The Physics of MOS Insulators

CONFERENCE CHAIRMAN
Gerald Lucovsky
North Carolina State University, Raleigh, NC

ORGANIZING COMMITTEE

G. Lucovsky, Chairman	NCSU
D. E. Aspnes	Bell Labs
R. S. Bauer	Xerox PARC
H. C. Casey	Duke Univ.
B. E. Deal	Fairchild
D. Emin	Sandia Labs
F. L. Galeener	Xerox PARC
D. L. Griscom	NRL
T. W. Hickmott	IBM
J. D. Joannopoulos	MIT
M. A. Littlejohn	NCSU
R. K. MacCrone	RPI
S. T. Pantelides	IBM
D. E. Sayers	NCSU
P. J. Stiles	Brown Univ.
H. H. Wieder	NOSC

INTERNATIONAL ADVISORY COMMITTEE

I. Lundstrom	Linköping, Sweden
N. F. Mott	Cambridge, UK
G. G. Roberts	Durham, UK
M. Schultz	Erlanger, W. Germany
W. E. Spicer	Stanford, USA
J. Stuke	Marburg, W. Germany
T. Sugano	Tokyo, Japan

PROGRAM COMMITTEE

G. Lucovsky, Chairman	NCSU
D. E. Aspnes	Bell
R. S. Bauer	Xerox
T. W. Hickmott	IBM
S. T. Pantelides	IBM
H. H. Wieder	NOSC

THE PHYSICS OF MOS INSULATORS

Proceedings of the International Topical Conference on
THE PHYSICS OF MOS INSULATORS
held at the Jane S. McKimmon Conference Center
North Carolina State University
Raleigh, North Carolina
June 18-20, 1980

Editors

GERALD LUCOVSKY
North Carolina State University

SOKRATES T. PANTELIDES
IBM, Yorktown Heights, NY

FRANK L. GALEENER
Xerox, Palo Alto, CA

PERGAMON PRESS
New York/Oxford/Toronto/Sydney/Frankfurt/Paris

Pergamon Press Offices:

U.S.A.	Pergamon Press Inc., Maxwell House, Fairview Park, Elmsford, New York 10523, U.S.A.
U.K.	Pergamon Press Ltd., Headington Hill Hall, Oxford OX3 0BW, England
CANADA	Pergamon of Canada, Ltd., Suite 104, 150 Consumers Road, Willowdale, Ontario M2J 1P9, Canada
AUSTRALIA	Pergamon Press (Aust.) Pty. Ltd., P.O. Box 544, Potts Point, NSW 2011, Australia
FRANCE	Pergamon Press SARL, 24 rue des Ecoles, 75240 Paris, Cedex 05, France
FEDERAL REPUBLIC OF GERMANY	Pergamon Press GmbH, Hammerweg 6, Postfach 1305, 6242 Kronberg/Taunus, Federal Republic of Germany

Copyright © 1980 Pergamon Press Inc.

Library of Congress Cataloging in Publication Data

International Topical Conference on the Physics
  of MOS Insulators, North Carolina State University, 1980.
   The physics of MOS insulators.

  Includes index.
  1. Metal oxide semiconductors—Congresses.
 2. Metal insulator semiconductors—Congresses.
 I. Lucovsky, G.   II.   Pantelides, Sokrates T.
 III.   Galeener, F.L.   IV. Title.
 QC611.8.M4I57   1980   537.6'22   80-22281
 ISBN 0-08-025969-3

All Rights reserved. No part of this publication may be reproduced, stored in a retrieval system or transmitted in any form or by any means: electronic, electrostatic, magnetic tape, mechanical, photocopying, recording or otherwise, without permission in writing from the publishers.

FOREWORD

An International Topical Conference on The Physics of MOS Insulators was held at the Jane S. McKimmon Conference Center, North Carolina State University, Raleigh, North Carolina, from June 18 to June 20, 1980. This was a follow-on to the International Topical Conference on The Physics of SiO_2 and Its Interfaces held at Yorktown Heights, New York, from March 22 to March 24, 1978. There were 195 participants from approximately one dozen countries in the Western Hemisphere, Europe, Asia and Africa. The Raleigh Conference was cosponsored by the American Physical Society, the American Vacuum Society, the Army Research Office, the National Science Foundation, the Office of Naval Research, and the Xerox Corporation, with financial support from the Army Research Office, the National Science Foundation, the Office of Naval Research, the Xerox Corporation, and the Schools of Engineering and Physical and Mathematical Sciences at North Carolina State University.

The conference program and abstracts have been published under separate cover in a booklet which was distributed at the time of the conference. These Proceedings have been organized into Chapters which reflect the structure of the conference sessions.

> Transport Properties
> Bulk Properties
> Bulk Defects
> Oxidation
> Interfaces
> Defects at Interfaces
> Device Physics

The success of this conference depended to a large extent on the participation of the Jane S. McKimmon Conference Center of North Carolina State University and, in particular, to Rosemary Jones of the McKimmon Center and Dale E. Sayers of the Department of Physics of North Carolina State University. Dale Sayers acted as Local Arrangements Chairman and was responsible for the effectiveness of all of the local functions, including the social events, the meals, and the transportation. In addition, he worked with the students from North Carolina State University who ran the projection booth and handled the microphones during the question periods. The students who are acknowledged for their assistance are Chuck Bennett, Jiann Liu, Azzam Mansour, Chris Roddy, and Ron Rudder. Secretarial assistance at the conference was provided by Becky B. Gates. Special thanks also go to Frank L. Galeener for the preparation of the Abstract Book, to Sokrates T. Pantelides for organizing the refereeing of the manuscripts, and to the members of the Program Committee, Dave Aspnes, Bob Bauer, Tom Hickmott, Sokrates Pantelides, and Harry Wieder.

> Gerald Lucovsky
> Conference Chairman
> Raleigh, North Carolina
> July 1980

CONTENTS

CHAPTER I: TRANSPORT PROPERTIES

High Current Injection into SiO_2 Using Si-Rich SiO_2 Films and Experimental Applications -
D. J. DiMaria 1

High Field Conduction in Thick Oxide MNOS Capacitors on p-Type Silicon -
I. Kashat and N. Klein 19

Dielectric Breakdown in Thermal SiO_2 Grown from Doped Polycrystalline Silicon Thin Films -
M. Berberian 24

The Effect of Diffusion on the Photoconductivity of Thin Films -
R. C. Hughes and R. J. Sokel 29

The Kinetic Behaviour of Mobile Ions in SiO_2 Studied with TSIC and TVS Measurements -
Ji Li-jiu, Wang Yang-yuan, Zhang Li-chun, Ni Xuie-win 34

Interactions Between Small-Polaronic Particles in Solids -
D. Emin 39

Small-Polaron Hopping Without Trap Participation in Dispersive Transient Transport in SiO_2 of MOS Structures -
K. L. Ngai, Xiyi Huang and Fu-sui Liu 44

Electron Transport in SiO_2 Films at Low Temperatures -
S. Othmer and J. R. Srour 49

Physical Effects in Lateral MIS Structures with Ultra-Thin Oxides -
J. Ruzyllo 54

Oxygen as a Two-Level Tunneling System in SiO_2 -
W. B. Fowler and A. H. Edwards 59

CHAPTER II: BULK PROPERTIES

Periodic Structural Models and Radial Distribution Functions of SiO_x: x = 0 to 2 -
W. Y. Ching 63

The Optical Absorption Edge of SiO_2 -
R. B. Laughlin 68

Band Structure and Density of States of β-Silicon Nitride -
S.-Y. Ren and W. Y. Ching 73

Electron Microscopy and Raman Spectroscopy of Nb_2O_5, Ta_2O_5 and Si_3N_4 Thin Films -
F. L. Galeener, W. Stutius and G. T. McKinley 77

Phonons and Submicrocrystallites in Amorphous SiO_2 -
K. Hübner, A. Lehmann and L. Schumann 82

Chemical Bonding in SiO -
G. Hollinger 87

Structural and Bond Flexibility of Vitreous SiO_2 Films -
A. G. Revesz and G. V. Gibbs 92

CHAPTER III: BULK DEFECTS

Electron-Transfer Model for E'-Center Optical Absorption in SiO_2 -
D. L. Griscom and W. B. Fowler 97

Assignment of the Optical Absorption of the E'_1 Center in SiO_2 -
O. F. Schirmer 102

Electronic Structure of Vacancies and Interstitials in SiO_2 -
F. Herman, D. J. Henderson and R. V. Kasowski 107

Surface and Bulk Vibrations in Ion-Implanted Amorphous Silica -
G. W. Arnold 112

Energy Distribution of Electron Trapping Defects in Thick-Oxide MNOS Structures -
V. J. Kapoor and S. B. Bibyk 117

Traps in SiO_2-Si Structure Determined by Electrochemical Method -
A. Wolkenberg 122

Charge Trapping and Associated Luminescence in MOS Oxide Layers -
C. Falcony-Guajardo, F. J. Feigl and S. R. Butler 127

Time Decay of Photoluminescence from Amorphous SiO_2 -
C. M. Gee and M. Kastner 132

Electron-Beam-Induced Luminescence in SiO_2 -
S. W. McKnight 137

Photoionization Cross Section of the 2.5 eV Electron Trap in SiO_2 -
D. D. Rathman, F. J. Feigl, S. R. Butler and W. B. Fowler 142

Hydrogenation of Amorphous Silicon Nitride -
H. J. Stein, P. S. Peercy and D. S. Ginley 147

CHAPTER IV: OXIDATION - Si

Initial Oxidation of Ion-Sputtered Silicon (100) -
D. Ellsworth and C. W. Wilmsen 152

Fixed Surface Charge Density Generation at the Interface of Anodic SiO_2-Si Systems -
J. L. Martinez and E. Gómez 157

Tracer Measurements of Network Oxygen Exchange During Water
Diffusion in SiO_2 Films -
R. Pfeffer and M. Ohring 162

An ^{18}O Study of the Oxygen Exchange in Silicon Oxide Films
During Thermal Treatment in Water Vapor -
S. Rigo, F. Rochet, A. Straboni and B. Agius................................167

X-ray Photoelectron Spectroscopy of Siloxene: A Model Compound Representing
Intermediate Oxidation States of Silicon and Interface Defect Sites -
J. A. Wurzbach...172

Effect of Annealing in O_2/N_2 Mixture on the MOS Characteristics -
A. K. AboulSeoud and S. Masoud 177

CHAPTER V: OXIDATION - COMPOUND SEMICONDUCTORS

Chemical Reactions in Native Oxide Films Formed on
III-V Semiconductors -
G. P. Schwartz 181

Anodic Oxide Insulators on InP and InAs -
D. A. Baglee, D. H. Laughlin, C. W. Wilmsen and D. K. Ferry..............191

Optical Properties and Interface Analysis of the GaAs-Anodic Oxide System -
D. E. Aspnes, G. J. Gualtieri, B. Schwartz, G. P. Schwartz
and A. A. Studna..197

XPS Study of GaAs(100) Surface Oxide Chemistry and Interface Potential -
R. W. Grant, S. P. Kowalczyk, J. R. Waldrop and W. A. Hill...............202

Germanium (Oxy)nitride Based Surface Passivation Technique as Applied
to GaAs and InP -
B. Bayraktaroglu, R. L. Johnson, D. W. Langer and M. G. Mier.............207

KrF-Laser Annealing of Native Oxides on GaAs -
R. K. Ahrenkiel, G. Anderson, D. Dunlavy, C. Maggiore, R. B. Hammond
and S. Stotlar 212

Anodic Oxidation of $Hg_{0.68}Cd_{0.32}Te$ -
B. K. Janousek, M. J. Daugherty and R. B. Schoolar........................217

CHAPTER VI: INTERFACES

Chemical Bonding at $Metal/SiO_2/Si(111)$ Interfaces -
R. S. Bauer, R. Z. Bachrach and L. J. Brillson 221

Dipole Layers at the $Gold-SiO_2$ Interface -
T. W. Hickmott 227

Measurement of Tunneling into Interface States -
W. E. Dahlke and D. W. Greve 232

Improved Experimental Characterization of the Si/SiO_2 Interface -
A. Sher, Y. H. Tsuo, P. Su and W. E. Miller 236

Gap States of Crystalline Silicon and Amorphous SiO$_2$ System -
    T. Sakurai and T. Sugano 241

Interface Width and Structure of the SiO$_2$ Layer on Oxidized Si -
    H. Frenzel and P. Balk..246

Chemical Composition and Kinetic Law of the SiO$_2$/Si Interface -
    A. Lora-Tamayo, E. Dominguez, E. Lora-Tamayo, A. Payo,
    F. Ferrer and J. Llabrés...250

Auger Analysis Coupled with Capacitance Studies of the Si-SiO$_2$ Interface -
    M. Hirose, S. Sakano, Y. Osaka and T. Hattori..255

Some Metal - Silicon Dioxide Interface Phenomena -
    C. M. Svensson..260

Field Effect Spectroscopy of Semiconductor-Insulator
    Interface States Using Thin Film Transistor Structures -
    L. J. Brillson, F. Luo, and F. Wysocki...265

The Properties and Applications of GaAs and InP MIS Structures -
    D. L. Lile and H. H. Wieder..270

A Study of the Electronic Structure of the GaAs/Natural Oxide Interface -
    E. W. Kreutz and P. Schroll 275

Interface States in GaAs/LaF$_3$ MIS Configurations -
    A. Sher, Y. H. Tsuo, J. E. Chern and W. E. Miller...280

CHAPTER VII: DEFECTS AT INTERFACES

Generation of Interface States in the Si-SiO$_2$ System by
    Photoinjection of Electrons -
    S. Pang, S. A. Lyon and W. C. Johnson 285

Reduced Oxidation States and Radiation-Induced Trap Generation
    at Si/SiO$_2$ Interfaces -
    F. J. Grunthaner, B. F. Lewis, R. P. Vasquez and J. Maserjian.....................290

Studies of Electron-Beam Radiation and Hydrogenation Effects
    on Si-SiO$_2$ Interface and SiO$_2$ by XPS -
    T. Hattori, T. Totsuka and T. Suzuki...296

A Microscopic Model for the Q_{ss} Defect at the Si/SiO$_2$
    Interface -
    G. Lucovsky and D. J. Chadi 301

EPR Defects and Interface States on Oxidized (111) and (100) Silicon -
    P. J. Caplan, E. H. Poindexter, B. E. Deal and R. R. Razouk 306

Characteristic Defects at the Si-SiO$_2$ Interface -
    N. M. Johnson, D. K. Biegelsen and M. D. Moyer 311

Impurity Segregation at the Si-SiO$_2$ Interface -
    R. W. Barton, J. Rouse, S. A. Schwarz and C. R. Helms...............................316

Investigation of Hydrogen and Chlorine at the SiO_2/Si Interface -
I. S. T. Tsong, M. D. Monkowski, J. R. Monkowski, P. D. Miller,
C. D. Moak, B. R. Appleton and A. L. Wintenberg..........................321

Surface-Potential Dependence of EPR Centers at the Si/SiO_2 Interface -
E. H. Poindexter, P. J. Caplan, J. J. Finnegan, N. M. Johnson,
D. K. Biegelsen and M. D. Moyer 326

Electron-Beam-Induced Defects at Si–SiO_2 Interface -
E. Rosencher, A. Chantre and D. Bois 331

CHAPTER VIII: DEVICE PHYSICS

Anomalous Gate Current on Avalanche Hot Electron Injection in MOS Structures -
K. Yamabe and Y. Miura...336

Noise from MOS Transistors at Weak and Uniform Inversion -
A. A. Walma 341

Polymerized Langmuir Film MIS Structures -
K. K. Kan, M. C. Petty and G. G. Roberts....................................344

MOS Wearout and Breakdown Statistics -
D. Wolters, T. Hoogestyn and H. Kraaij.......................................349

**Effect of Preparation Methods on Performance of
MOS Photovoltaic Solar Cell -**
F. Abou-Elfotouh and M. Al-Mass'ari...353

List of Participants 359
Author Index 368

CHAPTER I

TRANSPORT PROPERTIES

HIGH CURRENT INJECTION INTO SiO$_2$ USING Si-RICH SiO$_2$ FILMS AND EXPERIMENTAL APPLICATIONS

D.J. DiMaria
IBM Thomas J. Watson Research Center
Yorktown Heights, New York 10598

ABSTRACT

Although silicon dioxide has a high mobility for electrons ($\sim$ 30 cm^2/V-sec), it forms large interfacial energy barriers (> 2 eV) with all contacting metals or semiconductors because of its large energy bandgap of $\sim$ 9 eV which makes current injection into it very difficult. This results in the excellent insulating properties of SiO$_2$ capacitors. By depositing a thin layer of Si-rich SiO$_2$ between SiO$_2$ and a contacting electrode, such as Al or Si, high electronic current injection at moderate applied voltages can be obtained. This "apparent" reduction of the interfacial energy barrier is believed to be obtained by "local" electric field distortion associated with the highly conducting, two-phase (Si and SiO$_2$) nature of the deposited Si-rich SiO$_2$. This injection phenomenon has been shown to be limited by the interface of the Si-rich SiO$_2$ layer with the SiO$_2$ layer. The currents are Fowler-Nordheim-like and have weak temperature dependence. Internal photoemission has been used to show that uniform homogeneous grading of the energy bandgap of the Si-rich SiO$_2$ layers does not account for the high current injection. Injection currents are not strongly affected by high temperature (> 1000°C) annealing which is shown, using Raman spectroscopy, to convert amorphous Si regions (< 50 Å in size) into Si crystallites in the Si-rich SiO$_2$ layer.

Layered structures composed of Si-rich SiO$_2$ and SiO$_2$ have been shown to have several applications, particularly in the area of non-volatile semiconductor memory and can be used to improve voltage breakdown characteristics. Devices using Si-rich SiO$_2$ injectors to charge up floating polycrystalline-Si gate type memory structures are capable of being "written" or "erased" at low voltages in 5 msec at least 10^5 times with the excellent charge retention characteristics of a floating poly-Si structure. Low voltage breakdowns of SiO$_2$ layers, which are believed to be associated with the field at the cathode, are shown to be suppressed when a thin Si-rich SiO$_2$ layer is present. Reversible space charge build up in this thin layer relaxes high electric fields at the metal or silicon contacts. This is believed to result in the observed dramatic increase in breakdown voltage. Also, current as a function of ramped gate voltage characteristics are shown to be a rapid, easy means for studying trapping kinetics and centroid position of trapping sites in the SiO$_2$ layer of structures with Si-rich SiO$_2$ injecting layers.

I. INTRODUCTION

Silicon dioxide, SiO_2, is commonly used in the electronics industry for passivation because of its excellent insulating properties. These properties are due to the large energy barriers (> 2 eV) that SiO_2 forms at interfaces with contacting metal or silicon layers because of its large energy bandgap (≈ 9 eV [1]). However, once electrons have entered the SiO_2 layer from the contact at the lower potential, they rapidly move through this layer to the opposite contact with a rather high mobility, for an insulator, of ~ 30 cm^2/V-sec [1] with less than 1 out of every 10^5 carriers being captured by traps in the thermally grown SiO_2 [1,2]. These electronic trapping centers are believed by many researchers to be related to residual H_2O present in the SiO_2 layers [2,3]. Hole injection from the contacts into SiO_2 layers is much more difficult than electron injection due to an asymmetry of the interfacial energy barriers, with holes "seeing" a larger barrier [1]. Hole mobility is also typically $\approx 10^{-5}$ times less than the electronic mobility due to the presence of shallow traps in the SiO_2 forbidden bandgap near the top of the valence band [1]. In principle, the blocking effect of the contacts to SiO_2 could be overcome if an interfacial matching layer is inserted between the contact and the SiO_2. For example if the contact is Si, then a region with the bandgap graded from Si to SiO_2 should be appropriate; namely, a semiconductor to insulator heterojunction [4]. At low applied fields, this heterojunction would be blocking with a limiting energy barrier for electrons (holes) of ≈ 3 eV (5 eV) from the bottom of the Si conduction band to the bottom of the SiO_2 conduction band (top of the Si valence band to the top of the SiO_2 valence band) [1,5]. As the applied field is increased, the energy bands would be pulled down for electrons (up for holes) until the smaller energy barrier at the Si-contact–graded-insulator interface limits the injection process [5]. With this type of reasoning in mind, metal-oxide-semiconductor (MOS) structures, were initially fabricated with a thermal silicon dioxide layer grown on Si and various thin stepped or graded Si-rich SiO_2 layers [5] deposited on top of the SiO_2 before gate electrode deposition. Current as a function of gate voltage characteristics such as those in Fig. 1 indeed showed enhanced electron injection under negative gate voltage bias when compared to control structures without the stepped or graded Si-rich SiO_2 layers present. However as seen in Fig. 1, the current enhancement was not particularly sensitive to the nature of the grading as long as the last layer closest to the contacting electrode, which was Al for this case, was composed of the Si-rich SiO_2 material with the highest Si content used (46% atomic Si [5,6]). Also, no strongly enhanced hole injection under positive gate voltage bias was seen. These observations led to a questioning of whether or not the Si-rich SiO_2 layers were really graded in a homogeneous way, or whether the material was actually composed of separate phases (like Si and SiO_2).

This review will be concerned with experimental data correlating the material properties of the composite Si-rich-SiO_2–SiO_2 structures with the observed electrical properties. In section II, several experiments showing a two phase nature (Si and SiO_2) of the Si-rich SiO_2 materials will be discussed. Then a model depending on this two phase nature will be used to explain the high electronic current injection phenomenon shown in Fig. 1. The dependence of the injection process on temperature,

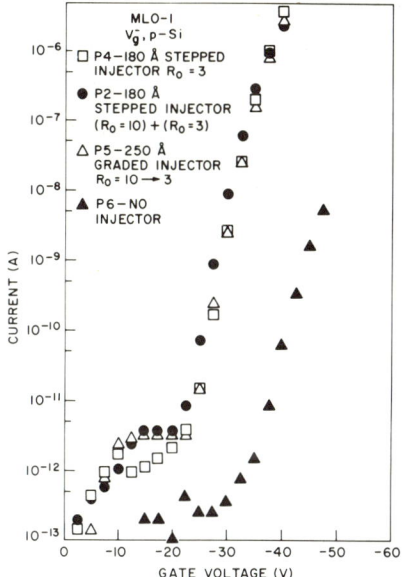

Fig. 1. Point-by-point measurements of the magnitude of the dark current as a function of negative gate voltage on composite MOS structures with various Si-rich SiO_2 layers stacked on top of thermal SiO_2. In this measurement, the gate voltage was stepped by -2.5 V starting from 0 V every 20 sec with the dark current being measured 18 sec after each voltage step increase. The Si-rich SiO_2 layer was either stepped or graded with R_o defined as $[N_2O]/[SiH_4]$ as an indicator of the Si content of this layer. R_o from 10 (40% atomic Si) to 3 (46% atomic Si) was used with Si content increasing towards the top metal gate electrode when several layers were stacked on top of the underlying 550 Å thick thermal SiO_2 layer ($R_o = 10 + 3$) or when a graded layer was used ($R_o = 10 \rightarrow 3$). Taken from Reference 5.

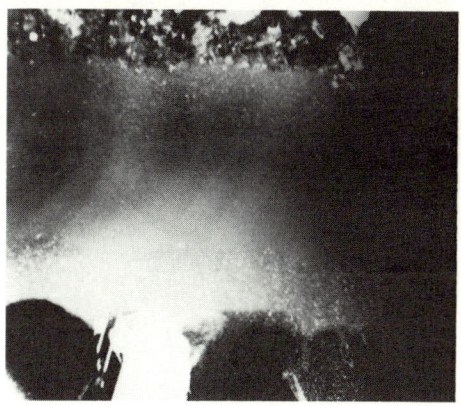

Fig. 2. High-resolution cross-section TEM of a dual electron injector structure (DEIS) stack on top of the first (floating) poly-Si layer with the lower part of the top (control gate) poly-Si layer showing in the uppermost part of the micrograph. The DEIS stack is composed of sequentially deposited layers of Si-rich SiO_2, SiO_2, and Si-rich SiO_2 with 46% atomic Si in the Si-rich SiO_2 layers. The entire structure was annealed at 1000°C in N_2 for 30 min.

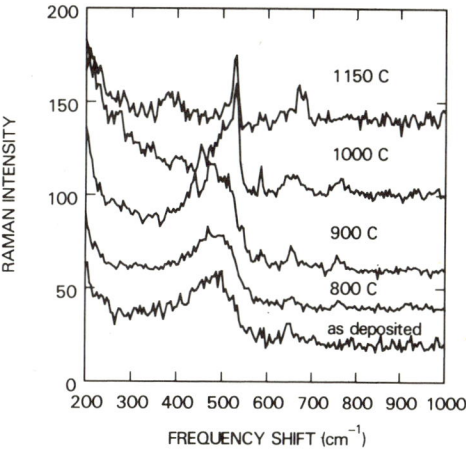

Fig. 3. Raman scattering spectra for a Si-rich SiO_2 film with 46% atomic Si which was sequentially annealed in N_2 for 30 minutes at 800°C, 900°C, 1000°C and 1150°C from its as-deposited state (fabricated at 700°C). Taken from Reference 14.

Si-rich SiO_2 thickness, gate electrode area, annealing, and sensitivity to Si content of the Si-rich SiO_2 layers will be described. In section III, experimental applications involving various stacked layers of Si-rich SiO_2 and SiO_2 will be shown for electrically-alterable read-only-memory (EAROM). Also, low voltage breakdowns in MOS structures which are believed to be associated with contact-SiO_2 interfaces are shown to be dramatically improved with the presence of the intervening Si-rich SiO_2 layer. Finally, the current–ramped-voltage characteristics of these structures will be discussed in terms of a rapid, easy way to study trapping kinetics and centroid position of sites in the SiO_2 layer which are either purposely or inadvertently introduced during fabrication.

II. PHYSICS AND MATERIAL CONSIDERATIONS

A. Two Phase Characteristics

Several researchers have recently shown that Si-rich SiO_2 (also called semi-insulating polycrystalline silicon (SIPOS) [7]) has Si crystallites in an oxide matrix made up of mostly SiO_2 at least after annealing at high temperatures (1000°C) in an inert ambient such as nitrogen [8-12]. These films are usually prepared by chemical vapor deposition (CVD) at 700°C-800°C using SiH_4 and N_2O [6]. Figure 2 shows a dark-field high-resolution cross-section transmission-electron-micrograph (TEM) of stacked layers of 1000°C annealed Si-rich SiO_2, SiO_2, and Si-rich SiO_2 [12]. The bright white spots in these micrographs are crystallites in the Si-rich SiO_2 layers. The largest crystallites are approximately 50 Å in size. The intervening, chemically-vapor-deposited oxide layer does not show any crystallites.

Further studies, initially by Goodman et al. [13] and more recently by Hartstein et al. [14] using Raman spectroscopy, have shown that the as-deposited unannealed Si-rich SiO_2 layers are also made up of at least two phases (Si and oxide) with the Si regions being amorphous. Figure 3 demonstrates how these amorphous Si regions are converted to Si crystallites by high temperature annealing. In this figure, the Raman spectrum of amorphous Si (α-Si) shows no sharp lines for frequency shifts above 200 cm^{-1} but rather a broad asymmetric continuum peaked near 480 cm^{-1}. This continuum arises from the relaxation of the normal Raman selection rules for scattering from a crystal due to the loss of translational symmetry in the amorphous state. As the Si-rich SiO_2 layer is progressively annealed at higher and higher temperatures, the continuum of the amorphous Si state disappears (completely by 1150°C) and the Raman spectrum of crystalline silicon with a single strong line at $\approx$ 525 cm^{-1} appears. Optical transmission measurements in the visible region show a similar behavior in comparing annealed and unannealed Si-rich SiO_2 layers [14].

Also, measurements of the infrared absorption using the attenuated-total-reflection (ATR) technique [15] of CVD Si-rich SiO_2 and control CVD SiO_2 films show absorption lines attributed to SiOH, H_2O, and SiH groups in as-deposited films [16]. The concentrations of the SiOH and H_2O

impurities are in the low 10^{21} cm^{-3} range and the concentration of the SiH impurity is in the mid 10^{18} cm^{-3} range with the Si-rich SiO$_2$ films containing about an order of magnitude more SiH than the SiO$_2$ films [16]. After annealing at 1000°C in N$_2$ no absorption lines are observed, and the concentration limit for SiOH and H$_2$O is $\lesssim$ low 10^{19} cm^{-3} while for SiH it is $\lesssim 10^{16}$ cm^{-3} [16].

B. Current-Voltage Characteristics

(i) Dark Currents

Electrical characterization of Si-rich SiO$_2$ films deposited on top of or underneath SiO$_2$ layers was performed by measuring the dark current using point-by-point or ramped gate voltage techniques [5,17]. Results from these measurements which will be discussed in this section together with the results from material characterization summarized in section II-A, will lead to the physical model proposed in section II-C.

Figure 1 shows the dark current as a function of negative gate voltage for various Si-rich SiO$_2$ layers stacked on top of a thermal SiO$_2$ layer incorporated into an MOS structure. These measurements were taken point-by-point, and they show a ledge in the low voltage region. This ledge is due to a reversible electronic space charge build-up in the Si-rich SiO$_2$ layer [5]. As the negative voltage is increased, the Al gate electrode injects electrons into the Si-rich SiO$_2$ layer which traps and holds this charge in the Si regions. This causes the observed ledge which is terminated when injection from the Si-rich SiO$_2$ layer into the SiO$_2$ layer becomes dominant at ≈ -22 V. At higher negative voltages ($\gtrsim -22$ V), most of the applied voltage is dropped across the underlying SiO$_2$ layer due to the low resistivity of the Si-rich SiO$_2$ film compared to SiO$_2$ [6]. As seen in Fig. 1, the current is strongly dependent on voltage after dominant injection into the SiO$_2$ layer appears, and this is consistent with Fowler-Nordheim tunneling [18]. The remainder of this review will be concerned mostly with this high electronic current injection into the SiO$_2$ layer. With the Si-rich SiO$_2$ layer present, the MOS shows essentially a switching action where electron injection from the Al electrode is blocked by a reversible space charge build-up in the Si-rich SiO$_2$ layer until a critical voltage is reached. Then injection (via Fowler-Nordheim tunneling) of some of the stored electrons in the Si-rich SiO$_2$ layer into the SiO$_2$ layer becomes dominant, and these electrons can rapidly flow through the SiO$_2$ to the opposite electrode which is the Si substrate for this case. This switching action is extremely useful for device applications as will be shown in section III-A.

MOS structures with Si-rich SiO$_2$ layers at either the Si substrate or both the Si substrate or gate electrode interfaces [19] can be fabricated using in situ CVD of Si-rich SiO$_2$ and of SiO$_2$ as opposed to using a thermal SiO$_2$ layer grown from the Si substrate. Annealed CVD SiO$_2$ and thermal SiO$_2$ are very similar in both their physical and electrical properties [1,17,19]. Such a stacked structure consisting of a Si-substrate − CVD Si-rich-SiO$_2$ − CVD SiO$_2$ − CVD Si-rich-SiO$_2$ − metal or polycrystalline Si (poly-Si) gate electrode is called a dual electron injector structure (DEIS) [19], because enhanced electron injection can occur from either the Si substrate for positive gate voltage V_g^+ or from the gate electrode for negative gate voltage V_g^-. Figures 4 and 5 show enhanced electron injection from the Si substrate for DEISs which were unannealed and annealed at 1000°C in N_2 for 30 min prior to Al gate metallization, respectively. These dark I-V measurements were taken under ramped gate voltage conditions using a constant rate of .47 V/sec. The first low voltage ledge is due to a capacitive displacement current $C\ dV_g/dt$ where C is the total series capacitance of the stacked structure [17]. The second broad ledge at $\approx 10^{-7}$ A in Fig. 4 for the unannealed structure is due to trapping on certain H_2O related sites [17] in the intervening CVD SiO_2 layer which are removed with annealing as shown by Fig. 5. The study of these trapping ledges can give useful information on trapping kinetics and will be discussed in section III-B. For the annealed structures in Fig. 5, the I-V characteristics show only a slight dependence on the thickness of the Si-rich SiO_2 layers from 100 Å to 1000 Å. This is consistent with the low resistivity of this material with respect to SiO_2 and consistent with a physical model which suggests that the enhanced electron injection into the SiO_2 layer is controlled by the Si-rich SiO_2 interface with this oxide layer. The ledge at $\approx 10^{-6}$ A in Fig. 5 is not due to trapping. It is caused by depleting minority carriers (electrons for the 2 Ωcm p-type Si substrates used here) from the substrate-Si–Si-rich-SiO_2 interface faster than they can be generated thermally (or optically) in the Si for enhanced injection into the SiO_2 layer. As this depletion occurs, any additional voltage is dropped across the Si substrate and the oxide fields are held approximately constant. On the unannealed structure in Fig. 4, there is a more pronounced dependence on Si-rich SiO_2 thickness. This again is probably due to H_2O in the oxide regions of the Si-rich SiO_2 layers which cause more trapped space charge build-up in this layer and make it appear more resistive to electronic carrier flow to the Si-rich-SiO_2–SiO_2 interface.

The dark I-V dependence on contacting electrode material was studied in detail for Al, Au, planar Si, and Si with a rough surface (poly-Si surfaces [20]). For the case of a Si-rich SiO_2 injecting layer deposited on SiO_2, poly-Si and Al gate electrodes give similar results, but Au electrodes regardless of the deposition technique (resistance heated W boat or electron-beam deposition [5]) have less injected current for comparable negative gate voltages. However, these differences between Si, Al, and Au were not as large as observed on control structures without the Si-rich SiO_2 layer [5]. The differences in dark I-V characteristics on the control structures are adequately predicted by differences in the work functions of Al, Si and Au with the work function of Au being ≈ 1 eV larger than that of Al or Si [5,18]. Since the current injection is controlled by Fowler-Nordheim tunneling at the metal-SiO_2 interface [18], a 1 eV difference in work function gives a 1 eV larger energy barrier at this interface and a larger tunneling distance into the SiO_2 for the same oxide fields. However when the injecting Si-rich SiO_2 layer is present, the Si islands and the space charge they build up tend to limit the electrode injection, as discussed previously. The energy barrier height differences are minimized further as the Si-rich SiO_2 layer is made thicker [5].

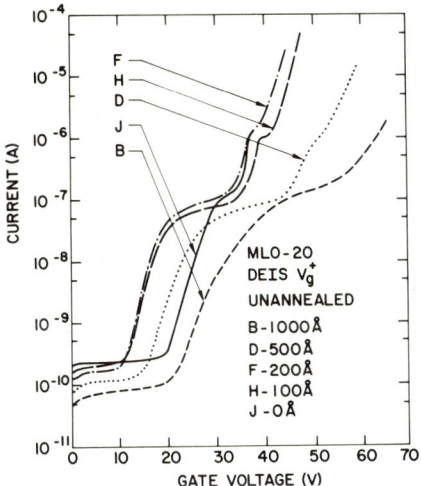

Fig. 4. Magnitude of the dark current as a function of positive ramped (0.47 V/sec) gate voltage on DEIS capacitor structures with various CVD Si-rich SiO_2 (46% atomic Si) injector thicknesses (0 Å, 100 Å, 200 Å, 500 Å, and 1000 Å) and a 400 Å intervening CVD SiO_2 layer. Structures were not annealed prior to Al gate electrode deposition.

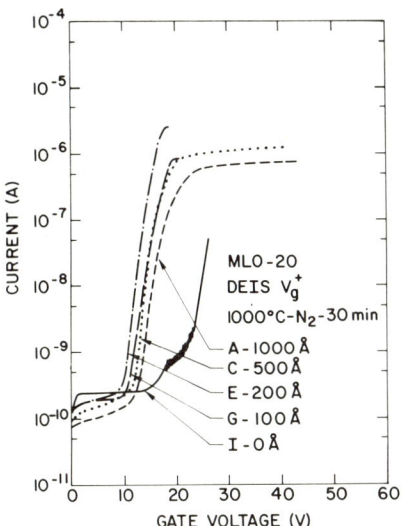

Fig. 5. Magnitude of the dark current as a function of positive ramped (0.47 V/sec) gate voltage on DEIS capacitor structures which are identical to those in Fig. 4 except that they were annealed at 1000°C in N_2 for 30 min. prior to Al gate metallization.

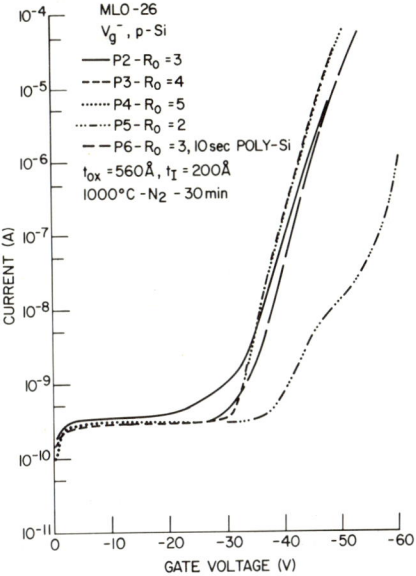

Fig. 6. Magnitude of the dark current as a function of negative ramped (−0.47 V/sec) gate voltage on single Si-rich SiO_2 injector structures with various amounts of incorporated Si in the vicinity of 46%. The ratio R_o of $[N_2O]/[SiH_4]$ is used as an indicator of the changes in Si content. The injectors were all 200 Å thick with an underlying thermal oxide of 560 Å in thickness.

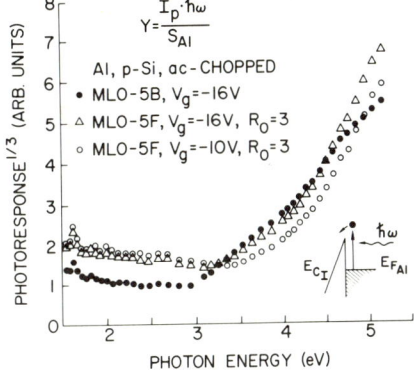

Fig. 7. Cube root of the a.c. photoresponse as a function of photon energy on Si-rich-SiO_2–SiO_2 composite MOS and SiO_2 MOS structures under negative gate voltage bias. $R_o = 3$ (46% atomic Si) Si-rich SiO_2 material with a thickness of 120 Å was used for the injector on MLO-5F, the gate electrode was Al (135 Å), and the thermal SiO_2 thickness was 290 Å. MLO-5B is the control structure with no injector. Taken from Reference 5.

For structures (such as the DEIS) where the Si-rich SiO$_2$ injector is deposited on either a planar-Si single crystal surface or a rough poly-Si surface and then the SiO$_2$ layer is deposited on top of it, an asymmetry in the dark I-V characteristics between this bottom injector and a Si-rich SiO$_2$ injector deposited on top of the SiO$_2$ layer (top injector) is observed [19]. The bottom injector gives a larger electron current injection for positive gate voltages than the top injector for negative gate voltages of the same magnitude for structures which have supposedly equivalent injectors and contacts [19]. This asymmetry is thought to be due to observed (by high-resolution cross-section TEM [12]) differences in the size and density (and possibly shape) of the Si islands at the two Si-rich-SiO$_2$–SiO$_2$ interfaces.

Dark current-voltage characteristics on composite structures of Si-rich SiO$_2$ and SiO$_2$ as a function of the following variables, were also investigated: area from 6×10^{-3} cm^2 to 8.4×10^{-6} cm^2; temperature in the range from 77°K to 473°K; annealing at 1000°C in various gaseous ambients such as N$_2$ and N$_2$-H$_2$ mixtures; and Si content of the Si-rich SiO$_2$ layers ranging from 33% to 46% atomic Si. No strong dependence of the currents of any of the composite structures was seen with variations in area [5,21], temperature [5], or annealing [5]; but a strong dependence with Si content of the Si-rich SiO$_2$ layers was observed [5]. However, it should be noted that Si-rich SiO$_2$ layers alone do show a strong dependence of their conductivity on thickness, temperature, and annealing conditions [22]. A lack of dependence of the currents on electrode area implies that the injection process is uniform over the scale of dimensions discussed here. The slight temperature dependence of the currents that was observed is consistent with Fowler-Nordheim tunneling at the Si-rich-SiO$_2$–SiO$_2$ interface limiting the electron injection into the SiO$_2$ layer [18]. Although the Si islands in the two phase Si-rich SiO$_2$ layers are converted from an amorphous to crystalline state with high temperature annealing in N$_2$ or N$_2$-H$_2$ ambients, the electronic injection process remains essentially unchanged [5]. This implies that if the Si-rich-SiO$_2$–SiO$_2$ interface controls the injection into the SiO$_2$, at least the Si islands at this interface are not drastically changing in size, shape, or density.

A dependence on the Si content of the Si-rich SiO$_2$ layer of the composite structures was observed for $\lesssim$ 40% atomic Si. This is consistent with an increase in the resistivity of the Si-rich SiO$_2$ layer with more of the applied voltage being dropped across it rather than across the intervening SiO$_2$ layer. However, once a Si content in the range of $\approx$ 46% atomic Si was reached, small changes (1% to 2%) in the Si content had very little effect on the I-V characteristics. This is shown in Fig. 6 where the ratio R_o of [N$_2$O] to [SiH$_4$] in the gas phase was varied from 2 to 5. There is a decrease in the injection efficiency only for $R_o = 2$. This was caused by the characteristics of the CVD conditions and reactor used here where less Si in the Si-rich SiO$_2$ layer is deposited at the wafer position in this reactor for less than an $R_o = 3$ condition [6].

(ii) **Photocurrents**

To investigate and measure interfacial energy barrier heights, photocurrent as a function of light energy was used [23]. Typical cube root of the photoresponse Y (photocurrent I_p normalized to the incident photon flux where S_{Al} is the light intensity transmitted through the Al electrode) as a function of light energy $\hbar\omega$ is shown in Fig. 7 for an MOS structure with a Si-rich SiO_2 injector on top of SiO_2 and for a control MOS structure with just an SiO_2 layer present, both with Al gate electrodes. These measurements were done using chopped light with a.c. detection techniques at voltages where high dark current electron injection from the Si-rich SiO_2 layer into the SiO_2 is occurring [5]. The photocurrent response from 1.5 eV to 3 eV is due to a 90° out of phase signal from the Si substrate while that from 3 eV to 5 eV is due to internal photoemission from the Si islands or Al gate electrode into the SiO_2 regions [5]. Both of these latter two processes have an $\approx$ 3 eV energy barrier [23], and the slight increase in photoresponse with voltage (comparing $V_g = -10$ V to -16 V) is consistent with ordinary Schottky barrier lowering [23]. Clearly no energy barriers $\lesssim$ 2 eV due to uniform homogeneous bandgap grading (as discussed in section I) which would be necessary to explain the dark current data are observed in Fig. 7. D.c. photocurrent measurements lead to the same conclusions and are very similar to the data in Fig. 7. Differences in the photoresponse-energy characteristics were observed when the metal contact was changed from Al to Au. Au has an $\approx$ 1 eV greater work function and therefore an $\approx$ 1 eV greater energy barrier when in contact with SiO_2 [23]. However, these changes were not as large as on the control MOS structures with just an SiO_2 layer present due to the photoconductivity of the trapped space charges (electrons) on the Si islands in the Si-rich SiO_2 layer [5] as discussed in section II-B-i.

C. Physical Model

From the experimental observations discussed in sections II-A and II-B, the large electron current injection into SiO_2 appears to be controlled by the Si-rich-SiO_2–SiO_2 interface. Since <u>uniform</u> bandgap grading or stepping seems unlikely from the two phase (Si and SiO_2) nature of the Si-rich SiO_2 and from the photocurrent measurements where no lowered energy barriers due to a homogeneous Si_xO_y material graded or stepped to SiO_2 are observed, a localized physical mechanism seems reasonable. Figure 8 schematically illustrates such a localized mechanism. For this physical model, electrons move easily through the Si-rich SiO_2 (because of its low resistivity) to its interface with the SiO_2 layer. At the Si-rich-SiO_2–SiO_2 interface, electrons are injected from the last layer of Si islands into the SiO_2 at lower applied average fields than from a continuous planar surface due to distortion of the local electric field. This enhancement of the local fields occurs because of the size, shape, and density of the Si islands [24] which must have some curved surfaces since they are non-planar (see Fig. 2). To explain the dark I-V data reported here only a field enhancement of $\approx$ 1.5 to 2 [17,19] is necessary due to the strongly field dependent nature of Fowler-Nordheim tunneling from Si into SiO_2 [18]. Fowler-Nordheim tunneling at this interface is consistent with the weak temperature dependence and the strong electric field dependence observed for the dark currents. The photocurrent results, which are only weakly dependent on the electric field, would still yield results which are not drastically different from those of the control structures. The lack of dependence on electrode area is also consistent with this localized model because the Si islands are much smaller than the electrode area and they are densely packed for the 46% atomic Si material.

The conductivity of the Si-rich SiO_2 layer <u>itself</u> is probably controlled by direct tunneling between the closely spaced Si islands for the 46% atomic Si material (see Fig. 8). For smaller percentages of Si in the Si-rich SiO_2 layer, the conductivity is probably controlled by Poole-Frenkel conduction [25] (field-assisted thermal-activation of carriers from the Si wells). For these lighter doped Si materials, more of the applied voltage is dropped across the Si-rich SiO_2 layer of the composite structures and degradation of the current enhancement is seen on dark I-V characteristics. Of course, the density of the Si islands will decrease and possibly their size and shape might change in these lighter doped films which will also affect the injection efficiency at the Si-rich-SiO_2–SiO_2 interface. For larger amounts of Si, the Si regions become connected and percolation [26] controls the conduction to the Si-rich-SiO_2–SiO_2 interface. For nearly 100% atomic Si, the Si-rich-SiO_2–SiO_2 interface would look planar and the electric field distortion would disappear.

There are other localized mechanisms at the Si-rich-SiO_2–SiO_2 interface which might also be operating in addition to or instead of field-enhanced tunneling. One such mechanism could be localized energy bandgap grading or stepping occurring in the transition region from Si to SiO_2 surrounding each of the Si islands in the last layer at the Si-rich-SiO_2–SiO_2 interface. In a similar fashion, many localized states across the SiO_2 bandgap might be introduced in this same transition region and lead to a localized mechanism involving trap-assisted tunneling [1].

III. EXPERIMENTAL APPLICATIONS

In this section, three experimental applications using structures containing stacked layers of Si-rich SiO_2 and SiO_2 are described. The first application discussed in section III-A uses a DEIS stack in a non-volatile memory device. In section III-B, the improved breakdown characteristics of MOS-type structures with Si-rich SiO_2 layers acting as buffers between the contacting electrode and the SiO_2 layer will be demonstrated. The third application presented in section III-C involves using Si-rich SiO_2 injectors as a large electronic current source at moderate applied electric fields for charge trapping studies in SiO_2 layers.

A. DEIS EAROM

A novel non-volatile memory device using a DEIS stack for putting electrons on or taking them off a floating poly-Si storage layer is depicted schematically in Fig. 9 [19,21]. This device is essentially an MOS field effect transistor (MOSFET) [25,27] with an energetically deep storage well for electrons formed by the isolated poly-Si electrode. When electrons are put on this floating poly-Si layer, the internal electric field they generate affects the field at the substrate-Si–SiO$_2$ interface. For the n-channel structure (electrons are the mobile charge carriers in the Si surface channel when the device is turned on) shown in Fig. 9, the charge state of the floating poly-Si layer is sensed when an electron current flows in the channel region between the source and drain contacts. The channel is turned on by putting a large enough positive voltage on the control gate to invert the Si surface by pulling minority carriers (electrons in this case) to the Si–SiO$_2$ interface. When electrons are trapped on the floating poly-Si layer, more positive gate voltage is required to invert the Si surface by overcoming the opposing electric field introduced by the trapped negative charge. This voltage at which the channel becomes conducting is called the turn-on or threshold voltage, V_T [25,27]. Determining the charge state of the floating poly-Si layer by sensing V_T is called the "read" operation of the non-volatile memory.

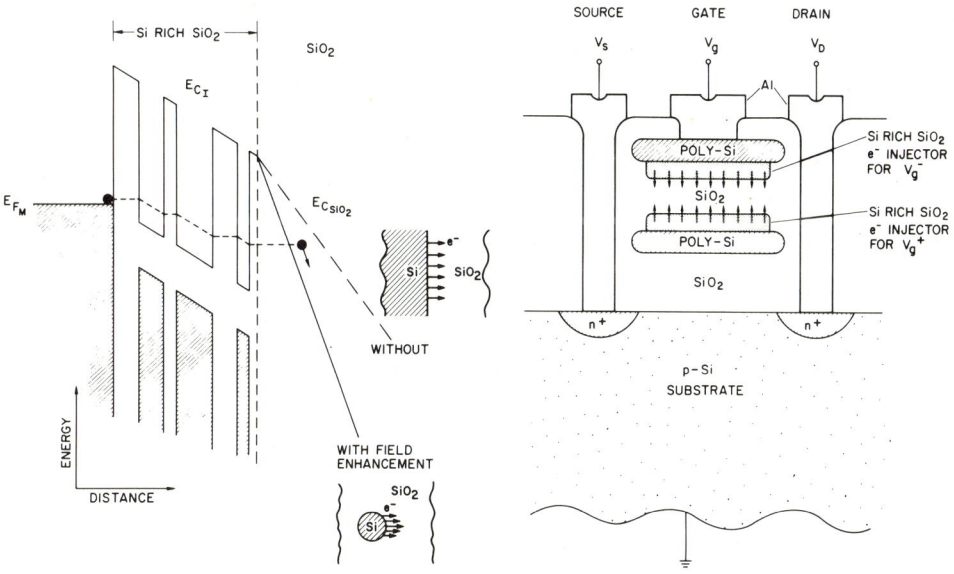

Fig. 8. Schematic energy band representation of conduction in the Si-rich SiO$_2$ layer via direct tunneling between isolated Si regions in the SiO$_2$ matrix of this two phase system and subsequent high field injection into the underlying SiO$_2$ region due to local electric field enhancement caused by the curved surfaces of the Si regions. Electronic Fowler-Nordheim tunneling into SiO$_2$ from a planar Si surface is shown for comparison. Taken from Reference 5.

Fig. 9. Schematic representation of a non-volatile n-channel field-effect-transistor (FET) memory using a DEIS stack between a control gate and a floating poly-Si layer. Writing (erasing) is performed by applying a negative (positive) voltage, V_g^- (V_g^+), to the control gate which injects electrons from the top (bottom) Si-rich SiO$_2$ injector to the floating poly-Si storage layer (back to the control gate). Structure is not drawn to scale. Taken from Reference 19.

The difference between the memory described here and others using a similar floating poly-Si storage layer [28,29] is the way in which electrons are put on (called the "write" operation) or taken off (called the "erase" operation) the poly-Si. Here a DEIS stack is used to give write/erase operations at moderate gate voltages and low power because of the enhanced electron injection phenomenon associated with the Si-rich-SiO_2–SiO_2 interfaces. Most commercially available floating poly-Si gate non-volatile memories are written by avalanching a Si junction to get "hot" electrons over the highly blocking 3 eV barrier at the Si–SiO_2 interface [28,29]. This requires much more power consumption because the Si currents are $\gtrsim 10^5$ times the SiO_2 currents. These devices are typically very difficult to erase, usually requiring photo-discharge of the trapped electrons off the poly-Si with ultraviolet (UV) light which takes $\gtrsim 10$ minutes [28,29]. The DEIS EAROMs described here can be written and <u>erased</u> at comparable voltages ($\lesssim 30$ V) and speeds ($\lesssim 5$ msec) at least 10^5 times.

Figure 10 demonstrates the write/erase cycling of DEIS EAROMs like that depicted in Fig. 9. The turn-on voltage V_T is used as an indication of the charge state of the floating poly-Si where a positive value of V_T implies a written condition of stored electrons and a negative value of V_T implies an over-erased condition of ionized donors. A virgin as-fabricated device, which is approximately in a neutral charge state initially, is cycled using the voltages indicated by similar symbols (solid characters for write and open characters for erase). A control structure with just an SiO_2 layer between the control gate and floating poly-Si layer could not be written for similar voltages. However, the control structure can be over-erased somewhat because of the rough nature of the top surface of the poly-Si layer [20,30,31] which gives field-enhanced local injection near the tips of the asperities on this Si surface into the SiO_2 layer between the control and floating gates [30]. A collapse in the V_T window is observed to occur after $\approx 10^4$ to 10^5 write/erase cycles. This is due to a permanent trapped electronic space charge build-up in the intervening CVD SiO_2 layer due to H_2O related trapping sites [1,2,17,21]. This negative trapped charge generates an internal electric field which reduces the field at either Si-rich-SiO_2–SiO_2 injecting interface, and therefore reduces the injected current from the control gate or off the floating poly-Si layer which in turn determines V_T. The effect of this trapped charge can be overcome by increasing the write/erase gate voltages so that the electric fields at the injecting interfaces are returned to their initial values [21]. Also, the trapped electrons can be discharged thermally by heating the device to temperatures $\gtrsim 200°C$ [2,3,21]. Recently, write/erase voltages between 10 V and 15 V have been obtained on DEIS EAROMs like that depicted in Fig. 9 with a thinner oxide layer (100 Å in thickness) between the single crystal Si substrate and the floating poly-Si layer than that used for the DEISs in Fig. 10 [21].

Although the DEIS EAROM described here has excellent write/erase characteristics, it must also retain the stored electronic information for long periods of time (10-100 years) when no power is applied which gives it a "non-volatile" nature. Figure 11 shows that the DEIS EAROM for a grounded gate electrode condition has the same retention characteristics as a commercially available floating poly-Si gate device with an SiO_2 layer between the control and floating gates, at least for the times indicated. The discharge rate for $\approx 5 \times 10^{12}$ stored electrons/cm^2 for temperatures between 25°C and 300°C is similar for a control device, DEIS EAROM, or a single electron injector structure (SEIS) EAROM with only a Si-rich SiO$_2$ injector under the control gate. The electron discharge which becomes significant after $\approx 10^5$ sec and 10^4 sec at 200°C and 300°C, respectively, can be shown to be thermally activated with a barrier energy of ≈ 3 eV for zero electric field [21]. This is consistent with electron discharge from a Si well surrounded by SiO$_2$ [23]. At normal device operating temperatures between 25°C and 80°C, this thermal activation process would give only 5% charge loss in $\gtrsim 3 \times 10^7$ years [21]. The DEIS structure in Fig. 11 shows a similar retention characteristic compared to the control structure because the local fields are not high enough to cause significant charge loss for the times considered by the enhanced current injection mechanism via tunneling into the SiO$_2$ layer which is only weakly dependent on temperature [5].

B. Low Voltage Breakdown

Numerous measurements on many large area (.006 cm^2) capacitors with DEIS or SEIS stacks have shown that low voltage breakdowns are drastically reduced compared to MOS controls with just an SiO$_2$ layer present. A low voltage breakdown is one that causes destructive shorting of the SiO$_2$ layer at average electric fields which are less than the "intrinsic" maximum dielectric strength of SiO$_2$. This intrinsic value is regarded as usually between 8-12 MV/cm depending on oxide thickness [32]. Figure 12 demonstrates this for a DEIS structure which only shows 3 capacitors out of approximately 100 breaking down at low average fields after being sequentially ramped to current levels of 2×10^{-6} A and then 5×10^{-4} A. The portion of this histogram (in 0.5 MV/cm bins) between 8.5 and 10.5 MV/cm is simply a measure of the reproducibility across a wafer of the gate voltage needed to draw a dark current of 5×10^{-4} A ($\approx .1$ A/cm^2). The scatter in the histogram is due to variations in the intervening CVD SiO$_2$ thickness which is of the order of 10% for the CVD reactor used. Ramping the voltage to the same current level again would produce essentially the same distribution as shown in Fig. 12. The breakdown distribution on the control structure fabricated in an identical fashion except for the absence of the Si-rich SiO$_2$ injecting layers and subjected to the same ramped voltage sequence is shown in Fig. 13. Clearly most of the 100 capacitors of the control wafer have broken down at lower voltages than expected for a 5×10^{-4} A Fowler-Nordheim current injected from the Al gate electrode into the SiO$_2$ layer [18].

Low voltage breakdowns are usually associated with localized defects or asperities at the contacting-electrode–SiO$_2$ interface where the injection current increases locally in a drastic manner compared to the current from the remaining electrode area causing shorting of the Si substrate to the gate electrode [32]. The DEIS structure (or SEIS for the polarity favoring injection into SiO$_2$ via the Si-rich SiO$_2$ layer) minimizes these localized contact effects that occur at low average electric fields by building up reversible space charge layers in the Si-rich SiO$_2$ layer opposite to the defects or asperities at the contacting-electrode–Si-rich-SiO$_2$ interface. This field screening effect due to the trapped space

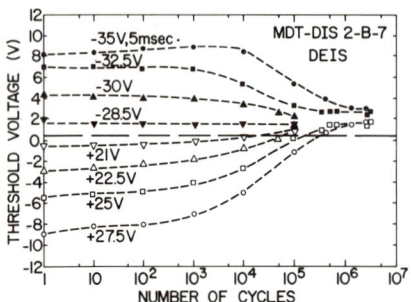

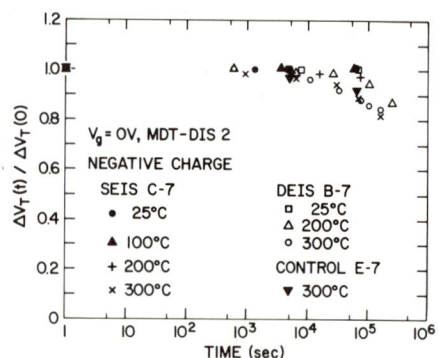

Fig. 10. Threshold voltage after writing and erasing as a function of the number of write/erase cycles for various $V_{W/E}$ conditions on DEIS FETs from wafer MDT-DIS 2-B (340 Å gate oxide from the floating poly-Si layer to the Si substrate and a CVD DEIS stack of 150 Å Si-rich-SiO$_2$ – 100 Å SiO$_2$ – 150 Å Si-rich-SiO$_2$ from floating poly-Si to control gate poly-Si with the Si-rich SiO$_2$ having 46% atomic Si). Solid and open symbols correspond to the threshold voltage after writing and erasing for 5 msec, respectively. The horizontal dashed line indicates the initial threshold voltage of the as-fabricated FETs before cycling. Taken from Reference 21.

Fig. 11. Stored electronic charge (initially $\approx 5 \times 10^{12}$ electrons/cm^2) loss as a function of time on DEIS, SEIS, and the control FETs from the MDT-DIS 2 series for a grounded gate condition $V_g = 0$ V at temperatures of 25°C, 100°C, 200°C, and 300°C. Charge loss is calculated in normalized units of $\Delta V_T(t)/\Delta V_T(0)$ where $\Delta V_T(0)$ and $\Delta V_T(t)$ are the threshold voltage shifts (from the uncharged virgin state) initially after charging and after time t, respectively. The SEIS and control FETs had stacks of 100 Å SiO$_2$ – 150 Å Si-rich-SiO$_2$ and of 100 Å SiO$_2$, respectively, between floating poly-Si and control gate poly-Si. The DEIS FETs are described in Fig. 10. Taken from Reference 21.

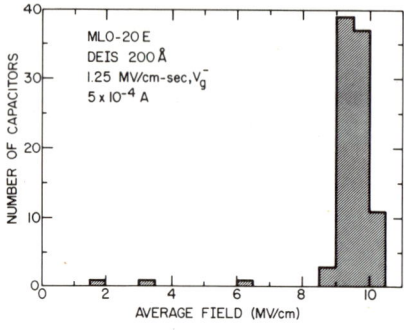

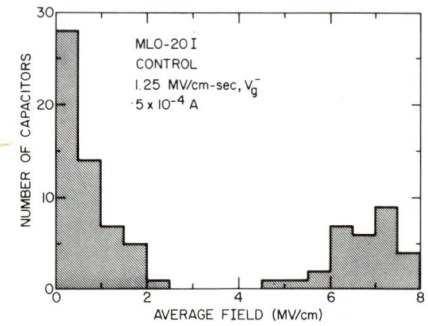

Fig. 12. Histogram (in 0.5 MV/cm bins) of the number of capacitors (.006 cm^2 area) on a DEIS wafer to draw a current of 5×10^{-4} A as a function of the average oxide field, performed after first ramping to 2×10^{-6} A. The CVD Si-rich SiO$_2$ injectors were 200 Å thick and contained 46% atomic Si, the intervening CVD SiO$_2$ was 400 Å thick, and the structure was annealed at 1000°C in N$_2$ for 30 min prior to Al metallization. Samples were ramped with negative gate voltage bias from 0 V at a rate of 1.25 MV/cm-sec. Taken from Reference 21.

Fig. 13. Histogram (in 0.5 MV/cm bins) of the number of capacitors (.006 cm^2 area) on a control wafer with just a 400 Å thick CVD SiO$_2$ layer to draw a current of 5×10^{-4} A as a function of the average oxide field, performed after first ramping to 2×10^{-6} A. The structure was annealed at 1000°C in N$_2$ for 30 min prior to Al metallization. Same experimental conditions as in Fig. 12 were used. Taken from Reference 21.

charge has been demonstrated previously for minimizing injection from a rough poly-Si surface into SiO$_2$ [20,30,31] by using a "W" trapping layer ($\approx$ 70 Å from the injecting interface) which will build up a stable trapped electron distribution [31].

C. SiO$_2$ Trapping Characterization

Charge trapping ledges in dark current as a function of ramped gate voltage characteristics on DEIS or SEIS structures can yield useful information on trapping kinetics and trapped charge location. Examples of such ledges were seen in Fig. 4 at a current level of $\approx 10^{-7}$ A due to certain H$_2$O related sites in the intervening CVD SiO$_2$ layer. The ledge appears, during a dark-current–ramped-gate-voltage measurement, when a current level is reached at which the SiO$_2$ traps start to fill significantly for the constant ramp rate used and build up an internal electric field opposed to the increasing applied field [17,33]. The field near the injecting interface is held approximately constant until the traps are filled which therefore keeps the electron injection current nearly constant. The ledge is much more pronounced on the DEIS structures as compared to the control because of the lower average electric field in the bulk of the SiO$_2$ layer where the electrons are trapped [33]. This lower field minimizes field ionization of trapped charges and/or a reduction in the capture rate of the charge carriers [1]. Similar ledges are also seen for injection from the top Si-rich SiO$_2$ injector for negative gate voltage bias V_g^- [17,19]. From the current position of the ledge I_l and its voltage width ΔV_{g_L}, it can easily be shown that the SiO$_2$ trapping parameters (the electron capture cross section σ_{c_o} and the total number of traps per unit area N_{∞_o}) can be determined, assuming the Si-rich SiO$_2$ layers are highly conductive compared to the SiO$_2$ layer, from the relationships [17,33]

$$\sigma_{c_o} = \frac{qA(dV_g/dt)}{\Delta V_{g_L} I_l} \tag{1}$$

and

$$N_{\infty_o} = \frac{\epsilon_o \Delta V_{g_L}}{q(\ell_o - \bar{x}_o)} \tag{2}$$

where A is the electrode area, q is the charge on an electron (-1.6×10^{-19} coul), dV_g/dt is the voltage ramp rate, ℓ_o is the total SiO$_2$ layer thickness, $\bar{x}_o$ is the centroid of the trapped electron distribution measured from the appropriate injecting Si-rich-SiO$_2$–SiO$_2$ interface, and ϵ_o is the low frequency permittivity of SiO$_2$ ($3.9 \times 8.86 \times 10^{-14}$ F/cm). Furthermore, if partial ledge widths $\Delta V_g^{\pm}$ are measured for both polarities of DEIS structures [17,19] or if ΔV_g^- is measured for a SEIS with only a top Si-rich SiO$_2$ injector and the flat-band voltage shift ΔV_{FB} is deduced from capacitance-voltage measurements [17], it can be shown that the charge centroid $\bar{x}'_o$ measured from the top Si-rich-SiO$_2$–SiO$_2$ interface and the number of bulk trapped SiO$_2$ charges per unit area N_o can be determined from [17]

$$\frac{\bar{x}'_o}{\ell_o} = \left[1 - \frac{\Delta V_g^-}{\Delta V_g^+ (\Delta V_{FB})}\right]^{-1} \quad (3)$$

and

$$N_o = \frac{\varepsilon_o}{q\ell_o}[\Delta V_g^- - \Delta V_g^+(\Delta V_{FB})] \quad (4)$$

where it is assumed that the Si-rich SiO_2 is highly conductive and thin compared to the SiO_2 layer [5,6,17] with the ΔV_{FB} term replacing ΔV_g^+ when a SEIS is used. These relationships given by Eqs. 3 and 4 are the same as the photocurrent-voltage (photo I-V) equations [34] with essentially the same physics applying; that is, $\Delta V_g^\pm$ are a measure of the reduction in field at the appropriate injecting interface due to the trapped charge build-up in the bulk of the SiO_2 layer.

Purposely added "W" traps, CVD SiO_2 traps, and thermal SiO_2 traps have been studied using these ramp I-V techniques with DEIS or SEIS capacitors [17,19]. The results of these measurements have been shown to be in agreement with the determination of the trapping parameters on MOS structures without Si-rich SiO_2 injectors using the standard techniques of avalanche-injection with flat-band voltage tracking [2,35] and photo I-V [34]. The ramp I-V technique has the advantage that it is simpler and faster than the standard techniques, although probably less accurate. For instance, it would be appropriate when large numbers of structures must be monitored.

IV. CONCLUSIONS

Si-rich SiO_2 layers incorporated as an intervening layer between contacting electrodes and an SiO_2 layer have been demonstrated to give high electronic current injection at moderate applied average electric fields. This electron injection phenomenon is believed to be controlled by electric field distortion at the Si-rich-SiO_2–SiO_2 interface associated with the two phase nature of the Si-rich SiO_2 material. Several electrical and physical measurements support this model.

Structures based on the "insulator engineering" discussed here (such as the DEIS with its stacked layers of Si-rich SiO_2 and SiO_2) have been shown to have importance in the area of non-volatile memory, the area of minimizing low voltage breakdowns in MOS structures, and the area of rapidly measuring charge trapping parameters of impurities either purposely or inadvertently introduced into SiO_2 films. The possibility of even faster switching times (perhaps ≤ 1 μsec) for EAROM structures might be demonstrated as devices using DEIS stacks are optimized further through increased electron injection efficiency of the injectors or through design considerations where most of the applied voltage is dropped across the DEIS stack rather than any of the other SiO_2 layers.

In the future, other materials (such as Si-rich Si_3N_4) may also give enhanced electron injection into SiO_2 at even lower applied fields. Another interesting possibility in the realm of this new field of insulator engineering is the use of DEIS stacks with Si_3N_4 replacing the SiO_2 layer.

ACKNOWLEDGEMENTS

The author would like to acknowledge helpful discussions with A. Hartstein, J.M. Gibson, C.M. Serrano, D.R. Young, D.W. Dong, K.M. DeMeyer, E.A. Irene, and N.J. Chou; the use of the unpublished high-resolution cross-section TEM data of J.M. Gibson; the experimental assistance and data interpretation of K.M. DeMeyer on most of the device and voltage breakdown measurements; the materials work and sample preparation of D.W. Dong; the technical assistance of F.L. Pesavento and J.A. Calise; the assistance in device fabrication by the Si facility at the IBM T.J. Watson Research Center and by C.M. Osburn; the support of this work by D.R. Young, M.I. Nathan, and M.H. Brodsky; the critical reading of this manuscript by D.R. Young, K.M. DeMeyer, and M.H. Brodsky; and the partial financial support by the Defense Advanced Research Projects Agency monitored by the Deputy for Electronic Technology, RADC, under Contract F19628-78-C-0225.

REFERENCES

1. D.J. DiMaria, in *The Physics of SiO_2 and its Interfaces*, ed. by S.T. Pantelides (Pergamon, New York, 1978), p. 160 and references contained therein.
2. D.R. Young, E.A. Irene, D.J. DiMaria, R.F. DeKeersmaecker and H.Z. Massoud, *J. Appl. Phys.* 50, 6366 (1979).
3. E.H. Nicollian, C.N. Berglund, P.F. Schmidt, and J.M. Andrews, *J. Appl. Phys.* 42, 5654 (1971).
4. D.J. DiMaria, *J. Appl. Phys.* 50, 5826 (1979).
5. D.J. DiMaria and D.W. Dong, *J. Appl. Phys.* 51, 2722 (1980).
6. D.W. Dong, E.A. Irene, and D.R. Young, *J. Electrochem. Soc.* 125, 819 (1978).
7. T. Matsushita, T. Aoki, T. Otsu, H. Yamoto, H. Hayashi, M. Okayama, and Y. Kawana, *IEEE Trans. Electron. Devices* 23, 826 (1976).
8. M. Hamasaki, T. Adachi, S. Wakayama, and M. Kikuchi, *J. Appl. Phys.* 49, 3987 (1978).
9. J.H. Thomas, III, and A.M. Goodman, *J. Electrochem. Soc.* 126, 1766 (1979).
10. T. Adachi and C.R. Helms, presented at 1979 IEEE Semiconductor Interface Specialists Conference, November 29-December 1, 1979, New Orleans, La., *(unpublished)*.
11. E.A. Irene and N.J. Chou, *private communication*.
12. J.M. Gibson, *private communication*.
13. A.M. Goodman, G. Harbeke, and E.F. Steigmeier, in *Physics of Semiconductors 1978*, ed. by B.L.H. Wilson (The Institute of Physics, Bristol and London, 1979), p. 805.
14. A. Hartstein, J.C. Tsang, D.J. DiMaria, and D.W. Dong, *Appl. Phys. Lett.* 36, 836 (1980).
15. K.H. Beckmann and N.J. Harrick, *J. Electrochem. Soc.* 118, 614 (1971).
16. A. Hartstein, D.J. DiMaria, D.W. Dong and J.A. Kucza, *J. Appl. Phys.*, July 1980.
17. D.J. DiMaria, R.A. Ghez and D.W. Dong, *J. Appl. Phys.*, July 1980.
18. M. Lenzlinger and E.H. Snow, *J. Appl. Phys.* 40, 278 (1969).
19. D.J. DiMaria and D.W. Dong, *Appl. Phys. Lett.* 37, 61 (1980).
20. R.M. Anderson and D.R. Kerr, *J. Appl. Phys.* 48, 4834 (1977).
21. D.J. DiMaria, K.M. DeMeyer, C.M. Serrano, and D.W. Dong, *(unpublished)*.
22. S.K. Lai and D.W. Dong, *(unpublished)*.
23. B.E. Deal, E.H. Snow, and C.A. Mead, *J. Phys. Chem. Solids* 27, 1873 (1966).
24. T.J. Lewis, *J. Appl. Phys.* 26, 1405 (1955).

25. S.M. Sze, *Physics of Semiconductor Devices* (Wiley-Interscience, New York, 1969), Chapters 9 and 10.
26. B. Abeles, *RCA Rev.* 36, 594 (1975).
27. A.S. Grove, *Physics and Technology of Semiconductor Devices* (Wiley, New York, 1967), Chapters 9 and 11.
28. D. Frohman-Bentchkowsky, *Solid State Electronics* 17, 517 (1974).
29. H. Iizuka, F. Masouka, T. Sato, and M. Ishikawa, *IEEE Trans. Electron Devices* ED-23, 379 (1976).
30. D.J. DiMaria, and D.R. Kerr, *Appl. Phys. Lett.* 27, 505 (1975).
31. D.J. DiMaria, D.R. Young, and D.W. Ormond, *Appl. Phys. Lett.* 31, 680 (1977).
32. P.L. Solomon, *J. Vac. Sci. Technol.* 14, 1122 (1977).
33. P.L. Solomon, *J. Appl. Phys.* 48, 3843 (1977).
34. D.J. DiMaria, *J. Appl. Phys.* 47, 4073 (1976).
35. E.H. Nicollian and C.N. Berglund, *J. Appl. Phys.* 41, 3052 (1970).

HIGH FIELD CONDUCTION IN THICK OXIDE MNOS CAPACITORS ON
P-TYPE SILICON

I. Kashat and N. Klein, Dept. Electrical Engineering,
Israel Institute of Technology, Haifa, Israel

ABSTRACT

Observations on electrical conduction are interpreted with transport processes of electrons. A near constant current step in the I-V characteristic of the inversion mode is due to limitation of electron supply from the silicon. The step is terminated by breakdown of a deep depletion layer in the substrate, which is influenced by electron-electron interactions at the inversion layer. Positive charges formed at fields above 8 MV/cm are trapped at the silicon-silicon dioxide interface. Currents in the accumulation mode are orders of magnitude smaller than in the inversion mode, or in single layer insulators owing to low fields for current injection into the nitride, electron trapping in the nitride and absence of hole conduction.

INTRODUCTION

Many investigations on electrical transport in silicon dioxide-silicon nitride dual layers have been reported since the basic studies of Frohman-Bentchkowsky and Lenzlinger[1]. Reviews show [2-4] that most of the work centered on memory elements with very thin oxides. Transport in dual layers with thicker oxides, in which electrons were shown to be the dominant charge carriers[5] were studied less and we report here on transport properties, which have so far received relatively little attention. We will describe briefly studies on electrical conduction with p-type silicon substrates in the accumulation and also in the inversion mode[6], investigations on the effect of positive charges produced at very high fields and on the influence of the oxide-nitride thickness ratio on transport.

SAMPLES AND MEASUREMENTS

MNOS capacitor samples were prepared by conventional techniques on p-type 0.4 Ω-cm <100> silicon wafers. The oxide was thermally grown in dry oxygen and the nitride by chemical vapor deposition. The counterelectrodes were aluminum dots of 0.015cm^2 area and 1 μm thickness. The dual layer thicknesses and the type of sample will be denoted by pairs of numbers, the first number giving the oxide and the second the nitride thickness in nanometer. The five types of samples investigated were: 10-20; 20-10; 25-110; 45-50 and 70-25. Observations consisted of current and capacitance measurements as function of voltage and time and of determination of the flat-band voltage V_{FB}, within less than 25 msec after a voltage pulse. A tungsten filament lamp was used to examine the effect of illumination on transport in the sample.

OBSERVATIONS IN THE INVERSION MODE

Approximate steady state I-V characteristics of a 20-10 sample in the inversion mode were recently discussed in ref.[6]. Three such characteristics, marked L_1, L_2 and L_3 are presented in fig. 1 with the data of open marks. The characteristics consist of (a) a steeply rising lower voltage range, (b) a nearly constant current step and (c) a steeply rising upper voltage range. The current of the step increases with increase of illumination $L_1 < L_2 < L_3$. Data of the characteristics are current values measured 20 sec after the application of a step voltage, when inversion layers were already formed. The observations of fig. 1 are completed by V_{FB} measurements, presented by solid marks, the same type of open and solid mark be-

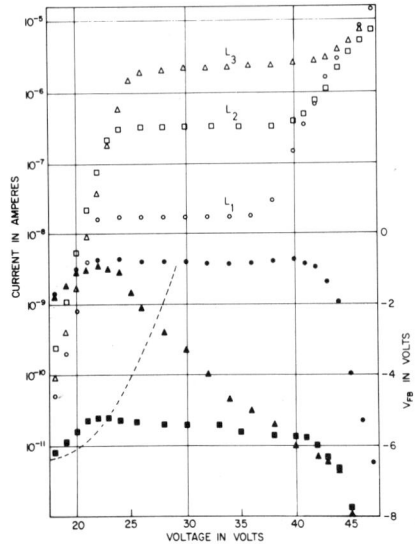
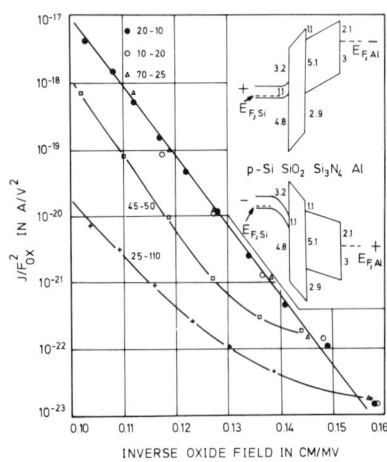

FIG. 1 I-V characteristics (open marks) and V_{FB} measurements (solid marks) of a 20-10 sample in the inversion mode. Dashed line represents I-V characteristic for opposite polarity.

FIG 2 Fowler-Nordheim plots of five types of MNOS samples in the inversion mode. Schematic energy band diagrams in the inset.

longing to one experiment.

For the discussion of conduction schematic energy band diagrams are shown in the inset of fig. 2 for both polarities with energies in eV. Fowler-Nordheim plots were drawn in fig. 2 for the interpretation of conduction in the lower voltage range. The oxide field of the plot was calculated disregarding charges in the insulator with relative permittivities 3.9 and 7.6 for the oxide and nitride respectively. Fig. 2 shows that the data of the 20-10 and approximately those of the 10-20 and 70-25 samples fall on a straight line with a slope of 240 MV/cm. This value is in close agreement with slopes quoted for electron tunneling from silicon into silicon dioxide[8,9], indicating that this mechanism determines conduction in the three samples.

The plots of the 45-50 and 25-110 samples fall much below the straight line in Fig. 2. The current is assumed to be due to electron tunneling in these cases too and the current lowering is attributed to the effect of electron trapping in the nitride. This assumption was checked by calculating flat band voltage shifts, V_{FB}, with the current lowering data of Fig. 2. Calculated and measured V_{FB} data were in better than 5 % agreement. These results show that, when the nitride layer is thick, currents in the inversion mode can be much smaller in dual than in single layer insulators.

The current in the step of the characteristics in Fig. 1 was found to be limited by electron supply from the silicon by diffusion and by thermal and optical generation[9,6]. The voltage increase in the step serves to form a deep depletion layer in the substrate. This is proven by measuring decrease of the sample capacitance, in good agreement with calculations. Breakdown of the deep depletion layer removes the limitation on electron supply. The current resumes its rapid rise with voltage beyond the step and electron tunneling determines the current as in the lower voltage range[6].

Discussion of the electron injection relationship indicated little difference in the lower and upper voltage ranges of the characteristic[6]. It was assumed therefore that the breakdown voltage of the deep depletion layer V_b is the voltage difference across the step defined approximately by intersections of linear extensions of the characteristic. V_b=20 V was observed earlier in a 0.4 Ω-cm sample on the application of a fast ramp voltage pulse [10]. V_b was found to increase with illumination in the range of 15 to 20 V in most samples tested here and the slope of the characteristics was smaller in the upper than in the lower voltage range.

The increase in V_b with illumination can be attributed to the effect of the inversion layer on impact ionization by the electron current, noting that the bulk of the ionization takes place near the inversion layer[6]. The expected effect of the inversion layer is a decrease in carrier multiplication by the electron current due to electron-electron interactions. The larger the electron density in the inversion layer and the oxide field, the larger the decrease in multiplication. Since the current of the step and the oxide field increase with the intensity of illumination an increase in breakdown voltage V_b with illumination and a decrease in the slope of the characteristic is expected.

This can explain the observations in Fig. 1 and those of ref.[6] on a 20-10 sample in which V_b was 14.8 V for an experiment in darkness, but 18.2 V when the illumination was strong. An increase in V_b with illumination was found also for the samples 10-20, 45-50 and 70-25. No change with increasing illumination was found however, in the 20 V step voltage V_b of 25-110 samples and the reason for this is not clear.

It is interesting to examine now changes in the flatband voltage V_{FB} illustrated by the experiments of Fig. 1. The weak and strong illumination V_{FB} curves, denoted by ● and ▲ dots respectively, show a small initial rise with increasing voltage and also a range strongly decreasing at widely differing voltages. The currents though at which the V_{FB} curves begin to decline are approximately equal, 10^{-7} A at an oxide field of about 8.5 VM/cm. Impact ionization is expected in the oxide at this field[11] and trapping of holes very close to the silicon-silicon dioxide interface[12]. Practically no effect is expected on the I-V characteristics, when the positive charge is at the interface and this is indeed shown by the unchanged slopes in Fig. 1. A further check on the location of the positive charge was obtained by applying a large negative pulse to the counterelectrode of a sample, reducing the initial V_{FB} to -6.2 V. Subsequently measured data presented by ◻ and ◼ in Fig. 1 show that the initially trapped positive charge had no significant effect on the lower voltage range of the I-V characteristic and was located at the interface. The effect of the inversion layer on carrier multiplication can be tested also by trapping positive charge at the interface. The positive charge increases the charge in the inversion layer and a decrease in carrier multiplication

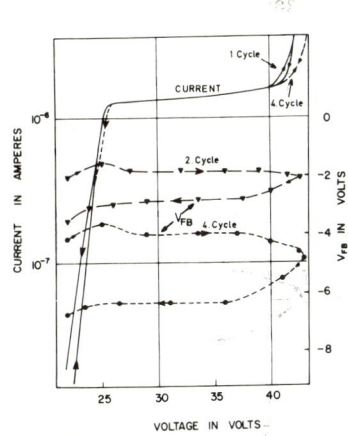

FIG. 3 Changes in I-V characteristics and V_{FB} measurements on the application of four cycles of ramp voltages to 20-10 sample.

and an increase in V_b is expected. The presence of an effect was examined on 20-10 samples with experiments, which consisted of four up and down ramp voltage cycles at a voltage change rate of 0.5 V/sec. Voltage was applied in these experiments for much shorter times than in those of Fig. 1 and the trapped positive charge was found to increase in the course of the experiments. Results are presented in Fig. 3, which for better legibility gives only significant current and V_{FB} data of the cycles. The first cycle produced little positive charge and marks ▼ and ● for the second and fourth cycle indicate the increase in positive charge. The I-V characteristic of the first cycle is represented by the full line. The increase in positive charge was nearly $2 \times 10^{12}/cm^2$ by the end of the fourth cycle, the upper range of the I-V characteristic shown by the broken line shifted by nearly 1 V to the right and V_b increased with increase in positive charge trapping.

OBSERVATIONS IN THE ACCUMULATION MODE

Current changes become small 10 seconds after the application of a negative step voltage to the aluminium electrode and quasi-steady I-V characteristics were obtained with measurements 20 seconds after voltage application. Such character-

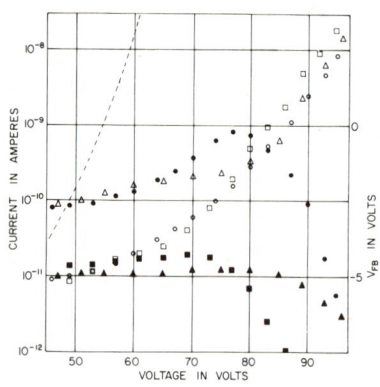

FIG. 4 I-V characteristics (open marks) and V_{FB} measurements (solid marks) of a 45-50 sample in the accumulation mode. Dashed line represents I-V characteristic for opposite polarity.

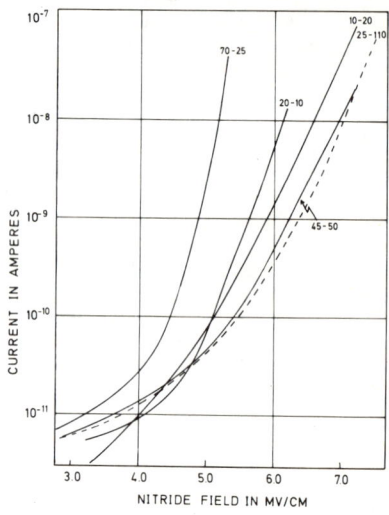

Fig. 5 Current versus nitride field plots of five types of MNOS samples in the accumulation mode.

istics are presented in Fig. 4 for 45-50 samples by open marks, the solid marks giving V_{FB} data obtained after a voltage pulse. Fig. 5 presents current versus a nitride field curves for a comparison of the five types of samples, data dots being omitted for better legibility. Fig. 5 offers only a rough comparison, since nitride fields were calculated neglecting charge trapping and these fields are larger than the injection fields at the aluminum.

The characteristics consist of a slowly rising lower voltage and a steeper higher voltage range. The current is assumed to be due to electron injection from the aluminum electrode; injected hole current is insignificant owing to the thickness of the oxide layer[5]. The slow rate of current rise in the lower voltage range is ascribed to increasing electron trapping, indicated in Fig. 4 by the rise of V_{FB} with voltage for the sample marked by o ●. Application of voltage pulses of opposite polarities producing positive charges prior to measurements had differing effects: For a positive pulse on the sample ▫ ▪ the initial V_{FB} decreased to -5V. Since comparison of the I-V data ▫ and o shows no change in the characteristics, the positive charge produced by the initial pulse appeared to be trapped at the silicon interface. A negative pulse to the sample ▲▴ produced again a decrease in V_{FB} but also an increase in the current in the lower voltage range shown by comparison with the characteristic marked o. The effect is likely to be due to hole trapping in the nitride.

The V_{FB} curves in Fig. 4 begin to decrease rapidly roughly at voltages where the slope of the characteristics becomes steeper. Fields in the oxide vary at these voltages between 8 and 10 MV/cm. Assuming impact ionization in the oxide the decrease in V_{FB} and the increase of the slope of the I-V characteristics can in part be ascribed to the effect of hole trapping in the silicon nitride. The decrease in V_{FB} at the bend of the I-V characteristics was pronounced also with 10-20 and 25-110 samples.

Dashed lines in Fig. 1 and 4 present I-V characteristics for a polarity opposite to that of the characteristics given by open marks. Comparison of these characteristics shows that currents in MNOS dual layers are several orders of magnitude smaller in the accumulation than in the inversion mode; they are very much smaller than in single nitride, or oxide layers of the same thickness and voltage. This is due to the absence of hole conduction, to electron trapping in the nitride and due to differences in electron injecting fields, which are relatively large in the oxide in the inversion mode, but low in the nitride in the accumulation mode. These results show that suitable combination of dual layer thicknesses, permittivities and band structure properties can produce a considerable decrease in electrical conduction.

The authors express their thanks to R. Gdula for the supply of samples and to P. Balk for discussions.

REFERENCES

1. D. Frohman-Bentchkowsky and M. Lenzlinger, J.Appl.Phys. $\underline{40}$, 3307 (1969).
2. J.J. Chang, Proc. IEEE $\underline{64}$, 1039 (1976).
3. J.F. Verwey, Adv. Electr. Electron. Phys. (ed.L. Marton), Acad. Press, Vol.41, 249-309 (1976).
4. M. Pepper, Proc. Conf. Insulating Films on Semiconductors, ed. G.G. Roberts and M.J. Morant, Inst. Physics Bristol, pp. 193-205, 1979.
5. E. Suzuki and Y. Hayashi, J.Appl.Phys. $\underline{50}$, 7001 (1979).
6. S. Davidoff, I. Kashat and N. Klein, Appl.Phys.Lett. $\underline{34}$, 782 (1979).
7. M. Lenzlinger and E.H. Snow, J.Appl.Phys. $\underline{40}$, 278 (1969).
8. Z.A. Weinberg, Solid-State Electron. $\underline{20}$, 11 (1977).
9. P.M. Solomon, Appl.Phys.Lett. $\underline{30}$, 597 (1977).
10. A. Goetzberger and E.H. Nicollian, J.Appl.Phys. $\underline{38}$, 4582 (1967).
11. P. Solomon and N. Klein, Solid State Comm. $\underline{17}$, 1397 (1975).
12. P.M. Solomon and J.M. Aitken, Appl. Phys. Lett. $\underline{31}$, 215 (1977).

DIELECTRIC BREAKDOWN IN THERMAL SiO$_2$ GROWN FROM DOPED

POLY CRYSTALLINE SILICON THIN FILMS

Michael Berberian
American Microsystems Inc., Pocatello, Idaho 83201

INTRODUCTION

The understanding and characterization of dielectric breakdown in thermal SiO$_2$ is recognized as an important concern in the physics of MOS insulators. A large body of work has accumulated on this subject particularly for thermal SiO$_2$ layers grown from single crystal silicon substrates. Considerably less is known when doped polycrystalline silicon (polysilicon) is used as the substrate. In particular, the effects of the polycrystalline structure and the dopant type on field induced breakdown are of interest because of the increasing use of polysilicon thin films in semiconductor processing technology. In the physics of MOS processing the properties of doped polysilicon have found wide use as a high temperature electrical interconnecting layer requiring oxide insulation, and in the near future SiO$_2$ grown from polysilicon will provide oxide isolation for VLSI (Very Large Scale Integration) processing schemes involving self-aligned contact technologies. The purpose of this report is to evaluate the field breakdown characteristics of SiO$_2$ thin films grown at high temperatures from polysilicon (hereafter referred to as polyoxide) doped with the most commonly used elements (boron, phosphorus, arsenic) and sourcing methods.

SAMPLE PREPARATION AND MEASUREMENT PROCEDURE

Polysilicon films were deposited on 75mm diam. single crystal wafers (n type, <100>orientation, 3-5Ω-cm prime starting material) on which a thermal oxide (SiO$_2$) isolation layer of approximately 2500-3000 Å was first grown. The polysilicon films were deposited in a low pressure chemical vapor deposition (LPCVD) furnace tube system. The reaction used was the pyrolysis of silane and oxygen carried out at 700°C. The film thickness was 4000±100 Å and this measurement was made using optical reflectometry. Following deposition of polysilicon the samples were precleaned for dopant deposition/sourcing. Following dopant sourcing an anneal cycle was carried out to drive and activate the dopant consisting of a soak in an N$_2$ atmosphere for 15min. at 1000°C. The samples were then deglazed by immersion in HF and V/I readings recorded for each wafer. In Table 1 the various dopants and sourcing methods are outlined. The oxidation preclean used was of a well known type used in MOS processing consisting of immersions in 10:1 (H$_2$O:HF) and 5:1 (H$_2$SO$_4$:H$_2$O$_2$) baths. The oxidations were carried out in a fused silica tube furnace using a dry, flowing oxygen atmosphere at a temperature of 1075±1°C for periods of time between 12 and 24 min. A metal deposition followed (Al-2%Si, rinse preclean only). Masking then proceeded with first defining capacitors in a metal mask, then in a second masking step, opening contact windows to the polysilicon substrate. Finally, all samples received a H_2 heat treatment at 490°C for 20 min. In Fig. 1 the sample geometry is given.

TABLE 1 Elements Used as Dopants and Sourcing Methods

Dopant	Sourcing Method	Typical V/I	Field Strength (MV/cm)
P	Solid phase (Vapox-8%P)	3-5	1.7
	Vapor phase ($POCl_3$ 950°C)	5-6	4.3
	Ion Implant*	20-25	3.8
B	Vapor Phase (BBr_3 1070°C)	7-10	4.5
	Ion Implant*	15-20	3.6
As	Vapor Phase (Capsule)	5-6	1.8
	Ion Implant*	50-55	2.0

*Dose 5×10^{15}, range 0.1µm

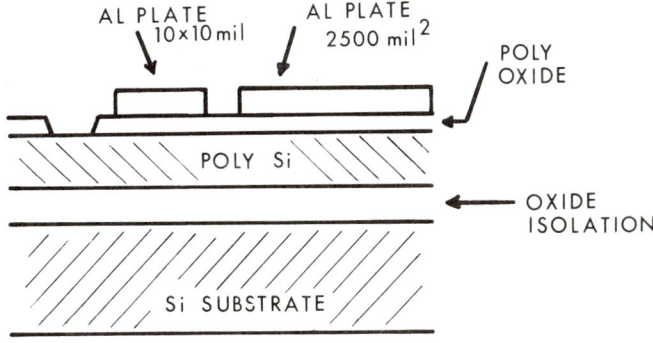

Fig. 1. The sample geometry.

The electrical testing was carried out using a capacitance meter (Boonton 72b) to derive dielectric thickness values. The breakdown voltage measurements were carried out using a computer aided automatic test system (LOMAC ATE system). This system utilized an auto prober/stepper to consecutively probe and step to the next capacitor structure, thus sampling the entire wafer area in a relatively short period of time. The sampling could be modified to skip every other row of die, thus allowing a check of results using a bench probe station and curve tracer. The test programs detected breakdown voltage by measuring a threshold current (1µa or 10µa) at an applied voltage.

RESULTS AND DISCUSSION

The procedure outlined above was carried out over a period of approximately one year. Phosphorus and boron doped polysilicon substrates were produced most frequently since they represented production cycles. Wafer lot sizes were typically 20-30 wafers.

The breakdown voltage distributions show systematic differences between the dopant types and the sourcing methods. In Fig. 2 typical breakdown voltage distributions for phosphorus (sourced via vapox layer, and vapor phase) and boron (vapor phase) are given. In Fig. 3 the derived breakdown field is plotted versus oxide thickness for one experimental run. The samples sourced via vapor phase deposition show similar behavior in contrast to the vapox sourced samples which had lower breakdown fields. In Table 1 results of representative runs are shown. The similarity between samples with phosphorus and boron doped polysilicon substrates continues when ion implantation is used as the sourcing method. For the case of arsenic, however, the ion implantation and the capsule sourcing methods give lower breakdown fields than for dielectrics grown with phosphorus or boron doped polysilicon using either ion implantation or vapor phase sourcing.

There is no simple explanation for the apparent differences in breakdown characteristics between sourcing methods or between arsenic and phosphorus and boron. Recently, the morphology of polyoxide has been shown to be more complex than previously believed (1). Enhanced grain growth results when polysilicon is doped and oxidized at high temperatures. Grain growth is diffusion controlled and may be dependent on dopant type and concentration. The grain structure of polysilicon resulting from heat treatments appears to be one of normal grain growth (1) rather than a duplex structure. Presently, in the description of field breakdown in polyoxide, no connection exists between polysilicon microstructure (doped or undoped), and the breakdown strength of the high temperature thermal oxide grown from it. Asperities occurring in polysilicon have been mentioned and some evidence for their detection has been given (2). These asperities appear to flatten out at temperatures of approximately 1150°C with accompanying increases in the field breakdown strength of the polysilicon. In future studies the connection between this phenomenon and grain growth, and the influence of doping element and grain growth need to be determined.

Finally, attention should be drawn to the measurement technique. The effects of charge trapping have been shown to occur in polyoxide and may be eliminated by the use of a floating gate capacitor structure (3). In the measurements described here, reversal of the applied voltage polarity produced no significant changes in the breakdown distribution. Nevertheless, the two types of sample geometries could produce field breakdown strength differences of approximately 1 MV/cm (3) for current densities of 10^{-6} A/cm^2.

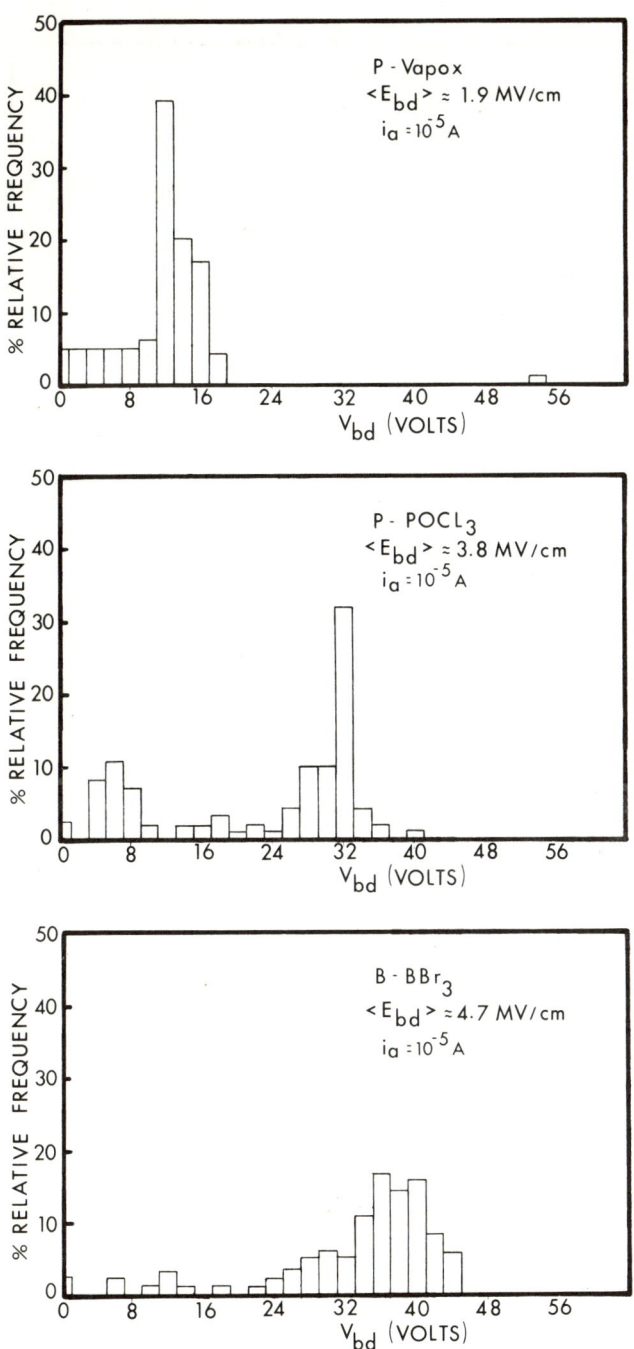

Fig. 2. Breakdown voltages for polyoxide

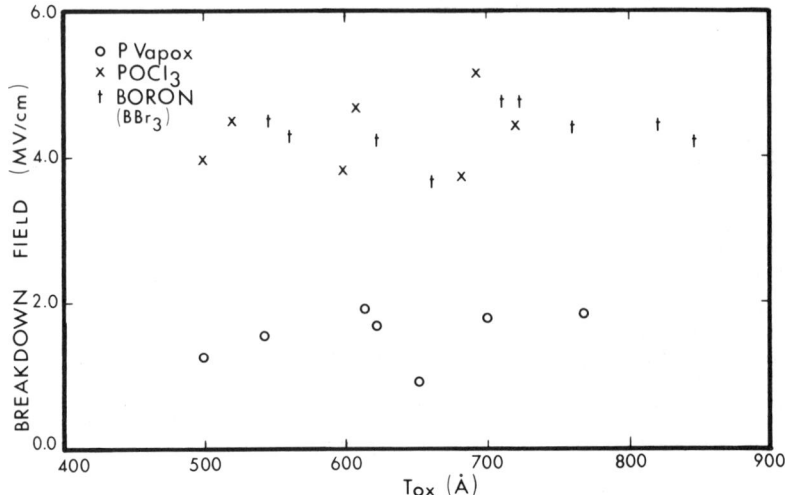

Fig. 3. Breakdown field versus oxide thickness for P and B doped polysilicon.

REFERENCES

(1)   E. A. Irene, E. Tierney, and D. W. Dong, Silicon Oxidation Studies: Morphological Aspects of the Oxidation of Polycrystalline Silicon, J. Electrochem. Soc., 127, 705 (1980).

(2)   R. M. Anderson and D. R. Kerr, Evidence for Surface Asperity Mechanism of Conductivity in Oxide Grown on Polycrystalline Silicon, J. Appl. Phys. 48, 4834 (1977).

(3)   C. Hu, Y. Shum, T. Klein, and E. Lucero, Current-field Characteristics of Oxides Grown from Polycrystalline Silicon, Appl. Phys. Lett. 35(2), 189, (1979).

THE EFFECT OF DIFFUSION ON THE PHOTOCONDUCTIVITY OF THIN FILMS*

R. C. Hughes and R. J. Sokel
Sandia National Laboratories†
Albuquerque, New Mexico 87185

ABSTRACT

The equations governing time dependent and steady state photoconductivity are solved analytically for some approximations and numerically in exact form. The diffusion terms are shown to be important especially in thin films and some data of Farmer and Lee on photoconductivity in MOS structures is reinterpreted with losses due to diffusion to the contacts included.

INTRODUCTION

The simplest photoconductivity problem involves two metal plates, in plane-parallel geometry, separated by an insulator layer with no free carriers in the dark. Electron-hole pairs are created in the insulator by band gap or other ionizing radiation and a photocurrent is observed in an external circuit when a voltage is applied. The first serious efforts to analyze the current-voltage behavior of such a device occurred in the early 20th century with the interest in air and liquid ionization chambers.(1) Even without carrier trapping which we often face in our solid photoconductors, the complete analysis of the photoconductivity is quite complicated, and without approximations, results in differential equations which are only soluble by numerical methods. One of the first approximations that is usually made is to ignore the gradients in carrier concentrations which lead to diffusion currents in favor of using electric fields which are high enough to mask diffusion effects and also neglect variations in electric field (spatially) caused by concentration gradients in the charge carriers. A good summary of precomputer solutions and their application to ionization in liquids has been given by Hummel and Schmidt.(1)

The approximations usually work fine for very thick insulating layers, but in thin film photoconductors and insulators, diffusion of the charge carriers to the metal (or semiconductor) contact cannot be ignored in many experimental situations. In this paper, we will discuss the results of approximate analytical solutions and exact numerical solutions to the photoconductivity equations including diffusion. We will compare the results to some published data on radiation-induced conductivity in MOS devices where diffusion was ignored in the analysis. Diffusion is also important in thicker photoconductors when trapping of one sign of carrier causes large distortions in the electric field inside of the sample.

PHOTOCONDUCTIVITY EQUATIONS

The current density in the one-dimensional case is

$$J = J_n + J_p + \varepsilon \varepsilon_o \frac{\partial E}{\partial t}(x,t) \quad (1)$$
$$J_n = eD_n(dn/dx) + \mu_n enE(x) \quad (2)$$
$$J_p = -eD_p(dp/dx) + \mu_p epE(x) \quad (3)$$

---

*This work was supported by the U. S. Department of Energy under Contract DE-AC04-76-DP00789. †A U. S. Department of Energy facility.

where D_n and D_p are the diffusion constants for electrons and holes respectively, μ_n and μ_p are the mobilities of the electrons and holes respectively (related by the Einstein relation), n and p are the carrier densities for electrons and holes, and E is the electric field. It is useful to note that if the system is in steady state, J is a constant for all values of x, even though J_p and J_n may vary quite a bit with x. This relation comes from the continuity equations which govern the transient response of the system:

$$\frac{\partial p}{\partial t} = D_p \frac{\partial^2 p}{\partial x^2} - \mu_p \frac{\partial(E \cdot p)}{\partial x} + g - kp(N_t^o - N_t^+) \tag{4}$$

$$\frac{\partial n}{\partial t} = D_n \frac{\partial^2 n}{\partial x^2} + \mu_n \frac{\partial(E \cdot n)}{\partial x} + g - \gamma n N_t^+ \tag{5}$$

$$\frac{\partial N_t^+}{\partial t} = kp(N_t^o - N_t^+) - \gamma n N_t^+ \tag{6}$$

where g is the rate of production of electron-hole pairs, assumed for the present to be independent of time, electric field and space; k is the trapping rate of holes into a uniform distribution of neutral traps, N_t; N_t^+ is the concentration of holes in those traps; and γ is the "recombination coefficient" of electrons with those trapped holes. Other trapping or recombination rates could be included in the numerical solutions, but for the present this seems like a good approximation for insulators like SiO_2,(2) and PbO.(3) Similar approximations were made by Churchill, et al.(4) in their numerical solutions for charging of MOS capacitors. The correct solutions to these differential equations requires that Poisson's equation be satisfied

$$\nabla^2 V(x) = \frac{-e}{\varepsilon \varepsilon_o}(p - n + N_t^+) = \frac{-e}{\varepsilon \varepsilon_o} \rho(x) \tag{7}$$

where V is the potential (including any applied bias) and ε is the dielectric constant for the insulator and ε_o is the permitivity of free space.

Some of the approximate solutions are given in Ref. 1, including the simple high field limit, J = geL, with L the thickness of the insulator. Another important specification is the boundary conditions at x = 0 and x = L, where the carriers are extracted at the electrodes. The simplest boundary condition is to let n(0,L) = p(0,L) = 0 which has been used in most plasma physics and ionization chamber problems. The rationale is that the electrons, holes or ions are unstable energetically with respect to the metal or semiconductor contact, so the recombination velocity should be high. It is possible to use a recombination rate constant (often called the surface recombination velocity in semiconductor problems), but all values of the rate constant above a certain level give the same results as the simple condition and without more specific knowledge of the rate constant it does not seem necessary to introduce a more complicated boundary condition.

A solution of interest which has not appeared anywhere to our knowledge is the one in which the steady state space charge is not sufficient to disrupt the initial field, E = V/L where V is the applied bias. Also letting k,γ = 0 and assuming steady state, the current-voltage characteristic is

$$\frac{J}{geL} = \frac{e^{V/kT} + 1}{e^{V/kT} - 1} - \frac{2kT}{V} \tag{8}$$

given as the ratio of the current to the saturation current. It should be noted that the solution is independent of the magnitudes of the diffusion constants, D_n and D_p. Of course, if the difference in D's leads to a significant space charge distortion of E, the solution will no longer be valid. However, for thin films, extraordinarily high values of g have to be used to generate sufficiently high concentrations of carriers in thin films to cause a deviation from Eq. 8. For example, in a 5000 A SiO_2 film with D_n = .5 cm^2/s and D_p = 10^{-7} cm^2/s and g = 10^{14} and V = .05 volt, the numerical solution with no trapping is quite close to the prediction of Eq. 8.

NUMERICAL SOLUTION

If transient or pre-steady state solutions are desired [or steady state with appreciable space charge distortion of E(x)], then numerical analysis is required. Specifically, the particle density and current density for both electrons and holes, the trapped charge density, and the electric field as functions of position and time constitutes a complete solution to Eqs. 4 through 6. In addition, the total current, which is the sum of the particle currents and the displacement current (which is only a function of time) is calculated. The current voltage characteristic is derived from the value of the total current in steady state.

A standard procedure for solving partial differential equations numerically is followed. The continuity equations are discretized with respect to the spatial variable according to a scheme introduced to model transport in a p-n junction.(5) This scheme avoids potential unphysical results which can occur if conventional linear difference approximations are used. Spatial discretization of the continuity equations reduces the partial differential equations into a set of first order ordinary differential equations in the time variable which can then be easily solved with techniques for solving initial value problems. Because the electron mobility and the hole mobility can differ by many orders of magnitude, different time constants are introduced and the system is said to be "stiff." Thus, special care must be taken to obtain accurate transient solutions. A differential equation solver based on the work of Gear has been employed for this purpose.(6)

Several approaches for solving Poisson's equation numerically are available. The method here takes advantage of the fact that in one dimension Poisson's equation can be integrated analytically to obtain E in terms of the charge density ρ,

$$E(x) = \frac{1}{\varepsilon\varepsilon_o} \int_0^x \rho(x')dx' - V/L - \frac{1}{\varepsilon\varepsilon_o} \int_0^L (1 - x'/L)\rho(x')dx' \quad (9)$$

where V is the applied voltage. This result is used to eliminate E in favor of p and n in the continuity equations.

A simple expression for the total curent, J, is obtained by differentiating Eq. (9) with respect to t and comparing the result to the definition of the total current. It is found that

$$J = L^{-1} \int_0^L (J_p + J_n) dx + \frac{\varepsilon\varepsilon_o d}{dt} (V/L) \quad (10)$$

The total current is the spatial average of the particle currents plus the displacement current due to a time varying external field. This result follows from the fact that the total current is a function of t only. It can be extended to three dimensions if the integral is understood to be replaced by a path integral.

DISCUSSION

Thin SiO_2 films provide a difficult test of the numerical solutions since the electron and hole mobilities are known from transient experiments to be orders of magnitude different. Our code has had no trouble giving the correct predictions for cases which can be solved analytically like Eq. 8, despite the "stiffness" of

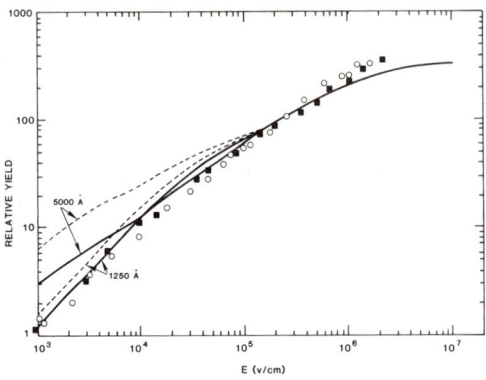
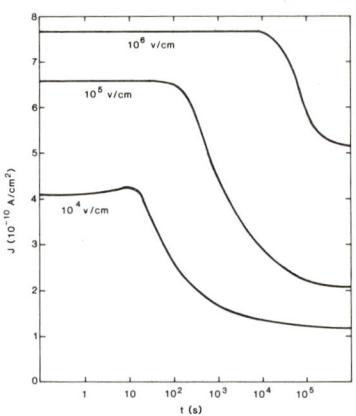

Fig. 1. The relative photoconductivity yield for a 1250 A oxide (■) and a 5000 A oxide (○) vs. applied field. The dashed lines are the theory predictions for geminate recombination and diffusion alone, and the solid lines are the numerical solutions for trapping with parameters given in the text.

Fig. 2. The numerical solutions for the time dependence of the photocurrent for the 5000 A oxide for the given fields. Parameters are given in the text.

the difference equations thus giving us confidence in the time dependent and non-linear regions. A few years ago Farmer and Lee(7) published data for steady-state photocurrents in MOS devices and had problems interpreting the data at low-electric field strengths. Their data, normalized to gel or relative yield, is given in Fig. 1 for two oxides of thickness 1250 and 5000 A. They did not take diffusion into account, and their only loss mechanisms were geminate recombination and trapping. The dashed lines in Fig. 1 contain the effect of diffusion and the known geminate recombination yield (which is field dependent).(2) Note that the thickness dependence appears in this method of plotting yields because the abscissa is in field rather than voltage units. Even the most perfect oxide, without any traps, would be expected to have a relative yield no larger than the dashed lines, unless "g" was so large that bulk recombination between free electrons and holes could occur. The g in the Farmer and Lee experiments is about 10^{14} cm^{-3} s^{-1}, and it would require a g of about 10^{18} to bring the steady state concentrations up to the point where bulk recombination would be important.

The dashed curve is a pretty good fit to the thinner (1250 A) oxide, which indicates that trapping is not too important. However, the fact that the thicker oxide yield falls far below the diffusion curve indicates that some trapping needs to be included in the model. The solid line gives the numerical solutions for the yield for the following parameters: hole trapping - k = 10^{-14} cm^3 s^{-1}, hole trap density - 10^{18} cm^{-3}, $\gamma = 10^{-7}$ cm^3/s, D_n = .5 and $D_p = 10^{-7}$ cm^2/s. The data in Fig. 1 was taken from the integrated charge during a 50 ms exposure to x-rays and is thus only steady state in the sense that 50 ms is long enough for the sweep out and diffusion of the free carriers to come to steady state. The numerical solutions show that 50 ms is long enough if one ignores the dispersive transport of the holes, which our previous data shows is not too bad an approximation. The trapped hole concentration is of course not in steady state but is building up a constant rate for all the fields and thicknesses shown in Fig. 1. At a time later than 50 ms the trapped holes cause a severe distortion in the electric field and

at a much later time a real steady state is reached. The numerical solutions for
the time dependence of the total current are given in Fig. 2 for the thicker oxide,
showing the transition from the early "steady" state to the real steady state.
The trapping brings the two thicknesses much closer together in relative yield,
emphasizing the fact that trapping and diffusion have the opposite effect on
yield vs. thickness. An increase in trapping rate for holes does not bring the
5000 A curve any lower in the low-field regime and some electron trapping must be
added to make the curve coincide with the data. An electron trapping rate of
about 4×10^9 s^{-1} is needed. There is no other evidence for a short electron
lifetime in these oxides, although the fact that the thicker oxide was produced by
a different company than the thin oxide make it plausible that the electron- and
hole-lifetimes are different in the two kinds of oxide.

The decay in current as steady state is approached has been seen in a number of
films including SiO_2,(7) PET, polystyrene and other polymer films,(8) but data has
not been taken over sufficient time to accurately check the model (note the log
time abscissa). The reported current decays have the general shape predicted in
Fig. 2, and we plan future studies to see if consistency can be found between
transient transport parameters and the long term steady state current. Other
areas where the full photoconductivity code will be helpful include predicting
flatband shifts in field effect devices, photocurrents and efficiencies in thin
film solar cells, thermally stimulated currents in insulating films and behavior
of electrets. There are additional physical effects which will be incorporated
into the code. These include: (a) geminate recombination which implies that the
value of g will change in different parts of the sample as the space charge dis-
torts the electric field; (b) field induced injection from the metal-insulator
contact - a new source of carriers in the problem; (c) the dispersive transport
of carriers which is common in amorphous materials; and (d) non-uniform trap
distributions.

In summary, we have shown that diffusion can have important effects on the photo-
condutivity of thin films and that numerical solutions to the photoconductivity
equations can give new insight into complicated behavior of MIM and MIS structures
under excitation.

REFERENCES

1. A. Hummel and W. F. Schmidt, Radiation Research Review 5, 199 (1974).
2. R. C. Hughes, Solid State Electronics 21, 251 (1978).
3. A. M. Goodman and A. Rose, J. Appl. Phys. 42, 2823 (1971).
4. J. N. Churchill, F. E. Holmstrom and T. W. Collins, J. Appl. Phys. 50, 3994 (1979).
5. D. L. Scharfetter and H. K. Gummel, IEEE Trans. Electron. Devices 16, 64 (1969).
6. A. C. Hindmarsh, Computer Code GEAR (Lawrence Livermore Lab, Livermore, CA, 1974).
7. J. W. Farmer and R. S. Lee, J. Appl. Phys. 46, 2710 (1975).
8. H. Maeda, M. Kurashige and T. Nakakita, J. Appl. Phys. 50, 7247 (1979).

THE KINETIC BEHAVIOUR OF MOBILE IONS IN SiO₂ STUDIED WITH TSIC AND TVS MEASUREMENTS

Ji Li-jiu, Wang Yang-yuan, Zhang Li-chun and Ni Xuie-win

Department of Computer Science, Beijing University.
Beijing, China

ABSTRACT

The measurement results about trapping energy level of the mobile ions in SiO₂ at both interfaces for poly Si-SiO₂-Si system are reported with TSIC method. A new approximation method to calculate the initial energies distributions of ions among the interface trapping states is presented. It is intuitively simpler and clearer than one's suggested by Boudry and Stagg. We have pointed out that is unfeasible to calculate trapping energy levels with the relation between $\ln\sigma$ and $1/T$ presented by Kuhn and Silversmith in 1971.

INTRODUCTION

Since it was recognized the migration of mobile ions is one of the important sources of instabilities in silicon device, especially, in MOS devices, the kinetic behaviour of mobile ions in SiO₂ has been extensively studied by various methods, such as TVS by Chou[1] and Kuhn[2] and TSIC by Hickmott[3]. It is generally recognized the kinetic behaviour of mobile ions in MOS system is not only dominated by traps at the two interface, but also has an initial distribution related to energy among the trapping states. Recently, an approximation method to calculate this distribution was suggested by Boudry and Stagg[4]. But all these results are based on Al-SiO₂-Si system. In this paper, we report the measurement results for poly Si-SiO₂-Si system and propose a new approximate method to calculate directly the initial energies distribution of ions among the trapping states at interface from the TSIC experimental curve. It is simpler and clearer than Boudry and Stagg's.

Our results are different from Osburn[5] and Kuhn's on several points. We will discuss the possible sources.

THE TRAPPING ENERGY LEVELS OF MOBILE IONS AT POLY Si/SiO₂ AND SiO₂/Si INTERFACES MEASURED WITH TSIC

The principles about TSIC had been described in detail in[4]. The mode of temperature, T, is linearly increased with time, t, have been taken in our experiment. The relevant equations for single energy level are:

$$\frac{dQ}{dt} = -Qse^{-\frac{E}{KT}} \quad (1)$$

Where Q is the charge density of mobile ions still trapped at time t; S is frequency, and equal to $\beta\sqrt{E}$, $\beta = 4\times 10^{11}$ [sec⁻¹ ev⁻¹/²] [4]. We obtain

$$Q = Q_0 \exp(-\frac{\beta\sqrt{E}}{b}\int_{T_0}^{T} e^{-\frac{E}{KT}} dT) \quad (2)$$

and

$$J(T) = -\frac{dQ}{dt} = Q_0 f(E,T) \quad (3)$$

where

$$f(E,T) = \beta\sqrt{E}\exp[-(\frac{E}{KT} + \frac{\beta}{b}\sqrt{E}\int_{T_0}^{T} e^{-\frac{E}{KT}} dT)] \quad (4)$$

$$b = \frac{dT}{dt} = const.$$

It may be proved that the f(E,T) is a function which contains a peak with respect to T. Using $\partial f(E,T)/\partial T=0$, we obtain

$$\frac{\beta E}{KT^2} = \frac{\beta}{b} \exp(-\frac{E}{KT}) \qquad (5)$$

Equation (5) gives the corresponding relation of T and E when f(E.T) takes the maximum. We can obtain the level of trapping states from the TSIC curves from Eq.(5)

The MOS samples used in the experiments are polySi-SiO$_2$-Si structures. A SiO$_2$ film, about 2000Å, was grown in wet O$_2$ on a p-type (100) oriented si wafer whose resistivity is about 6~8Ωcm. Polysi film is about 5000Å into which the dopant impurities were doped from vapor-deposition and diffusion respectively. Phosphorus concentration is about 10^{20}/cm^3. The measurement conditions are, bias voltage: +5v or -5v; initial temperature 280°K; b=1.4°K/sec. The results are shown in Fig. 1.

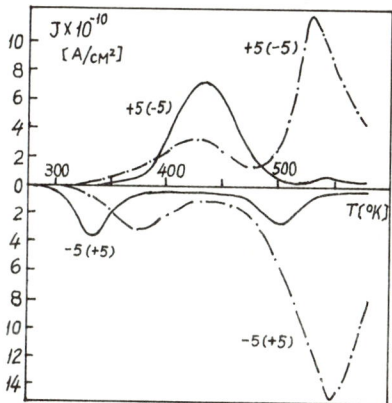

Fig. 1. Poly Si-SiO$_2$-Si system's TSIC curves.
——·——, ———— vapor deposited; ———— diffused;

+5(-5) represent +5v TSIC after the gate is applied -5v T-V stress.

The trapping energy level at poly Si/SiO$_2$ interface is about 1. 1ev and 1. 3ev calculated from positive direction TSIC (upper in Fig. 1) and about 0.8-0.9ev and 1. 3ev at SiO$_2$/Si interface Calculated from other one (lower in Fig.1)

DIRECT CALCULATING METHOD ABOUT INITIAL ENERGY DISTRIBUTION OF MOBILE IONS AMONG THE INTERFACE TRAPPING STATES

It is generally recognized that mobile ions at interface trapping states are distributed in different energy levels. (A well-test described in[4] is repeated in our experiment). So that TSIC curve which is obtained in experiment must be the sum of all ions detrapping from various energy levels. Formulated on this analysis, Eq.(3) may be written in

$$J'(E.T)=Q'_0(E) f(E.T.) \qquad (6)$$

Where $Q'_0(E)$ [coul cm^2 ev^{-1}] is charge density of mobile ions at temperature T=To and in unit energy range with energy E; J'(E.T.) [Acm^2 ev^{-1}] is contributed to TSIC by $Q'_0(E)$ at temperature T, let us denote the TSIC with J(T),

$$J(T)= \int_0^\infty Q'_0(E) f(E,T) dE \qquad (7)$$

Taking T as parameter, the relationship between f(E.T) and E may be calculated and can be seen that it is well approximated by δ-function, i.e. f(E.T) will only take a certain value when E is equal to E_T which corresponds to temperature, T, so we can rewrite Eq. (7)

$$J(T)=Q'(E_T) \cdot F(T) \qquad (8)$$
where $\qquad F(T)= \int_0^\infty f(E,T)dE \qquad (9)$
is a known function.

All the same, we can obtain the relation between E and T when f(E,T) is taking maximum with respect to E,

$$\frac{1}{KT}-\frac{1}{2E} = \frac{\beta\sqrt{E}}{Kb}\int_{T_0}^{T}\frac{1}{T}e^{-\frac{E}{KT}}dT - \frac{\beta}{2b\sqrt{E}}\int_{T_0}^{T}e^{-\frac{E}{KT}}dT \qquad (10)$$

Since function $e^{-E/KT}$ depends on T rapidly, we can approximationally write Eq.(10) as

$$\frac{1}{KT}-\frac{1}{2E} = (1-\frac{KT}{2E})\frac{\beta T}{b\sqrt{E}}\exp(-\frac{E}{KT}) \qquad (11)$$

and $\qquad \frac{\sqrt{E}}{KT^2} = \frac{\beta}{b}\exp(-\frac{E}{KT}) \qquad (12)$

Eq.(12) has a form identical to Eq.(5). The relation between E and T can be determined from Eq.(12) and is a approximate linear one. So far, we can obtain the initial energy distribution of mobile ions at interface trapping states from experimental TSIC curve. For example, taking the +5(-5)V curve in Fig. 1 as J(T), according the steps stated above, we show the relations between $n_0(E)= Q'_0(E)/q$ and E, as well as between J(T) and T in Fig.2. It can be seen that the two curves are similar in shape. It is because of that F(T) depends on T weakly

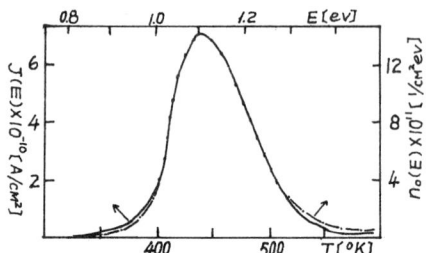

Fig.2. $n_0(E)=Q'_0(E)/q$ curve and TSIC experiment curve.

For verification purpose, we have calculated it from the experimental data given in [4], the $n_0(E)$ at peak obtained with our method is 1.6×10^{12} [$cm^{-2}ev^{-1}$] while that given by Boudry and Stagg is about 1.5×10^{12} [$cm^{-2}ev^{-1}$].

DISCUSSION
Comparasion With Osburn's Results

The activity energy of 2.7ev was reported by Osburn[5] for polySi-SiO_2-Si MOS system using the dependence of the longest breakdown time on temperatur. Seung et al[6] quoted the value as trapping energy. It is much larger than our's. But the factors which effect the longest breakdown time would be more complicated, it seems to us that the reduction of 2.7ev value to trapping energies at interface suggested in [6] is questionable.

Discussion About Kuhn's Method Of The Calculation Of Activity Energies And Analysis Of TVS Process

Kuhn'a results given in[2] are 0.20ev for Na^+ contaminated $10^{13}/cm^2$ and 0.60ev for

$10^{11}/cm^2$ using the relation between $\ln\sigma$ and $1/T$, it is rather low. We also measure the realtion of dependence of σ on T for Al-SiO$_2$-Si system as shown in Fig.3. It may be **observed that the areas under peaks will hardly continue to increase from 180° c to 220°c** in negative direction of TVS. This fact has been seen by other authors[1]. It is basic on $\sigma = \sigma_0 e^{-E/KT}$ to calculate trapping energies using the relation between $\ln \sigma$ and $\frac{1}{T}$. As E=0.62ev, T=500°k, $\frac{\sigma}{\sigma_0} \sim 10^{-6}$, if measured result is $\frac{\sigma}{g} \sim 10^{12}/cm^2$, σ_0/g must be $10^{18}/cm^2$, as E=0.8ev, σ_0 will be larger, it is unusually large in comparison with the value obtained by means of other methods including neutron activation[7], so that its validity is questionable.

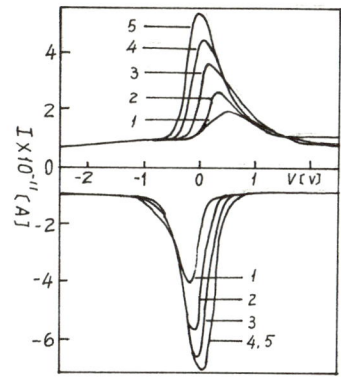

Fig.3. TVS curves at various temperature.
(1) 115C; (2) 135C; (3) 165 C; (4) 185C; (5) 220C.
Al-gate; oxide 2000A; $\alpha = dV/dt$ =55mv/sec.

Analysis of TVS Process. Let us think over the two extreme situations first. When gate voltage is negative, assuming total mobile ions density, Q_T, Q_t is trapping at traps with energy level E; $\sigma_1' = Q_T - Q_t$ is out of traps, the relations are:

$$\frac{dQ_t}{dt} = -Q_t S \exp(-\frac{E}{KT}) + R(Q_T - Q_t) = 0 \quad (13)$$

where R= retrapping frequency. We may obtain

$$Q_t = Q_T R[R + s\exp(-E/KT)]^{-1} \quad (14)$$
$$\text{and} \quad \sigma_1' = Q_T s \exp(-E/KT)[R + s\exp(-E/KT)]^{-1} \quad (15)$$

when gate voltage is positive and large enough, the retrapping process may be negligible, We obtain

$$dQ_1/dt = -Q_1 S \exp(-E/KT) \quad (16)$$
$$\text{and} \quad Q_1 = Q_{o1} \exp(-s \Delta t \, e^{-E/KT}) \quad (17)$$

where Q_1 is charge of mobile ions at traps, Q_{ol} is Q_1 value at the moment when the retrapping process begins to be negligible.

Because interface layer thickness is less than 100A°[8] and the mobility of Na$^+$ in SiO$_2$, μ, is about $10^{-8} cm^2/v.sec$ at 400°k[9], all mobile ions which are out of traps will depart from interface region in 0.3 sce for the sample of oxide $\sim$1000A° and α =50mv/sec while only less than 0.2 percent of mobile ions will be detrapped for 1ev trapping energy level, so that $Q_{ol} \doteq Q_t$. The charge of mobile ions excited from traps in time interval, Δt, which is required for voltage sweep to+Vo, σ_2', is given

$$\sigma_2' = Q_T R [1 - \exp(-s\Delta t \, e^{-E/KT})] \cdot [R + s e^{-E/KT}]^{-1} \quad (18)$$

All charges measured in TVS process are equal to

$$\sigma = \sigma_1' + \sigma_2' = Q_T g(E,T) \quad (19)$$
where
$$g(E,T) = 1 - R\exp(-s\Delta t \, e^{-\frac{E}{KT}}) \cdot (R + s e^{-\frac{E}{KT}})^{-1} \quad (20)$$

the relation, $\sigma = \sigma_o e^{-E/KT}$, will not hold except $s\Delta t e^{-E/KT} \ll 1$, but it would be only satisfied at temperature T<20°c for 0.85ev trapping energy states, however, it is at variance with actual measuremental situation. Up to now, we have only discussed the mode of single energy level. For the mode of multiple energy level,

$$\sigma = \int_0^\infty Q'_T(E) g(E,T) dE \qquad (21)$$

Taking T as parameter, The function of g(E.T) related to E is calculated and may be well approximated by step function. Eq.(21) can be rewritten as

$$\sigma = \int_0^{E_T} Q'_T(E) dE \qquad (22)$$

The relation, $\sigma = \sigma_o e^{-\frac{E}{KT}}$, does not hold as well. It is the physical meaning of Eq.(22) that at certain T, the charge density of mobile ions obtained from TVS measurement is the sum of the mobile ions at traps with energy E_T (as well as lower one) which corresponds to T. The increasing of $\sigma(T)$ with T is dependent on the number of ions at traps with higher energy.

The decay parts of TVS current at T,J,is mainly dominated by excited process of level E_T. We can obtain from Eq. (16)

$$J = Q_T R s e^{-\frac{E}{KT}} \cdot (R + s e^{-\frac{E}{KT}})^{-1} \cdot \exp(-s t e^{-\frac{E}{KT}}) \qquad (23)$$

Using the relation between ln J and t, E which corresponds to 480°K is about 1.18 1.22ev that is consistent with TSIC result approximately.

ACKNOWLEDGMENT

The authors wish to thank professor Li Zhi-jian for critical comments and valuable suggestions. We would also like to thank professor Huang Yong-bao, Han Ru-qi, Wu Guo-ying and Tan Chang-hua for useful discussion. Valuable assistance from Miss Li Xiang-lin and Gu Xiao-ling is also acknowledged.

REFERENCES

(1) N.J. Chou, J.Electrochem. Soc.118,601, (1971).

(2) M.Kuhn and D. J. Silversmith, ibid. 966, (1971).

(3) T. W. Hickmott, J. Appl. Phys. 46, 2583, (1975).

(4) M. R. Boudry and J.P. Stagg, ibid. 50, 942, (1979).

(5) C.M.Osburn and E.Bassous, J.Electrochem. Soc. 122, 89, (1975).

(6) Seung P.Li, E.T.Bates and J.Maserjian, Solid-State Electronics. 19, 235,(1976).

(7) E.Yon, W.H.Ko and A.B.Kuper, IEEE Tran. Electron Devices E-D 13, 276, (1966).

(8) J.van Turnhout and A.H.van Rheenen, Non-Crystalline Solids. 494, (1977).

(9) J.P.Stagg, Appl. Phys. Lett. 31, 532, (1977).

INTERACTIONS BETWEEN SMALL-POLARONIC PARTICLES IN SOLIDS*

David Emin

Sandia National Laboratories[†]
Albuquerque, New Mexico 87185

ABSTRACT

When a light particle in a solid composed of relatively heavy atoms is associated with substantial displacements of the equilibrium positions of the atoms immediately surrounding it, the composite entity may be regarded as being small-polaronic. Many instances of self-trapping of electronic charge carriers, excitons and light atoms such as hydrogen are known. A significant contribution to the interaction between such particles results from interference between their atomic displacement patterns. As a result oppositely charged small polarons may experience an intermediate-range repulsion, while both like-signed and neutral entities may have a tendency to cluster. These effects can be very important. As examples, the recombination kinetics of electron and hole small polarons and the ordering of defect atoms are discussed.

INTRODUCTION

The introduction of either interstitial atoms, vacancies, electronic charge carriers, or excitons into a solid is generally accompanied by alterations of the equilibrium positions of the atoms of the solid. These atomic displacements become especially significant if they exceed the amplitudes of the zero-point motion of the displaced atoms. This paper is concerned with that interaction between strongly coupled particles (or quasiparticles) which results from their mutual interaction with the atomic displacements of the host material.

APPROACH

A system of static entities which interact with each other and with the lattice containing them is described by a Hamiltonian which presumes a linear interaction between the added particles (or quasiparticles) and the atoms of the lattice, in addition to harmonic interactions between lattice atoms. Specifically, one has

$$H = \sum_i \sum_g \epsilon_g^i n_g^i + \tfrac{1}{2} \sum_{i\,j} \sum_{g\,g'} U_{g,g'}^{i,j} n_g^i n_{g'}^j + \sum_\lambda \sum_q \hbar\omega_{q,\lambda} (b_{q,\lambda}^+ b_{q,\lambda} + \tfrac{1}{2})$$

$$+ \sum_\lambda \sum_q \sum_i \sum_g n_g^i (V_{q,\lambda}^i e^{i q \cdot g} b_{q,\lambda}^+ + \text{c.c.}) , \qquad (1)$$

where n_g^i and ϵ_g^i are, respectively, the number and energy of each particle of type i located at a site designated by the position vector g, and $U_{g,g'}^{i,j}$ is the energy of direct interaction between a pair of static particles. Creation and annihilation operators for phonons of mode λ, wavevector q, and energy $\hbar\omega_{q,\lambda}$ are denoted by $b_{q,\lambda}^+$ and $b_{q,\lambda}$, respectively. The interaction between a particle of species i and a phonon is characterized by $V_{q,\lambda}^i$.

For any configuration of particles one can find the positions of the lattice atoms which minimize the energy of the system. Then the energy of the system, apart from that due to vibrational motion, is found to be

$$E = E_{isolated} + E_{direct} - E_b , \qquad (2)$$

where $E_{isolated}$ and E_{direct} represent the first and second terms of Eq. (1), respectively, and E_b is defined by

$$E_b \equiv \sum_\lambda \sum_{\underset{\sim}{q}} \sum_i \sum_j (v^{i*}_{\underset{\sim}{q},\lambda} \, v^{j}_{\underset{\sim}{q},\lambda}/\hbar\omega_{\underset{\sim}{q},\lambda}) \sum_{\underset{\sim}{g}} \sum_{\underset{\sim}{g}'} n_{\underset{\sim}{g}} \, n_{\underset{\sim}{g}'} \, e^{i\underset{\sim}{q}\cdot(\underset{\sim}{g}-\underset{\sim}{g}')} . \qquad (3)$$

If, for example, there were but a solitary particle of type i located in the solid, E_b simply reduces to the small-polaron binding energy of such a particle:

$$E_b = \epsilon^i_b = \sum_\lambda \sum_{\underset{\sim}{q}} |v^i_{\underset{\sim}{q},\lambda}|^2/\hbar\omega_{\underset{\sim}{q},\lambda} . \qquad (4)$$

However with two particles of the same type (species i) located at sites $\underset{\sim}{g}_0$ and $\underset{\sim}{g}_1$ one has that

$$E_b = \sum_\lambda \sum_{\underset{\sim}{q}} \left[|v^i_{\underset{\sim}{q},\lambda}|^2/\hbar\omega_{\underset{\sim}{q},\lambda} \right] \{ 2 + 2 \cos[\underset{\sim}{q}\cdot(\underset{\sim}{g}_0 - \underset{\sim}{g}_1)] \}$$

$$\equiv 2\epsilon^i_b + U^{i,i}_{\underset{\sim}{g},\underset{\sim}{g}'}(\text{indirect}) . \qquad (5)$$

Here the second term represents that portion of the interaction between the two particles which results from their mutual interaction with the displacements of the atoms of the lattice. Finally, if the particle at sites $\underset{\sim}{g}_0$ and $\underset{\sim}{g}_1$ are of different types (i and j, respectively) the energy of their indirect interaction is given by

$$U^{i,j}_{\underset{\sim}{g},\underset{\sim}{g}'}(\text{indirect}) = 2 \sum_\lambda \sum_{\underset{\sim}{q}} \text{Re}\left[v^{i*}_{\underset{\sim}{q},\lambda} \, v^j_{\underset{\sim}{q},\lambda} \, e^{i\underset{\sim}{q}\cdot(\underset{\sim}{g}_0-\underset{\sim}{g}_1)} \right]/\hbar\omega_{\underset{\sim}{q},\lambda} . \qquad (6)$$

In the present discussion the particle-lattice interaction is taken to be of short range. That is, the energy of a charge carrier on an atomic site depends on the proximity of the nearest-neighbor atoms, while the energy of an interstitial is a function of its distance from the atoms immediately adjacent to it. In these cases $v^i_{\underset{\sim}{q},\lambda}$ has a rather weak nonmonotonic dependence on $\underset{\sim}{q}$ with maxima occurring far from the center of the Brillouin zone. Concomitantly, one finds that the indirect interparticle interaction falls off with separation as the oscillatory terms of Eqs. (5) and (6) give rise to increasingly efficient cancellations in the $\underset{\sim}{q}$-summation.

EQUIVALENT INTERSTITIALS

In solids one is often concerned with the strain fields surrounding defects. Here the displacements about neutral interstitials are considered. Presuming a negligible direct interaction between these interstitials, the net interaction energy is simply $-E_b$. As the simplest example consider interstitials placed in a monatomic linear chain. Minimimizing the energy, Eq. (1), yields the displacement patterns depicted in Fig. 1. Namely, each interstitial occupies a space (in this case, a linear dimension) equal to A/k, where A is the (constant) repulsive force exerted

Fig. 1 One and three interstitials added to a monatomic chain. The strained bonds are shown with dashed lines.

between the interstitial and each of the two adjoining atoms, and k is the stiffness constant of the monatomic lattice. This constitutes a one-dimensional microscopic derivation of Vegard's theorem of elasticity theory: an elastic material expands by an amount equal to the extra volume of an inclusion.

In systems in which the added particle interacts with both optical and acoustic modes of the solid the situation becomes more complex.[1] A common situation is that in which light atoms form cages about relatively heavy atoms. A one-dimensional analogue of such a structure is that of a backbone of heavy masses to which light masses are attached. In essence the motions of the heavy atoms are associated with acoustic vibrational modes and the movements of the light masses relative to the heavy masses involve the optical modes. As illustrated in Fig. 2, if the interaction of an interstitial is mainly with the light atoms (optical modes), a local type of deformation pattern is produced which does not extend far from the interstitial. Furthermore, with a collection of three adjacent interstitials the asso-

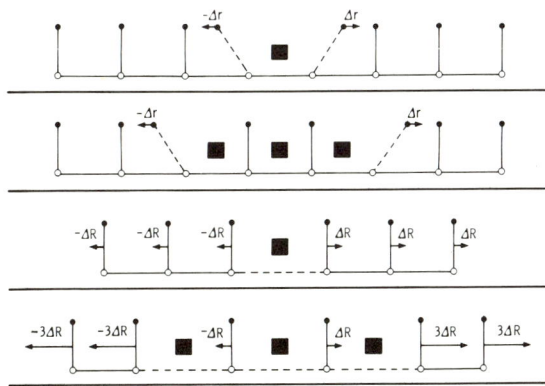

Fig. 2. Interstitials in a diatomic system. The optical-type deformations with one and three interstitials are shown in the top two lines. The respective acoustic-type deformations are shown in the bottom two lines for a diatomic chain. Dashes indicate strained bonds.

ciated deformation patterns tend to cancel. However, if the interaction of the interstitial is mainly with the heavy atoms, acoustic modes predominate and a long-range displacement pattern is established. Here Vegard's theorem may be applied.

Several comments are now in order. First, since Vegard's theorem is an outgrowth of elasticity theory (which involves only long-wavelength acoustic phonons) its in-applicability to some instances where optical-mode displacements play a role does not pose a contradiction for the macroscopic theory. Second, it is both obvious and well-known that in systems of higher dimensionality the displacement patterns at long range are altered so as to reduce the magnitude of the displacements of individual atoms at the expense of involving a greater number of atoms. Third, dimensionality plays a major role in the energetics of clustering of added particles. For example, it can be seen from the optical-mode portion of Fig. 2 that it is energetically unfavorable for interstitials to cluster. However in three-dimensional models involving optical-type displacements clustering can be energetically favorable. Similarly, three-dimensional strain fields associated with acoustic-type displacements favor clustering.

Although the preceding discussion considered only neutral interstitials, analogous results apply to the strain fields and clustering of excitons, and (with the addition of the coulomb interaction) excess charges. Indeed, it is the possibility that the atomic-displacement-induced tendency of like charges to cluster may overcome their coulombic repulsion that has led to the consideration of bipolaron formation in both solids and liquids.

ELECTRON-HOLE INTERACTIONS

The lattice-mediated interaction of electron and hole small polarons can be of critical importance in their direct recombination. For the ideal monatomic system illustrated in Fig. 3a the deformation (here a contraction) about an electron has an opposite sense to that about a hole (here an expansion). For illustration the distortions are taken to be local as with light masses harmonically coupled to a rigid frame. The central point is that when the electron-hole separation is reduced sufficiently so that the two distortion patterns overlap substantially the binding energy associated with each small polaron is reduced. In other words, as shown by the curve labeled small polaron in Fig. 3b, the lattice-mediated interaction provides a _repulsive_ component to the interaction between an electron and a hole small polaron. Furthermore, as shown in Fig. 3b, with sufficiently strong polaronic binding the combination of the small-polaron and coulombic terms yields a net energy which contains a repulsive barrier to recombination. This in effect screens out the strong portion of the coulombic attraction.

The luminescence associated with recombination in such a polaronic solid has been described elsewhere.[2] Hence it will only be noted that such a system (potentially) exhibits three luminescence processes: the recombination of excitons prior to lattice deformation, of self-trapped excitons, and of pairs of separated small polarons. These processes are associated with the positions labeled as E, STE, and GS in Fig. 3b.

The photoconductivity of a small-polaronic solid is proportional to the average mobility of the photogenerated carriers and their lifetime. Since a finite time is required before an optically generated charge forms a small polaron, each carrier's mobility is an average of its nonpolaronic (precursor) mobility $\tilde{\mu}_e$ or $\tilde{\mu}_h$, and its self-trapped mobility, μ_e or μ_h, weighted by that fraction of the lifetime each exists in the precursor state, f_e or f_h. The average photoconductive mobility is $[\tilde{\mu}_e f_e + \mu_e(1 - f_e)] + [\tilde{\mu}_h f_h + \mu_h(1 - f_h)]$. Since often $\tilde{\mu} >> \mu$ the obser-

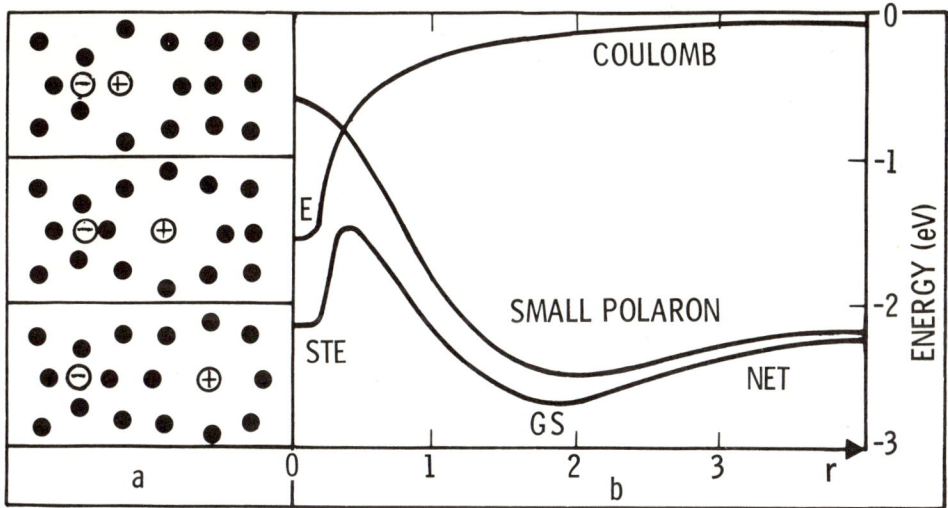

Fig. 3. The deformation patterns for an electron and a hole in a monatomic lattice are shown for three separations in (a). In (b) the net energy of the system as well as the coulombic and distortional (small-polaron) components of the energy are plotted against the separation, r, measured in units of the lattice constant.

vation of a low photoconductive mobility requires that $f_e, f_h << 1$.

For direct recombination of low-mobility charge carriers it has generally been assumed that the room-temperature recombination is diffusion limited.[3] That is, electrons and holes hop together spatially and then recombine. Specifically, the probability that, once within their mutual coulomb capture radius, a pair will separate rather than recombine has been assumed to be insignificant. Nonetheless, there are previously unexplained situations in which the photoconducting carriers have very low (hopping) mobilities for which the recombination coefficient does not display the temperature dependence characteristic of diffusion controlled recombination.[4] The presence of a repulsive barrier provides a mechanism to resolve this dilemma. Namely, the barrier keeps the carriers apart and thereby reduces their overlap and recombination rate. In addition, by shielding the carriers from the steepest portion of their attractive coulomb potential, the probability of their separation is enhanced. Thus such a barrier to the recombination of small polarons may be significant in understanding the photoconducting properties of insulating and semiconducting glasses and crystals in which the charges form small polarons.

REFERENCES

*This work was supported by the U. S. Department of Energy, DOE, under Contract DE-AC04-76-DP00789.
†A U. S. Department of Energy facility.
1. David Emin, Sandia Technical Report, SAND78-0165.
2. David Emin, J. Noncrystal. Solids, 35 & 36, 969 (1980).
3. N. F. Mott, E. A. Davis, and R. A. Street, Phil. Mag. 32, 961 (1975).
4. T. D. Moustakas and K. Weiser, Phys. Rev. B 12, 2448 (1975).

SMALL POLARON HOPPING WITHOUT TRAP PARTICIPATION IN DISPERSIVE
TRANSIENT TRANSPORT IN SiO_2 OF MOS STRUCTURES*

K. L. Ngai
Naval Research Laboratory, Washington, D.C. 20375

Xiyi Huang and Fu-sui Liu
Dept. of Electrical Engineering, University of Hawaii,
Honolulu, Hawaii 96822

ABSTRACT

Analyses of the dispersive transient transport data of a-SiO_2 indicate that the temperature dependences of the transit time and of the prompt mobility of small hole polaron hopping are strongly correlated. This correlation is in quantitative agreement with the predictions of a recently advanced model. The implications are that dispersive transient transport occurs also by small polaron hopping and that no traps or defects are involved.

Recently there has been considerable interest in time dependent carrier transport that is observed in time-of-flight experiments. Experimental studies on the chalcogenide glasses a-Se and a-As_2Se_3[1] and on a-SiO_2[2,3,4] have indicated dispersive non-Gaussian transport. The dispersive hole transport in a-SiO_2 was first observed by Boesch, et al.[2] at times longer than 10^{-4} sec. Later Hughes[3] performed high time resolution hole-transport. He has shown that after an initial very short time interval following the X-ray pulse, the hole relaxes to form a small polaron[3,4] (self-trapped) in the non-bonding 2p orbital of an oxygen. Transport at times shorter than 10^{-6} sec. takes place by the hopping of the small hole polaron from one oxygen to another. Its mobility has magnitude, temperature dependence and electric field dependence all in agreement with the predictions of small polaron theory.[5] In particular, the mobility is low, thermally activated (with activation energy $E_\mu \cong 0.14$ eV, see Fig. 1) at temperatures above one third the Debye temperature and breaking to a non-activated behavior at lower temperatures. For $t \gtrsim 10^{-5}$ sec., dispersive transient current is observed with $I(t) \sim t^{-(1-\alpha)}$ for $t < t_\tau$ and $I(t) \sim t^{-(1+\alpha)}$ for $t > t_\tau$ where $0 < \alpha < 1$ and t_τ is the "transit time." The dependence of t_τ on electric field E and sample thickness L through $t_\tau \propto (E/L)^{-1/\alpha}$ is also observed. These are in accord with the general result of the continuous time random walk (CTRW) model of Scher and Montroll[1] obtained for non-Gaussian waiting time distribution function $\psi(t) \propto t^{-(1+\alpha)}$. It has been hypothesized that the CTRW occurs in a-SiO_2 because the hole, after

---

*Work supported in part by ONR.

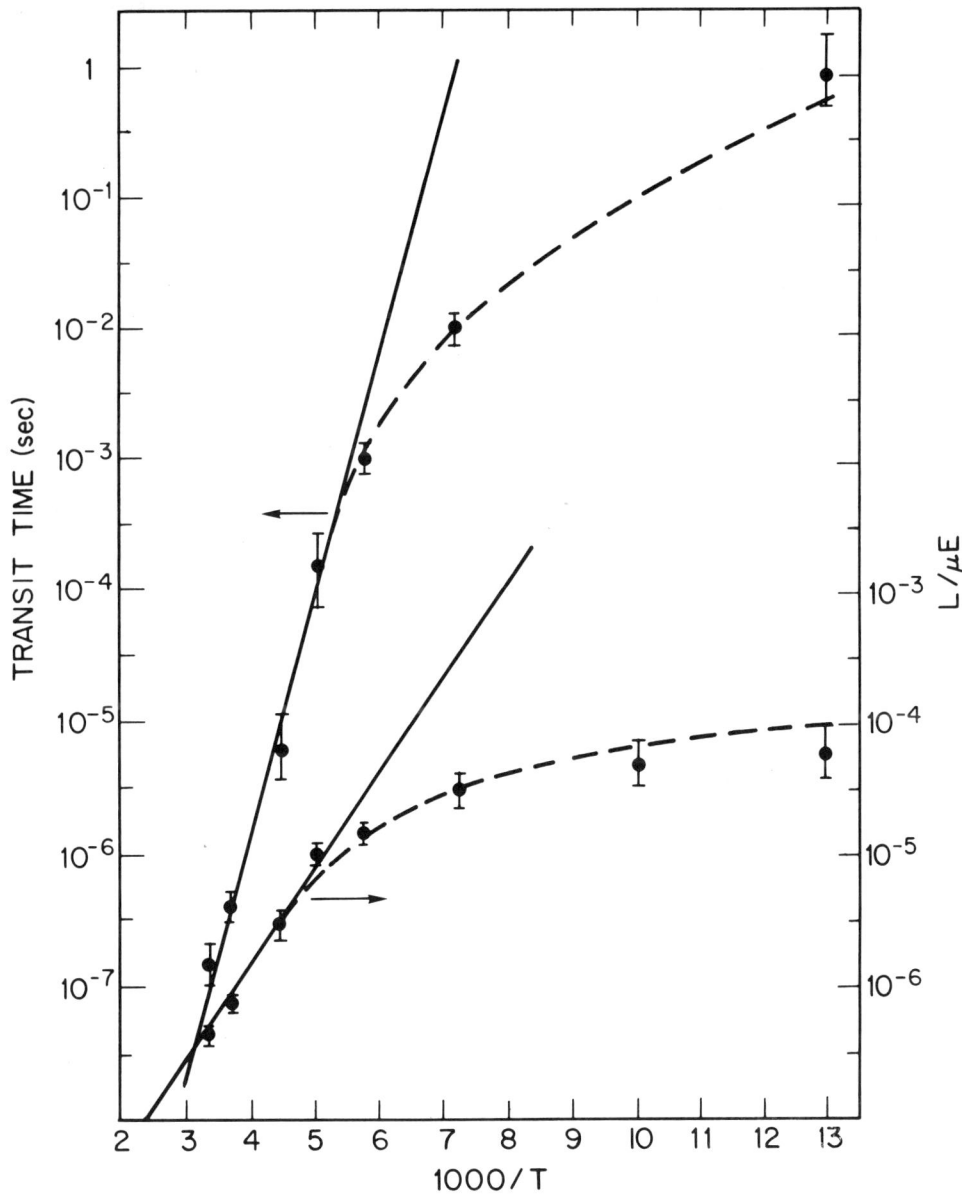

Fig. 1. Transit time t_τ and $L/\mu E$ data of Hughes replotted. The solid lines indicate Arrhenius behavior for T > 200K with activation energies of 0.37 eV and 0.14 eV. The value of 0.14 eV is slightly different from the value of 0.16 eV obtained by Hughes. At every T the ratio of the local slope of the lower dashed curve to that of the upper dashed curve is $\alpha \equiv 0.3$ (see text).

initially diffusing by small polaron hopping, becomes trapped at some structural defect. Further transport is by tunneling from defect to defect randomly located in the lattice,[3] leading to a non-Gaussian waiting time distribution function that is assumed to have the form of $t^{-(1+\alpha)}$. The temperature dependence of t_τ (Fig. 1) is thermally activated above 200K and can be fit to a simple Arrhenius plot with activation energy (AE) $E_A^* = 0.37$ eV. The large difference between E_A^* and E_μ would seem to indicate that the transient transport mechanism is different from the earlier transport mechanism of small polaron hopping; and it is the primary reason for invoking traps[3] in transient transport. The hole transport dispersion in a-SiO$_2$ is[2] remarkably stable with temperature. The dispersion parameter α as appears in $I(t) \sim t^{1-\alpha}$ for $t < T_\tau$ is shown by McLean, et al. for their own oxide films to have the constant value of 0.22 for the temperature range of 87-377K. The same is true for data of Hughes where he found for his own oxide samples $\alpha = 0.3$ for 140K $< T <$ 298K. This property rules out the mechanism of a distribution of trap depths for dispersive transport.

There are some remarkable correlations between the small polaron transport (SPT) and the dispersive transient transport (DTT). Both SPT and DTT are markedly non-Arrhenius below 200K (Fig. 1) with a decrease in both activation energies (AEs) as T decreases. The AEs above 200K are related, $E_A^* \cong E_\mu/(1-n)$ where $n \equiv (1-\alpha) = 0.7$ and most remarkably this relation continues to hold for $T \leq$ 200K which now takes the form $\partial(\log t_\tau)/\partial(1/T) = [\partial(\log(1/\mu))/\partial(1/T)]/(1-n)$. That is, in an Arrhenius plot, the local slopes of DTT and SPT differ by a factor of $(1-n)$, and n characterizes the dispersive current, i.e. $I(t) \sim t^{-n}$ for $t < t_\tau$. This is quantitatively illustrated in Fig. 1 by the two non-Arrhenius dashed curves which interpolate the (L/µE) and the t_τ data respectively for $T <$ 200K, and at any T the ratio of the local slope of the former to the local slope of the latter has the constant value of $(1-n) = 0.3$. Such correlations are not expected from the tunneling/hopping among traps model for DTT. The $T <$ 200K data also contradicts the "trap-controlled" hopping model[6] designed for a-As$_2$Se$_3$ which predicts the AE for t_τ to be $E_\mu + E_t$ with E_t the trap depth. The value of α depends on the preparation of the oxide films. McLean et al. have $\alpha = 0.22$ with a measured $E_A^* = 0.6$ eV, while Hughes has $\alpha = 0.3$ and $E_A^* = 0.37$ eV. The large discrepancy between the two AEs of t_τ imposes some difficulties in the trap models. Now what Hughes has measured for E_μ is for hole hopping in the non-bonding 2p oxygen orbitals, an intrinsic quantity which should be the same for both samples. Then, if we use the relation $E_A^* = E_\mu/\alpha$ for McLean et al.'s sample we find $E_\mu/\alpha = (0.14/0.23)$ eV $= 0.61$ eV which[10] is almost identical to the measured value of E_A^*. It is worthwhile to point out that this empirical relation holds not only for a-SiO$_2$ but also for a-As$_2$Se$_3$ and a-Si.[7]

Our approach to dispersive transient transport follows from a general and unified treatment[7] of several low frequency (or long time) fluctuation, dissipation and relaxation phenomena including dielectric and mechanical relaxations, deep level transient capacitance in semiconductors, NMR relaxations, flicker 1/f noise, G-R noise and diffusion. The fundamental and common ingredients of these unified theories are (a) the correlated states and their excitations[8] in condensed matter, (b) the application of the Wigner's random matrices statistical theory of energy levels to condensed matter[7] and (c) the infrared divergent excitations and deexcitations of correlated states.[7,8,9] For details, the reader is referred to the references quoted and expecially a recent article[7] where we have presented a version of our model on dispersive transient transport. Our theory[7] provides a distribution of hopping times from a fundamental process without assuming it. A constant transition rate $W = W_o \exp(-E_A/kT)$ that describes through $dQ/dt = -WQ$ the decay of the probability $Q(t)$ that the hole remains on a particular site caused by transfer to surrounding sites is modified by infrared divergent excitations and deexcitations of correlated states to be $W \exp(-n\gamma) (E_c t)^{-n}$, where $\gamma = 0.577$, E_c is the upper cut-off energy for correlated state excitations[7,8,9] and n is a constant and $0 \leq n < 1$. We have then $dQ/dt = -W\exp(-n\gamma) (E_c t)^{-n} Q$, and $\psi(t) = -dQ/dt$ can be written as $\psi(t)/\tilde{W} \propto \tau^{-n} \exp(-\tau^{1-n})$ with $\tilde{W} = \{W_o \exp(-n\gamma)/(1-n)E_c^n\}^{1/(1-n)}$ $\exp\{-E_A/(1-n)kT\}$ and $\tau = \tilde{W}T$ a dimensionless reduced time. In earlier treatment of DTT[7] we exploited some similarity of our $\psi(t)$ with the calculation of $\psi(t)$ for hopping through a random medium,[1] and noticed that our $\psi(t)$ will have the form of $t^{-(1+\alpha)}$ with $\alpha \cong (1-n)$ for $\tau > 1$ but not too large. We derived, in the same mathematical framework of Ref. 1, the transit time $t_\tau \propto (L/E)^{1/\alpha} \exp(E_A/(1-n)kT)$. Note that the apparent activation energy $E_A^* = E_A/(1-n)$, and not the true hopping activation energy E_A governs the transit time. Identifying E_A with E_μ and α with $(1-n)$, we can see that our model predicts the observed empirical relation $E_A^* = E_\mu/\alpha$ (Fig. 1), and implies that the DTT occurs from small polaron hopping also. No traps are involved. It can reconcile the large difference of E_A^* between measurements on two samples since it is only an apparent AE, while the product (αE_A^*) is nearly equal for both samples as it should be, because it is the intrinsic small polaron mobility AE, E_μ. We have studied also the early times τ^{-n} portion of our $\psi(t)$. The Laplace transform (LT) $\psi^*(s)$ of $\psi(t)$ is $\tilde{W}s^{-1+n}$ for $s \gg 1$. The LT $\phi^*(s)$ of the relaxation function $\phi(t)$ that appears in the master equation of the non-Gaussian transport[1] and related to $\psi^*(s)$ by $\phi^*(s) = s\psi^*(s)/[1-\psi^*(s)]$ has the form of $\tilde{W} s^n$ for $s \gg 1$. Using the notations of Ref. 1, the mean position $<\ell(t)> \sim \bar{\ell}L^{-1}$ $\{\psi^*(s)/\tilde{W}s(1-\psi^*(s))\} \sim \bar{\ell}Wt^{1-n}$. Hence the τ^{-n} portion of $\psi(t)$ gives rise also to the experimental transient current for $t < t_\tau$ and its AE given through W is the true E_A (or E_μ), which is smaller than E_A^*, the AE for t_τ, by a factor of α, in agreement with the observation of Hughes.[3]

The infrared divergent response that has modified a constant rate W to $\tilde{W} \tau^{-n}$, introduces in a fundamental way a distribution of rates or time constants. We illustrate this through a reconsideration of the multiple-trapping model[10] of transient transport. The model we shall discuss is not appropriate for a-SiO$_2$ since we believe no traps are involved. Nevertheless we consider it to demonstrate how our mechanism enters into trapping and detrapping. For a compact presentation, we adopt the notations in Ref. 10. Instead of a distribution of traps characterized by different capture rates w_i and release rates r_i, we consider only a single trap with a fixed capture rate W and release rate R. $q(\vec{x},t)$ is the concentration of carriers localized by the trap at $\vec{x}$. The capture of mobile carriers with concentration $p(\vec{x},T)$ by the trap at $\vec{x}$ is accompanied by the excitations and deexcitations of correlated states in an infrared divergent manner,[7,8,9] and is described by $\partial p(\vec{x},t)/\partial t = -We^{-n\gamma}(E_c t)^{-n} p(\vec{x},t)$. For earlier times $\tau \equiv \tilde{W}t < 1$, it can be shown that this process can equivalently be described as the capture by a distribution of fictitious traps. The distribution function of traps with capture rate $w_i = 1/t_i$ is $\sim t_i^{-n}$. Similarly the release of carriers from the trap at $\vec{x}$ is described by $\partial q(\vec{x},t)/\partial t = -\text{Re}^{-m}\gamma(E_c t)^{-m} q(\vec{x},t)$, where $0 \leq m < 1$ is the infrared divergent exponent for the release process. This release process is equivalent to the release from a fictitious distribution of traps of density $\sim t_i^{-m}$ with release rate of $r_i = 1/t_i$. Through a simple transformation, the capture process can be recast as the capture by this latter distribution of traps of density $\sim t_i^m$ but the release rate is modified as $w_i = \text{const.} \times t_i^{-\beta}$ with $\beta=(1-m)/(1-n)$. In this common fictitious trap distribution picture, the key relation $w_i \propto r_i^\beta$ follows as a consequence and need not be assumed as in Ref. 10. The quantity $\Sigma(s) = s\Sigma_i [M_i/(s+r_i)]$ can be evaluated by integration over t with weighting function $\sim t^{-m}$. We then arrive at the result $\Sigma(s) \sim s^\alpha$ with $\alpha = n\beta$. Then, it follows that the transient current varies as $t^{-1+\alpha}$ for $t < t_\tau$ and as $t^{-1-\alpha}$ for $t > t_\tau$. Thus, we have shown that our mechanism when considered in the multiple-trapping model leads also to the observed behavior even though no explicit distribution of trap parameters is assumed.

REFERENCES

1. H. Scher and E.W. Montroll, Anomalous transit time dispersion in amorphous solids, Phys. Rev. B 12, 2455 (1975).
2. F.B. McLean, H.E. Boesch and J.M. McGarrity, Hole transport and recovery characteristics of SiO_2 gate insulators, IEEE Trans. Nuc. Sci. NS-23, 1506 (1976); and private communication from F.B. McLean.
3. R.C. Hughes, Time-resolved hole transport in a-SiO_2, Phys. Rev. B 15, 2012 (1977); and G. Lucovsky, Defect-controlled carrier transport in a-SiO_2, Phil. Mag. B 39, 531 (1979).
4. R.C. Hughes and D. Emin, Small polaron formation and motion of holes in a-SiO_2, Proc. Int'nal Conf. Physics of SiO_2 and its Interfaces, ed. by S. Pantelides, Pergamon (1978), p. 14.
5. D. Emin, Phonon-assisted transition rates I. Optical-phonon-assisted hopping in solids, Adv. Phys. 24, 305 (1975).
6. G. Pfister and H. Scher, Time-dependent electrical transport in amorphous solids: As_2Se_3, Phys. Rev. B 15, 2062 (1977).
7. K.L. Ngai, Universality of low frequency fluctuation, dissipation and relaxation properties of condensed matter. I, Comments Solid State Phys. 9, 127 (1979); K.L. Ngai,---II, Comments Solid State Phys. 9, 141 (1980).
8. K.L. Ngai and C.T. White, Frequency dependence of dielectric loss in condensed matter, Phys. Rev. B 20, 2475 (1979).
9. K.L. Ngai, A.K. Jonscher and C.T. White, On the origin of the universal dielectric response of condensed matter, Nature 277, 185 (1979).
10. F.W. Schmidlin, Theory of trap-controlled transient photoconduction, Phys. Rev. B 16, 2362 (1977); and J. Noolandi, Multiple-trapping model of a-Se, Phys. Rev. B 16, 4466 (1977).

ELECTRON TRANSPORT IN SiO$_2$ FILMS AT LOW TEMPERATURES*

Siegfried Othmer and J. R. Srour
Northrop Research and Technology Center, One Research Park
Palos Verdes Peninsula, CA 90274

ABSTRACT

Electron transport has been investigated in thermally grown SiO$_2$ films over the range 4-300 K. Charge buildup was measured subsequent to steady-state ionizing radiation, and charge transport was measured subsequent to pulsed ionizing excitation. Results of charge buildup measurements were consistent with sweepout of nearly all generated electrons at all temperatures. Under pulse conditions, electron transport was found to be complete within the experimental time resolution of 1 μsec. This indicates a lower limit to the drift mobility of 6×10^{-4} cm^2/V-sec at 4.7 K. If the low-temperature mobility is thermally activated, and under certain other assumptions, an upper limit to the activation energy of 4.7 meV is indicated.

INTRODUCTION

Several years ago, hole transport in SiO$_2$ films was the subject of extensive investigation. Studies of charge transport and charge buildup as a function of temperature, field, and oxide thickness revealed complicated transport behavior (1,2) that was successfully described both in terms of the continuous-time random-walk (CTRW) model of Scher and Montroll (3,4,5) and in terms of a multiple trapping model.(6) It was established that after a brief period, of the order of picoseconds, (7) in which holes undergo transport with their intrinsic mobility, formation of small polarons takes place. Subsequent transport then occurs with polaronic mobilities until the carriers are trapped in states of successively longer time constant, giving rise to the dispersive transport that has been modeled in such detail.

In all of these studies, results were consistent with essentially complete sweepout of electrons generated within the oxide in a time short compared to experimental resolution ($\geqslant 10^{-5}$ sec). This is consistent with expectations on the basis of electron transport studies performed by Hughes on amorphous SiO$_2$ (Suprasil). (8) Hughes found that the electron mobility was about 20 cm^2/V-sec at room temperature, and showed a temperature dependence characteristic of scattering-limited conduction. The mobility attained a value of 40 cm^2/V-sec at low temperature, where it was found to be temperature-independent down to the lowest temperature measured, about 114 K.

It has been shown that in amorphous materials the absence of long-range order, and variations in local potential, should lead to the existence of localized states at the extremes of the allowed energy bands. (9) Transport is then expected to take place either by excitation to extended states above the "mobility edge," or by variable range hopping among localized states. The present investigation was undertaken to determine the time scale, and the mechanism, of electron transport at low temperatures. The vehicle of SiO$_2$ films thermally grown on silicon was chosen for

―――――――――
*Supported by Defense Nuclear Agency under Contract DNA001-79-C-0190.

its technological importance. Charge buildup in irradiated oxide films presents a problem because of resulting threshold shifts in MOS transistors. At room temperature, such charge buildup at times long compared to the hole transit time is largely a function of the quality of the SiO_2-Si interface, which determines the fraction of hole charge trapped near the interface. At temperatures less than about 90 K, however, generated holes are trapped essentially in place at moderate field strengths ($\leq 10^6$ V/cm), giving rise to quite severe radiation sensitivities. Consequently, it is of interest whether at sufficiently low temperatures electron trapping yields ideal compensation for such hole trapping.

The concern here is not with those electron traps which have been found to be important at higher temperature. Such traps have been studied in detail by Ning (10) and by Young et al. (11) They have relatively small cross sections and thus trap only a small fraction of generated electrons, offering no effective compensation for trapped holes. In fact the experiments to be reported here were not sensitive to small fractional trapping of generated electrons.

EXPERIMENTAL PROCEDURE AND RESULTS

Electron trapping was investigated initially by means of charge buildup measurements under ionizing radiation (Co-60 gamma rays) in the range 4-77 K. Subsequently, electron transport was investigated with higher time resolution using pulsed ionizing radiation (Febetron 705 flash X-ray). Charge buildup measurements were made on capacitors with oxides grown pyrogenically at 925°C on 3 ohm-cm phosphorus-doped ⟨100⟩ material. Oxidation was followed by N_2 anneal at 925°C. Aluminum metallization was followed by an alloy performed at 500°C in N_2. These devices were fabricated for us by Hughes Aircraft. Charge buildup was determined from C-V measurements of the shift in flatband voltage. Capacitance was sensed either from the quadrature response to 3-Hz modulation or via quasistatic techniques. Results of such measurements on a number of samples are shown in Fig. 1a and 1b. In Fig. 1a, comparison is made between flatband voltage shifts obtained under 4 K irradiation with those obtained under 77 K irradiation. The latter have been successfully modeled on the assumption of complete electron sweepout. (12) At low bias, charge buildup is governed by the field dependence of charge yield, i.e., the field dependence of geminate or columnar recombination. (13) At high bias, sweepout of holes is observed on the experimental time scale of 20 minutes, the time required to accumulate 10^4 rads(Si). The maximum shift in flatband voltage, under assumption of 100% charge yield and complete hole trapping, is 1.7 V at this dose, taking the pair generation energy in SiO_2 to be 18 eV. (13) Under the combined influence of finite charge yield and hole sweepout, a maximum shift of 80% of the above value is expected at 2.5×10^6 V/cm, in agreement with the observed value of 1.45 V.

The results obtained at 4 K agree with those obtained at 77 K, leading to the conclusion that essentially all of the electrons have been swept out even at 4 K. C-V measurements were made at this temperature using optical excitation to generate carriers in the silicon. Since this excitation could have depopulated electron traps in the oxide as well, the experiment was repeated using thermal excitation of carriers in silicon at 19 K, with results displayed in Fig. 1b. A scale factor error is evident in the ordinate of Fig. 1b, which is not of consequence to the conclusion that, once again, agreement with 77 K data was obtained. Finally, to investigate whether thermal depopulation of electron traps might have occurred between 4 K and 19 K, charge transport measurements were made upon Co-60 irradiation while the sample was raised from 4-19 K. Charge transport of 7 pC was measured, amounting to 1.4% of the charge expected if all electrons had been trapped at 4 K and swept out between 4 and 19 K. It is therefore apparent that the fractional trapping of electrons is small on the experimental time scale at the field strengths employed.

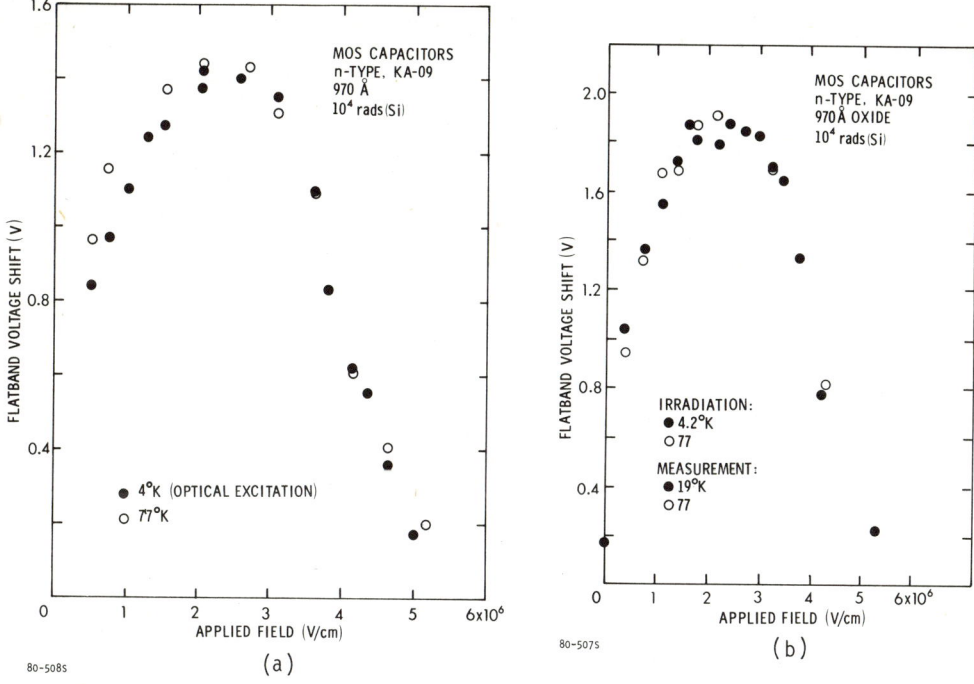

Fig. 1. Flatband voltage shift in MOS capacitors after 10^4 rad(Si), as a function of bias applied during irradiation. Results for two irradiation temperatures are shown: (a) Measurement at 4 K under optical excitation; (b) measurement at 19 K with thermally generated carriers.

In order to examine electron trapping on shorter time scales, charge transport studies were made utilizing the 30-nsec pulse of a flash X-ray. Large area (100-mil diameter) capacitors were fabricated for this purpose. A dry-wet-dry cycle was used to grow a 7600 Å oxide on 0.1 ohm-cm P-doped silicon with a degenerately doped surface layer, obtained by phosphorus diffusion at 1000°C. Direct contact to the surface layer was made in order to eliminate effects of transport within the silicon. Samples were equilibrated thermally using helium exchange gas, which had to be removed during measurement to avoid effects of ionization of the gas. Secondary electron emission from the capacitor metallization was dealt with by using two facing capacitors in close proximity, so that charge emitted by one would be absorbed by the other. In this manner, sensitivity to 1 pC of charge was achieved. To increase the signal amplitude, a total of four capacitors was irradiated simultaneously. Upon preamplification in a lead-shielded chamber near the dewar, charge transport was monitored using a Nicolet 1090 digital oscilloscope, which limited time resolution to 0.5 μsec. The flash X-ray trigger was random within this time window. Results of charge transport observed at 10 μsec are shown in Fig. 2 for a number of temperatures. On this time scale, holes do not contribute significantly to transport for such a large oxide thickness even at room temperature. (14) Thus, electron transport is observed exclusively. The field dependence is that of charge yield, as in the low-field portion of Figs. 1a and 1b. The temperature independence of electron charge transport manifested in Fig. 2 indicates that electron trapping is insignificant on this time scale even at 4.7 K. The only data points at which a temperature dependence is discernible are those at zero bias. At higher temperatures,

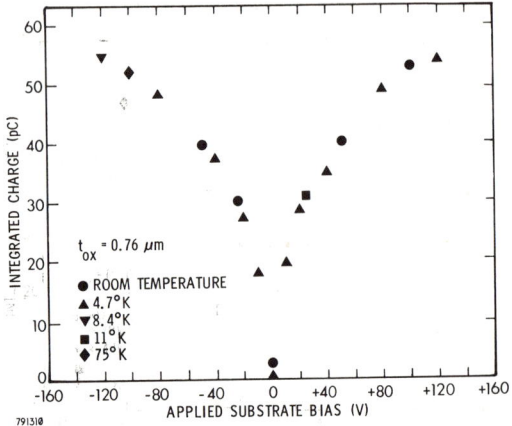

Fig. 2. Integrated charge as a function of applied substrate bias for MOS capacitors with an oxide thickness of 7600 Å irradiated at various temperatures. Charge integration time is 10 μsec.

small residual fields due to charge residing in the oxide or at the interface give rise to finite charge yield. This is not seen at 4.7 K, which could be explained by reduced low temperature mobility. Charge transport as a function of time at 4.7 K is shown in Fig. 3 for various biases. Charge collection appears to be essentially complete within 1 μsec at all biases. Due to the finite time resolution available, only at 10 V is there any evidence of signal current on that time scale.

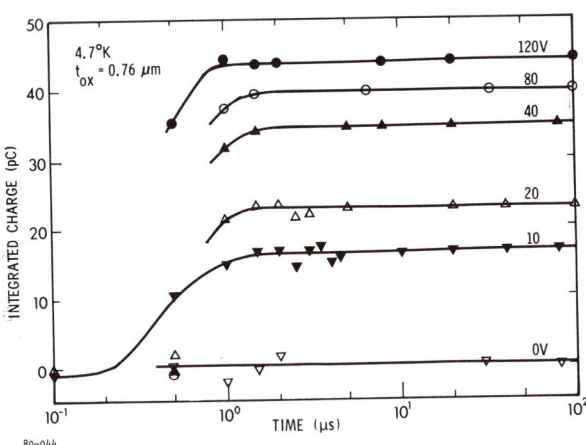

Fig. 3. Integrated charge versus time observed at 4.7 K following a 30-nsec X-ray burst for the same MOS capacitor treated in Fig. 2. Data are shown for various values of applied substrate bias.

DISCUSSION

It has been demonstrated that relatively little shallow trapping of electrons occurs in thermally grown oxide layers at cryogenic temperatures under the present experimental conditions. Using the apparent transit time of about 1 μsec at a bias of 10 V, a minimum drift mobility of 6×10^{-4} cm^2/V-sec is determined. Such a low mobility could arise either via thermal excitation to extended states above the mobility edge from localized states below, or via hopping conduction among localized states. At high temperatures, it is likely that electron drift mobilities should be comparable to those previously observed in amorphous SiO_2 (40 cm^2/V-sec at about 114 K). (8) If this is assumed, then an upper limit to the activation energy governing the mobility is 4.7 meV. One would like to interpret this as defining the width of the tail of the conduction band in this amorphous material. However, a number of questions remain. It is not clear whether the carriers thermalize prior to sweepout. The field strengths employed thus far have been rather large, perhaps leading to hot carrier effects such as reduction in capture probabilities for shallow traps. Nevertheless, one is tempted to conclude that the conduction band density of states must terminate rather abruptly, even if the above 4.7 meV cannot be taken at face value. Certainly no conclusive evidence has been offered here that demonstrates the existence of a mobility edge in the conduction band of thermally grown SiO_2. Hartstein et al. found evidence of a tail in the conduction band density of states amounting to about 0.1 eV using photon-assisted tunneling measurements. (15) Sensitivity of that technique is limited to within 20 Å of the interface. Moreover, dependence of this value on post-oxidation treatment of the oxide tends to raise doubts as to whether it is characteristic of the bulk. It would be desirable to extend the present studies to thicker oxides, and to establish more firmly whether electron transport is thermally activated at low temperature in thermally grown SiO_2.

REFERENCES

1. H.E. Boesch, Jr., F.B. McLean, J.M. McGarrity, and G.A. Ausman, Jr., IEEE Trans. Nucl. Sci. 22, 2163 (1975).
2. J.R. Srour, S. Othmer, O.L. Curtis, Jr., and K.Y. Chiu, IEEE Trans. Nucl. Sci. 23, 1513 (1976).
3. F.B. McLean, G.A. Ausman, Jr., H.E. Boesch, Jr., and J.M. McGarrity, J. Appl. Phys. 47, 1529 (1976).
4. R.C. Hughes, Phys. Rev. B15, 2012 (1977).
5. H. Scher and E.W. Montroll, Phys. Rev. B12, 2455 (1975).
6. O.L. Curtis, Jr. and J.R. Srour, J. Appl. Phys. 48, 3819 (1977).
7. R.C. Hughes and D. Emin, in The Physics of SiO_2 and Its Interfaces, S.T. Pantelides, Ed. (Pergamon Press, New York, 1978), p. 14.
8. R.C. Hughes, Phys. Rev. Lett. 30, 1333 (1973).
9. N.F. Mott and E.A. Davis, Electronic Processes in Non-Crystalline Materials (Clarendon Press, Oxford, 1979).
10. T.H. Ning, J. Appl. Phys. 49, 5997 (1978).
11. D.R. Young, E.A. Irene, D.J. DiMaria, R.F. DeKeersmaecker, and H.Z. Massoud, J. Appl. Phys. 50, 6366 (1979).
12. J.R. Srour and K.Y. Chiu, IEEE Trans. Nucl. Sci. 24, 2140 (1977).
13. G.A. Ausman, Jr. and F.B. McLean, Appl. Phys. Lett. 26, 173 (1975).
14. H.E. Boesch, Jr., F.B. McLean, J.M. McGarrity, and P.S. Winokur, IEEE Trans. Nucl. Sci. 25, 1239 (1978).
15. A. Hartstein, Z.A. Weinberg, and D.J. DiMaria, in The Physics of SiO_2 and Its Interfaces, S.T. Pantelides, Ed. (Pergamon Press, New York, 1978), p. 51.

PHYSICAL EFFECTS IN LATERAL MIS STRUCTURES WITH ULTRA-THIN OXIDES

Jerzy Ruzyllo[1]
Department of Electrical Engineering, The Pennsylvania
State University, University Park, PA 16802

ABSTRACT

Lateral conduction in silicon MIS structures with ultra-thin (10Å-60Å) thermal oxides is considered, and various resulting effects are investigated. Devices of this type with oxide thickness from about 20Å to 40Å are found to be featured by the distinct lateral conduction mechanism. The paper outlines their general properties and also suggests their use in certain useful device structures.

INTRODUCTION

In conventional thick oxide MIS structures made on silicon the current flowing between the two electrodes formed on the surface of the silicon dioxide depends only on the conductivity of the oxide outer surface. Such a surface conduction, being ionic in nature, is independent of the silicon substrate, but on the other hand, depends strongly on the composition of the surrounding atmosphere, especially on the content of water in this atmosphere.
The situation is changed both qualitatively and quantitatively when the oxide thickness in the same system is reduced to a few tens of angstroms (Fig. 1). In this case charge transport occurs via the silicon substrate, and the tunneling controls the charge transfer between the metal and silicon.[2] Considering this, it is

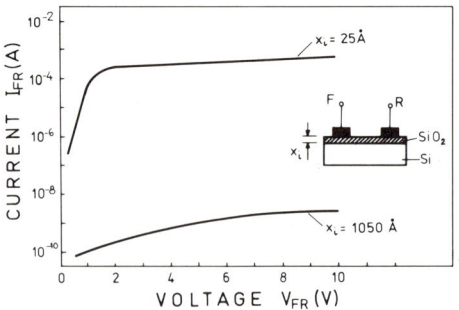

Fig. 1. I_{FR} vs. V_{FR} in the structure with thick and ultra-thin oxide (distance F-R = 25 μm).

appropriate to describe such structures as lateral MISIM (Metal-Insulator-Semiconductor-Insulator-Metal) tunnel devices.
The peculiarity of the conduction mechanism in lateral MISIM devices made on high resistivity ($\rho \geq 10\Omega$cm) silicon covered with 20Å-40Å thick thermal SiO_2 films was already pointed out previously.[2] The purpose of this work is to study in more detail the physical effects in such structures. In addition, the novel structure of the lateral MIS tunnel diode is introduced. The paper also suggests the use of certain properties of the lateral MISIM structures in useful devices such as a novel transistor structure.

EXPERIMENTAL

In Fig. 2 the cross-sectional views of the devices fabricated for this study are presented. In most cases they were made on n-type, high resistivity ($\rho \geq 10\Omega$cm) silicon, although in some cases p-type substrates were also used.

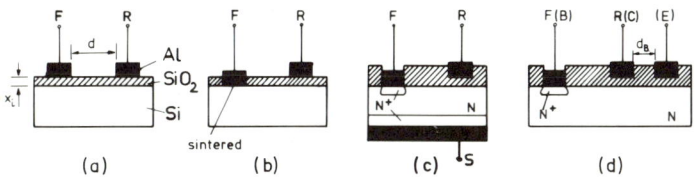

Fig. 2. Cross-sectional views of devices fabricated for this study.

The ultra-thin SiO_2 was grown during brief consecutive treatments in dry N_2 and dry O_2 at 820°C. In each case, aluminum was used for electrodes, and whenever an ohmic contact to the substrate was required the aluminum was sintered (Fig. 2.b) or an n^+ region was formed beneath the electrode (Fig. 2.c). The oxide thickness x_i was measured by ellipsometry and was varied from below 10Å to 60Å.
All measurements were performed in a low humidity ambient, and unless otherwise stated, in the dark and at room temperature, and on the devices with distance d=25μm.

RESULTS

Transport Mechanism in Lateral MISIM Tunnel Structures
Let us consider the lateral MISIM structure shown in Fig. 2.a. In order to make the discussion of any physical effects occurring in such a structure meaningful, it has to be demonstrated that the presence of the deliberately grown ultra-thin oxide, rather than the native oxide, accounts for the entirely different physical effects. In Fig. 3 the I-V characteristics of the lateral MISIM device with an ultra-thin

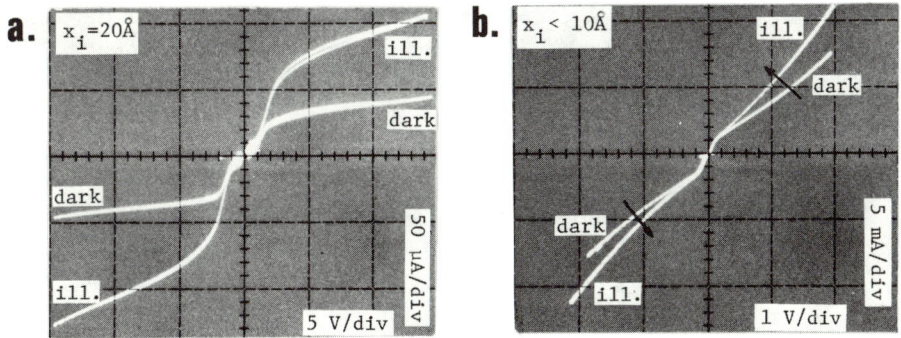

Fig. 3. I_{FR} vs. V_{FR} for the lateral MISIM structure with (a) thermal ultra-thin oxide, and (b) native oxide.

oxide and with a native oxide are shown. The shape of the curves presented, the level of the current, the different sensitivity to illumination, and the high instability of current in the latter case (arrows show the direction of current changes as a function of time at the given voltage) reflect the differences in the conduction mechanism in the structures considered. It is clear that the surface leakage current to a large extent dominates the conduction in the latter case. In the former case, however, the grown oxide, although very thin, provides enough insulation over the oxide surface between the electrodes to force the charge carriers to tunnel from one metal electrode through the oxide to the substrate, and then back to the other electrode. Therefore, we deal in this case with an unusual system of two oppositely biased MIS tunnel contacts linked by a conduction path through the substrate.

The most characteristic feature of the I-V characteristics of the lateral MISIM structures represented in Fig. 3.a and Fig. 4 is saturation of I_{FR} current. Such saturation was not observed in either the devices with native oxide (Fig. 3.b) or in the devices with oxides thicker than a certain critical value (Fig. 4). This is very similar to the behavior described and analyzed by Shewchun et al.[3,4] of the conventional (vertical) reverse biased non-equilibrium MIS tunnel diodes which is due to the depletion of the minority carriers (holes) at the surface by their tunneling to the metal electrode. This in turn imposes a limit on the number of electrons tunneling from the metal to the semiconductor conduction band and causes the diode current to saturate.

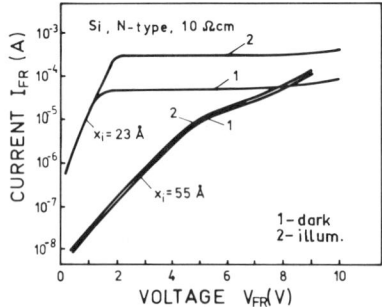 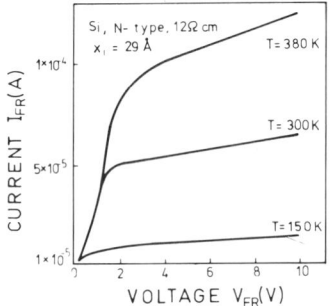

Fig. 4. I_{FR} vs. V_{FR} for two different oxide thicknesses.

Fig. 5. I_{FR} vs. V_{FR} as a function of temperature.

Several experimental results point out that virtually the same mechanism is responsible for the current saturation in the case of the lateral MISIM structures considered in this work. First, current saturation was observed in the lateral MISIM structures in this work and in the reverse biased vertical MIS tunnel diodes[3] for an oxide thickness x_i within the same range (~ 15Å to ~ 40Å). Second, sensitivity of the I_{FR} current to illumination (Fig. 3.a and Fig. 4) and temperature (Fig. 5) demonstrates its dependence on the minority carrier supply to the depletion region of the reverse biased MIS tunnel contact of the lateral MISIM structure. Electrical injection of minority carriers has also been shown to result in the same behavior of the current.[2]

The above data show that the saturation of the current in the lateral MISIM structures results from the properties of that MIS tunnel contact which is reverse biased, and which in fact controls the current flow in the whole system. In Fig. 6 the resistances are ascribed to the reverse biased MIS tunnel contact (R_{DR}) in the MISIM structure, to the forward biased MIS tunnel contact (R_{DF}), to the region of the substrate through which current is flowing (R_S), and also to the oxide surface (R_{ox}). In terms of these resistances the above conclusion can be presented as $R_{DR} \gg R_{DF} + R_S$ (although still $R_{ox} \gg R_{DR}$).

The resistance R_S is naturally related to the distance d defined in Fig. 2.a. An increase of this distance, however, does not cause a decrease in the I_{FR} current, but instead it clearly increases (Fig. 7) in the range of d variations considered.

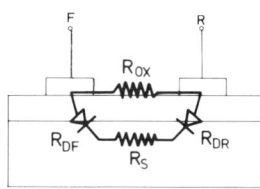

Fig. 6. Simplified equivalent circuit of the lateral MISIM structure.

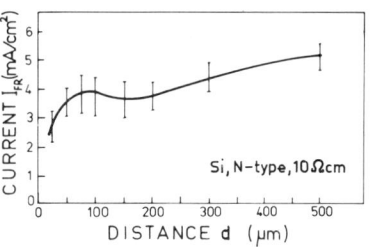

Fig. 7. I_{FR} vs. d at V_{FR}=4V.

It is assumed to be due to the change of the voltage drop on resistances R_S and R_{DR} resulting in a smaller portion of the V_{FR} voltage drop on R_{DR}. In the reverse biased non-equilibrium MIS tunnel diode in saturation this narrows the depletion region[3] which accounts for the higher probability of electrons tunneling from the metal to the semiconductor conduction band. This effect supports the conclusion that in the lateral MISIM tunnel structures the limitation to the current flow comes from only one (reverse biased) MIS tunnel contact incorporated into this structure.

Lateral MIS Tunnel Diode
The lateral MISIM structure can be transformed into a novel structure, the lateral MIS tunnel diode, by replacing one of the MIS tunnel contacts by an ohmic contact to the substrate(Fig. 2.b and c). In this way a rectifying device can be obtained which under reverse bias displays all of the features of the lateral MISIM tunnel structure discussed in the previous section, e.g. sensitivity to illumination. This is illustrated in Fig. 8 which shows the I-V characteristics of the lateral MIS tunnel diode taken in dark and under two different levels of illumination. These characteristics clearly display the photo-diode-like behavior of the lateral MIS tunnel diode. In the forward direction the current is controlled by the tunneling of majority carriers and is not affected by illumination.
The characteristics shown in Fig. 8 suggest a similarity between the lateral MIS tunnel diodes studied in this work and conventional vertical MIS tunnel diodes. The device shown in Fig. 2.c made possible a direct comparison between these two structures (configuration R-F and R-S respectively). The I-V characteristics,

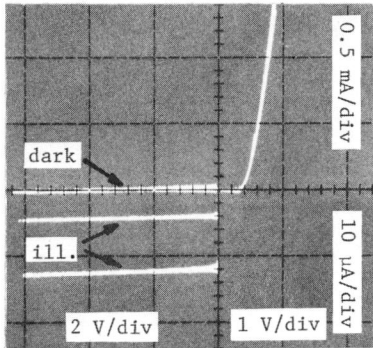

Fig. 8. I-V characteristics of the lateral MIS tunnel diode (x_i = 22Å, Si p-type).

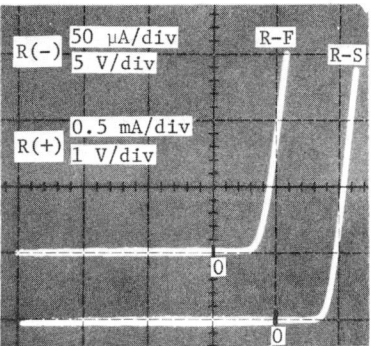

Fig. 9. I-V characteristics of the lateral (R-F) and vertical (R-S) MIS tunnel diodes (x_i = 25Å).

shown in Fig. 9, are virtually identical. This demonstrates that the characteristics of the whole device are controlled only by the properties of the MIS tunnel system, and that the position of an ohmic (reference) contact on the wafer has no influence on the device performance.

SOME APPLICATIONS OF THE LATERAL MIS TUNNEL STRUCTURES

According to the data presented in the previous section the lateral MIS tunnel diode can probably be applied wherever the vertical MIS tunnel diode can be used including use of the lateral configuration to exploit the photovoltaic effect. The preliminary experiments performed in the course of this study have shown that the lateral MIS tunnel structure (Fig. 2.b) might perform as a solar cell as well as does a vertical MIS tunnel structure. It offers at the same time new possibilities as far as the exposure of the device to the light is concerned.
In addition, the mechanism of conduction in the lateral structures under study can be used to fabricate a novel device called the lateral MIS tunnel transistor in the form shown in Fig. 2.d.[5] Such a device was found to be fully operable with current gain in the common-emitter configuration up to 15 at the distance $d_B = 7 \mu m$.

SUMMARY

The conduction in the lateral MISIM structures with oxide 20Å-40Å thick is featured by the specific properties related to the phenomena occurring in the non-equilibrium MIS tunnel diodes. It is the reverse biased MIS tunnel contact incorporated into the lateral MISIM tunnel structure which controls the current flow and other properties of such device. The lateral MISIM structure can be transformed into the novel structure of lateral MIS tunnel diode which displays the same properties as the vertical MIS tunnel diode, but due to the electrode configuration, it offers new possibilities in some applications, e.g., in the field of solar cells. The described effect can be also exploited to form a novel transistor structure.

REFERENCES

1. On leave from Institute of Electron Technology, Technical University of Warsaw, Koszykowa 75, 00-662 Warsaw, Poland.
2. J. Ruzyllo, K. Kucharski and J. Jakubowski, Effect of minority carrier injection on lateral current in MIS tunnel structures, Solid State Electronics (in press).
3. R. A. Clarke and J. Shewchun, Non-equilibrium effects on metal-oxide-semiconductor tunnel currents, Solid State Electronics, 14, 957 (1971).
4. V. A. K. Temple, M. A. Green and J. Shewchun, Equilibrium-to-nonequilibrium transition in MOS (surface oxide) tunnel diode, J. Appl. Phys., 45, 4934 (1974).
5. J. Ruzyllo, Lateral MIS tunnel transistor (to be published).

OXYGEN AS A TWO-LEVEL TUNNELING SYSTEM IN SiO_2

W. Beall Fowler and Arthur H. Edwards
Department of Physics and Sherman Fairchild Laboratory
Lehigh University, Bethlehem, Pa. 18015

ABSTRACT

Evidence has accumulated that oxygen atoms are responsible for the two-level tunneling systems (TLS) which result in a number of low-temperature acoustic anomalies in amorphous SiO_2. By means of MINDO/3 and MNDO computer codes we have found off-centered oxygen atoms in sufficiently elongated Si-O-Si bonds, in agreement with Strakna. We can fit this behavior to a pseudo-Jahn-Teller effect with plausible parameters. By considering a distribution of strained bonds, we can account for the small electric dipole moments of TLS, a major difficulty in earlier attempts to relate the anomalies to displacements of a single oxygen atom.

INTRODUCTION

The anomalous low-temperature properties of amorphous insulators have been widely studied during the past decade[1]. The many anomalous properties include specific heat, thermal conductivity, acoustic attenuation and velocity, dielectric properties, thermal expansion, and optical properties. These effects have been studied below 1 K, although most of them persist to higher temperatures. Many amorphous materials have been shown to exhibit anomalies, including SiO_2, GeO_2, B_2O_3, Se, As_2S_3, and even Ge No. 7031 varnish!

We have concentrated on effects pertinent to SiO_2, which has been the most studied and most widely discussed material exhibiting anomalous behavior. Microscopic mechanisms which provide plausible explanations in SiO_2 should of course be investigated in other materials. The review article by Pohl and Salinger[1] is an excellent general reference.

The anomalous behavior referred to in this paper involves effects which cannot be attributed to electronic excitations or atomic vibrations in an idealized model of an amorphous solid. That is, when one cools a glass to $\sim$ 1 K and below and freezes out most of the vibrations, one finds that the glass retains some "anomalous" degrees of freedom through which it can interact with external probes. Such effects are not seen in high-quality crystals. Low-temperature data on specific heat and thermal conductivity[2] for several glasses indicate that the number of these degrees of freedom is $\sim 10^{17}$ cm^{-3}. For SiO_2 this would correspond to one for every $\sim 10^4$ atoms.

THEORY AND MODELS

Several authors[3,4] have pointed out that a two-level tunneling model can be used to describe much of the data. The anomalies are presumed to be associated with two nearly equivalent atomic configurations with a barrier between them, as shown in Fig. (1). Here "d" is a generalized coordinate; the actual motion may involve groups of atoms, not just one. Furthermore, there is a nearly uniform spread in energy separation ϵ, at least up to ~ 1 K, and barrier characteristics are expected to vary as well. One thus describes the anomalies as being associated with a two-level system (TLS) or, more accurately, a number of slightly different TLS.

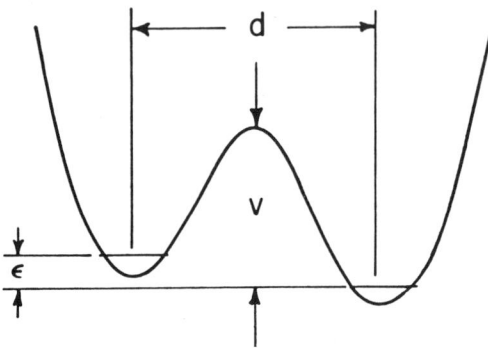

Fig. 1. A two-level tunneling system. The potential wells are separated by a distance d and a barrier of height V. Their depths differ by an amount ϵ.

The most widely discussed microscopic models for TLS involve transverse motion of oxygen atoms[5,6]. In view of the flexibility of the Si-O-Si bond and the well-known spread[7] in this bond angle in amorphous SiO_2, this model appears attractive. In one version of this model[6] there are postulated two values of the Si-O-Si angle, each of which has a local equilibrium with respect to small displacements.

A second model, first suggested by Strakna[8], and recently invoked by Laermans[9], involves Si-O-Si bonds which are sufficiently elongated to yield a double-well potential for the oxygen (Fig. 2). Strakna supported this suggestion with a calculation using a Morse potential, which indicated that the central oxygen would move off center if the silicon-silicon distance exceeds ~ 3.8 A.

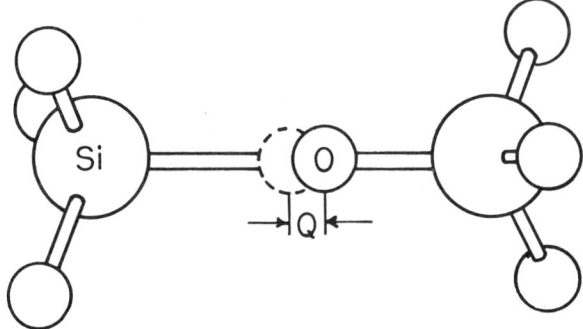

Fig. 2. An off-centered oxygen associated with an elongated Si-O-Si bond.

One of the chief difficulties with any model involving displacement of just one atom is the small observed dipole moment associated with the anomalies in SiO_2. Recent measurements[10], which are consistent with earlier reports, indicate an average dipole moment of ~ 0.5 Debye or ~ 0.1 eÅ, with considerable spread. Thus, e.g. an oxygen "ion" with charge ~ -1 could move only ~ 0.1 Å, a small value indeed. On the other hand, one can conceive of angular motions which would result in a small dipole moment. Several authors[1,3] have recognized this problem.

Other recent experiments[11] on thin films of Si and SiO_2 indicate that oxygens are indeed involved in the anomalies. These authors suggest that TLS do not exist in amorphous Si, and present evidence that in SiO_2 the relaxing particle is a single oxygen atom.

PRESENT WORK

We are using MINDO/3 and MNDO semiempirical computer codes, written by Dewar and associates[12,13] and supplied by the Quantum Chemistry Program Exchange, to investigate potential energy surfaces associated with clusters of atoms in SiO_2. At present we have done only preliminary work on angular variations related to the TLS problem. We have studied the elongated bond, or Strakna, mechanism in more detail and feel that it has considerable promise in explaining the TLS.

For the most part we have investigated an $(Si_2O_7)^{6-}$ cluster. We find that the central oxygen goes off center at a Si-Si separation of $\gtrsim 4.2$ Å, slightly larger than predicted by Strakna's Morse potential but in general agreement. In our calculation as in Strakna's the energy is the same with the oxygen on either side of center; the small inequivalences which are part of the tunneling model[3,4] are readily explained by the differences in the two sides associated with the amorphous structure, and we will not pursue this point further. We find that the amount of off-centered displacement is approximately proportional to the Si-Si distance; at 4.3 Å the oxygen is $\sim .2$ Å off-center, at 4.5 Å it is $\sim .45$ Å off-center. The barrier height V varies more rapidly, going from .03 eV at 4.3 Å to .29 eV at 4.5 Å.

We can fit this behavior to a pseudo-Jahn-Teller (PJT) formalism, much as we have done for the E_1' center in alpha quartz[14]. The parameters which work[15] are consistent with those which fit the E_1' center. In this formalism the ground-state energy is given by

$$E(Q) = \frac{1}{2} \mu\omega^2 Q^2 - \frac{1}{2} [E_{sp}^2 + 8 E_{JT} \mu\omega^2 Q^2]^{1/2} \quad (1)$$

where Q is the displacement of the oxygen from the center, E_{sp} is an electronic excitation energy at Q=0 ($\gtrsim$ band gap), the Jahn-Teller energy E_{JT} is a measure of the coupling strength, and $(\mu\omega^2)$ is a force constant for oxygen motion.

While it appears at first sight that the elongated bond model predicts displacements which are too large (leading to large dipole moments), we believe that the data may be reconciled if we consider the distribution of Si-Si distances in the glass. The equilibrium Si-Si distances ℓ should be characterized by a distribution function which peaks at the mean Si-Si separation ℓ_o and which falls off monotonically with ℓ. For example, one might assume a Gaussian distribution of the form

$$P(\ell) = \sqrt{\frac{\alpha}{\pi}} e^{-\alpha(\ell-\ell_o)^2} \quad (2)$$

Thus highly strained bonds which would produce off-centered oxygens are relatively improbable, and of these the most probable are those with small dipole moments.

Quantitatively, if the oxygen displacement from center, Q, is given by

$$Q(\ell) = 0 \qquad \ell < \ell_1 \\ = Q(\ell) \qquad \ell > \ell_1 \quad (3)$$

where ℓ_1 is the Si-Si separation for which oxygen goes off center, then

$$\bar{Q} = \int_{\ell_1}^{\infty} Q(\ell) \, P(\ell) \, d\ell \qquad (4)$$

$\bar{Q}$ may be rather small, depending on the chosen parameters. Our computations indicate that $Q(\ell)$, for $\ell > \ell_1$, goes approximately as

$$Q(\ell) = \gamma(\ell - 4.2 \text{ Å}), \qquad \ell > 4.2 \text{ Å} \qquad (5)$$

where γ is ~ 2.

If ℓ_0 is 3.2 Å, a value $\alpha \sim 7$ Å^{-2} will yield a fraction of bonds $\sim 10^{-4}$ whose length is greater than $\ell_1 = 4.2$ Å, which therefore have off-centered oxygens which can provide the desired tunneling systems. This is consistent with the number cited in ref. (2). Evaluation of $\bar{Q}$ (Eq. 4) under these conditions yields $\bar{Q} \sim .09$ Å, consistent with measured dipole moments.

These results are of course rather crude. More detailed calculations will certainly yield somewhat different results. For example, α should probably be larger than 7 Å^{-2} and ℓ_1 smaller than 4.2 Å. (The distribution function used here falls off too slowly[7] for $\ell \sim \ell_0$.)

CONCLUSIONS

The elongated bond mechanism has been shown to be a very plausible candidate for the TLS in amorphous SiO_2, and (by extension) perhaps in other amorphous systems as well, provided that the distribution of strained bonds is taken into account. The off-centeredness of the oxygen atom, computed by semiempirical MO techniques, can be modeled as a pseudo-Jahn-Teller effect. Further implications of this result are being investigated.

While we have by no means ruled out variations in the Si-O-Si angle as a mechanism for TLS, it should be noted that such angular variations involve motion of large numbers of atoms if, as expected, the O-Si-O angle is held constant. The existence of local minima of energy vs. angle under such conditions, we feel, is still open to question.

ACKNOWLEDGEMENT

This research was supported by the ONR, Contract No. N00014-76-C-0125.

REFERENCES

(1) R. O. Pohl and G. L. Salinger, Annals N.Y. Acad. Sci. 279, 150 (1976).
(2) R. B. Stephens, Phys. Rev. B8, 2896 (1973).
(3) P. W. Anderson, B. I. Halperin and C. M. Varma, Phil. Mag. 25, 1 (1972).
(4) W. A. Phillips, J. Low Temp. Phys. 7, 351 (1972).
(5) O. L. Anderson and H. E. Bömmel, J. Am. Chem. Soc. 38, 125 (1955).
(6) M. R. Vukcevich, J. Noncryst. Sol. 11, 25 (1972).
(7) R. L. Mozzi and B. E. Warren, J. Appl. Cryst. 2, 164 (1969).
(8) R. E. Strakna, Phys. Rev. 123, 2020 (1961).
(9) C. Laermans, Phys. Rev. Lett. 42, 250 (1979).
(10) Brage Golding, M. v. Schickfus, S. Hunklinger, and K. Dransfeld, Phys. Rev. Lett. 43, 1817 (1979).
(11) M. Von Haumeder, U. Strom, and S. Hunklinger, Phys. Rev. Lett. 44, 84 (1980).
(12) R. C. Bingham, M.J.S. Dewar, and D.H. Lo, J. Am. Chem. Soc. 97, 1285 (1975).
(13) M.J.S. Dewar and W. Thiel, J. Am. Chem. Soc. 99, 4899 (1977).
(14) K. L. Yip and W. B. Fowler, Phys. Rev. B11, 2327 (1975); D. L. Griscom and W. B. Fowler, this volume.
(15) A good (though not unique) fit is obtained with E_{sp} = 8.0 eV, $\mu\omega^2$ = 16.6 eV/Å^2, E_{JT} = 2.9 eV.

CHAPTER II

BULK PROPERTIES

PERIODIC STRUCTURAL MODELS AND RADIAL DISTRIBUTION
FUNCTIONS OF SiO_x: x=0. to 2.*

W. Y. Ching
Department of Physics
University of Missouri-Kansas City
Kansas City, Mo. 64110

A series of random network structural models with periodic boundary conditions for SiO_x for x range from 0. (pure amorphous Si) to 2. (pure vitreous silica) within the context of microscopic random bond mixing model have been constructed. A Keating type of potential is used in the computer relaxation process. These cubic models which have 54 Si atoms and variable number of O atoms have no internal voids or dangling bonds. Each O atom binds to two Si atoms in a non-linear bridging position while each Si atom binds to four other atoms (Si or O) in a tetrahedral configuration. The densities, radial distribution functions (RDF) and partial RDF are studied as a function of x and are compared with available x-ray scattering data. It is found that for cases of x ≠ 0. or 2., the Si-Si bond lengths tend to be larger and Si-O bond lengths tend to be shorter than their respective crystalline bond lengths. Utilization of these structural models in the electronic and vibrational calculations is also discussed.

INTRODUCTION

In the past few years, the continuous random network (CRN) theory[1] has been generally accepted as the appropriate theory to describe the structure of glasses, amorphous semiconductors and other covalently bonded non-crystalline solids. Cluster models (CM) of CRN which contain no internal voids or dangling bonds has been constructed for vitreous silica or amorphous SiO_2 (a-SiO_2) as early as mid-sixties.[2] Later, many tetrahedrally bonded network models were also constructed for amorphous silicon (a-Si).[3,4] These physical models enable us to calculate theoretical radial distribution functions (RDF) and compare with the RDF deduced from scattering experiments. Since agreement with experimentally derived RDF is a necessary but not sufficient condition of being a good model. A more stringent test is to calculate the electronic[5,6] or vibrational properties[7] of the material based on the atomic coordinates of the model structure. It is expedient to have a CRN with periodic boundary conditions (PBC). Since then, the mathematical formulism of electronic or vibrational calculations will be similar to the crystalline cases with perhaps an enormous increase in the complexity of the calculation because of the large number of atoms in the quasi unit cell. The first of such quasi-periodic model (QPM) has been constructed by D. Henderson for a-Si.[8] In contrast to the CM which always associate itself with a finite surface, the QPM represents a truely infinite array of network from which the density of the material can be accurately determined and no surface effect will appear. Of course, the unit cell should be sufficiently large such that the residual long range order of quasi-crystallinity is negligible. The only drawback of QPM is that the constraint of PBC make the construction of model much more difficult. In this paper, I present the first QPM for vitreous silica which has a RDF and density in excellent agreement with experiment. Similar models are also constructed for SiO_x with x = 1.5, 1.0, and 0.5 within the context of microscopic random bond mixing model.[9] These models should be very useful in elucidating the atomic structures of SiO_x films and Si-SiO_2 interfacial regions and provide a means of deeper microscopic understanding about the interfacial electronic states.

*Supported by DOE Contract DE AC02 79ER10462

MODEL CONSTRUCTION

The construction of QPM of CRN for an amorphous solid is a rather difficult and extremely time consuming process. Fortunately, reasonably good QPM for a-Si is already in existance and can be utilized to construct QPM for a-SiO$_2$ and SiO$_x$. We start with the QPM originally constructed by Guttman,[6] which has 54 Si-atoms in a cubic cell and insert O atoms between each pair of Si atoms. We then rescale the size of the cell such that the density of the model is equal to 2.2 gm/c.c. for vitreous silica. A two dimensional illustration of the process is shown in Figure 1. From this initial configuration in which the topology of the network and the bond pattern is determined, a computer relaxation process[4] is applied which reduce the overall elastic energy of the network by successively moving each atom to the position of zero force under a Keating[10] type of potential.

$$V = \sum_\ell V_\ell^{Si} + \sum_n V_n^O$$

$$V_\ell^{Si} = \frac{3}{16} \frac{\alpha}{d^2} \sum_{i=1}^{4} (\vec{r}_{\ell,i} \cdot \vec{r}_{\ell,i} - d^2)^2 + \frac{3}{8} \frac{\beta_1}{d^2} \sum_{(i,i')}^{6} (\vec{r}_{\ell,i} \cdot \vec{r}_{\ell,i'} - d^2 \cos^{-1} \phi)^2$$

$$V_n^O = \frac{3}{16} \frac{\alpha}{d^2} \sum_{j=1}^{2} (\vec{r}_{n,j} \cdot \vec{r}_{n,j} - d^2)^2 + \frac{3}{8} \frac{\beta_2}{d^2} (\vec{r}_{n,j} \cdot \vec{r}_{n,j'} - d^2 \cos^{-1} \theta)^2$$

where $\vec{r}_{\ell,i}$ is the radius vector from atom ℓ to its nearest neighbor atom i; ϕ is the ideal O-Si-O angle (109° 28'); θ is the average Si-O-Si angle (147.° in this work) and d is the Si-O or Si-Si bond length as the case may be. The first term in the above expressions sums over nearest neighbors, while the second term sums over the nearest neighbor pairs. The ratio of bond bending to bond streching force constants are set to be $\beta_1/\alpha = 0.17$,[8] $\beta_2 = 1/3 \beta_1$. The model structure is found to be rather insensitive to a reasonable range of choice of the constants α, β_1, β_2.

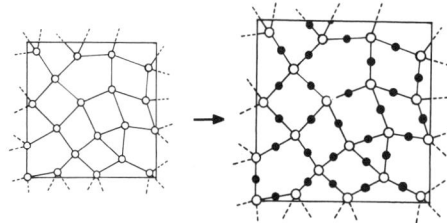

Fig. 1. Two dimensional illustration for obtaining initial atomic configurations of a periodic model of a-SiO$_2$ from periodic model of a-Si. ○ Si atom; ● O atom.

If only a random fraction of the available O sites in the above described process is occupied, we have the initial configuration of a random bond mixing model of SiO$_x$.[9] Each O atom binds to two Si atoms and each Si atom binds to four atoms which may be O or Si, depends on the random statistics. The unit cell sizes are rescaled appropriately according to x. The densities are found to be 2.45, 2.50, 2.44 gm/c.c. for x = 1.5, 1.0, 0.5, respectively. The structures are then relaxed as before where we assumed the bond streching constant α for Si-Si and for Si-O are the same, so are the bond bending constants β_1 for Si-Si-Si, Si-Si-O and O-Si-O angles. For each value of x three inequivalent models with different random statistics were constructed.

RESULTS

In Figure 2 (a) and (b) we present the calculated RDF of a-SiO$_2$ for x-ray and neutron scattering respectively. The atomic scattering factors used for x-ray (neutron) scattering are F^x_{Si} = 14.75 electrons (f^n_{Si} = 0.4210 x 10^{-12} cm) and f^x_O = 7.625 electrons (f^n_O = 0.577 x 10^{-12} cm) respectively.[12] The corresponding experimental curves as presented in Ref. 12 are also shown as dotted lines. It is apparent that the agreement is very good up to a radial distance of 8 Å. Not only the position of the major peaks are in match, but also the smaller structures

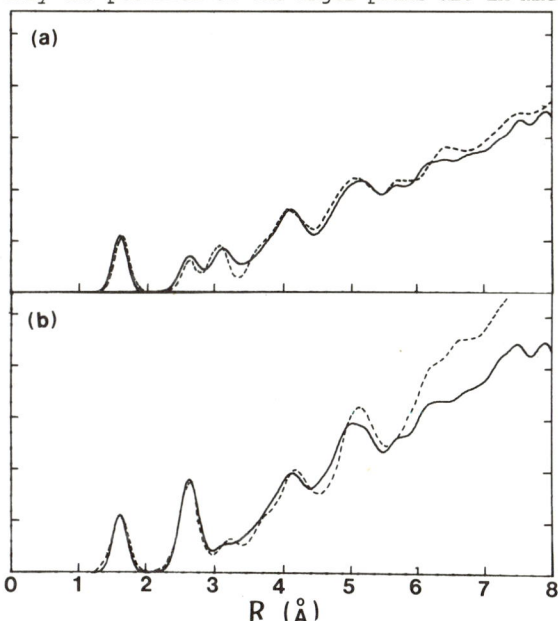

Fig. 2. RDF of vitreous silica for: (a) x-ray scattering, (b) neutron scattering. Solid line, computed from the present model; dotted line, experimental curve as presented in Ref. 12.

are quite well reproduced. The resolved Si-Si, O-Si and O-O partial RDF are shown in Fig. 3 (a) to (c). These results are also in good agreement with those from CM of a-SiO$_2$ by Bell and Dean.[12] The only difference being the absence of a shoulder at 3.7 Å which is attributed to Si-O pairs in four membered rings. The present model, being derived from a model for a-Si[6] does not contain four membered rings. The distributions of Si-O bond lengths R, the O-Si-O angles ϕ and the Si-O-Si angle θ are shown in Figure 4 (a) to (c). The average value is indicated by an arrow. It is obvious that the distortion of SiO$_4$ units from a perfect tetrahedron is very small. The root mean square deviations of R and ϕ are only 0.017 Å and 4.8° respectively. The distribution of θ is non-symmetric and ranges from 116° to 180° with an average value $\bar{\theta}$ = 147°. This is to be compared with the value of $\bar{\theta}$ = 153° in the model of Bell and Dean[12] and also $\bar{\theta}$ = 144° in α-quartz crystal.

The RDF for x-ray scattering of SiO$_x$ for x from 2.0 (vitreous silica) to 0. (a-Si) are shown in Figure 5 (a) to (e). For x = 1.5, 1.0 and 0.5, the curves represent the average of three statistically inequavalent models. Going from x = 2.0 to 0. the Si-O peak diminishing and the Si-Si peak growing and moving from 3.12 Å in a-SiO$_2$ to 2.33 Å in a-Si. The interpretations of other peaks above 2.5 Å is less straight forward because in the microscopic random bond mixing model, all types of atomic pairs contribute. It was found that for x not equal to 2.0 or not equal to 0. the Si-O (Si-Si) bond distances tend to be slightly shorter (longer) than the corresponding crystalline bond length. This is a consequence of additional short range disorder present in the cases of x ≠ 2.0 or 0. in contrast to the presence of short range order in vitreous silica or a-Si.

Experimental x-ray RDF for the specific x values of 1.5, 1.0, and 0.5 are not aware to us at this time. The closest is the x = 1.08 measurement of Yasaitis and Kaplan.[13] The RDF below 5 Å have four prominent peaks and agrees quite well with the calculated RDF for x = 1.0 of Figure 5 (c).

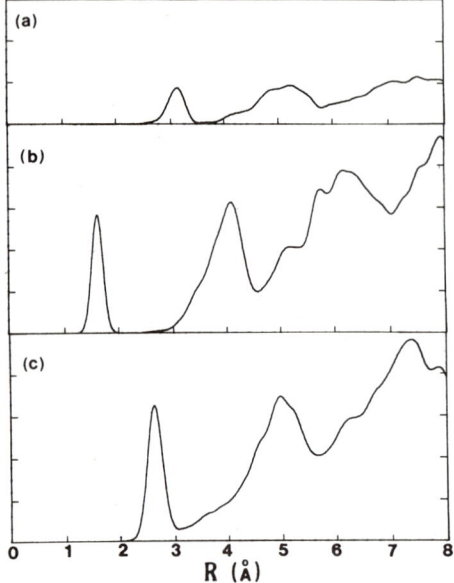

Fig. 3. Partial RDF of vitreous silica from present model: (a) Si-Si, (b) Si-O, (c) O-O.

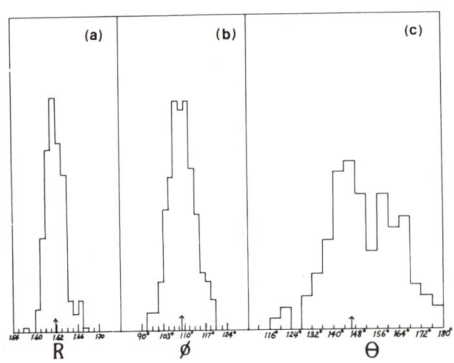

Fig. 4. Distribution of (a) Si-O bond lengths, (b) O-Si-O bond angles, (c) Si-O-Si bond angles in the present model. The average value is indicated by an arrow.

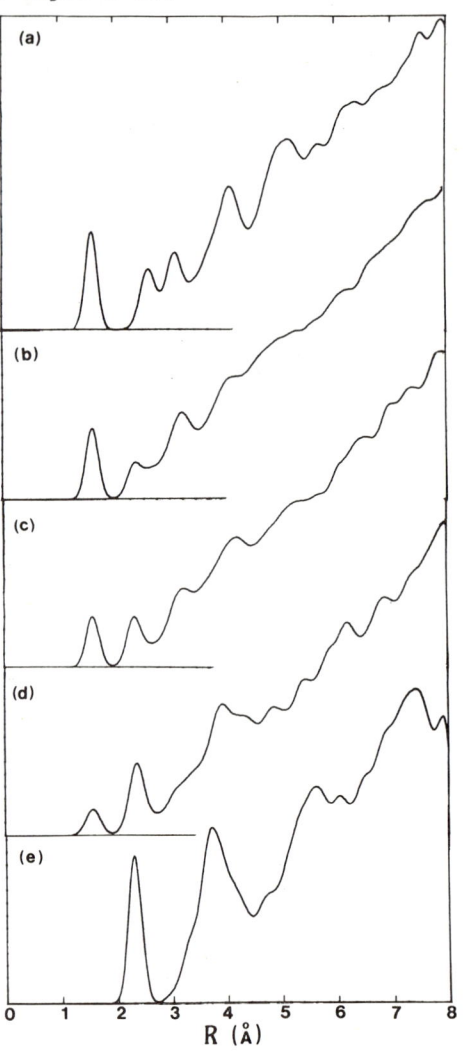

Fig. 5. Calculated x-ray scattering RDF for SiO_x, (a) x = 2.0 (vitreous silica), (b) x = 1.5, (c) x = 1.0, (d) x = 0.5, (e) x = 0. (amorphous Si).

DISCUSSION

We have successfully constructed QPMS which represent an infinite array of random network structure for a-SiO$_2$ and SiO$_x$. The calculated RDF is in good agreement with experimental measurement and that of earlier CM. Because of limitation of size of the present paper, detailed analysis of these models with respect to the ring topology, random bond statistics, partial RDF, effects of presence of four member rings, effects of varying relaxation parameters etc., will be presented elsewhere. It should be pointed out that the flexibility of Si-O-Si angle plays an important role in model construction. This accounts for the fact that the distortion of SiO$_4$ tetrahedral coordination in the case of a-SiO$_2$ is much smaller than the corresponding distortion in a-Si case.[6]

The models for SiO$_x$ should be of particular interest because they yield microscopic information about atomic scale structures of Si-SiO$_2$ interfacial regions. At present there are two competing models for the structure of SiO$_x$: (1) the microscopic random bond mixing model,[9] base on which the present models are constructed and (2) the randomly dispersed mixture of tetrahedrally bonded a-Si and a-SiO$_2$ model[13] in which the domain of Si-like or SiO$_2$-like regions are of few tetrahedral units. Although the present result based on the idea of first model is not inconsistent with experiment, we should be able to construct similar QPMS based on second model and make critical comparisons. Similar models with specific bonding configurations of defects can also be constructed. Using the atomic positions of the QPM constructed, one should be able to attempt realistic quantum mechanical calculations of electronic and vibrational states of SiO$_x$ for various values of x. In conjunction with photoemission or vibrational spectroscopy data, we are then in position to discriminate between competing models for SiO$_x$ and achieve a deeper understanding about the microscopic origin and nature of interfacial states. Work in this direction is in progress and will be reported in future publication.

REFERENCES

1. R. J. Bell and P. Dean, Physics Chem. Glasses, 9, 125, 1968.
2. F. Ordway, Science, 143, 800 (1964);
   D. L. Evans and S. V. King, Nature, Lond., 212, 1353 (1966);
   R. J. Bell and P. Dean, Nature, Lond., 212, 1354 (1966).
3. D. E. Polk, J. Non-Crystalline Solids, 5, 365 (1971);
   D. E. Polk and D. S. Bondreaux, Phys. Rev. Lett., 31, 92 (1973);
   G. A. N. Connell and R. J. Temkin, Phys. Rev. B, 9, 5323 (1974).
4. P. Steinhardt, R. Alben and D. Weaire, J. Non-Crystalline Solids, 15, 199 (1974).
5. W. Y. Ching, C. C. Lin, Phys. Rev. Lett., 34, 1223 (1975).
   W. Y. Ching, C. C. Lin and D. L. Huber, Phys. Rev. B, 14, 620 (1976).
   W. Y. Ching, C. C. Lin, Phys. Rev. B, 18, 6829 (1978).
6. W. Y. Ching, C. C. Lin, L. Guttman, Phys. Rev. B, 16, 5488 (1977).
7. P. Dean, Review of Modern Physics, 44, 127 (1972).
8. D. Henderson, J. Non-Crystalline Solids, 16, 317 (1974).
9. H. R. Phillips, J. Non-Crystalline Solids, 8-10, 627 (1972).
10. P. N. Keating, Phys. Rev., 145, 637 (1966).
11. F. L. Galeener, Phys. Rev. B, 19, 4292 (1979).
12. R. J. Bell and P. Dean, Phil. Mag., 25, 1381 (1972).
13. J. A. Yasaitis and R. Kaplow, J. Appl. Phys., 43, 995 (1972).
14. R. J. Temkin, J. of Non-Crystalline Solids, 17, 215 (1975).

THE OPTICAL ABSORPTION EDGE OF SiO_2

R. B. Laughlin
Bell Laboratories, Murray Hill, New Jersey 07974

ABSTRACT

The optical absorption spectrum of SiO_2 is calculated using tight-binding single-particle Hamiltonian and a screened coulomb electron-hole interaction. The results of the calculation suggest strongly that all four features in ε_2 are excitonic resonances, that the gap in SiO_2 is 8.9eV as indicated by internal photoemission, and that a Rydberg series of disallowed excitons exists beginning at 8.4eV.

INTRODUCTION

Despite the technological importance of SiO_2 and the extensive experimental and theoretical work performed on it in recent years[1,2], its optical absorption spectrum is still not completely understood. The reasons for this, which are numerous, may be traced to the unfortunate confluence in this material of three factors antagonistic to simple analysis of optical spectra: the large gap and small conduction band effective mass SiO_2 has in common with alkalai halides[3] and rare-gas solids; the structural complexity of SiO_2 in all its naturally occurring allotropes, particularly its glassy state; and, most serious of all, its lone-pair nature, i.e. the complexity and density of its upper valence bands, and the consequent tendency of these to be thoroughly mixed by a strong electron-hole interaction. The difficulty in interpreting the absorption spectrum of this material has had a number of serious consequences, the most significant being uncertainty as to the size of its gap.

In this paper an attempt is made to shed some light on the nature of electronic excitations in SiO_2 via a model calculation of ε_2. The calculation is motivated by the observation[4] that matrix elements of the single-particle Green's function tend to be the same in *any* structure in which the integrity of the SiO_4 tetrahedron is preserved. An obvious corollary of this is that inclusion of the electron-hole interaction via a Green's function technique[2] must also be insensitive to structure if the excitons are sufficiently small. That this is probably the case in SiO_2 is indicated by the remarkable similarity of the crystalline and amorphous absorption spectra. Insensitivity of the Green's function to structure has been used in this calculation as justification for adopting the highly symmetric but non-existent β-Cristobablite as a prototype structure and exploiting the favorable symmetry to make tractable the inclusion of a screened Coulomb electron-hole interaction. The calculation leads to sufficiently small excitons to justify to original approximations and suggests very strongly that

    1) All four features in ε_2 are excitonic resonances

    2) The gap in SiO_2 is 8.9 eV.

    3) There is a Rydberg series of disallowed excitons beginning at 8.4eV.

THE MODEL

In the presence of the electron-hole interaction ε_2 may be calculated by solving Dyson's equation

$$G = \frac{1}{G_o^{-1} - V} \tag{1}$$

for the 2-body Green's function G using the single-particle spectrum to generate G_o and using a screened coulomb interaction for V. Five major approximations have been used in the present calculation:

TABLE 1

Holes	Electrons
$\epsilon^{O_{2p}}_{Bonding} = -6.98$	$\epsilon^{O_{2s}} = 19.952$
$\epsilon^{O_{2p}}_{Non-bonding} = -1.739$	
$V_1^0 = -0.625$	$V = -1.842$
$V_2^0 = 0.0$	
$V_3^0 = -0.616$	
$V_4^0 = 0.108$	

The notation for the oxygen-oxygen interaction is that of Ref. 4.

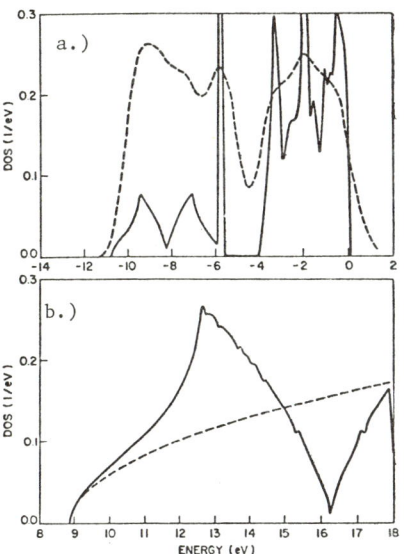

Fig. 1: a.) Solid: Density of valence states of β-Cristobalite. Dashed: Experimental XPS of amorphous silica. b.) Solid: Density of conduction states. The fine structure above 13 eV is computer noise. Dashed: Free electron density of states for $m^* = 0.3\ m_e$.

1. A simplified tight-binding Hamiltonian has been used to describe the single-particle excitations. The applicability of tight-binding Hamiltonians to this system and the significance of the parameters in them has been discussed in detail in Ref. 4. The present calculation was performed using oxygen 2p and oxygan s orbitals to represent the valence and conduction states, respectively. The tight binding parameters are listed in Table 1.

2. The β-Cristobalite structure, which is that of silicon with oxygen atoms centered in the bonds, has been adopted. This is justified on the ground that neither the single-particle spectrum[4] nor ϵ_2 is very sensitive to structure, and it has the important effect of making the symmetry of the tetrahedron the symmetry of the problem. Fig. 1 shows the valence and conduction states of β-Cristobalite calculated using the Hamiltonian of Table 1. The XPS of amorphous SiO_2[5] and the free electron ($m^* = 0.3m_e$)[6] densities of states are shown for comparison.

3. A dipole matrix has been adopted in which the electric field connects only oxygen s and oxygen p orbitals on the same atom. This mechanism, which has been suggested before,[2] is physically reasonable, and it tends to give the same result are more complicated mechanisms involving crossing transitions.

4. The electron-hole interaction is assumed not to distort the tight-binding basis set, and to take the value $\frac{e^2}{\epsilon r}$ between a pair of orbitals, where ϵ is the high-frequency dielectric constant of SiO_2 (2.38) and r is the separation of the orbital centers. If the valence and conduction orbital are not on the same atom, r is the separation of the orbital centers. If the valence and conduction orbital are on the same atom, r is taken to be a bond length, the assumption being that most of the conduction wavefunction lies on the neighboring silicon atoms.[6] As the numerical

solution of equation (1) requires a finite number of degrees of freedom, the electron-hole interaction is cut off beyond a separation of 8 Å. This distance corresponds to two effective Bohr radii, and is selected for three reasons: 1.) Numerical considerations make this a convenient place to stop. The number of degrees of freedom grows as the cube of the cutoff radius. 2.) This is the distance at which the crystal and amorphous topologies begin to differ. (The first six-fold ring forms at 6 Å in β-Cristobalite.) 3.) This is the point at which the n=1 exciton wavefunction and energy begin to stabilize. Most of the optical oscillator strength into parabolic excitons is into the n=1 state. The most serious effect of the cutoff is to prevent the binding of excitons with primary quantum number n=2 and higher. It also induces a small upward shift in the n=1 binding energy which may be taken into account using perturbation theory. The introduction of a cutoff is consistent with the notion that the features in ε_2 are local in origin.

5. The number of degrees of freedom is restricted by adopting a basis of Bloch states involving a valence orbital in the zeroth unit cell combined with the nth (n=1,12) spherically symmetric shell of conduction orbitals about that valence orbital. This approximation corresponds physically to disregarding excitons of high angular momentum and is justified on the ground that these are not generally visible in optical spectra of wide-gap materials.[3] Further reduction is achieved at q=0 by decomposing this basis into the irreducible representations of the holes.

RESULTS

Consider first the case of model electrons in the presence of an infinitely massive hole. It suffices to solve Eq. (1) in the presence of a point charge on an oxygen atom, using as G_0 the single-particle Green's function generated by the conduction bands alone. The imaginary part of the Green's function matrix element connecting the orbital at the center with itself is proportional to ε_2. Its value before and after inclusion of the electron-hole interaction is shown in Fig. 2a. In the presence of the interaction, two new features appear near the band edge: the ordinary parabolic exciton at 8.3 eV and an excitonic resonance, or hyperbolic exciton, near 10.5 eV. Both of these are well known[7] in the spectra of alkalai halides and solid noble gases. The hyperbolic exciton is usually portrayed as an electron-hole pair at the L-point bound in two dimensions and unbound in the third. The oscillator strengths of the hyperbolic and parabolic excitons are comparable, the hyperbolic being about 1,6 times as strong. The calculated binding energy of the parabolic exciton is 0.58 eV, compared to a theoretical hydrogenic value of

$$E_b = \frac{R}{\varepsilon_2}(\frac{m^*}{m}) = 0.72 \text{eV}. \tag{2}$$

This disparity disappears when the effect of the missing coulomb tail is taken into account by perturbation theory, a somewhat surprising result in light of the small exciton size. The agreement is probably fortuitous since approximation errors are expected to be the order of 10%. The parabolic exciton is roughly the size of the ideal hydrogenic exciton, which has a radius of 4 Å. 83% of the state lies within 8 Å of the hole, compared with an ideal value of 76%. About 6% of the electron is on the hole, 24% is on nearest-neighbor oxygen atoms, 18% is on second neighbors, and 13% is on third neighbors.

Now consider the case of β-Cristobalite. In figure 2b the contribution to ε_2 from transitions between the non-bonding (-5 eV to 0 eV in figure 1a) valence band and the conduction band with and without the electron-hole interaction is shown. The obvious similarity between this and Fig. 2a results from the tendency of the upper valence bands to behave as though they were composed of three heavy holes. The three

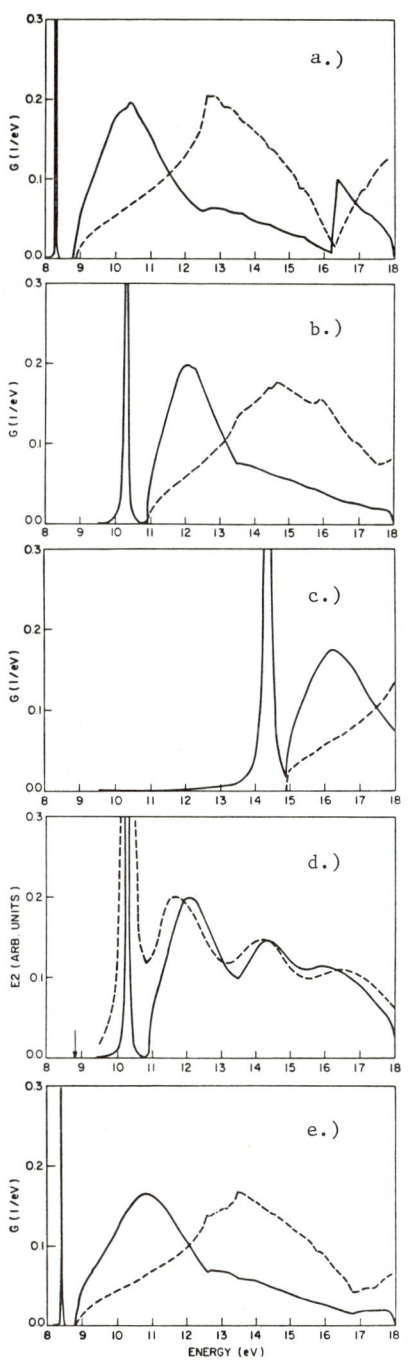

structure-insensitive peaks (0 eV, -2 eV and -4 eV) in the non-bonding density of states are centered on the three energy eigenvalues at Γ of β-Cristobalite, and thus give rise to two excitons apiece. However, the phase factors between O_p orbitals cause the excitons formed about the 0eV and -4eV holes to be dipole forbidden. One important difference between Figs. 2a and 2b is the width of the parabolic exciton in the latter, attributable to the slightly dissipative nature of the F_2 part of the Green's function between 8.9 eV and 11 eV. This width corresponds physically to a lifetime for decay of the exciton (an A_1 electron bound to an F_2 hole) into a free A_1-F_1 electron-hole pair. The width of the peak is smaller in the model than it is in experiment, probably because of the excess symmetry in β-Cristobalite.

In Fig. 2c, the contribution to ε_2 from transitions from the bonding (-6 eV to -11 eV in Fig. 2a) valence band into the conduction band with and without the electron-hole interaction is shown. The transitions giving rise to the unperturbed Green's function involve primarily the peak at -6 eV, which contains most[4] of the F_2 character. The spectrum in Fig. 2c has been phenomenologically broadened to simulate decays of excitons into free electron-hole pairs involving a hole in the non-bonding band. The absence of such decays is a pathology of the model stemming from the ability to decompose it exactly into bonding and non-bonding parts. Including a factor of 2 for spin degeneracy one has for the electronic contribution to ε

$$\varepsilon_{electrons} = \frac{-8\pi d^2}{V} \left[\left(\frac{8}{3}\right) G^{Nonbonding} + \left(\frac{4}{3}\right) G^{Bonding} \right] , \quad (3)$$

where V is the effective unit cell volume as determined from the density of α-quartz and d is the effective dipole matrix element between p and

Fig 2: a.) Model ε_2 for the case of infinitely heavy holes with (solid) and without (dashed) the electron-hole interaction. b.) Contribution to ε_2 of transitions from the non-bonding valence band into the conduction band with (solid) and without (dashed) the electron-hole interaction. c.) Contribution to ε_2 of transitions from the bonding valence into the conduction band with (solid) and without (dashed) the electron-hole interaction. d.) Superposition of the solid curves in 2b and 2c (solid) compared with the experimental ε_2 (dashed) of amorphous SiO_2. e.) Imaginary part of the Green's function matrix element between the on-site (n=1) F_1 state and itself with (solid) and without (dashed) the electron-hole interaction.

s orbitals on-site. Assuming an experimental width for the 10.3 eV peak of 0.5 eV and a height of 7, and knowing the integral

$$\int_{10.3eV peak} -Im(G^{Nonbonding})dE = 0.19 \quad , \tag{4}$$

one estimates a value for d of 1.2 e-Å, which is of the expected order. The function ϵ_2 as given by Eq. (3) with arbitrary normalization is compared with experiment[8] in Fig 2d. The bonding contribution has been further broadened to better agree with experiment. With the bandgap fit to 8.9 eV, it is clear that a realistic electron-hole interaction reproduces the four peaks of the experiment at approximately the correct energies. The theory yields the correct relative strengths for all four peaks, indicating that in this energy range the oscillator strength ratio given by Eq. (3) is satisfactory. As the binding mechanisms of the non-bonding and bonding excitons are similar, the separation of these energies must be that of the bonding and non-bonding F_2 peaks in the valence bands, seen in Fig 1a to be 3.7 eV. Not readily visible in Fig 2d, but indicated by an arrow, is a precipitous drop in the theoretical ϵ_2 corresponding to the band edge.

In Fig 2e, the on-site matrix element for the optically-forbidden F_1 part of the Green's function with and without the electron-hole interaction is shown. The F_1 analog of the visible excitonic resonance at 10.3 eV is thus a disallowed exciton at 8.4 eV. This is only the first in a Rydberg series, visible in this plot by virtue of its large amplitude for finding the electron and hole on the same atom. These excitons are probably the cause of the tail of absorption below 10.3 eV reported by Appleton[9]. They have the additional feature of being approximately transverse at $q \neq 0$, even though their disallowed nature arises from a cooperative effect between transitions on all four atoms in a tetrahedron. The likely explanation for the failure[10] to observe this exciton in ELS is thus that it contributes only to the transverse part of ϵ_2 at $q \neq 0$.

REFERENCES

1. See D.L. Griscom, J. Non-Cryst. Solids 24, 155 (1977).

2. S.T. Pantelides, The Physics of SiO$_2$ and its Interfaces, S.T. Pantelides, ed. (Pergamon Press, New York, 1978), p. 80.

3. J.C. Phillips, Phys Rev. 136, A1705 (1964); Phys Rev. 136, A1714 (1964).

4. R.B. Laughlin, J.D. Joannopoulos and D.J. Chadi, Phys. Rev. B20, 5228 (1979).

5. B. Fischer et. al, Phys. Rev. B15, 3193 (1977).

6. J.R. Chelikowsky and M Schlüter, Phys. Rev. B15, 4020 (1977).

7. J.C. Phillips "Fundamental Optical Spectra of Solids", vol. 18 of Solid State Physics, (Academic Press, New York, 1966).

8. H.R. Philipp, J. Non-Cryst. Solids 8-10, 627 (1972).

9. A. Appleton, et. al, ref. 2, p. 94.

10. A.E. Meixner, et. al, ref. 2, p. 85.

BAND STRUCTURE AND DENSITY OF STATES OF β-SILICON NITRIDE*

Shang-Yuan Ren[+] and W. Y. Ching
Department of Physics, University
of Missouri-Kansas City,
Kansas City, Mo. 64110

ABSTRACT

The electronic energy band structure of β-Si_3N_4 has been calculated using the first principles LCAO method. The bottom of the Conduction Band (CB) is at Γ and the top of the valence band (VB) is located along ΓA line. The very flat top VB along ΓA accounts for a large hole effective mass. The indirect band gap obtained is very close to the experimental value of 5.2 eV. The density of states (DOS) and partial DOS are also obtained and are in good agreement with photoemission data. In the VB region from -20. to -14. eV the states are entirely composed of N 2s states while in the range from -10.5 eV up, the states are predominately N 2p in character. In the CB region, the DOS is dominated by Si 3s and 3p orbital components. These results are consistant with charge analysis which indicates that on average, 0.56 electron is transferred from Si to N per Si-N bond.

INTRODUCTION

Silicon nitride has been an important material in ceramic industry with many diversified applications.[1] In recent years, silicon nitride has also become a very important material in microelectronic technology.[2] Strange enough, no energy band structure of Si_3N_4 has ever been calculated and little is known about its electronic structures.[3] Silicon nitride exists in two crystalline phases called α and β and possibly in amorphous forms. The crystal structure of β-Si_3N_4 is a hexagonal lattice with fourteen atoms (two Si_3N_4 units) per unit cell.[4] The α phase has similar crystal structure with a unit cell twice as large.[5] The short range order in both α and β phases are quite similar. Each Si atom bonds to four N atoms in a tetrahedral configuration and each N atom bonds to three Si atoms. In β-Si_3N_4, the S-N bond distances distribute between 1.73 Å and 1.75 Å and Si-N-Si angles range from 114.5° to 122.5°. The fairly complicated crystal structure may account for the lack of detailed electronic structure calculations in this material as compared to SiO_2 crystals where many band structure calculations have been attempted in recent years.[6]

In this paper, we present the first band structure calculation of β-Si_3N_4 using first principles orthogonalized linear combination of atomic orbital method (OLCAO). The method had been adequately described in the literature[7] and will not be repeated other than emphasizing that this method is particularly suitable for detailed electron state calculations of disordered systems or crystals of high complexity. We use a minimal basis set of contracted atomic-like orbitals of 1s, 2s, $2p_x$, $2p_y$, $2p_z$, 3s, $3p_x$, $3p_y$ and $3p_z$ for silicon atoms and 1s,

*Supported by DOE Contract, DE AC02 79ER10462

[+]On leave from University of Science and Technology of China.

2s, 2p$_x$, 2p$_y$, 2p$_z$ for nitrogen. The crystal potential is a superposition of spherically symmetric atomic potentials constructed from the free atom Hartree Fock atomic wave functions[8] with a Slater-type of local density approximation for exchange. The exchange parameter is set to be 2/3. All the multicenter integrals occuring in the Hamiltonian matrix are evaluated exactly and the lattice sum is carried to full convergence. The energy eigenvalues and wave functions are then obtained by diagonalizing the secular equations at various points in the Brillouin Zone (BZ).

RESULTS

In Figure 1, we display the calculated band structures along the various symmetry lines. The zero of energy is set at the top of the valence band (VB) which is located at 0.35 of ΓA along the ΓA line. The difference between the top of VB and that at Γ is very small (0.07 eV). The bottom of conduction band (CB) is at Γ, so in this calculation, an indirect band gap of 5.4 eV is obtained for β-Si$_3$N$_4$. This value is in good agreement with experimental gap value of about 5.2 eV.[9] The upper VB has a width of about 10.5 eV and the lower VB has a width of about 4.3 eV. These two bands are separated by a gap of about 4.1 eV. The CB is about 11.2 eV wide.

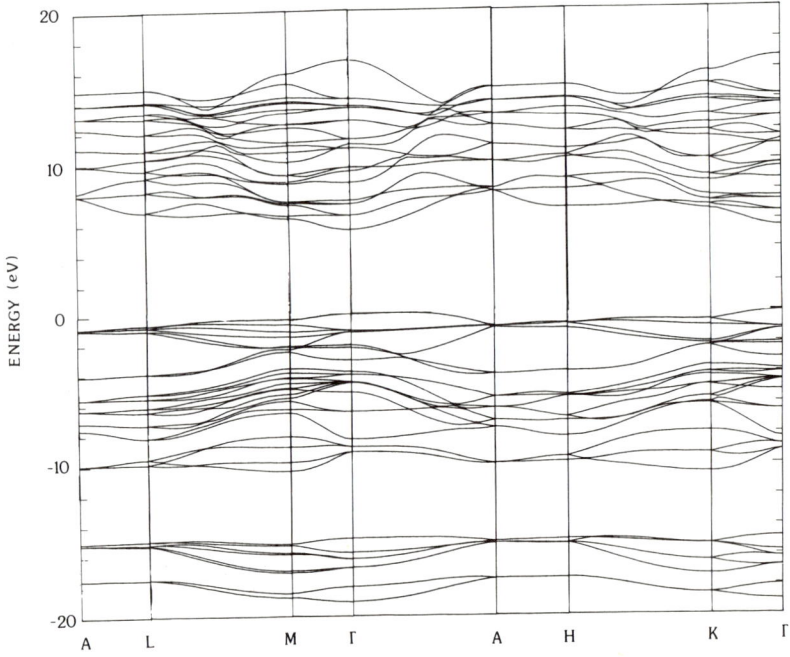

Fig. 1 Band structure of β-Si$_3$N$_4$ along symmetry lines.

The density of states (DOS) of β-Si$_3$N$_4$ is also calculated based on the eigenenergies obtained at eighteen $\vec{K}$-space points in the irreducible part of the BZ. Since the basis functions in our calculation are atomic orbitals, the DOS can be easily resolved into their partial components of Si 3s, 3p, and N 2s, 2p. These results are shown in Figure 2.

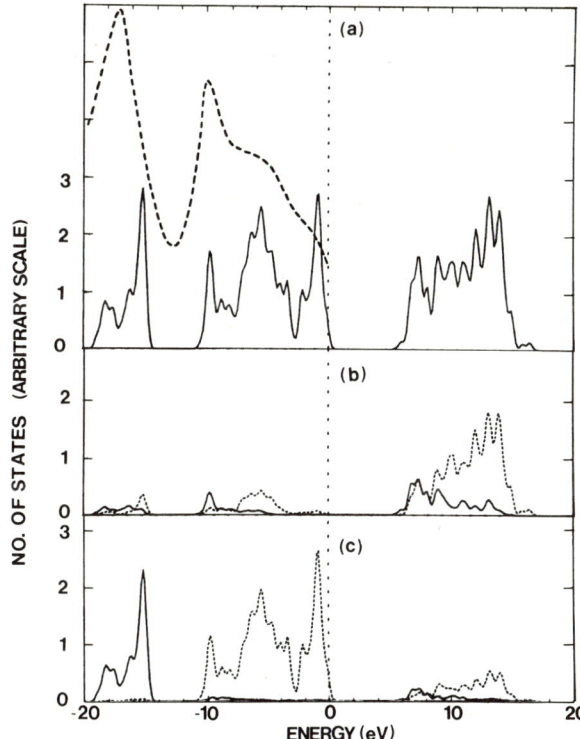

Fig. 2 (a) Total DOS of β-Si₃N₄, a Gaussian broadening of 0.25 eV is applied to each eigenenergy, dotted line: Experimental curve from Ref. 10.
(b) Partial DOS, full line: Si 3s; dotted line: Si 3p.
(c) Partial DOS, full line: N 3s; dotted line: N 3p.

The upper VB DOS has three prominent structures at -0.8, -5.5 and -9.7 eV. The lower VB DOS has peak at -15.2 eV. These are in good agreement with experimental XPS results[10] which is also shown in Figure 2 as dotted lines. The width of the VB and the positions of the four major structures in the DOS are well reproduced. On examining the partial DOS in Figure 2(b) and (c) we recognize that the lower VB is almost entirely of N 2s in character and the upper VB is dominated by N 2p components. In VB region, there is very little Si 3s and 3p mixing. In the CB region Si 3s and Si 3p are the major orbital constituents with the bottom of CB being Si 3s in character. Using the method of Mulliken's population analysis, we have calculated the average effective charge in Si and N atoms to be 1.77 and 6.67 electrons respectively reflecting a partially ionic character of Si-N bond. On the average, 0.56 electron is transferred from Si to N per Si-N bond. This number is less than the corresponding value in quartz crystals[11] indicating that Si_3N_4 has less ionic character than SiO_2.

From the band structures, we have estimated the effective mass of electrons (holes) to be about 0.8 (8.0) electron mass along the C axis and about 1.0 (2.0) electron mass on the plane perpendicular to the C axis. The relatively large hole effective mass is the consequence of very flat band at the top of VB. Since holes are found to be dominant charge carriers in Si_3N_4,[12] the relative heaviness of hole effective mass as compared to electron effective mass then suggest that, in Si_3N_4, the hole traps must be much shallower than the electron traps. This is totally consistant with recent transport measurement on thin films of Si_3N_4 prepared by chemical vapor deposition.[12]

DISCUSSION

We have presented the first full band structure calculation for β-Si_3N_4 using a first principles method. In spite of the non-self consistant nature of the calculation and the use of a minimal basis set, the calculated results are in good agreement with experimental information. Several improvements can be made such as including the Si d orbitals and use of overlap of atomic charge density model instead of overlap of atomic potential model. These efforts as well as the computation of electronic structures of α-Si_3N_4 are in progress and will be reported elsewhere. The major differences in electronic structures of SiO_2 and β-Si_3N_4 are: (1) In SiO_2, the upper VB splits into two portions with a well defined gap as the result of O lone pair orbitals; while in Si_3N_4, the upper VB is a continuous band of about 10.5 eV in width with three major structures, (2) the O 2s band is SiO_2 is about 2 to 3 eV lower than the N 2s band in β-Si_3N_4 relative to their respective top of the VB. Since explicit Bloch wave functions are obtained from the present first principles OLCAO calculation, it is possible to use these wave functions to calculate other microscopic and macroscopic properties of this material.

REFERENCES

1. R. N. Katz, Science, vol. 208, 841 (1980).
2. J. J. Chang, Proc. IEEE, 64, 1039 (1976);
   J. F. Verwey, Adv. Electron. Electron Phys., 41, 249 (1976);
   J. Wong, J. of Electronics Materials, 5, 113 (1976).
3. R. J. Sokel, (to be published) has recently discussed electronic structures of silicon nitride use a simplified LCAO method based on the energies at one point in the center of the BZ.
4. S. Wild, P. Grieveson and K. H. Jack in Special Ceramics, 5 edited by P. Popper (British Ceramic Research Association, Stoke-on-Trent (1972) p. 385.
5. I. Kohatsu and J. W. McCauley, Mat. Res. Bull., vol. 9, 917 (1974).
6. M. Schluter and J. R. Chelikowsky, Solid State Commun. 21, 381 (1977);
   J. R. Chelikowsky and M. Schluter, Phys. Rev. B 15, 4020 (1977).
   S. Ciraci and I. P. Batra, Phys. Rev. B 15, 4923 (1977).
   P. M. Schneider and W. B. Fowler, Phys. Rev. Letter, 36, 425 (1976).
   E. Calabrese and W. B. Fowler, Phys. Rev. B, 18, 2888 (1978).
   R. B. Laughlin, J. D. Joannopoulos and D. J. Chadi, Phys. Rev. B, 20, 5228 (1979).
7. W. Y. Ching and C. C. Lin, Phys. Rev. B, 12, 5536 (1975).
   W. Y. Ching and C. C. Lin, Phys. Rev. B, 16, 2989 (1977).
   W. Y. Ching, J. G. Harrison and C. C. Lin, Phys. Rev. B, 15, 5975 (1977).
8. E. Clementi and C. Roetti, Atomic Data and Nuclear Data Tables, vol. 14, Nos. 3-4, (1974).
9. A. M. Goodman, Appl. Phys. Lett. 13, 275 (1968).
10. Z. A. Weinberg and R. A. Pollak, Appl. Phys. Lett. 27, 254 (1975).
11. W. Y. Ching, unpublished.
12. J. M. Andrews, B. G. Jackson and W. J. Polito, Appl. Phys. Lett. 34, 785 (1979), J. Appl. Phys. 51, 495 (1980).

ELECTRON MICROSCOPY AND RAMAN SPECTROSCOPY OF
Nb_2O_5, Ta_2O_5 and Si_3N_4 THIN FILMS

Frank L. Galeener, Wolfgang Stutius, and Grady T. McKinley
Xerox Palo Alto Research Center
Palo Alto, CA 94304, U.S.A.

ABSTRACT

We report transmission electron micrographs and preliminary measurements of the Raman spectra of thin films of Nb_2O_5, Ta_2O_5 and Si_3N_4. Results indicate that the films are amorphous, on a scale above ~15 Å. Simple molecular models are used to discuss the spectra.

INTRODUCTION

Thin films of Nb_2O_5, Ta_2O_5 and Si_3N_4 are of interest in various electronic devices because these materials are high bandgap insulators of superior quality. They are also of interest in integrated optics applications because each has a very large index of refraction for visible light ($n_D \geq 2$), and because waveguiding in thin films requires that the film index exceed that of the materials on either side [1]. The materials are interesting scientifically because (as we shall show) they correspond to amorphous (a-) structures that have not been studied, and whose elucidation may help in developing general techniques for understanding structure and excitations in disordered solids.

THIN FILM DEPOSITION AND CHARACTERIZATION

The Nb_2O_5 (and Ta_2O_5) films were about 1 μm thick, and were prepared by reactive RF sputtering from Nb (and Ta) targets onto 7059 Corning glass microscope slides mounted on a water cooled anode. The Si_3N_4 films were about 3200 Å thick and were deposited by low pressure CVD on a substrate held at about 1100°C. In this case, the substrate was an 800 nm thick layer of steam SiO_2 grown at 1100°C on a (100) Si wafer. This buffer layer was sufficiently thick to completely isolate the guided wave in the Si_3N_4 from the highly absorptive Si substrate. Procedures for fabrication of the Si_3N_4 waveguides and for waveguide characterization are given in Ref. 2.

The morphology of the films was inspected by optical and electron microscopy. The electron micrographs, made by replication, are shown in Figs. 1-3. They were taken at critical underfocus in a Philips EM-301 transmission electron microscope (TEM) with goniometer stage and instrument resolution of 3.5-5.0 Å. The replication medium was Bioden RFA, later coated with ~250 Å of evaporated carbon and then evaporation shadowed at 60° from normal with a mixture of carbon and platinum. The Bioden was dissolved away with acetone leaving the Pt-shadowed carbon replica, whose transmission micrograph is shown. The procedure is known to preserve structures down to at least 30 Å wide. Shadowing is from the left.

The films show no evidence of grain boundaries as might be expected in polycrystalline material. The Nb_2O_5 and, to a lesser extent, the Ta_2O_5 show evidence of "hillocks" with lateral dimensions of about 100 Å. It is conceivable that these are related to the "islands" separated by "voids" that have been predicted to form [3] and have been seen [4] in evaporated a-Ge, here made less observable by the reduced resolution of the replication procedure. These hillocks are not seen in Si_3N_4. The meaning of the structure on a 10 Å scale (black dots) is open to question, and may be due to phase contrast effects alone. Micrographs of replicas from the uncoated glass substrates show neither hillocks nor the 10 Å scale structure, so that the latter may be real and represent small clusters of atoms. On the other hand, similar structure is seen in direct TEM micrographs of a-Ge films [4].

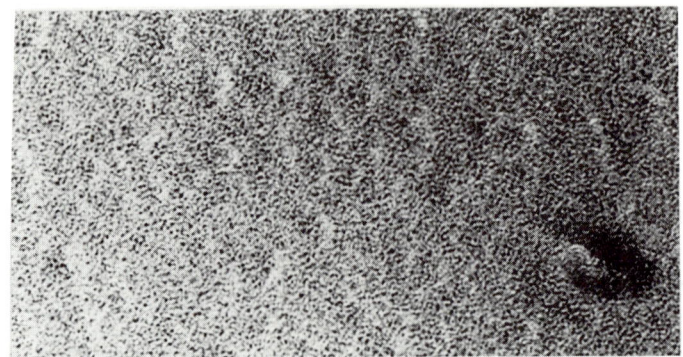

Fig. 1. TEM micrograph of replica of Nb_2O_5 film. Field width = 2500 A.

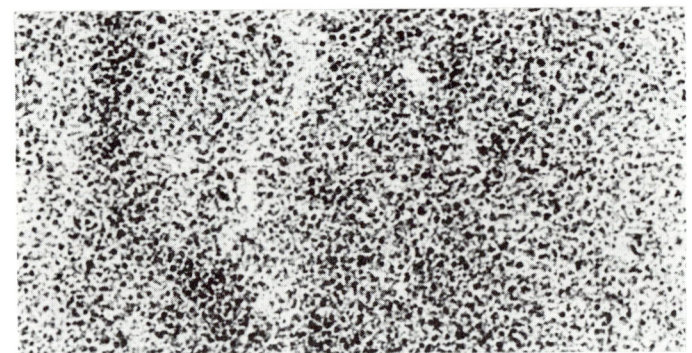

Fig. 2. TEM micrograph of replica of Ta_2O_5 film. Field width = 2500 A.

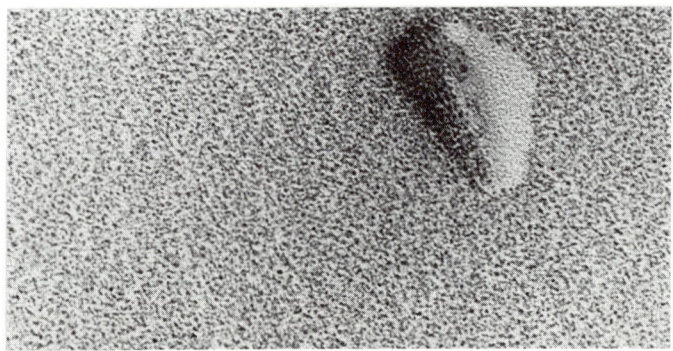

Fig. 3. TEM micrograph of replica of Si_3N_4 film. Field width = 2500 A.

RAMAN SPECTROSCOPY

Raman spectra of the films shown in the micrographs were obtained in the 90° scattering configuration using the physical arrangement illustrated in Fig. 4. Rutile (TiO_2) was used as the prism material since its refractive indices exceed those of the films to be studied and because the resultant angles were convenient [1,2]. The laser beam was polarized perpendicular to the plane of the figure.

The Raman spectra of the Nb_2O_5 and Ta_2O_5 films obtained at two different laser wavelengths (λ_L) are shown in Fig. 5. These are "unpolarized" spectra since no analyzer was used in the collection optical system. The "Bose-Einstein" peak, labeled B, is an artifact arising from the competition between a low frequency density of states that increases with wavenumber and a thermal population factor that decreases. The peaks labeled P are spurious. They are due to strong forward Raman scattering in the TiO_2 prism; this is waveguided along with the laser light then scattered with unshifted frequency into the collection system by the ~1 db/cm Rayleigh (and/or Tyndall) scattering characteristic of our films. We are investigating several methods of eliminating this problem.

The intrinsic Raman spectra of Nb_2O_5 and Ta_2O_5 are thus nearly identical, showing broad features at ~220 cm^{-1}, ~670 cm^{-1}, and ~900 cm^{-1}. Since the atomic weights of Nb (93) and Ta (181) are quite different, this suggests that the Raman active modes primarily involve oxygen motion, and that the interatomic force constants are similar. In crystalline forms of these materials, each Nb (or Ta) is octahedrally coordinated to six O atoms, so we can discuss the vibrations <u>approximately</u> by considering the replacements $(NbO_6) \rightarrow MoF_6$ and $(TaO_6) \rightarrow WF_6$, where the fluorides are molecules (liquids) whose vibrational frequencies have been

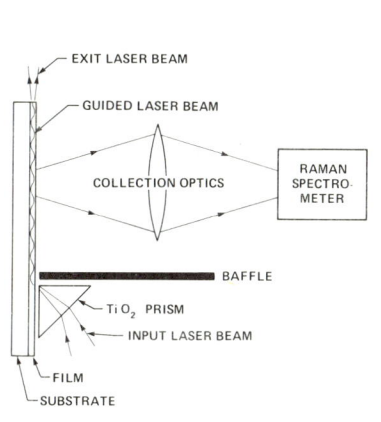

Fig. 4. Schematic diagram of the apparatus used to obtain the Raman spectra of a thin film material, as excited by a waveguided laser beam.

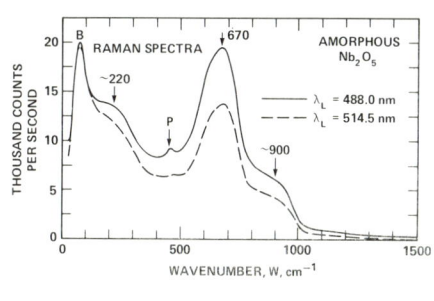

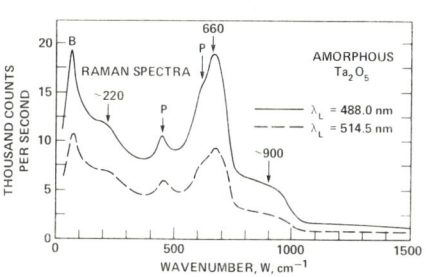

Fig. 5. Unpolarized Raman spectra of thin films of a-Nb_2O_5 and a-Ta_2O_5.

measured. This scheme of replacing cations in a glass by atoms of similar coordination and atomic weight [Nb → Mo (96), Ta → W (184)] and O (16) by F (19) has proven to be of heuristic value in discussing vibrations in several bulk glasses [5]. Since the experimental frequencies [6] of MoF_6 and TaF_6 are almost identical, the hypothesis regarding oxygen motion and similar force constants in the films is strongly supported. A central force network analysis like that employed for tetrahedral glasses [7] is being prepared and will be published elsewhere. The much greater breadth of the Raman features (≥ 200 cm^{-1}) than those of crystals (~2 cm^{-1}) identifies the materials as amorphous.

The spectra obtained from CVD thin film Si_3N_4 are the solid lines in the upper panel of Fig. 6. These are "polarized" spectra since an analyzer was used in the collection optics; HH indicates that incident and scattered electric vectors are parallel, while HV indicates they are perpendicular. The spectra slope upwards to the right because of a broad luminescence band which peaks at larger wavenumbers than shown. Efforts are underway to eliminate this unwanted luminescence from the experiment. In the interim, the background is estimated by linearly projecting the high frequency portion of the signal to zero at the laser line frequency (W=0), as shown by the dashed lines.

Upon subtraction of the dashed line data from the solid line data we obtain the polarized Raman spectra of Si_3N_4 shown in the lower panel of Fig. 6. Since the luminescence background is treated in an ad-hoc fashion, and better data is being sought, we will comment only on the most obvious features of the Raman response. First, it is even broader and more featureless than the spectra of a-Nb_2O_5 and a-Ta_2O_5, and thus Si_3N_4 is identified as amorphous. Secondly, the spectrum is only weakly polarized, with HH/HV ~ 1.3 throughout the more intense portion of the spectrum. Thirdly, the Raman response is about two orders of magnitude weaker than that seen in a-Nb_2O_5 and a-Ta_2O_5. (Counting rates can be directly compared, since laser powers, prism coupling efficiencies and refractive indices are quite similar).

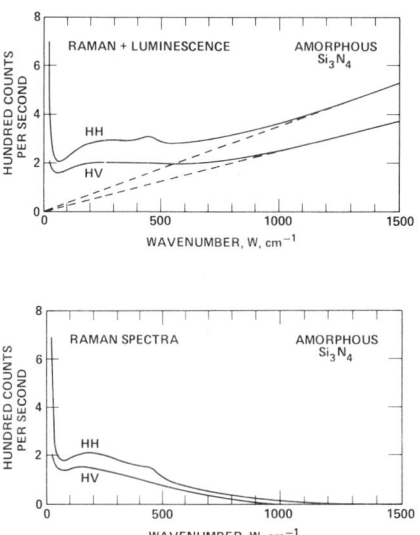

Fig. 6. The polarized Raman spectra of a thin film of CVD a-Si_3N_4 (lower panel), obtained by subtracting the luminescence (dashed line) from the experimental results shown in the upper panel.

This weak (and weakly polarized) Raman response is to be contrasted with the strong highly polarized response seen in fused silica and other bulk glasses, and associated by Galeener [7] with symmetric stretch (SS) motions. We infer that SS motions are not allowed or not possible in a-Si_3N_4. This hypothesis appears to be supported by a central force dynamical model, as will be elaborated elsewhere.

ACKNOWLEDGEMENTS

We are grateful to Dr. F. Zernike (Perkin-Elmer Corp.) who kindly supplied us with our first Nb_2O_5 thin film, and to Mr. W. J. Mosby for his assistance in the Raman spectroscopy. F. L. G. is also grateful to the Office of Naval Research for partial support of this work.

REFERENCES

1. P. K. Tien and R. Ulrich, J. Opt. Soc. Am. **60**, 1325 (1970).
2. W. Stutius and W. Streifer, Appl. Opt. **16**, 3218 (1977).
3. F. L. Galeener, Phys. Rev. Lett. **27**, 1716 (1971).
4. T. M. Donovan and K. Heinemann, Phys. Rev. Lett. **27**, 1794 (1971).
5. See, e.g., F. L. Galeener, in "Lattice Dynamics," ed. M. Balkanski (Flammarion, Paris, 1978), p. 345. Also see limitations noted in Ref. 7.
6. K. Nakamoto, "Infrared Spectra of Inorganic and Coordination Compounds (Wiley, New York, 1970), p. 122.
7. F. L. Galeener, Phys. Rev. **B19**, 4292 (1979).

PHONONS AND SUBMICROCRYSTALLITES IN AMORPHOUS SiO_2

Klaus Hübner, Annemarie Lehmann, and Lutz Schumann
Wilhelm-Pieck-Universität Rostock, DDR

ABSTRACT

The phonon density of states (DOS) of SiO_2 is calculated in dependence on elastic and Coulomb force constants and for different bond angles at the oxygen bridge between neighbouring SiO_4 tetrahedra, embedded in an effective medium. The different contributions to experimental phonon spectra from stretching, bending, and rocking motions of oxygen are identified by means of a corresponding local DOS. From the average DOS from submicrocrystallites with the structure of quartz, cristobalite, and tridymite within a random network of SiO_4 tetrahedra we determine the statistical weights of these different portions of the atomic structure of amorphous SiO_2. The explicit consideration of Coulomb forces due to dynamic effective charges leads to a longitudinal optical DOS and to corresponding LO-TO splittings.

CALCULATION OF THE PHONON DENSITY OF STATES

Equations of motion for the displacements of oxygen and silicon atoms in SiO_2 from their equilibrium positions are derived from the elastic valence-force potential for an Si-O-Si bridge (Fig. 1) with one central and one non-central force constant and a variable bond angle θ at the oxygen atom. The analytic elimination of the oxygen motion leads to a modified displacement correlation of neighbouring Si atoms and SiO_4 tetrahedra, respectively, and the resulting effective dynamic matrix involves frequency- and bond-angle-dependent force constants. Therefore, the cluster of five Si and four O atoms shown in Fig. 2 can be substituted by an effective tetrahedron of five Si atoms, which is embedded in a surrounding effective medium. The derivation of the Green functions for the effective Si vibration is presented elsewhere[2],

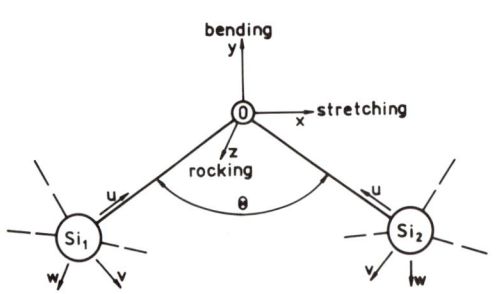

Fig. 1. Oxygen bridge between neighbouring SiO_4 tetrahedra of SiO_2 with an idealized classification of the oxygen motion

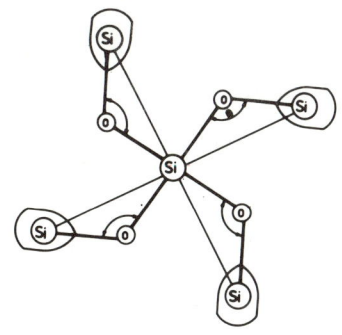

Fig. 2. Effective Si-Si$_4$ cluster within an effective medium

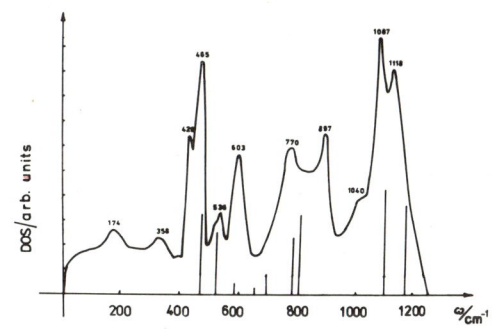

Fig. 3. Phonon density of states (DOS) of SiO_2 for $\theta = 144°$ (bond angle at oxygen) in comparison with positions and relative heights of the peaks of an IR spectrum of α-quartz

where it is shown that the isotropy condition for the effective medium leads to the diagonal elements of the Green function for the effective Si motion, from which the diagonal elements of the Green function for the O motion can be extracted with the help of the equations of motion mentioned above. The density of states (DOS)

$$\rho(\omega) = -\frac{2\omega}{\pi} \text{Im}\left\{ M_{Si}(\bar{G}_{uu}+\bar{G}_{vv}+\bar{G}_{ww})+2M_O(\bar{G}_{xx}+\bar{G}_{yy}+\bar{G}_{zz})\right\} \quad (1)$$

is calculated in dependence on the phonon frequency ω and for different angles θ at the oxygen bridge (M_{Si}, M_O atomic mass of Si and O, respectively). The DOS calculated for the special value of $\theta = 144°$ is shown in Fig. 3. We observe for this value of θ and corresponding force constants of SiO_2 in the quartz structure as well as for other crystalline forms of SiO_2 (cristobalite, tridymite) a surprisingly good agreement with experimental phonon spectra[6] of these SiO_2 polymorphs. For the comparison with experimental spectra the theoretical DOS is smoothed with the help of a bell-spline programme.

LOCAL DENSITY OF STATES

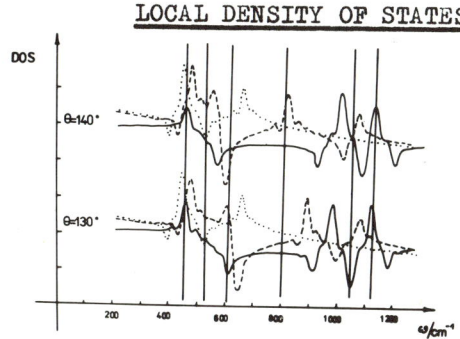

Fig. 4. Local DOS for the stretching(———),bending(- - -),and rocking (···)motion of oxygen in SiO_2

The local DOS is more useful for the comparison with experimental phonon spectra than the total DOS, since it reveals the different types of motion of the atomic constituents of a material. From the local DOS shown in Fig. 4 for two values of θ follows that the IR activity of SiO_2 at high frequencies comes primarily from the stretching motion of oxygen, whereas the bending and rocking displacements of oxygen determine the rest of the spectrum. Furthermore, for $\Theta = 140°$ the local DOS is in better agreement with experiment than for $\Theta = 130°$, as follows from the comparison with the peak positions of a corresponding Raman spectrum[7] (verticals in Fig. 4).

INTERMEDIATE-RANGE ORDER OF THE ATOMIC STRUCTURE OF AMORPHOUS SiO_2

The intermediate-range order of the atomic structure of amorphous SiO_2(a-SiO_2), which results from submicrocrystallites with the structure of SiO_2 polymorphs within a connective random network of regular SiO_4 tetrahedra[3], is taken into account by a corresponding distribution function of the bond angle at the oxygen bridge and

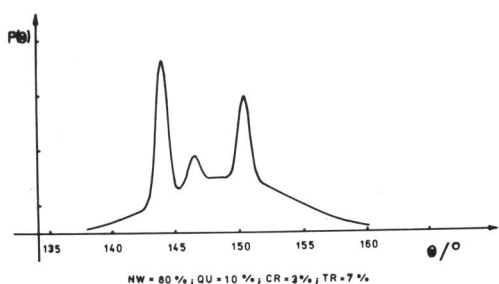

Fig. 5. Distribution function of the bond angle at oxygen for a special set of statistical weights of submicrocrystalline (quartz, cristabalite, tridymite) and random network parts in the structure of amorphous SiO_2

related force constants for the different structural parts of a-SiO_2. Figure 5, e.g., shows a broad distribution of the bond angle within the network (80% of the total structure) with peaks at the quasidiscrete angles of its submicrocrystalline portions. By means of a trial-and-error method for the comparison of the average DOS

$$\langle \rho(\omega) \rangle_\Theta = \int_{\Theta_i}^{\Theta_f} P(\Theta) \rho(\omega,\Theta) \, d\Theta \qquad (2)$$

with experimental spectra we determine the statistical weights of the different parts of the structure of a-SiO_2. The DOS for the empirically determined weights of the submicrocrystalline and network portions of an a-SiO_2 sample investigated in our laboratory is shown in Fig. 6 together with the DOS for a hypothetical distribution of the different substructures with equal weights. The latter reveals an unrealistic emphasis of some crystalline contributions to the spectrum.

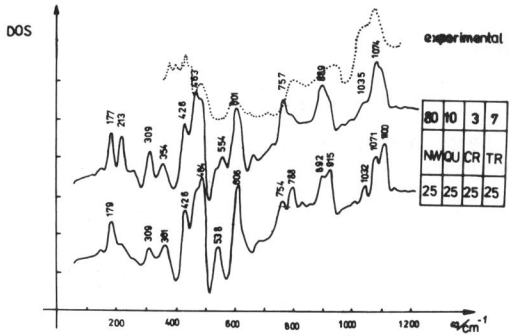

Fig. 6. DOS for two different structural compositions of amorphous SiO$_2$. The ratio of 80:10:3:7 of the different substructures was determined from the fit of the DOS to the experimental IR spectrum measured in our laboratory

COULOMB INTERACTIONS AND LO-TO SPLITTING

It can be shown that the phonon spectra resulting from the DOS calculated above have transversal optical (TO) character. The longitudinal optical (LO) DOS was determined by an explicit consideration of Coulomb forces due to dynamic effective atomic charges[4] and a corresponding redetermination of elastic bond-stretching and bond-bending force constants. An example for the resulting LO-TO splitting of the DOS of a-SiO$_2$ is shown in Fig. 7. It is in moderate agreement with corresponding experimental results[5], but can be used for seeking out the LO phonons dominating in electron-phonon interaction processes.

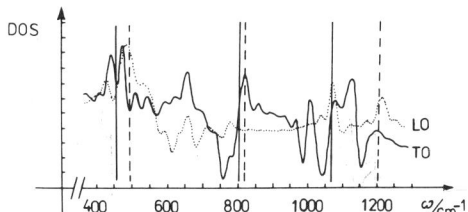

Fig. 7. Longitudinal and transverse optical DOS of amorphous SiO$_2$ for $\theta = 140°$. The theoretical LO-TO splitting should be compared with the marked positions of LO ($\vdots$) and TO ($|$) peaks identified in a reduced Raman spectrum by Galeener and Lucovsky[5]

REFERENCES

[1] P. N. Sen and M. F. Thorpe, Phys. Rev. B $\underline{15}$, 4030 (1977).

[2] L. Schumann, A. Lehmann, and K. Hübner, in: Physics of the Semiconductor Surface $\underline{11}$, Acad. Sci. GDR, Berlin 1980; in: SiO_2-Growth, Structure and Properties, I-III (ed. K. Hübner and G. Zuther), Wilhelm-Pieck-Universität Rostock 1979, II, p. 111.

[3] K. Hübner, in: The Physics of SiO_2 and its Interfaces (ed. S. T. Pantelides), Pergamon Press, New York 1978, p. 111; K. Hübner and A. Lehmann, phys. stat. sol. (a) $\underline{46}$, 451 (1978).

[4] S. T. Pantelides and W. A. Harrison, Phys. Rev. B $\underline{13}$, 2667 (1976).

[5] F. L. Galeener and G. Lucovsky, Phys. Rev. Letters $\underline{37}$, 1474 (1976).

[6] A. F. Vlassov and V. A. Florinski, Infrared Spectra of Inorganic Glasses and Crystals (in russian), Chemistry Press, Leningrad 1972.

[7] D. Heiman, R. W. Hellwarth, and D. S. Hamilton, J. non-crystalline Solids $\underline{34}$, 63 (1979).

CHEMICAL BONDING IN SiO

Guy Hollinger
Institut de Physique Nucléaire (and IN2P3), Université
Claude Bernard Lyon-I, 43, Bd du 11 Novembre 1918
69622 Villeurbanne Cedex France

ABSTRACT

High resolution SiKLL Auger spectra and Si2p X-ray photoelectron spectra are reported for amorphous SiO. Deconvoluted spectra show the occurrence of five main local atomic arrangements of Si atoms attributed to Si - ($Si_{4-x} O_x$) units whose distribution is determined experimentally. So for the first time the chemical structure of SiO is described directly from spectroscopic results and is found different from both the conflicting extreme models referred to as the Si - SiO_2 mixture model and the random bonding model.

INTRODUCTION

The chemical structure of the commercial SiO powder and of amorphous SiO_x films has received considerable attention in the past. Numerous experimental results obtained from X-ray diffraction, optical absorption, X-ray, Auger and photoelectron spectroscopies have been discussed on the basis of two extreme models, namely the Si - SiO_2 mixture model (1) and the random bonding model (2). This controversy can be related to two factors. First SiO is not a stable compound and is characterized by a temperature dependent chemical structure (3). Secondly informations on the local order cannot be directly extracted from the experimental results ; generally experimental data are compared to theoretical spectra calculated on the basis of a supposed structural model and a method of data analysis. For example previous XPS results (4) have shown that the Si2p core level spectra of SiO is characterized by an unsymmetrical maximum with a shoulder in the low binding energy side. This shape is incompatible with either the mixture or random bonding model predictions, but the experimental resolution in these experiments was not good enough to separate all the components and to propose another model. Consequently no experimental results have yet been able to clearly define the structure of SiO.

In this paper the results of high resolution SiKLL Auger spectra and high resolution Si2p X-ray photoelectron spectra of SiO are reported. From numerical resolution-enhancement procedures and an analysis of the core level chemical shifts the local order in SiO has been estimated as a function of the preparation temperature.

EXPERIMENTAL METHODS

Auger $SiKL_{23}L_{23}$ (1_D) spectra were measured with a Vacuum Generators Spectrometer (resolution : 1 eV) using $CrK\alpha$ radiation. In this apparatus the SiO films were prepared in situ by evaporation for different substrate temperatures (3). The photoelectron Si2p spectra were measured with a Hewlett Packard HP 5950 A spectrometer (resolution : 0.6 eV). In this case the films were evaporated on a substrate maintained at room temperature. From the intensity ratios of the O1s and Si2p lines the "SiO" films are expected to have the composition of $SiO_{1.1 \pm 0.1}$.

The observed XPS and Auger spectra can be understood as a convolution of an initial energy distribution by a broadening function with addition of a random noise. The initial energy distribution is directly related to the chemical structure of the solid. Deconvolution techniques were used to tentatively recover the initial energy distribution. Taking into account that deconvolution is often viewed with suspicion (occurrence of spurious unphysical features in the deconvoluted spectra) the following procedures have been adopted : starting with spectra of sufficient resolution, smoothing and background subtraction, comprehensive use of both the Van Cittert and Fourier transform methods (5, 6), tests by computer simulation. Respecting these conditions, it has been found that deconvolution restore the main features of the initial energy distribution.

RESULTS AND DISCUSSION

The Auger spectra for a SiO_2 - Si macroscopic mixture, SiO powder and three films prepared at 25°C, 200°C and 500°C are shown in Fig. 1. Modifications in the film spectra as the temperature increases, give evidence of the metastable character of SiO (3). The corresponding deconvoluted spectra are presented in Fig. 2. In Fig. 3 a high resolution Si2p experimental spectrum for a film prepared at 25°C is reported together with the results of the deconvolution process. In this latter case the deconvoluted spectra correspond to the average values of six measurements. From these results it is shown that the core level spectra consist of at least five peaks. However only four peaks are resolved in the lower resolution Auger spectra. In order to define more precisely the characteristics of the components, each experimental spectrum was resolved into five components using a non-linear least-squares curve-fitting procedure. The values of the intensities and the energies of the $SiKL_{23}L_{23}$ (1_D) peaks are reported in Fig. 4 and are compared with the predictions of both the mixture and random bonding models.

It is now well known that the photoelectron core level shifts must be interpreted in terms of the sum of three effects : the true chemical shift $\Delta \epsilon$, the extra-atomic relaxation energy shift ΔE_R, the shift of the Fermi level ΔE_F, as follows :

$$\Delta E_\ell = \Delta \epsilon + \Delta E_R + \Delta E_F$$

In a crude model we can express the Auger-shift in the same manner (6, 7) :

$$\Delta E_c = \Delta \epsilon + 3 \Delta E_R + \Delta E_F$$

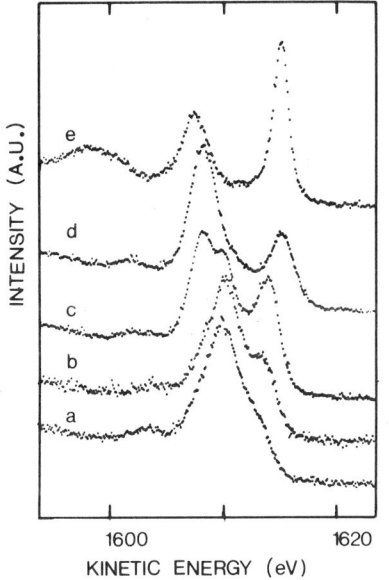

Fig. 1 SiKLL Auger spectra of SiO films prepared at 25°C (a), 200°C (b), 500°C (c), of commercial SiO powder (d) and of a thermally grown SiO_2 film on Si (e)

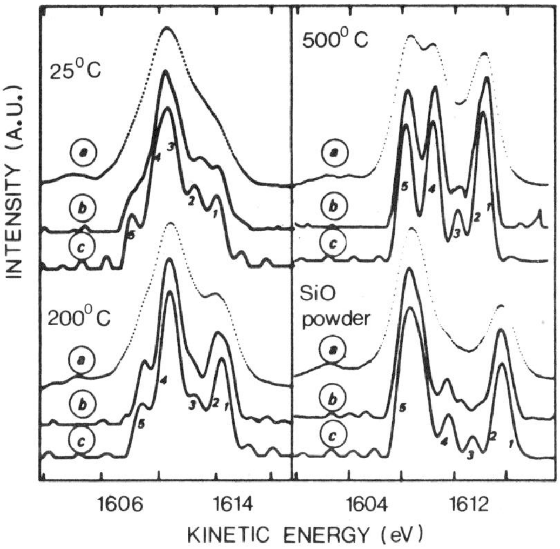

Fig. 2 Experimental (a) and deconvoluted Auger spectra of SiO
b : Van Cittert method
c : Fourier transform method

but the relaxation contribution is of the order of three times that for photoelectron energies. This effect leads to stronger shifts in Auger energy than in photoelectron energy and the different components are better resolved. So, a combined study of Auger and photoelectron spectra yields informations on the relaxation energy and allow to estimate the true chemical shift $\Delta \epsilon$, which is directly related to the change of chemical environment. In table 1 are listed the results for SiO prepared at room temperature, the differences between the K and L binding energy shifts were taken into account.

In the framework of a model in which the Si atoms are tetrahedrally bonded the five peaks are naturally associated to five kind of tetrahedron of the type Si - $(Si_{4-x}O_x)$ $0 \leq x \leq 4$. This assignment is confirmed first by the good values $(SiO_1 - SiO_{1.2})$ of the average composition which have been calculated from the component relative intensities. Secondly the values of the $\Delta \epsilon$ true chemical shift are found nearly similar to what we could expect in a model where the shift is correlated to the number of Si - O bonds, on the basis of the results of Si and SiO_2 (6).

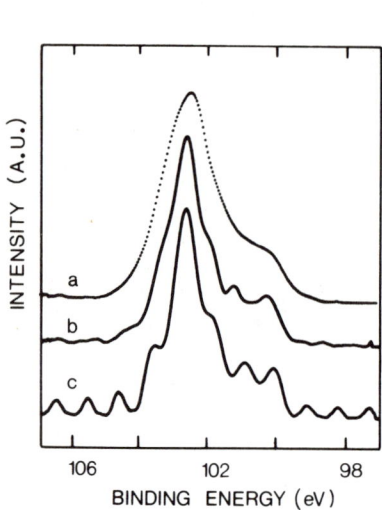

Fig. 3 Experimental (a) and deconvoluted (b and c) Si2p photoelectron spectra of SiO prepared at 25°C.

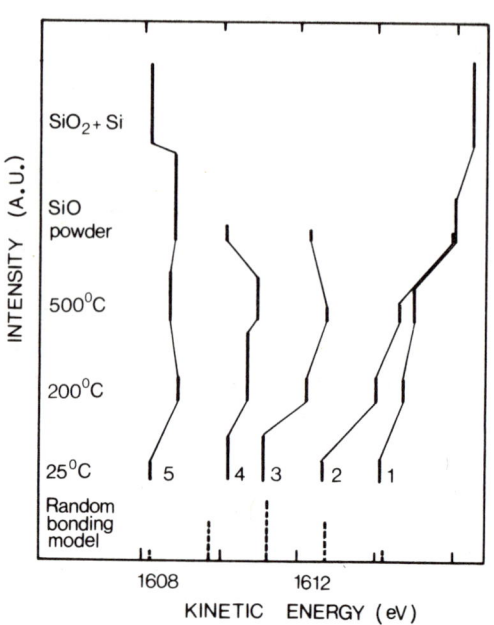

Fig. 4 The relative peak intensities and kinetic energies for the five tetrahedron types in SiO films and SiO powder are compared with the predictions of the macroscopic mixture model and of the Random Bonding model.

	$Si-O_4$	$Si-SiO_3$	$Si-SiO_2$	$Si-Si_3O$	$Si-Si_4$
Tetrahedron type	5	4	3	2	1
ΔE_c (eV)	-8.5	-6.2	-5.3	-3.8	-2.3
ΔE_ℓ (eV)	4.1	3.3	2.55	1.6	0.75
$\Delta \varepsilon$ (eV)	1.75	1.6	0.99	0.38	0.08

Table 1 Auger shifts, photoelectron shifts, and true chemical shifts for the five tetrahedron types in SiO(25°C) referenced to the bulk silicon values.

From the considerations described above and the results shown in Fig. 4 several points are noteworthy :

- SiO is a metastable chemical compound in which five coordinations ($Si - Si_{4-x} O_x$) of Si atoms are possible as also predicted by the Random Bonding model. The tetrahedra distribution is found different from Philipp predictions (2) and varies with the preparation temperature ; in particular $Si - O_4$ and $Si - Si_4$ units become more and more numerous as the temperature increases. This observation is in agreement with the thermodynamic properties of the Si-O system (8).

- In both SiO film and powder the values of the true chemical shift found for the $Si - Si_4$ and $Si - O_4$ units are similar to those calculated for Si and SiO_2 (6). Thus any modification of the Δ difference between the Auger components attributed to the $Si - Si_4$ and $Si - O_4$ units, in SiO compounds is indicative of a variation of the extra-atomic energy and may be used to give evidence of phase separation. Δ values of $\sim$ 6, 7.2 and 8.3 eV are found for the SiO films, SiO powder and the macroscopic mixture, respectively. These values support a microscopic mixture model for the SiO powder as discussed previously (3). In SiO films the same value $\sim$ 6 eV is found when the temperature increases. This fact gives evidence of local rearrangements of Si - Si and Si - O bonds in $Si - Si_{4-x}O_x$ units without nucleation or beginning of phase separation.

- FWHM values of 2.5 and 1.3 eV are found for the Auger and photoelectron lines respectively ; the larger Auger linewidths give evidence of the occurrence of a continuous distribution of sites around each main site in SiO films.

REFERENCES

(1) R.J. Temkin, J. Non Cryst. Solids 17, 215 (1975).

(2) H.R. Philipp, J. Phys. Chem. Solids 32, 1935 (1971).

(3) G. Hollinger, Y. Jugnet and Tran Minh Duc, Solid State Comm. 22, 277 (1977).

(4) G. Hollinger, J. Tousset and Tran Minh Duc, Tetrahedrally bonded Amorphous Semiconductors, M.H. Brodsky, S. Kirkpatrick and D. Weaire eds. American Institute of Physics, New York, (1974), p. 102.

S.I. Raider and R. Flitsch, J. Electrochem. Soc. 123, 1754 (1976).

J.R. McCreary, R.J. Thorn and L.C. Wagner, J. Non Crystall. Solids 23, 293 (1977).

(5) G. Wertheim, J. Electron Spectrosc. 6, 239 (1975).

(6) G. Hollinger, Thesis Lyon (1979) and to be published.

(7) C. Wagner, Faraday Discuss Chem. Soc. 60, 291 (1975).

(8) L. Brewer and R.K. Edwards, J. Phys. Chem. 58, 351 (1954).

STRUCTURAL AND BOND FLEXIBILITY OF VITREOUS SiO_2 FILMS

A. G. Revesz
COMSAT Laboratories, Clarksburg, Md. 20734

G. V. Gibbs
Virginia Polytechnic Institute and State University,
Blacksburg, Va. 24060

ABSTRACT

Thermal oxidation of Si is unique as several μm thick SiO_2 films can be grown with a vitreous rather than (poly)crystalline structure. The reason for this unusual behavior is that the energy of the Si-O-Si bond angle, ϕ, and the charge on the Si, obtained from ab initio MO calculations on pyrosilicic acid molecule, vary very little (by 3.4 kcal mole^{-1} and 0.03e, respectively) within the range of $120° \leq \phi \leq 180°$. Accordingly, the lack of long range order in v-SiO_2 arises from the wide distribution of ϕ rather than broken Si-O bonds.

INTRODUCTION

Many properties of silica polymorphs, silicates, and siloxane polymers are comparable and can be discussed in terms of the quantum chemistry of the Si-O bond because the bonding forces are localized.[1] Various methods of molecular orbital (MO) calculations have been used for molecules or clusters containing Si-O bonds, particularly Si-O-Si bridging bonds which are characteristic of SiO_2 polymorphs having four-fold coordinated Si atoms ($SiO_{4/2}$); these works have been recently summarized.[2,3] This bond approach can be contrasted with the band approach, which has been mostly used to calculate the electronic transitions.[4] The validity of the bond approach is justified, e.g., by the fact that the calculated electronic properties of the pyrosilicic acid, $(HO)_3SiOSi(OH)_3$ or $H_6Si_2O_7$ molecule[3] and SiO_4^{-4} cluster[5] are remarkably close to those of solid $SiO_{4/2}$ polymorphs.

Although the electronic and several other properties of SiO_2 are essentially determined by the $SiO_{4/2}$ tetrahedra, there are several properties which do depend on their arrangement (configuration), such as, density, compressibility, chemical reactivity, etc. The important fact, that thermal oxidation of silicon results in vitreous (i.e., noncrystalline but with high degree of short range order) rather than (poly)crystalline SiO_2 film, is also intimately tied to the configuration of $SiO_{4/2}$ tetrahedra; this film plays a crucial role in silicon devices. These aspects are discussed in this paper in terms of recent ab initio MO calculations performed on $H_6Si_2O_7$.[2]

Si-O BOND OVERLAP POPULATION AND ENERGY

The ab initio SCF-MO calculations were based on the s and p Slater type orbitals (STO) expanded as a linear combination of three Gaussian-type orbitals (GTO); this basis set is designated as STO-3G. The basis set supplemented with the five 3d orbitals of silicon is designated as STO-3G*. In these calculations the HO-Si-OH angles were fixed at 109.5°, the O-H bond lengths at 0.96Å, and the mean Si-O bond length at 1.62Å. The most important results of the calculations are as follows.

As the Si-O-Si bond angle, ϕ, decreases from 180° to 110°, the electron overlap population of the Si-O bridging bond, n(Si-O), calculated from the STO-3G basis set, decreases from 0.579 to 0.502. This change is linear with -sec ϕ, the hybridization index of the bridging oxygen AO's indicated by the superscript $sp^{-sec\,\phi}$, just as the change in the Si-O bond length, d(Si-O), in SiO_2 polymorphs: d(Si-O) = 1.526 - 0.068 sec ϕ Å.[6] The minimum total energy of the $H_6Si_2O_7$ molecule is at d(Si-O) = 1.60Å (corresponding to d(Si-OH) = 1.66Å) and $\phi \approx 140°$. These values are very close to the corresponding mean values of crystalline $SiO_{4/2}$ polymorphs: $\bar{d}$(Si-O) = 1.605Å and $\bar{\phi}$ = 147°,[6] as well as to those of v-SiO_2: d(Si-O) = 1.60Å and $\bar{\phi}$ = 144°[7] or 152°.[8]

The relationship between the energy of $H_6Si_2O_7$ and ϕ is shown by the dotted line in Fig. 1 which also includes the histograms of the ϕ values characteristic of v-SiO_2 and the 80 SiO_2 units in the primitive cell of α-tridymite. The steep energy barrier at lower angles is responsible for the fact that the lowest value of ϕ in crystalline SiO_2 polymorphs is 137.3°[9] and about 120° in v-SiO_2.[7] Using the STO-3G* basis set does not change the essential features of this behavior but lowers the minimum energy by 0.536 a.u. (335.8 kcal mole^{-1}). It is clear from this figure that the distribution of ϕ simply reflects the dependence of the energy on ϕ. This is a crucial point in understanding the properties of v-SiO_2 and details are discussed below.

These results are at variance with the findings that the total energy of SiO_2 has its minimum at ϕ = 100° and that the difference in energy values at the minimum and ϕ = ~168° (where it peaks) is about 58 kcal mole^{-1}; even after an empirical adjustment for Si-Si repulsion at the distance of 3.06Å (corresponding to ϕ = 146°), the difference between the maximum and minimum energy values is still about 20 kcal mole^{-1}.[11]

Inclusion of Si 3d orbitals increases the bond overlap population by 0.276 at ϕ = 140°. This increase is due to σ and π bond components, $n[Si(3d)-O(2s,2p)]_\sigma$ = 0.113 and $n[Si(3d)-O(2p)]_\pi$ = 0.163, respectively. The π bond is associated with 20 percent of the total overlap population and increases by 4.7 percent as ϕ increases from 140° to 180° while the total bond overlap population increases by 3.6 percent. Variations in the bond polarizability and other properties of SiO_2 polymorphs have been attributed to the increasing π character of Si-O bond going from coesite to v-SiO_2.[12]

THE IONICITY OF THE Si-O BOND

The effective negative charge on the bridging O atom calculated from the STO-3G basis set is 0.66e at ϕ = 140° and decreases to 0.423e when the STO-3G* basis set is used. Assuming that the charge on oxygen in SiO_2 is the same as on the bridging oxygen in $H_6Si_2O_7$, the average charges can be calculated as 0.56e and 0.88e with and without d-orbitals, respectively. According to the electroneutrality principle[13] the electronic configuration corresponding to the lowest value of the average charge is the favored one.

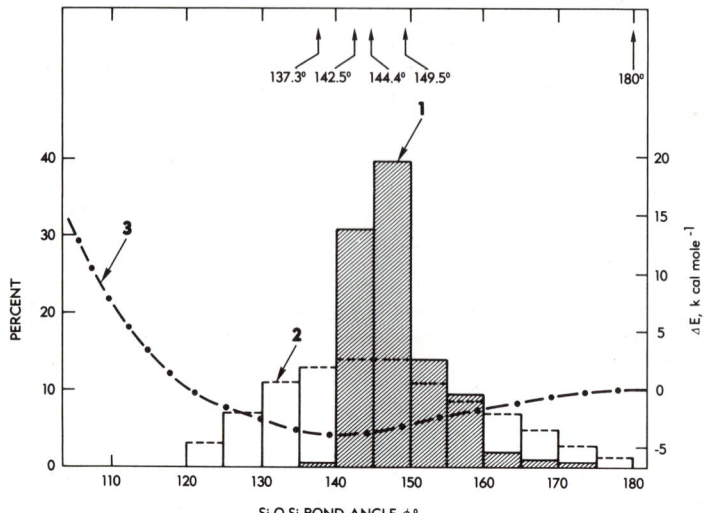

Fig. 1. The differential energy of $H_6Si_2O_7$ and distribution of the Si-O-Si bridging bond, ϕ. Curves 1 and 2 are histograms of ϕ for tridymite crystal and v-SiO_2 based on Refs. (10) and (7), respectively. The five values of ϕ in coesite crystal[9] are shown by the arrows. Curve 3 is the differential energy of $H_6Si_2O_7$; the absolute value at 180° and 120° is -1092.29474 a.u.

The effective negative charge on oxygen increases by 5 percent as ϕ increases from 140° to 180°. This increase is about the same as the increase in the overlap population of the $d\pi$-$p\pi$ bond. This increase in the effective charge was not considered in an earlier paper;[14] there the increase in π bonding with ϕ was incorrectly associated with a decrease in the ionicity of the Si-O bond.

A detailed analysis of the lattice dynamics of α-quartz also resulted in an effective negative charge on oxygen as 0.48e,[15] a value close to that obtained in a monopole and radial least square refinement of X-ray diffraction data for α-quartz (0.50e).[16] According to a classification based on the dielectric dispersion behavior, SiO_2 is a typical covalent solid as its value of the normalized macroscopic oscillator strength (5.1eV) is in the range of values characteristic of covalent solids: 4.3 - 5.1eV.[17] These considerations agree with the results that the ionicity of the Si-O bond is 0.25,[16] and emphasize the importance of the Si(3d) orbitals in lowering the effective charge.

In contrast, the ionicity of the Si-O bond as derived from dielectric considerations is 0.57, indicating that the bond is somewhat more ionic than covalent.[18] It should be noted that in derivation of this value the radii of the Si and O atoms were taken as half of the Si-O length in quartz (~1.61Å) rather than the covalent radii (1.17Å for Si and 0.68Å for O). It can be shown that the use of the covalent radii results in unrealistic values of the average band gap, 8.63eV (instead of 18.eV when $r_{Si} = r_O = 0.805$Å is used) and the ionicity, 0.0005. However, the same type of calculation which resulted in 0.57 as the ionicity of

$SiO_{4/2}$ leads to 0.67 as the ionicity of the Si-O bond in stishovite, $SiO_{6/3}$, in which the Si is six-fold coordinated and d(Si-O) = 1.78Å. This result is in accordance with classification of stishovite as ionic based on dielectric dispersion dispersion behavior.[17] These considerations demonstrate that the unique and decisive property of $SiO_{4/2}$ polymorphs is the unusually short Si-O bond length. Some of the values of the effective negative charge on oxygen as obtained from band structure calculations also appear to be too high, e.g., ~1.0e⁴ and their variation with ϕ unrealistically large, e.g., ~0.9e at ϕ = 120° and ~1.25e at ϕ = 180°.[11]

PROPERTIES OF VITREOUS SiO_2

The small change of the energy of the Si-O bond with ϕ is responsible for many unique properties of SiO_2 polymorphs, especially v-SiO_2. First of all, it is now clear why thermal oxidation of silicon results in a vitreous rather than crystalline SiO_2 film even when the oxidation temperature is as high as ~1300°C. Since the glass transition temperature of SiO_2 is 1000-1100°C,[19] one would expect that the SiO_2 film grown at high temperature would be crystalline; particularly since the growth rate in O_2 ambient is rather low. However, the thermal energy at 1000°C, ~2.5 kcal mole⁻¹, is comparable to 3.6 kcal mole⁻¹ that is the difference in energy of $H_6Si_2O_7$ at its minimum and at the ϕ values corresponding to the extremes of the distribution of the bridging bond angles in v-SiO_2, i.e., 120° and 180°. Accordingly, the proportions of Si-O-Si angles with ϕ = 160° and 120° relative to those with ϕ = 140°, estimated on the basis of Boltzmann statistics combining with sin ϕ as a Jacobian multiplier, are ~0.30 and ~0.34, respectively, at 1000°C and 0.34 and 0.44, respectively, at 1300°C. These values are reasonable since the ratios of the 160°-165° and 120°-125° ranges of ϕ to the range of 140°-150° in v-SiO_2 (see Fig. 1) are 0.50 and 0.25, respectively. Thus, it is understandable that the SiO_2 film is vitreous since the lack of long order results from the wide distribution of ϕ rather than broken Si-O bonds.

The vitreous structure is essential to obtain low density of interface states because no epitaxy requirement exists; the lack of grain boundaries coupled with the high degree of short range order are important regarding the insulator properties of the SiO_2 film in MOS structures.[20] The structural flexibility of v-SiO_2 results from the flexibility of the Si-O-Si bridging bond. This flexibility arises from subtle changes in the bond overlap population so that a wide range of ϕ values can be accommodated with very little difference in energy. In this respect, SiO_2 is unique as, for instance, thermal oxidation of metals in a comparable temperature range invariably results in (poly)crystalline oxide films.

The Raman absorption in v-SiO_2 at 606 cm⁻¹ (Ref. 21) may be closely related to the wide distribution of ϕ. It is suggested here that Si-O bonds associated with bond angles less than ~120° are responsible for this effect. These bonds are strained since their length is larger than ~1.66Å, but they are not broken. On the basis of Fig. 1, the proportion of Si-O-Si bonds with ϕ = 110° relative to those with ϕ = 140° (corresponding to the minimum energy in $H_6Si_2O_7$) is estimated as 0.015 at 1000°C and 0.037 at 1300°C. These values are comparable with the equilibrium fractional areas (relative to the 450 cm⁻¹ peak characteristic of the SiO_2 network) of the 606 cm⁻¹ peak in v-SiO_2: ~0.012 and ~0.024 corresponding to fictive temperatures of 1000° and 1300°C, respectively.[21] The suggestion that three-fold coordinated oxygen is responsible for this Raman absorption[22] appears to be unlikely since this configuration would correspond to that of oxygen in stishovite in which the Si-O bond length (1.78Å) is longer than the Si-O bond length in v-SiO_2 (1.60Å). A cursory ab initio calculation of the OSi_3H_9 configuration resulted in a bond length of 1.72Å; it is unlikely that a defect in v-SiO_2 (where d(Si-O) = 1.60Å) is associated with such a long Si-O bond.

ACKNOWLEDGMENT

This paper is based on work performed at COMSAT Laboratories under the sponsorship of Communications Satellite Corporation and under the support of NSF Grant EAR 77-23114. Thanks are due to S. H. Wemple for providing some unpublished data.

REFERENCES

1. W. Noll, Z. Angew. Chemie 75, 123 (1963).
2. M. D. Newton and G. V. Gibbs, Phys. Chem. Minerals, in press.
3. B. H. W. S. DeJong and G. E. Brown, Geochim. Cosmochim. Acta 44, 491 (1980).
4. For a recent review see D. L. Griscom, J. Non-Cryst. Solids 24, 155 (1977).
5. J. A. Tossel and G. V. Gibbs, Phys. Chem. Minerals 2, 21 (1977).
6. R. J. Hill and G. V. Gibbs, Acta Cryst. B35, 25 (1979).
7. R. L. Mozzi and B. E. Warren, J. Appl. Cryst. 2, 164 (1969).
8. J. R. G. DaSilva, D. G. Pinatti, C. E. Anderson, and M. L. Rudee, Phil. Mag. 31, 713 (1975).
9. G. V. Gibbs, C. T. Prewitt, and K. J. Baldwin, Z. Krist. 145, 108 (1977).
10. J. H. Konnert and D. E. Appelman, Acta Cryst. B34, 391 (1978).
11. S. T. Pantelides and W. A. Harrison, Phys. Rev. B B13, 2667 (1976).
12. A. G. Revesz, Phys. Rev. Letters 27, 1578 (1971).
13. L. Pauling, The Nature of the Chemical Bond, Cornell University Press, Ithaca, New York, 1960.
14. A. G. Revesz, Phys. Stat. Sol(a) 57, 235 (1980).
15. M. E. Striefler and G. R. Barsch, Phys. Rev. B 12, 4553 (1975).
16. R. F. Stewart, M. A. Whitehead, and G. Donnay, Am. Mineral. 65, 324 (1980).
17. S. H. Wemple, J. Chem. Phys. 67, 2151 (1977) and private comm.
18. B. F. Levine, J. Chem. Phys. 59, 1463 (1973).
19. R. Brueckner, J. Non-Cryst. Solids 5, 123 (1970).
20. A. G. Revesz, J. Non-Cryst. Solids II, 309 (1973).
21. J. C. Mikkelsen and F. L. Galeener, J. Non-Cryst. Solids 37, 71 (1980).
22. G. Lucovsky, Phil. Mag. B 39, 513 (1979).

CHAPTER III

BULK DEFECTS

ELECTRON-TRANSFER MODEL FOR E'-CENTER OPTICAL ABSORPTION IN SiO_2

David L. Griscom and W. Beall Fowler*
Naval Research Laboratory
Washington, DC 20375

ABSTRACT

A model is proposed to account for the 5.85 eV optical absorption of the E' center in SiO_2. It is assumed with Feigl, Fowler, and Yip that the E' center comprises an O^- vacancy with an asymmetric relaxation which localizes the unpaired electron in a dangling sp^3 orbital of one neighboring silicon (I) while the other silicon (II) has relaxed into the plane defined by its three oxygen neighbors. The suggestion that the 5.85 eV transition involves electron transfer from Si(I) to Si(II) is considered here in terms of the energy surfaces involved in the pseudo-Jahn-Teller effect. It is found that a plausible parameterization leads to a good fit to the experimental peak position and oscillator strength but not to the band width. Potential modifications to the present approach are discussed in light of the band width discrepancy and also constraints imposed by electron spin resonance data.

INTRODUCTION

The E' center[1] is perhaps the best known and most thoroughly characterized defect center in both the α-quartz and amorphous forms of SiO_2 (for reviews, see, e.g., Refs. 2 and 3). Evidence derived from electron spin resonance (ESR) has shown this defect to comprise an unpaired electron in a dangling sp^3 orbital of a single silicon[4-6] which is bonded to only three oxygens[7] in the glass or crystal framework. In the crystallographic context of α-quartz, Feigl, Fowler, and Yip[8,9] proposed and generally supported a model of the E' center as an "F^+ center", or simple oxygen vacancy, which is stabilized by an asymmetric relaxation at the two neighboring silicons such that the unpaired electron is trapped on one of them (I), while the second silicon (II) relaxes into the plane of the three oxygens to which it remains bonded. In effect, this model pictures the E' center as a hole trapped at a neutral oxygen vacancy. It is appropriate to note, however, that isolated positively-charged three-coordinated silicons are also thought to exist in the amorphous forms of SiO_2 and that these may serve as electron trapping sites with "E'-like" ESR signatures.

It has been firmly established[10] that the E' center (the variant with no neighboring proton) has an optical absorption centered at 5.85 eV with a full width at half height of 0.60 ± 0.06 eV and an oscillator strength of 0.14 ± 0.04. This careful work by Weeks and Sonder was performed on an oxygen-deficient fused silica (Corning 7943) and several more stoichiometric silicas which had been exposed to γ-ray or neutron fluences large enough to produce displacement damage (see, e.g., Ref. 11). Thus, there are strong grounds for the belief that the optical data cited here pertain to the oxygen-vacancy hole trap of Feigl, Fowler, and Yip[8,9] --

_____
*Permanent address: Lehigh University, Bethlehem, PA

as opposed to the isolated three-coordinated-silicon electron trap whose possible existence in amorphous SiO_2 has been mentioned above.

In view of the existence of both theoretical models and good quality optical data, surprisingly little attention has been devoted to the mechanism whereby the E' center absorbs light. Griscom[3] has recently proposed that the 5.85 eV transition is due to an electron transfer between Si(I) and Si(II) as schematically illustrated in Fig. 1. A similar suggestion was independently put forward by Schirmer.[12] It was recognized by Schirmer[12] and by Fowler[13] that this type of situation could be treated within the formalism of the pseudo-Jahn-Teller effect. The present paper analyzes the optical data in terms of this relatively simple theoretical model and discusses both the strengths and weaknesses of the approach. The implications of other available data inputs, notably luminescence and ESR, are also considered.

PSEUDO-JAHN-TELLER APPROACH

In a detailed molecular orbital (MO) theory investigation, Yip and Fowler[9] computed the ground-state energy of the E' center as a function of atomic position. To gain additional insights they fitted the results of their calculation by a simple pseudo-Jahn-Teller theory. Figure 2 illustrates the model, the symmetric A_1 and asymmetric B_2 normal modes, and A_1 ("s") and B_2 ("p") electronic wave functions. The adiabatic energy eigenvalues of the coupled system as a function of Q, the amplitude of the B_2 mode, are given by[14]

$$E = \tfrac{1}{2} \mu \omega^2 Q^2 \pm \tfrac{1}{2} [E_{sp}^2 + 4 G^2 Q^2]^{\tfrac{1}{2}}, \tag{1}$$

where E_{sp} is the electronic energy difference at Q = 0, ω is the angular frequency of the mode, μ is its effective mass, and G is an electron-phonon coupling constant. The Jahn-Teller energy E_{JT} is defined by

$$E_{JT} = G^2/2\mu\omega^2, \tag{2}$$

and if $4 E_{JT} > |E_{sp}|$ the ground state will have a minimum away from Q = 0. The MO calculations have shown that such an off-center minimum does indeed occur in the case of the E' center in α-quartz.[9]

The coefficients of the A_1 and B_2 wave functions, a_s and a_p, are given by

$$\frac{a_s}{a_p} = \frac{E_{sp} \pm [E_{sp}^2 + 4 G^2 Q^2]^{\tfrac{1}{2}}}{2 GQ} \tag{3}$$

The ground-state curve is approximately harmonic about its minimum, Q_{min}, and if the vibrational states are well localized one may treat optical absorption as a Franck-Condon process, as is seen in molecules and color centers.[15,16] In this case a nearly Gaussian absorption is predicted whose peak position ΔE, half-width W, and oscillator strength f are simply related to the parameters characterizing the curves.[17] One obtains

$$\Delta E = 4 E_{JT}, \tag{4}$$

$$W = 4 [2\ln 2 \, \hbar\omega_g E_{JT}]^{\tfrac{1}{2}} [\coth \hbar\omega_g/2kT]^{\tfrac{1}{2}}, \tag{5}$$

$$f = (m/6\hbar^2) E_{JT} \gamma^2 |\langle s|\vec{r}|p\rangle|^2, \tag{6}$$

where $\gamma = E_{sp}/E_{JT}$, m is the electron mass, $\hbar$ is Planck's constant, and ω_g is the frequency associated with the ground-state curve in the vicinity of Q_{min}.

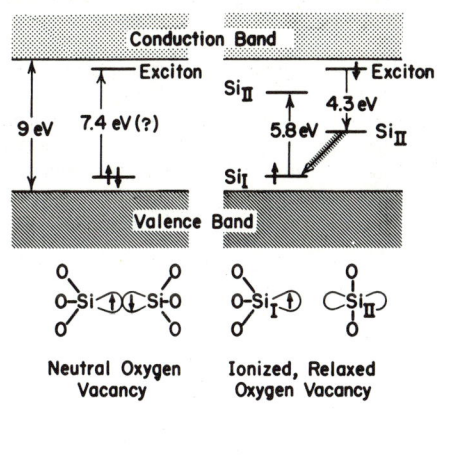

Fig. 1. Proposed model for optical absorption and luminescence at oxygen vacancy sites in SiO_2 (after Ref. 3). Si(II) is assumed to be in planar trigonal coordination. Cross-hatched arrow represents a radiationless transition.

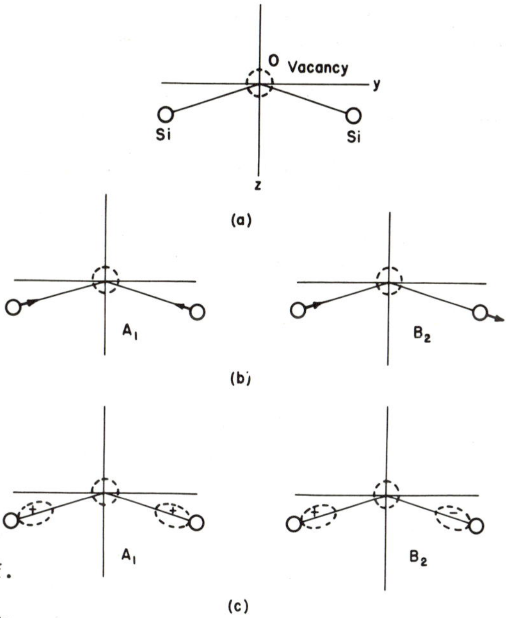

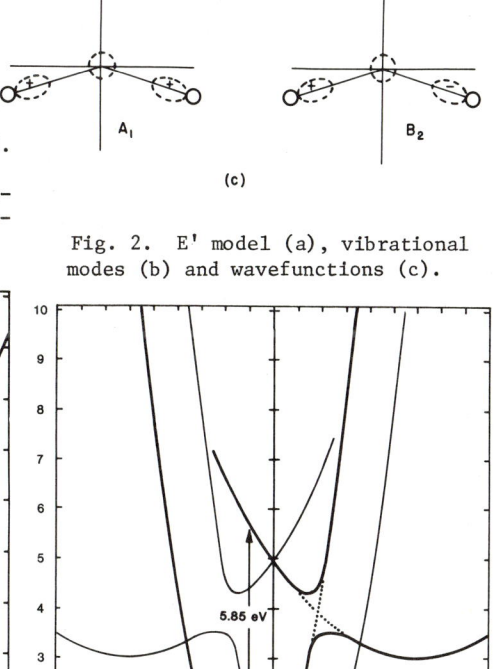

Fig. 2. E' model (a), vibrational modes (b) and wavefunctions (c).

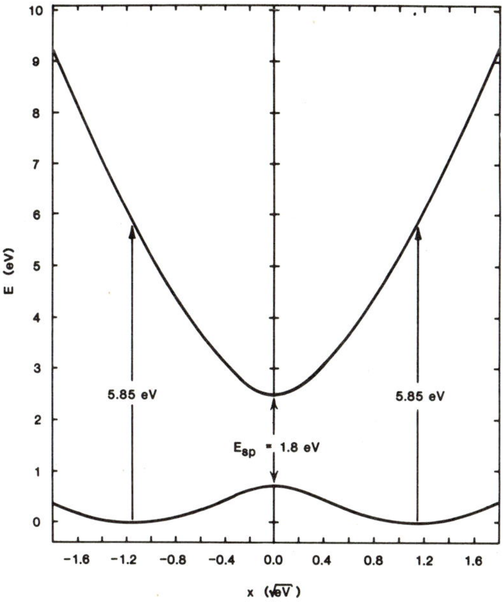

Fig. 3. Energy levels for E' center B_2 vibrations, calculated in a simple pseudo-Jahn-Teller model. Configuration is symmetric at X = 0.

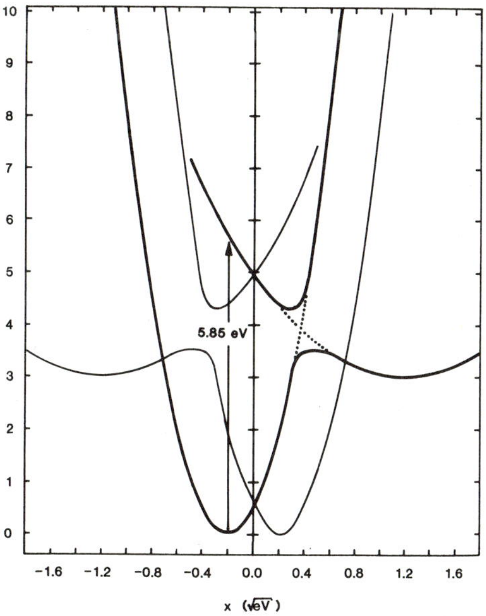

Fig. 4. Intuitive energy scheme for E' center, assuming no vibronic coupling between Si(I) and Si(II) and taking account of hybridization.

Using a simple electron transfer model for the optical transition,

$$f = (m/24\hbar^2)E_{JT}\gamma^2 d^2, \qquad (7)$$

where d is the distance between silicons.

In principle, eqs. (4), (5), and (7) can be solved simultaneously for E_{JT}, $\hbar\omega_g$, and γ, after substituting the experimental values of ΔE W, and f, together with a reasonable estimate for d. In fact, using Weeks and Sonder's values for ΔE and f, eqs. (4) and (7) yield E_{JT} = 1.46 eV and γ = 1.23 if d = 3.4 Å.[9] Figure 3 is a plot of eq. (1) for this particular choice of parameters (where $x \equiv Q\sqrt{\frac{1}{2}\mu\omega^2}$). However, eq. (5) is not satisfied for any value of $\hbar\omega_g$ when E_{JT} is constrained to be as large as 1.46 eV. A value $E_{JT} \sim 0.25$ eV is required to obtain W $\sim$ 0.6 eV. In addition, if it is accepted that W is essentially independent of temperature[18] between 300 and 78K, then $\hbar\omega_g \gtrsim 0.05$ eV.

DISCUSSION

The electron-transfer model for optical absorption by the E' center is supported by, or consistent with, a number of pieces of experimental evidence. First and foremost, the E' center is not photobleached[10,19] nor is there a photoluminescence induced by illuminating into the 5.85 eV band.[19] These findings strongly suggest that the state which terminates the 5.85 eV absorption is localized and not degenerate with either the conduction or valence bands. The only state which is obviously available and meeting these requirements is the unfilled 3p orbital on Si(II) in Fig. 1. A prediction of this model is that neither the surface E' center[20] nor the isolated three-coordinated-silicon electron center should absorb light at 5.85 eV.

It should be noted that the 4.3 eV luminescence which can be excited in oxygen-deficient or neutron-irradiated silicas by steady-state[21] or pulsed electron irradiation[19] is evidently due to electron-hole recombination at the E' site.[3] As indicated to the right of Fig. 1, it is believed that this emission results from the decay of a trapped exciton and hence is not described by the energy surfaces generated in the present study. It has been further suggested[3] that a 4.3 eV photoluminescence (7.4 eV excitation)[22] is also due to the formation of an exciton at a neutral oxygen vacancy.

The simple pseudo-Jahn-Teller theory employed here has been shown to be capable of fitting the experimental peak position and oscillator strength of the E' center optical absorption. The value of Q_{min} obtained in this fit was only $\sim$10% larger than that derived in a MO calculation,[9] though the ground-state well was a factor of $\sim$3 deeper in the present treatment. However, the present failure to reproduce the experimental band width hints that a more sophisticated approach may ultimately be necessary. Indeed, the vibrational amplitude associated with the zero-phonon level of the ground-state well in Fig. 3 is calculated to be $\sim$0.045 Å using $\frac{1}{2}\mu\omega^2$ = 16 eV/Å^2 (Ref. 9) and $\hbar\omega_g$ = 0.06 eV. By contrast, ESR data suggest that the vibrational amplitude of Si(I) should be substantially smaller than this estimate.[23] The simple theory assumes identical orbitals and vibrational motions for Si(I) and Si(II), whereas the former is hybridized sp^3 and undoubtedly undergoes more restricted motions than the latter whose dangling orbital is pure p. Figure 4 presents a more-or-less intuitive scheme which qualitatively incorporates these insights; the two silicons are envisioned to vibrate independently except when the electron is excited near the level crossing (dotted)--at which time the electron-phonon interaction may tend to contort the energy surfaces into their mirror image (lightly drawn curves) if the electronic relaxation is not sufficiently rapid.

REFERENCES

1. R. A. Weeks, J. Appl. Phys. 27, 1376 (1956); C. M. Nelson and R. A. Weeks, J. Am. Ceram. Soc., 43, 396 (1960); R. A. Weeks and C. M. Nelson, J. Am. Ceram. Soc. 43, 399 (1960); R. A. Weeks and C. M. Nelson, J. Appl. Phys. 31, 1555 (1960).
2. D. L. Griscom, The Physics of SiO_2 and Its Interfaces, S. T. Pantelides, ed. (Pergamon Press, Elmsford, NY, 1978) p. 232.
3. D. L. Griscom, Proc. Annual Frequency Control Symp., 33rd (Electronic Industries Assn., Washington, 1979) p. 98.
4. R. H. Silsbee, J. Appl. Phys. 32, 1459 (1961).
5. D. L. Griscom, E. J. Friebele and G. H. Sigel, Jr., Solid State Commun. 15, 479 (1974).
6. D. L. Griscom, Phys. Rev. B 20, 1823 (1979).
7. D. L. Griscom, (submitted to Phys. Rev. B).
8. F. J. Feigl, W. B. Fowler, and K. L. Yip, Solid State Commun. 14, 225 (1974).
9. K. L. Yip and W. B. Fowler, Phys. Rev. B 11, 2327 (1975).
10. R. A. Weeks and E. Sonder, Paramagnetic Resonance, Vol. 2, W. Low, ed. (Academic Press, NY, 1963) p. 869.
11. E. J. Friebele and D. L. Griscom, Treatise on Materials Science and Technology, Vol. 17, M. Tomozawa and R. Doremus, eds. (Academic Press, New York, 1979) p. 257.
12. O. F. Schirmer, (to be published in Journal de Physique).
13. W. B. Fowler, (unpublished work, 1979).
14. F. S. Ham, Phys. Rev. B 8, 2926 (1973).
15. D. L. Dexter, Solid State Physics, F. Seitz and D. Turnbull, eds. (Academic Press, New York, 1958), Vol. 6 p. 353.
16. C. C. Klick, D. A. Patterson, and R. S. Knox, Phys. Rev. 133, A1717 (1964).
17. This type of absorption process has been extensively treated by Schirmer for a number of hole centers in oxides. See, e.g., O. F. Schirmer, Z. Phyzik B24, 235 (1976).
18. E. W. J. Mitchell and E. S. G. Paige, Phil. Mag. 1, 1085 (1956).
19. G. H. Sigel, Jr., J. Non-Cryst. Solids 13, 372 (1973/74).
20. G. Hochstrasser and J. F. Antonini, Surf. Sci. 32, 644 (1972).
21. C. E. Jones and D. Embree, J. Appl. Phys. 47, 5365 (1976).
22. C. M. Gee and M. Kastner, (submitted for publication).
23. The observed ^{29}Si hyperfine coupling constant can be related to the O-Si-O bond angle at the defect site and hence to the Si position (see Ref. 5). In glassy silica there is a range of bond angles due to static disorder corresponding to a distribution of Si positions having a full width at half maximum of 0.032Å. E' center hyperfine linewidths are evidently an order of magnitude narrower in α-quartz (F. J. Feigl and J. H. Anderson, J. Phys. Chem. Solids 31, 575(1970)), implying dynamic variations in the position of the apex atom ≤ 0.003Å.

ASSIGNMENT OF THE OPTICAL ABSORPTION OF THE E_1' CENTER IN SiO_2

O. F. Schirmer
Fraunhofer-Institut für Angewandte Festkörperphysik, 7800 Freiburg, West Germany

ABSTRACT

The optical absorption of $SiO_2:E_1'$ is treated as a transition between the symmetric and antisymmetric states of the electron in the defect, modified by rather large coupling to an asymmetric vibration of the lattice and to a small asymmetric static potential.

INTRODUCTION

Fig. 1 shows the model of the E_1' center in SiO_2, as established by Feigl, Fowler and Yip (1,2). It consists of an oxygen vacancy occupied by one electron. This is localised in an sp^3 orbital pointing towards the vacancy, alternatively at one of the Si ions neighboring the vacancy, stabilised there by an asymmetric lattice relaxation. The electron thus has a twofold site degeneracy. Strictly speaking, both sites are not quite equivalent, since in the unperturbed SiO_2 lattice (no oxygen vacancy) the two SiO bonds differ in length by about 1 %.

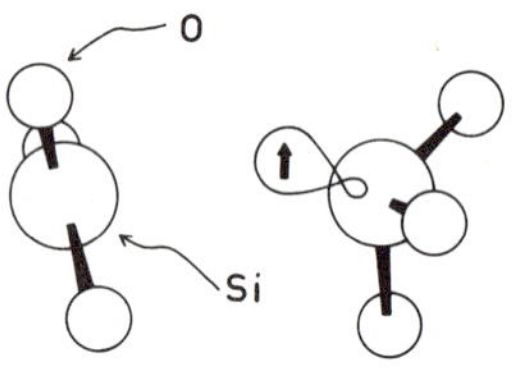

Fig. 1. Model of $SiO_2:E_1'$, adapted from Ref. 1

This center constitutes a simple member of a class of defects, especially numerous in oxide materials, which are characterised by site degeneracy of trapped electrons or defect electrons (3). A well known example among them is $SiO_2: [Al]^0$ (2). The symmetry of these centers generally is broken by electron phonon coupling and/or fluctuating fields due to structural irregularities of the crystal, localising the respective carriers at one of the degenerate sites. This mechanism can be understood as a Pseudo-Jahn-Teller effect (1,4) or as the formation of a small polaron bound to a defect (5). Such centers give rise to strong optical absorption, which can consistently be described as a light induced transfer of the carrier trapped at one site to an equivalent site (5,6). It has been proposed recently, independently by Griscom (7) and the present author (3), that the optical absorption of $SiO_2:E_1'$, which has remained unexplained for almost two decades, might be due to such a transfer. It is the aim of this paper to show that such a mechanism can indeed account for the optical absorption of $SiO_2:E_1'$ within the model of Feigl, Fowler and Yip, thereby further supporting this model, and to derive numerical values for the parameters describing the dynamics of this center. The problem is also appealing because it represents one of the most simple aspects of the Jahn-Teller effect. At the same time this example exhibits the essentials necessary for the treatment of the optical absorption of free small polarons (8).

METHOD

The simplicity of $SiO_2:E_1'$ consists in its (approximately) two-fold (C_{2v}) site symmetry. The degeneracy was higher in all other cases studied so far (3). Stationary electronic states of this system are symmetric with respect to reflection at the mirror plane, $|s\rangle$, and antisymmetric, $|a\rangle$. The corresponding eigenvalues are separated by twice the resonance integral J. Symmetry breaking is introduced by coupling these states to asymmetric vibration modes of the surroundings of the center. As is generally done in treating Jahn-Teller problems and as proposed in Ref. 1, we replace the multitude of possible phonon modes by a single representative one. In the basis of the eigenstates $|s\rangle$ and $|a\rangle$, the Hamiltonian will have the form:

$$H = \begin{pmatrix} J & VQ \\ VQ & -J \end{pmatrix} + H_{ph} ; \quad H_{ph} = \frac{1}{2m}(p^2 + m\omega^2 Q^2) \quad (1)$$

(V coupling parameter; Q displacement coordinate; H_{ph} phonon Hamiltonian, diagonal in $|s\rangle$ and $|a\rangle$, conventional notation). A slight energy inequivalence between both sites by 2 D as proposed in Ref. 1, can be considered in the C_{2v} approximation by adding D to both off-diagonal elements in H. In all systems studied so far (3), the experimental optical absorptions could be described by taking J as a perturbation. According to a LCLO-MO (linear combination of localised orbitals - molecular orbitals) calculation by Yip and Fowler (1), it is unlikely that this will be so in the present case. Therefore J and VQ will have to be treated on equal footing. We do so by expressing H in the basis $|s\rangle|n\rangle$, $|a\rangle|n\rangle$, where $|n\rangle$ are eigenstates of the n-th excitation of the asymmetric coupling vibration. The part $|s0\rangle$, $|a1\rangle$, $|s2\rangle$, ... of this basis is invariant under the mirror symmetry of the system, while $|a0\rangle$, $|s1\rangle$, $|a2\rangle$, ... changes sign. The Hamiltonian spanned by this basis can thus be reduced into two submatrices having the following appearance:

$$H^+ = \begin{pmatrix} J & \alpha & 0 & 0 \\ \alpha & -J+\hbar\omega & \sqrt{2}\alpha & 0 \\ 0 & \sqrt{2}\alpha & J+2\hbar\omega & \sqrt{3}\alpha \\ 0 & 0 & \sqrt{3}\alpha & -J+3\hbar\omega \end{pmatrix} \quad (2)$$

($\alpha = V(\hbar/2m\omega)^{1/2}) = (E_{JT} \hbar\omega)^{1/2}$. H^- is spanned by $|a0\rangle$, $|s1\rangle$, $|a2\rangle$, .. and differs from H^+ by the sign of J. These matrices were diagonalised numerically after truncation to sizes large enough to avoid truncation errors. Matrix sizes of 100 x 100 for each sub-Hamiltonian have been found to be sufficient.

Assuming for simplicity that the absorption takes place at zero temperature, transitions are allowed from the lowest eigenstate of H^+, E_g^+, to all eigenstates of H^-, for instance the l-th one, E_l^-. The intensity of this transition is given by (9)

$$I_l = (E_l^- - E_g^+) \cdot F_l \cdot C \qquad (3)$$

(C constant, independent of the transition energy (9)), where F_l is the value of the shape function for the transition

$$F_l = (\sum_i Z_{ig}^+ Z_{il}^-)^2 |\langle s|\vec{P}|a\rangle|^2 \qquad (4)$$

(Z_{ig}^+ i-th component of the eigenvector of the groundstate of H^+, Z_{il}^- corresponding component of the eigenvector to the l-th eigenstate of H^-, $\vec{P}$ electric dipole operator). The transitions implied by eq. 3 lead to a sequence of sharp lines (stick diagram). In order to account for the width of the distribution of coupling phonons, represented here by a single phonon mode, the separated transitions were convoluted with a Gaussian wide enough to cover all underlying structure. A FWHM of 4 $\hbar\omega$ was sufficient for that purpose.

RESULTS AND DISCUSSION

Our aim is to find parameters J, α and D, which lead to reasonable agreement with the experimentally observed absorption. It is reported to be peaked in the range between 5.85 eV (7,10) and 6.2 eV (11). The most narrow width reported is 0.62 eV (10). Fig. 2 shows an example. It also gives a fit to the experiment obtained with J = 2.7 eV, $\hbar\omega$ = 0.05 eV, E_{JT} (= $\alpha^2/\hbar\omega$) = 1.15 eV, D = 0.15 eV. Further parameter values yielding agreement are shown in Table 1 for vibration energies 1/3 to 1/2 of the Si-O vibration energy in an unperturbed SiO_2 lattice. It is seen: 1) These values are rather independent of the phonon energies assumed. 2) The value of J is somewhat smaller than that predicted in Ref. 1, 3.8 eV. 3) Also E_{JT} is smaller than the prediction (1), 2.6 eV. 4) The inclusion of D is essential to reach agreement

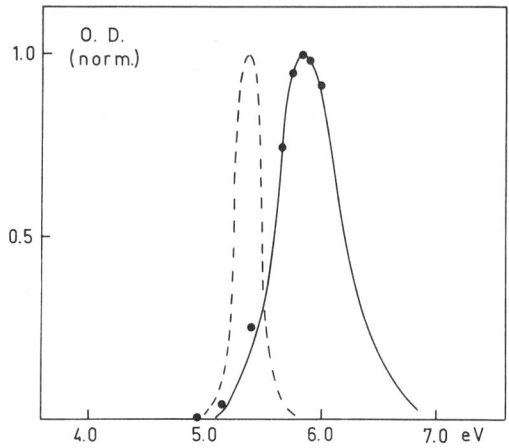

Fig. 2. Comparison of experimental E_1' absorption (10) (●) with the theoretical lineshape (full curve) derived from the model. See text for fitting parameters. Dashed line: same parameters, except D = 0.

with experiment. The deduced asymmetry energy, D, is of the order assumed in Ref. 1, 0.1 eV. Taking D = 0 leads to asymmetric lineshapes rising more steeply at their low energy sides and decreasing more gently at high energies than the experimental one, when the peak energy and band width had been adjusted.

TABLE 1 Fitting parameters (in eV)

$\hbar\omega$	J	E_{JT}	D
0.05	2.70	1.15	0.15
0.06	2.76	1.18	0.11
0.07	2.80	1.28	0.09
0.08	2.76	1.28	0.10

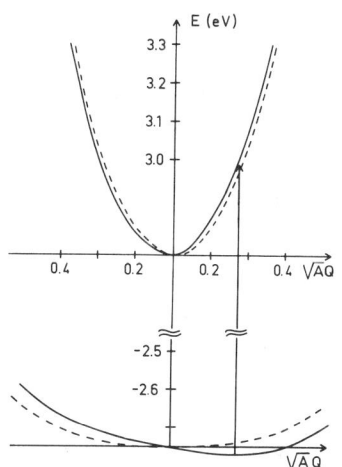

Fig. 2 also shows the sensitivity of line position and shape to the value of D. A change of D from 0.15 eV to zero reduces the peak energy by 0.5 eV and the width by 0.4 eV for the example given. Fig. 3 demonstrates that the fitting parameters are such that the system appears to be only at the verge of being localised at one side, if D is zero. This figure contains the eigenvalues of eq. 1 in the static limit (P = 0):

$$E = AQ^2 \pm 0.5((2J)^2 + 4(V^2Q^2 + D)^2)^{1/2}$$

(A = 0.5 mω^2). Only the inclusion of a small but finite D can induce the necessary localisation. This sensitivity of the system to asymmetric perturbations shows that slight structural changes in the neighborhood of E_1' can have strong influences on the optical absorption.

Fig. 3. Energies of symmetric (lower part) and asymmetric (upper part) electron states as modified by coupling to an asymmetric vibration (coordinate Q) (dashed). Static limit (P = 0). Full curve: additional coupling to an asymmetric static potential. (J = 2.75 eV, E_{JT} = 1.25 eV, D = 0.1 eV).

All such perturbation described by D will shift the absorption peak to higher energies, however. The lower peak energy of E_2', 5.5 eV (12,13), an E_1' center associated with a proton, can only be explained by a reduction of J. This may be attributed to decreased overlap between both sites caused by the distortion introduced by the perturbation.

The preceding arguments demonstrate that the optical absorption of SiO_2:E_1' can consistently be explained within the model of Feigl, Fowler and Yip (1). Since the model implies an electric dipole allowed transition, it is also in accord with the high oscillator strength, 0.14 (10), of the absorption.

I thank L. S. Cederbaum, W. Domcke and H. G. Reik for suggestions and discussions, and U. Kaufmann for a careful reading of the manuscript.

REFERENCES

(1) F. J. Feigl, W. B. Fowler, and K. L. Yip, Sol. State Comm. <u>14</u>, 225 (1974); K. L. Yip and W. B. Fowler, Phys. Rev. B <u>11</u>, 2327 (1975)

(2) for a recent review see: D. L. Griscom, in <u>The Physics of SiO_2 and its Interfaces</u>, edited by S. T. Pantelides (Pergamon Press, Elmsford, NY, 1978), p. 232

(3) O. F. Schirmer, in <u>Proceedings of the 3rd Europhysical Topical Conference on Lattice Defects in Ionic Crystals, 17. - 21. Sept. 1979, Canterbury, U.K.</u>, J. de Phys. (in print)

(4) see, e.g., G. D. Watkins, in <u>Radiation Defects in Semiconductors</u>, edited by J. E. Whitehouse (Institute of Physics, London, 1973), p. 228

(5) O. F. Schirmer, P. Koidl, and H. G. Reik, phys. stat. sol. (b) <u>62</u>, 385 (1974)

(6) O. F. Schirmer, Z. Phys. B <u>24</u>, 235 (1976)

(7) D. L. Griscom, in <u>Proc. Annual Freq. Control. Symp. 33rd</u> (Electronic Industries Assn., Washington, DC, 1979), p. 98

(8) see, e.g., P. Gerthsen, E. Kauer, and H. G. Reik, in <u>Festkörperprobleme</u>, Vol. 5, edited by F. Sauter (Vieweg, Braunschweig 1966), p. 4

(9) see, e.g., C. H. Henry and C. P. Slichter, in <u>Physics of Color Centers</u>, edited by W. B. Fowler (Academic Press, New York, 1968), p. 378

(10) R. A. Weeks and E. Sonder, in <u>Paramagnetic Resonance</u>, Vol. 2, edited by W. Low (Academic Press, New York, 1963), p. 869

(11) C. M. Nelson and R. A. Weeks, J. Am. Ceram. Soc. <u>43</u>, 369 (1960), R. A. Weeks and C. M. Nelson, ibid. <u>43</u>, 399 (1960)

(12) R. A. Weeks, J. Appl. Phys. <u>27</u>, 1379 (1956)

ELECTRONIC STRUCTURE OF VACANCIES AND INTERSTITIALS IN SiO_2

Frank Herman and Douglas J. Henderson
IBM Research Laboratory, San Jose, California 95193[1]

Robert V. Kasowski
Central Research and Development Department
E.I. du Pont de Nemours & Co., Wilmington, Delaware 19898

ABSTRACT

After discussing some of our recent studies of idealized Si/SiO_2 interfaces, we report new results for the electronic structure of Si and O vacancies and interstitials in SiO_2. The energy levels of these and other defects were obtained by carrying out extended muffin-tin-orbital band structure calculations for periodic arrays of non-interacting defects embedded in an idealized diamond-like SiO_2 crystal whose supercell normally contains 24 atoms.

INTRODUCTION

A primary obstacle to the development of a fundamental theory of the electronic structure of the Si/SiO_2 interface is lack of knowledge of the atomic arrangements in the neighborhood of the interface, particularly the distribution of structural and chemical defects[2]. In order to gain some insight into the nature of the Si/SiO_2 interface, some of us recently introduced an idealized model[3] designed to represent the average contact between silicon and its native oxide. Here we will discuss this model further, as well as a closely related model simulating the presence of Si and O vacancies and interstitials in SiO_2.

IDEALIZED Si/SiO_2 INTERFACES

It is expedient to begin by ignoring structural disorder at the actual interface. Let us then represent the Si/SiO_2 interface by the boundary between two crystalline domains, the first being the Si substrate, and the second an idealized crystalline form of SiO_2, diamond-like beta cristobalite. Except as otherwise noted, we will regard SiO_2 as having a diamond-like structure. Because the ratio of the linear Si-O-Si bond length in SiO_2 to the Si-Si bond length in crystalline Si is very nearly $\sqrt{2}$ (actually 1.39^4), we can obtain virtually perfect registry between SiO_2 and crystalline Si by placing the (100) face of one next to the $45°$-rotated (100) face of the other. This construction[3] leads to a remarkably simple interface model for which half the Si atoms at the interface are four-fold coordinated, and the remaining half are two-fold coordinated. Thus, we can describe the interface as a checkerboard with Si atoms common to Si and SiO_2 occupying the red squares, and the unsaturated Si atoms (two dangling bonds each) occupying the black squares. Because of the $45°$ rotation, the four bonds emanating from the common (red) Si atoms deviate from ideal tetrahedral geometry. This means that there is some residual lattice strain in our model. We will bear the existence of this strain in mind for the time being, and later take steps to relieve it.

We investigate the electronic structure of this idealized interface in terms of a superlattice composed of alternating Si and SiO_2 slabs which are sufficiently thick to isolate adjacent interfaces from one another. Our current studies are based on a 21 atom per unit cell superlattice consisting of 5 Si layers (2 atoms per layer) alternating with 3 tiers of SiO_4 tetrahedra (11 atoms). This superlattice, with 2 additional O atoms forming Si=O double bonds with the black Si atoms, is shown in Ref. 3 (cf. Fig. 1c). Because of the repeating slab geometry, we are actually dealing with a three-dimensional periodic structure, so we can exploit the well-developed methods of band theory to study localized electronic states at the interfaces[5].

In the present study we use the first-principles extended muffin-tin-orbital (MTO) method[6]. It is important to note that there are two types of MTO's: exp-MTO's having exponentially-damped tails, and osc-MTO's having oscillatory-damped tails. Both types are atomic-like in character, but the former are more highly localized than the latter. If we include both types of MTO's in our calculations, successive interfaces become coupled to one another through the long range of the osc-MTO's, and it becomes necessary to use considerably thicker slabs to eliminate this undesirable coupling. On the other hand, if we confine ourselves to the spatially compact exp-MTO's, successive interfaces in our 21 atom per unit cell superlattice do not interact with one another, and we can obtain numerical solutions that are sufficiently accurate for our purposes. In short, we use a minimal basis set composed of sp^3 exp-MTO's for each of the atoms in the supercell, where sp^3 denotes one s and three p functions.

The essential results for the unrelaxed superlattice can be summarized as follows:
(a) localized interface states occur in the thermal gap of Si if the dangling bonds are left unsaturated;
(b) these localized states are removed from the thermal gap if the dangling bonds are saturated by OH groups, hydrogen atoms, oxygen atoms, etc.

To relieve the lattice strain arising from the $45°$ rotation, we relaxed the superlattice using Monte Carlo techniques and a Keating-type microscopic atomic force constant model[7]. However, in accordance with our assumption of a diamond-like SiO_2 crystal, we used a $180°$ (straight) Si-O-Si bond as the norm. For this condition, there is little opportunity for the lattice to relax, so that the relaxed geometry is nearly identical to the unrelaxed geometry, implying negligible changes in our numerical results and no changes in our overall conclusions.

The simplest way to increase the realism of the above model is to retain the diamond-like topology of the SiO_2 region, but to relax the superlattice in accordance with a preferred Si-O-Si bond angle of $140°$ rather than $180°$. With this assumption we would expect the width of the SiO_2 slab to shrink slightly, and, more important, the interface to become slightly buckled as the local strain is relieved. Studies of such more fully relaxed Si/SiO_2 interfaces are currently in progress.

To recapitulate, our idealized Si/SiO_2 interface model provides a simple physical picture of this interface, as well as specific atomic arrangements in terms of which we can carry out detailed electronic structure calculations. The essential physical problem is the study of localized states associated with a buckled Si interface layer having half its atoms attached to the SiO_2, and the remaining half unattached. The unsaturated Si atoms have dangling bonds unless these bonds are saturated by hydrogen atoms, OH groups, etc. Because the Si atoms not attached to the SiO_2 are next-nearest rather than nearest neighbors, the interaction between adjacent dangling bonds is reduced relative to that for an ideal free (100) Si surface. It will be interesting to see whether a more realistic treatment of lattice relaxation at the interface will shift localized dangling bond states out of the thermal gap or not.

Finally, we note that the above 45° rotated (100) model is the simplest of a large class of idealized interface models connecting crystalline Si to crystalline SiO_2. Other members of this class involve low-index faces such as (110) and (111). In general, these other models lead to considerably larger supercells than the one we are using, making them less attractive for detailed numerical studies.

PERIODIC ARRAYS OF DEFECTS IN SiO_2 SUPERCELLS

A further extension of our idealized Si/SiO_2 interface model would involve the introduction of structural and chemical defects in the SiO_2 region. Within the framework of a band structure approach, we would naturally introduce a periodic array of defects at sites that are sufficiently far apart to insure negligible interaction among defects. How far apart should we place the defects? By way of reply, let us first consider the electronic structure of an infinite diamond-like SiO_2 crystal having 6 atoms per unit cell.

Using the extended muffin-tin-orbital method[6], neutral constituent atoms, Slater exchange, and a double basis set (sp^3 exp-MTO + sp^3 osc-MTO), we obtain a state-of-the-art SiO_2 band structure including a forbidden band width of 9.0 eV. We find essentially the same band structure whether we use both types of Si MTO's or either of these types alone. It is sufficient, therefore, to use just the compact Si exp-MTO's. Regarding the oxygen MTO's, we find that the compact O exp-MTO's are essential for the proper description of the valence bands. Omission of the O osc-MTO's does not affect the valence bands significantly, but does lead to a more or less rigid upward shift of the conduction bands by about 4.5 eV. If we use a minimal basis set consisting only of Si and O exp-MTO's, instead of the double basis set, the upper valence and lower conduction bands remain essentially unchanged, except that the separation between them is increased by 4.5 eV, leading to a forbidden band width of 13.5 instead of 9.0 eV.

Experimenting with various basis sets and SiO_2 supercells, we find that a satisfactory compromise between computational feasibility and physical realism involves the use of the minimal basis set (only exp-MTO's) and a diamond-like SiO_2 supercell containing 24 atoms (4 normal unit cells). This supercell is in fact a unit cube, as is the corresponding reduced zone. This 24-atom supercell has Si atoms at (0,0,0), (1/4,1/4,1/4), etc., and O atoms at (1/8,1/8,1/8), etc., where distances are measured in units of the cube edge. The perfect SiO_2 supercell contains 128 valence electrons, so there are normally 64 filled bands. If we work out the band structure for the supercell at the center of its reduced zone, we obtain the energy levels for the Γ and X points of the usual reduced zone, as we should. We can test the dispersion properties of defect-related bands by studying the band structure for the supercell at its zone center and also at its zone corner: negligible dispersion implies negligible interaction between adjacent defects. Introducing selected defects into the supercell, we indeed find negligible dispersion (order of 0.1 eV)[8] if the separation distance is the unit cube edge, which for our model is 7.68 Å. In other words, we introduce one defect of a given type into the supercell.

Let us now introduce a <u>neutral interstitial silicon atom</u> at lattice position (1/2,1/2,1/2). The 64 filled valence bands are perturbed only slightly. More important, we now have a filled interstitial band 6.0 eV above the <u>valence band maximum</u> (VBM) arising from the extra Si 3s level. There are also unfilled interstitial bands many eV above the <u>conduction band minimum</u> (CBM) arising from the extra Si 3p levels. The lowest (perturbed) conduction band is also filled, accounting for the remaining 2 of the 4 valence electrons contributed to the supercell by the Si interstitial. The same 6.0 eV interstitial band also occurs if we carry out the analogous calculation for the usual (6 atom) SiO_2 unit cell.

We can compensate for the shortcomings of our minimal basis set calculation by scaling the forbidden band width and all levels within its range arising primarily from conduction band states by a factor of 2/3. By construction, the forbidden band width of 13.5 eV reduces to 9.0 eV, the value given by our double basis set study (and also by experiment). Moreover, the Si 3s interstitial donor level is placed 2/3 x 6.0 eV = 4.0 eV above the VBM (or, equivalently, 5.0 eV below the CBM). The fact that we find a Si interstitial level roughly midway in the SiO_2 forbidden band suggests that this level may also lie within or close to the thermal gap of Si/SiO_2. Concerning the occupancy of the lowest conduction band, our model cannot distinguish between extended donor levels and delocalized conduction band states. The best we can say is that two electrons contributed by the interstitial Si atom occupy a mid-gap donor level, and the remaining two either an extended donor level just below the CBM, or low-lying conduction band states.

For a <u>neutral interstitial oxygen atom</u> located at the same position, (1/2,1/2,1/2), there is an occupied interstitial O 2s band 11.3 eV below the VBM, two occupied interstitial O 2p bands 2.1 and 2.3 eV above the VBM, and an empty interstitial O 2p band 3.0 eV above the VBM. The scaling factor of 2/3 should <u>not</u> be applied to these levels because they are adequately described by O exp-MTO's alone.

Returning to the normal supercell, let us now introduce a <u>neutral silicon vacancy</u>. The absence of a neutral silicon atom is equivalent to the removal of 4 electrons from the supercell. Hence the lowest 62 rather than the lowest 64 valence bands will be occupied. Because the uppermost valence bands in the normal supercell are built up from O 2p non-bonding orbitals, the removal of a Si atom has minor influence on the energies of these bands. Consequently, the two vacant levels (bands 63 and 64) remain essentially degenerate in energy with the topmost valence level (band 62). The absence of the Si atom and its orbitals reduce the width of the forbidden band from 13.5 to 13.0 eV, but this is an inconsequential effect.

Removing a neutral oxygen atom from the normal supercell to create a <u>neutral oxygen vacancy</u>, we have 128 - 6 = 122 electrons left in the supercell. These electrons fill the lowest 60 valence bands as well as an oxygen vacancy band located 1.9 eV below the CBM (at the zone center). The dispersion of this band is 0.2 eV. Because the vacancy orbitals are concentrated on the neighboring Si atoms, the scaling factor of 2/3 should be applied here to obtain a better estimate of the vacancy level position. The improved estimate places the oxygen vacancy level 2/3 x 1.8 eV = 1.2 eV below the CBM.

We have repeated the Si and O vacancy calculations relaxing the lattice in the same manner as already described for the Si/SiO_2 interface. Again, the assumption of straight (180°) Si-O-Si bonds leads to minor lattice relaxation, implying negligible changes in the numerical results listed above, In future studies we hope to deal with more realistic SiO_2 crystal structures and see whether the vacancy levels are strongly affected by more realistic lattice geometries and more complete lattice relaxations.

Replacing a neutral silicon atom in the normal supercell by a neutral aluminum atom merely removes one electron from the supercell, leading to a half-occupied valence band (band 64). For the <u>substitutional aluminum impurity</u>, our model cannot distinguish between a delocalized hole at the top of the valence bands and a shallow (extended) acceptor level. If, in addition to the Al impurity at lattice position (0,0,0) we place an <u>interstitial hydrogen atom</u> at position (-1/8,-1/8,-1/8), the topmost valence band (band 64) becomes fully occupied, and we find an empty hydrogen level 6.1 eV above the VBM. Further studies of complexes involving substitutional and interstitial impurities are in progress.

Having obtained estimates of Si and O vacancy and interstitial levels in an extended SiO_2 region, we now turn to the question of incorporating defects into idealized Si/SiO$_2$ interface models.

DEALING WITH DEFECTS CLOSE TO IDEALIZED INTERFACES

On the basis of the above experience with periodic arrays of defects in extended SiO$_2$ regions, we can readily imagine the construction of idealized Si/SiO$_2$ interface models which embody such arrays. The simplest of these would be the 21 atom per unit cell superlattice described earlier, but this would place adjacent defects rather close together, and would not be satisfactory for any but the most localized defects. A more acceptable model would be based on an extended supercell containing the same number of Si layers and SiO$_4$ tiers as before, but having twice as many atoms per layer or tier. This extended supercell would contain 42 atoms, and would place adjacent defects in SiO$_2$ just as far apart as in the infinite SiO$_2$ region discussed above (7.68 A, the unit cube edge).

Even more challenging than the study of defects at the interface is the study of interface models which include misfit dislocations.[8] While it is possible to carry out first-principles extended MTO calculations for as many as 50 or so atoms per unit cell in a reasonable time, using existing computer programs, it will undoubtedly become necessary to parameterize this method for more ambitious applications, such as the study of dislocation models involving hundreds of atoms.

REFERENCES

1. Supported in part by Office of Naval Research Contract N00014-79-C-0814.

2. For extensive references and many valuable papers on SiO$_2$, see: S. T. Pantelides, ed., <u>The Physics of SiO$_2$ and its Interfaces</u> (Pergamon Press, New York, 1978).

3. F. Herman, I. P. Batra, and R. V. Kasowski, in Ref. 2, p. 333.

4. L. E. Sutton, ed., <u>Interatomic Distances Supplement</u> (The Chemical Society, London, 1965). Using the data appearing on p. S 12S, we obtain for the bond length ratio of Si-O-Si to Si-Si: 2 x 1.633 A / 2.352 A = 1.39.

5. F. Herman, J. Vac. Sci. Technol. <u>16</u>, 1101 (1979).

6. R. V. Kasowski and E. Caruthers, Phys. Rev. <u>B21</u>, 3200 (1980).

7. P. N. Keating, Phys. Rev. <u>145</u>, 637 (1966); J. Bock and G. J. Su, J. Amer. Ceram. Soc. <u>53</u>, 69 (1970).

8. For our idealized diamond-like SiO$_2$ model we take as the unit cube edge the lattice constant of Si, 5.43 A, multiplied by $\sqrt{2}$, or 7.68 A. This is, of course, considerably larger than the experimental lattice constant for beta cristobalite, 7.16 A: cf. R. W. G. Wyckoff, <u>Crystal Structures</u> (Interscience, New York, 1963), Second Ed., Vol. 1. Because of the openness of the SiO$_2$ structure, and the dependence of the band structure on the local atomic arrangements, we would expect the Si and O vacancy and interstitial levels to come out nearly the same for idealized and actual beta cristobalite, particularly if we allow for more realistic lattice relaxation in the former. The most reasonable way to accommodate actual rather than idealized beta cristobalite in contact with a silicon substrate is to introduce misfit dislocations. But this would necessarily lead to quite large supercells.

SURFACE AND BULK VIBRATIONS IN ION-IMPLANTED AMORPHOUS SILICA*

G. W. Arnold
Sandia National Laboratories[†], Albuquerque, New Mexico 87185 USA

ABSTRACT

Infrared reflection spectroscopy (IRS) has been used to identify the Si-O vibrational mode and confirm previous assignments of Si-OH, and Si-OD vibrational modes in porous amorphous silica implanted with heavy ions and with H^+ and D^+ ions. The Si-O stretching mode (~ 1015 cm^{-1}) is produced by the damage cascade and is seen in all implanted bulk silicas as well as in porous silica. Implantation of porous silica with H^+ and D^+ ions produces bands at ~ 985 cm^{-1} and ~ 960 cm^{-1}, respectively. The position of all three bands is consistent with O, OH, and OD mass considerations. Implantation of D^+ ions into porous silica containing molecular water and OH^- groups results in D-H exchange. The Si-OH and Si-OD vibrations are also seen in bulk fused silica at low H/D fluences. These results suggest that intrinsic E'-type defects in bulk silica are dangling Si bonds at internal surface sites.

INTRODUCTION

Dangling-bond defects (e.g., Si-O) are formed in the near-surface region of ion-implanted silica and the vibrational modes of these defects can be readily identified using infrared reflection spectroscopy (IRS). Ion implantation further allows the incident ion to be chemically incorporated in the damage region formed by the stopping process. H and D implantations result in the formation of Si-OH and Si-OD modes in addition to the Si-O mode observed for heavy-ion implantation, and have been identified for the first time in this study. The identifications of the H and D modes agree with Raman studies of Hartwig, et al.[1] on gamma-irradiated fused silica loaded with H/D in overpressure soaking. We have also identified these modes in porous silica (CGW 7930). In this case, the H and D ions associate with dangling Si bonds at internal surface sites. The large internal surface area (200 m^2/g) greatly increases the number of these modes relative to bulk fused silica and makes investigation of such modes easy. Implantation of D ions into porous glass shows D-H exchange with surface species involving O, Si, and B. The results for Si-OH and Si-OD are in agreement with IR studies of Boccuzzi, et al.[2] on silica powders with and without H/D ambients, and with Raman studies of Murray, et al.[3] on porous glass exposed to H/D atmospheres. The removal of alkali from the near-surface region by inert-gas ion implantation and indiffusion of surface H further confirms the Si-OH assignment.

---

*This article sponsored by the U. S. Department of Energy under Contract DE-AC04-76-DP00789.

[†]A U. S. Department of Energy facility.

EXPERIMENTAL

Corning glasses 7940 and 7930 were used in this investigation. CFS 7940 is a high-purity silica glass (Type III) made by hydrolization of $SiCl_4$ in an oxyhydrogen flame. CGW 7930 is a porous silica glass ("thirsty" Vycor) in which a B_2O_3 phase has been leached out leaving a structure in which the pores are approximately 40 Å diameter. Implantations were made using an Accelerators, Inc. implanter (50-500 keV). IRS measurements were made using a Beckman IR 12 spectrophotometer with a N. J. Harrick reflectance attachment at an angle of incidence of 20°.

RESULTS AND DISCUSSION

Figure 1 shows the IRS spectra after implantation of 3.5×10^{16} 200 keV A^{++}/cm^2 into bulk fused silica in the spectral region of the Si-O-Si vibrations between 400 and 1300 cm^{-1}. Decrease in intensity and an energy shift of the Si-O-Si stretching vibration is caused by the disorder introduced by implantation. In particular, the decrease in intensity is proportional to the product of energy density into displacement processes and the ion range.[4] The reflectance peak induced by the A^{++} and other heavy ions occurs near 1015 cm^{-1}. It is worth noting that, (1) a peak at this energy can be seen in the reflectance data of Simon[5] for fused silica after a fast-neutron fluence of 2×10^{20} n/cm^2, and (2) that the vibrational mode ascribed to dangling Si-O bonds in the alkali oxide-silica glass system occurs at 935-950 cm^{-1}. Since the latter is clearly seen only at molar concentrations of alkali oxide exceeding about 20%,[6] we believe that this mode is typical of phase separated alkali-rich regions of the binary glasses and that the 1015 cm^{-1} band represents that of the dangling Si-O species in pure silica. Experiments on a $SiO_2:12\%K_2O$ glass, implanted with 7×10^{16} 250 keV Xe^+ ions/cm^2 in order to remove the K ions,[7] show that the 935 cm^{-1} band in the unimplanted glass is replaced by a 980 cm^{-1} Si-OH band as H from the surface replaces the K ions.

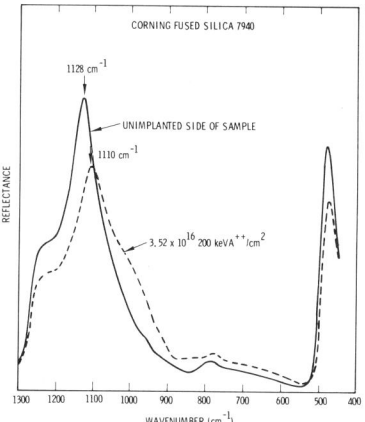

Fig. 1. Reflectance vs wavenumber for CFS 7940 after implantation of 3.5×10^{16} 200 keV A^{++}/cm^2 compared to the unimplanted side.

The results obtained by implantation of bulk fused silica with H- and D-ions are shown in Fig. 2. The reflectance maximum induced by H-implantation is at about 985 cm^{-1} which agrees with the expected value assuming that the 1015 value is that for the Si-O stretch. The D-ion also generates sufficient damage for this mode to be seen at 1015 cm^{-1}. Both the H and D implantations at these relatively low fluence levels produce dispersion in the reflectance curve. This dispersion is more clearly seen (different samples and fluences) in the difference spectra in Fig. 3. The positions of the two modes are clearly different even though the exact positions are difficult to determine in this case. These fluence levels cause little or no change in the intensity or position of the Si-O-Si band. At high fluence levels, the 1015 cm^{-1} damage band overwhelms the Si-OH and Si-OD modes.

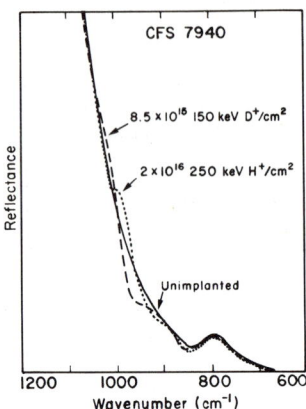

Fig. 2. Reflectance vs wavenumber for CFS 7940 after H- and D-implantation as shown compared to unimplanted sample.

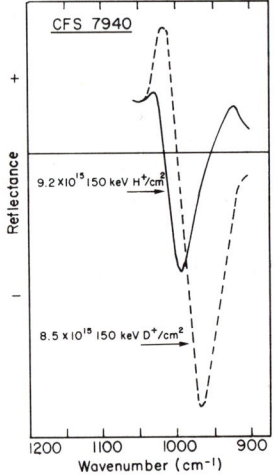

Fig. 3. ΔR vs wavenumber for CFS 7940 after H- and D-implantation as shown. ΔR is difference between implanted sample reflectance and that of the unimplanted side.

The use of porous silica glass (CGW 7930 Vycor) allows a much easier assignment to be made. In Fig. 4 are shown the results for Ne, H, and D implantations into 7930 glass. The samples were heated for 30 min. at 400°C prior to implantation. Induced bands at 1015, 985, and 960 cm^{-1} are clearly evident and are consistent with the assignment at Si-O, Si-OH, and Si-OD stretching modes, respectively. The effect of exchange of D for H is most evident at higher wavenumbers, as can be seen in Fig. 5 for the spectral region from 2000-3800 cm^{-1}. The easy exchange of D for H confirms the surface nature of the species. Before implantation the absorption at the H-bonded OH frequency of about 3650 cm^{-1} is the only apparent feature. After the D-ion implantation, however, the OD mode at about 2700 cm^{-1} is evident as well as the other features labeled on the figure. Also observed, but not shown, are modes due to H-O-D, B-O, B-OH, and B-OD in the spectral range of 1200-2000 cm^{-1}. A common feature of all implants into porous glass is the increased intensity of reflection after implantation, which is evident from the data in Figs. 4 and 5. This increase is due to the known action of ion-implantation in reducing the interstitial alkali concentration in the near-surface region which results in increased strength of the fundamental SiO$_4$ tetrahedral vibrations.

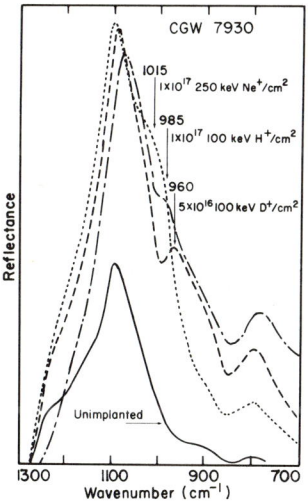

Fig. 4. Reflectance vs wavenumber for CGW 7930 porous glass after Ne-, H-, and D-implantations compared to an unimplanted reference sample.

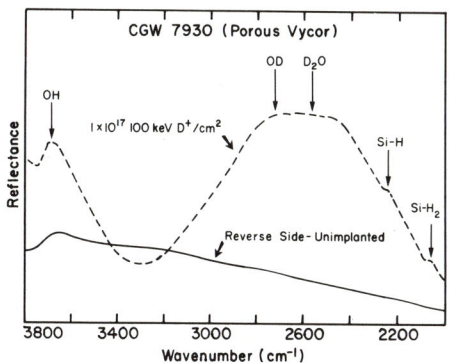

Fig. 5. Reflectance vs wavenumber for CGW 7930 porous glass after implantation with 1x10^{17} 100 keV D$^+$/cm^2 compared to the unimplanted reverse side.

CONCLUSIONS

The results reported here indicate that a basic feature of the damage in ion-implanted silica is the dangling-oxygen bond (Si-O), producing a vibrational mode at about 1015 cm^{-1}. The results also show that the surface vibrational modes Si-OH and Si-OD produced in porous glass by H- and D-ion implantation are seen, at nearly the same wavenumber, in bulk fused silica at low H- and D-fluences. This observation suggests that the H/D ions in fused silica are incorporated at sites which are similar to isolated surface sites. These internal surfaces represent sites for dangling Si valence bonds, i.e., intrinsic E´-centers, and the concentration of intrinsic defects is estimated to be from 10^{19}-10^{20} cm^3. The correspondence of irradiation produced E´-centers in bulk fused silica with those generated at the surfaces of ground silica powders has been noted by Arends, et al.,[8] and Hochstrasser and Antonini.[9] The present results and the earlier work of Boccuzzi, et al.[2] have also shown that the surface modes associated with Si-OH and Si-OD are readily seen by IR techniques in both porous silica glass and in silica powders. The bands were seen only in Raman scattering in the experiments of Murray, et al.[3] in which considerable care was taken to avoid atmospheric contaminants. Laughlin and Joannopoulos[10] used a Bethe lattice to model the surface sites in order to account for the non-appearance of this mode in the IR in these experiments.

The advantages of ion-implantation in glass studies have been exploited in the present experiments, i.e., (1) creation of a high-density damage region which allows the Si-O mode to be seen, (2) the incorporation of the incident ion (H/D) in the damaged region or at surface sites, and (3) the action of incident ions in sweeping out alkali network-modifier ions from the surface where the simultaneous indiffusion of surface H allows a different means of establishing Si-OH modes.

REFERENCES

1. C. M. Hartwig and L. A. Rahn, J. Chem. Phys. 67, 4260 (1977).

2. F. Boccuzzi, S. Coluccia, G. Ghiotti, C. Morterra, and A. Zecchina, J. Phys. Chem. 82, 1298 (1978).

3. C. A. Murray and T. J. Greytak, Phys. Rev. 20B, 3368 (1979).

4. G. W. Arnold, to be published.

5. J. Simon, in Modern Aspects of the Vitreous State, Vol. 1, edited by J. D. MacKenzie (Butterworth and Co., 1960), p. 120.

6. See, e.g., J. Simon and H. McMahon, J. Amer. Ceram. Soc. 36, 160 (1953).

7. G. W. Arnold and J. A. Borders, J. Appl. Phys. 48, 1488 (1977); J. A. Borders and G. W. Arnold, in Proceedings Int'l. Conf. on Ion-Beam Surface Layer Analysis, Karlsruhe, Germany, 1975, edited by O. Meyer, G. Linker, and S. Kappaler (Plenum Press, 1976), p. 415.

8. J. Arends, A. J. Dekker, and W. G. Perdok, Phys. Stat. Sol. 3, 2275 (1963).

9. G. Hochstrasser and J. F. Antonini, Surf. Sci. 32, 644 (1972).

10. R. B. Laughlin and J. D. Joannopoulos, Phys. Rev. 16B, 2942 (1977).

ENERGY DISTRIBUTION OF ELECTRON TRAPPING DEFECTS IN THICK-OXIDE MNOS
STRUCTURES

V. J. Kapoor and S. B. Bibyk
Department of Electrical Engineering and Applied Physics
Case Institute of Technology, Case Western Reserve University
Cleveland, Ohio 44106

ABSTRACT

The energy distribution of electrons trapped during internal photoemission in the Si_3N_4 layer of the thick-oxide $Al-Si_3N_4-SiO_2-Si$ structure has been investigated. Five well-defined electron trap levels were determined to be located at 2.50, 2.76, 3.03, 3.36 and 3.76 eV below the Si_3N_4 conduction band edge at room temperature. The data suggests an association of the trapped charge with oxygen impurity in the Si_3N_4 layer.

INTRODUCTION

It has been widely acknowledged that the MNOS (metal-Si_3N_4-SiO_2-Si) charge storage effects are associated with the charge capture in traps within the Si_3N_4 bulk and/or at the Si_3N_4-SiO_2 interface. The traps referred to are electron or hole energy levels, associated with impurities or defects in the Si_3N_4 layer of the MNOS structure. A detailed knowledge of the spatial and energy distribution of the memory traps is essential in understanding the charge storage, decay and fatigue characteristics of the MNOS devices.[1] This paper deals with the determination of the energy distribution of electrons trapped in the Si_3N_4 layer of a thick-oxide MNOS structure. The basic experimental technique used in this investigation was photoinjection and current-monitored trap photodepopulation. This is a greatly refined version of the early experiments of Williams,[2] as further elaborated by Thomas and Feigl.[3] We have studied the photocurrent response associated with optical release of electrons trapped in the Si_3N_4 insulator as a function of photon energy and report here the results of these measurements.

Our thick-oxide MNOS structures inhibited direct charge tunneling through the oxide since thick-oxide is essentially perfectly blocking at low applied fields. The thick-oxide also prevented photoinjection of holes from Si under negative gate bias[4] and thus eliminated simultaneous trapping of holes and electrons in the Si_3N_4 insulator during the photoinjection process.

EXPERIMENT

The MNOS structures were fabricated on p-type, <100> oriented silicon substrates with resistivities of 1-3 Ω-cm. The silicon wafers were thermally oxidized in a resistance heated furnace at 950°C in dry oxygen atmosphere. The Si_3N_4 film was deposited on the oxidized wafers by low pressure (0.5 torr) chemical vapor deposition at 800°C in a hot-wall LPCVD reactor from a gaseous SiH_2Cl_2/NH_3 mixture with various ratios. To allow light transmission into the MNOS structure, a semitransparent aluminum gate electrode (60% transmittance) of 100 A° thickness was deposited on the Si_3N_4 film. The thickness of SiO_2 was 0.1 μm for all the samples and the as-deposited Si_3N_4 thicknesses were 0.05 μm, 0.11 μm and 0.17 μm.

Figure 1 exhibits a schematic representation of the photoinjection-photodepopulation and C-V experimental arrangement. The optical system used was a high intensity 1000-W xenon arc lamp, mounted in a lamp housing and focused on the entrance slit of a Spex 1763, 1200-groove/mm grating monochromator, covering the wavelength range 900-200 nm. A water filter F_1 was placed between the lamp and the monochromator to remove the infrared part of the spectrum which can damage the monochromator and its grating. The entrance and exit slits of the monochromator were 2.5 mm wide and 20 mm tall, corresponding to a bandpass of 5 nm. The output of the monochromator was collimated and the image of the exit slit was focused on the electrode of the MNOS structure using a quartz lens system (L_1 and L_2) and a first surface flat reflector mirror with Al-MgF_2 overcoat (Oriel Optics A-33-262). Cut-off filters (F_2) were used to remove second and higher-order light wavelengths. An optical-bench mounted chamber with a vacuum-sealed quartz window W, a horizontal-plane sample holder and specially designed external lead configurations was used to enable execution of both photoinjection-photodepopulation and C-V measurements in the same facility. The measurements were executed at room temperature and in vacuum (5×10^{-5} torr) to eliminate surface contamination. The probe was positioned near the edge of the metal gate electrode to minimize shadowing effects in the photocurrent measurements.

The electrical system consisted of a series circuit composed of the MNOS device structure, a variable power supply and an electrometer. The background noise level maintained through out all experiments was in the 10^{-15} ampere range. Applied voltage V refers to the potential at the metal electrode of the MNOS structure referenced to ground. The C-V measurements were made on a Boonton model 72B capacitance meter, operated at 1 MHz.

The measurement program was developed around significant advances in instrumentation employing a computer based real time digital data acquisition, data processing and analysis system. This system consisted of a PET 2001 microcomputer, a 4 channel 12 bit Analog to digital converter (A/D), a 2 channel 10 bit D/A converter, operated on the IEEE-488 busline and a time sharing Harris/6 system computer.

The monochromator was driven by a computer controlled wavelength mini-drive (Spex 1673) as shown in Fig. 1. The mini-drive was connected to the microcomputer through its parallel output port to permit scanning the monochromator with externally programmed speed and direction. The microcomputer also recorded and stored digital signals corresponding to the wavelength data in its buffer memory. In this system the output signal of the Keithley electrometer and the digital signals corresponding to the wavelength of the monochromator were recorded simultaneously by the microcomputer in its buffer memory once in every 24 ms and then to a magnetic tape or to the memory of the main Harris computer. The entire measurement program was executed, analyzed and plotted using the microcomputer.

Photoinjection and current-monitored trap photodepopulation measurements were carried out to determine the energy distribution of electron trapped in the Si_3N_4 insulator of a MNOS device. In the photoinjection step, with the applied gate voltage bias of -10 V, electrons were photoinjected into the Si_3N_4 from the Al electrode with light of energy 4.3 eV. The photocurrent was measured by the electrometer as a function of time until the photocurrent reached a steady state and the positive shift in the flatband voltage saturated, as determined by C-V measurements. This indicated that a steady-state trap population had been attained. The net effect of photoinjection was the introduction of negative charge in the Si_3N_4 by capture of electrons into impurity/structural-related trapping states.

Trap photodepopulation measurements were carried out by continuously varying the monochromator wavelength from high to low wavelengths while simultaneously recording the photocurrent. The range of the wavelength covered was 900-305 nm and the applied gate voltage bias was +10.0V. The monochromator sweep rate was 0.1 nm/sec and the full intensity of the xenon lamp with sharp cutoff filters was used. It was determined that under the above operating conditions the trapping-state distribution was scanned so slowly that the trapped charges are essentially emptied by the complete scan. The successive scans produced no photocurrent or the photocurrent was of the order of the dark current. During the photodepopulation process the photoexcited electrons are collected by the aluminum electrode and the retrapping of the electrons in the Si_3N_4 was not found to be a significant factor. It was also determined that the spectra obtained during trap photodepopulation were independent of the photoinjection conditions such as applied voltage during photoinjection and time of photoinjection, as long as steady-state trap population was reached during the photoinjection process.

RESULTS AND DISCUSSIONS

Figure 2 exhibits the results of trap photodepopulation measurements where the photocurrent-vs-wavelength spectra was obtained after a photoinjection step for three groups of samples FSON1, FSON2 and FSON3. The Si_3N_4 film thickness was 0.05 μm for FSON1, 0.11 μm for FSON2 and 0.17 μm for FSON3. In all the samples used, the thickness of SiO_2 was 0.1 μm. No spectral response was observed prior to a photoinjection step and the photocurrent was found to be equal to the dark current ($+1.3 \times 10^{-14} A/cm^2$) during the virgin sample scan. At applied gate voltage bias of +10 V the wavelength scan was in the 900-305 nm region but the spectra for the 600-305 nm range only are shown in Fig. 2. No spectral response was observed and the photocurrent was of the order of the dark current above 600 nm.

Undulations were observed in the spectra. The amplitudes of maxima and minima observed in the spectra were found to be dependent on the Si_3N_4 thickness of the sample whereas the wavelength positions of these undulations did not change appreciably with Si_3N_4 thicknesses. Undulations which were observed in the photodepopulation spectra may be due to certain structures associated with the trapping levels in the Si_3N_4 insulator. These undulations may also be due to thin-film optical interference effects. A four-layer optical model was developed for the metal-nitride-oxide-semiconductor structure to analyze the interference effects and to calculate the light intensity in the Si_3N_4 film as a function of photon energy. The basic logic of the optical model was similar to that recently reported by Kapoor, Feigl and Butler.[5]

Each of the observed photocurrent-vs-wavelength curves, as shown in Fig. 2, were normalized to the light intensity $S(h\nu)$ in the Si_3N_4 film to obtain the "spectral response" as a function of photon energy. The spectral response $Y(h\nu)$ is defined as

$$Y(h\nu) = \frac{I/eA}{S(h\nu)/h\nu} \quad \text{(photoelectron/photon)}$$

where I is the measured photocurrent in amperes at the incident photon energy $h\nu$ in eV, e is the electronic charge in coloumbs, A is the electrode area in cm^2 and $S(h\nu)$ is the average light intensity in the Si_3N_4 film in mW/cm^2 determined from

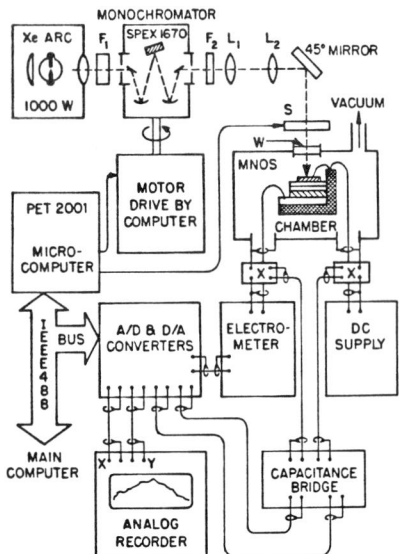

Fig. 1. Schematic of equipment used for photoinjection-photodepopulation and C-V measurements.

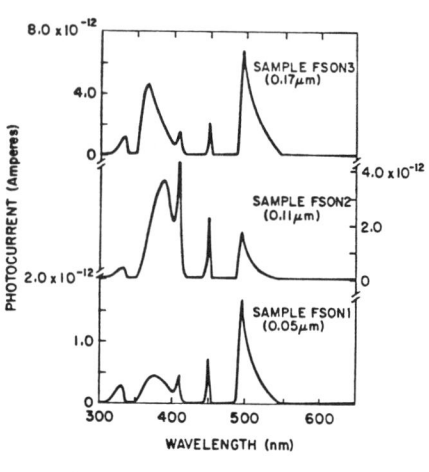

Fig. 2. Photocurrent response measured as a function of wavelength for samples FSON1, FSON2, and FSON3.

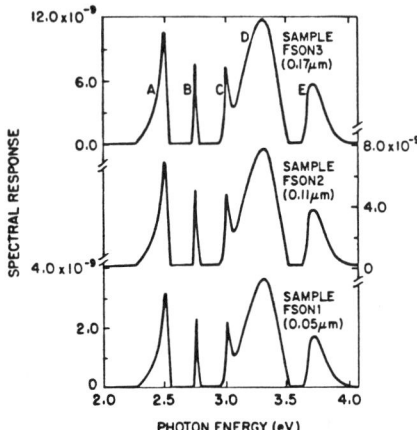

Fig. 3. Spectral response as a function of photon energy at room temperature for the MNOS samples FSON1, FSON2 and FSON3. A, B, C, D, and E designate the peak positions of the spectral response curves.

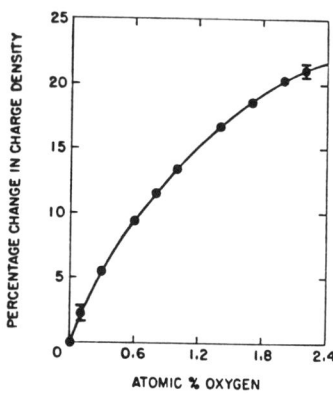

Fig. 4. Measured percentage change in photoinjected trapped charge density as a function of atomic % oxygen in the Si_3N_4 films of the MNOS structures.

the optical model calculations and the measured xenon arc lamp-monochromator intensity. Figure 3 shows the spectral response $Y(h\nu)$ as a function of photon energy $h\nu$ for FSON1, FSON2 and FSON3 samples.

The spectral response shown in Fig. 3 is interpreted to provide information about the energetically localized distribution of electron trapping states associated with chemical impurities and/or structural defects within the Si_3N_4 film of the thick-oxide MNOS structure. The trap depth (the peak position in energy) of five trap levels designated as A, B, C, D and E in Fig. 3 relative to the conduction band edge of Si_3N_4 were determined to be at 2.5, 2.76, 3.03, 3.36 and 3.76 eV with half width (full width at half maximum) of 0.08, 0.02, 0.05, 0.32 and 0.17 eV respectively.

In addition to photocurrent spectroscopy, Auger electron spectroscopy[6] has been used in conjunction with argon-ion sputtering to determine the chemical nature of the MNOS structures. The experiment involved bombardment of the specimen surface with a beam of 1-2 KeV argon ions and simultaneously scanning with the primary electron beam of 4.0 μA at 4 KeV over an area of $2.5 \times 10^{-2} cm^2$. The Auger spectrum showed the presence of nitrogen, carbon, iron, argon and oxygen together with silicon. We have obtained chemical depth profiles for MNOS devices with the Si_3N_4 grown at various SiH_2Cl_2 to NH_3 ratios. It was observed that the oxygen content in the nitride was sensitive and varied as a function of NH_3 to SiH_2Cl_2 ratio. This observed increase may be due to oxygen and water impurities in the NH_3 or in the carrier gases used for the deposition.

The density of trapped electronic charge during the photoinjection process was determined using C-V measurements as a function of the amount of oxygen in the Si_3N_4 layer of the MNOS structure. Figure 4 shows the results of such measurements. The atomic % oxygen was measured using Auger Depth Profiling of the MNOS samples.[6] The data indicates that the amount of oxygen in the Si_3N_4 film may be correlated to the trapped charge. The amount of oxygen is important with regard to electron trapping, since non-bridging oxygen atoms are electron acceptors. In addition the correlation of the photodepopulation spectra shown in Fig. 3 with the amount of oxygen in Si_3N_4 insulator indicated that the trap level A at 2.50 eV was associated with oxygen impurity in the film.

ACKNOWLEDGEMENTS

The authors would like to thank Drs. R. S. Singh, C. T. Kirk and P. C. Y. Chen for helpful discussions. We are grateful to Prof. F. J. Feigl for providing the sample chamber. Research supported by U.S. Navy Office of Naval Research.

REFERENCES

1. C. T. Kirk, Jr., J. Appl. Phys. 50, 4190 (1979).
2. R. Williams, Phys. Rev. 140, A569 (1965).
3. J. H. Thomas and F. J. Feigl, J. Phys. Chem. Solids 33, 2197 (1972).
4. B. H. Yun, Appl. Phys. Lett. 27, 256 (1975).
5. V. J. Kapoor, F. J. Feigl and S. R. Butler, J. Appl. Phys. 48, 739 (1977).
6. H. J. Stein, S. T. Picraux and P. H. Holloway, IEEE Trans. Electron Devices ED-25, 1008 (1978).

TRAPS IN SiO_2-Si STRUCTURE DETERMINED BY ELECTROCHEMICAL METHOD

Andrzej Wolkenberg

Institute of Electron Technology, 02 668 Warsaw, Poland

ABSTRACT

Physical and chemical properties of silicon dioxide in water solution of potassium chloride are discussed. It is shown that the electrochemically determined trap number as well as the space distribution of these traps can be obtained experimentally from anodic polarization curves. This trap number per unit area in the investigated thermal oxide was about $3.10^{16}/cm^3$. Implanted boron appeared to form either negatively or positively charged centers in silicon dioxide in contact with electrolyte. Boron implanted oxide shows that there is dependence between chlorine concentration and this traps number.

INTRODUCTION

The electrolyte side of the silicon dioxide-electrolyte interface consists of ions arranged in a double electric layer. When an electrode, say SiO_2 covered silicon electrode, is dipped in an electrolyte solution can happen that the electrons are swept out of the SiO_2 layer and the positive charge is generated by a significant negative charge in the double layer.

After polarization of this electrode, an electric field appears in silicon dioxide layer resulting in the electron injection from either the electrolyte or the semiconductor, depending on the field direction. This way traps become charged. We can thus imagine the following charges existing in the $Si-SiO_2$-0.1M KCl structure and being of considerable importance for measurements:

- charge in the space charge region of semiconductor,
- charge accumulated in SiO_2 layer, i.e., that present in the structure prior to dipping it in the electrolyte solution and the one

induced after dipping,
- charge in the double ion layer in the electrolyte solution.
This paper presents the experimental data on the interface model introduced above. The model was verified using the following structure: N-type silicon wafer covered with thermal oxide and dipped in water solution of potassium chloride.

EXPERIMENTAL TECHNIQUE

All details concerning the setup have been reported earlier[1] and so the peculiarities of Si-SiO$_2$ structures measured with it[2,3].
The present investigations were performed on /100/ and /111/ oriented N-type silicon wafers thermally oxidized in three different ambients: dry oxygen, oxygen with 5% HCl, and oxygen with 7% HCl at a temperature of 1100°C. Following oxidation, different doses of boron were implanted into the oxide layer at an energy of 30 keV. Next the samples were annealed at 920-1100°C for 30 minutes. An ohmic contact was prepared to the non-oxidized side of the sample to provide better accuracy and reproducibility of the results of electrical measurements. The distribution of implanted boron concentration has been assumed consistent with the LSS theory.

RESULTS AND DISCUSSION

The results of measurements of the number of electrochemically-detectable traps are collected in Table I.
The data for the flat-band potential, ϕ_{FB}, for the same samples in 0.1M KCl are illustrated in Fig.1.
The effect of crystal orientation on flat-band potential measured in the Si-SiO$_2$-0.1M KCl structure, compared to that in the Si-SiO$_2$-Hg structure,

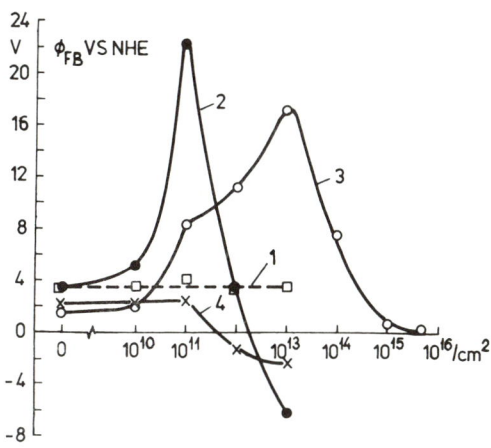

Fig.1. Flat-band voltage ϕ_{FB} versus implantation fluency
Si n ≅ 2·10^{15}/cm^3 <111>   S$_{SiO_2}$ ≅ 1525 ÷ 1710 Å
in 0,1M KCl.
 □ - without HCl   P implanted with 80 keV
 ● - without HCl
 ○ - HCl/O$_2$ = 5%   B implanted with 30 keV
 × - HCl/O$_2$ = 7%

has been explained in terms of the experimental results presented in Table II.

TABLE I Trap density

Si wafer N=2.10^{15}/cm^3 /111/		trap density N_t/cm^3					
		implantation dose/cm^2					
		0	1.10^{10}	1.10^{11}	1.10^{12}	1.10^{13}	1.10^{14}
s_{SiO_2}=1710 Å oxidation without HCl	B 30 keV	$4.5.10^{16}$	$7..10^{16}$	$3..10^{16}$	$4.5.10^{16}$	$-9..10^{16}$	–
	P 80 keV	$4.5.10^{16}$	$5..10^{16}$	$5.5.10^{16}$	$4..10^{16}$	$5.5.10^{16}$	–
s_{SiO_2}=1580 Å oxidation with 5% HCl	B 30 keV	$2.5.10^{16}$	$4.5.10^{16}$	$1.3.10^{17}$	$1.3.10^{17}$	$2.5.10^{17}$	$1.3.10^{17}$
s_{SiO_2}= 1525 Å oxidation with 7% HCl	B 30 keV	$4.5.10^{16}$	$5..10^{16}$	$5..10^{16}$	$-1..10^{16}$	$-4..10^{16}$	–

TABLE II Charge induced by 0.1M KCl

wafer number	Si N=2.10^{15}/cm^3 /100/ and /111/ oxidation with 7%HCl	flat band potential difference between the structures Si-SiO$_2$-0.1M KCl and Si-SiO$_2$-Hg	charge induced by 0.1 M KCl
I	s_{SiO_2} = 1135 Å /100/	0.25 V	$3.7.10^{10}$/cm^2
II		0.2 V	3.10^{10}/cm^2
1	s_{SiO_2} = 1525 Å /111/	1.6 V	2.10^{11}/cm^2
2		1.6 V	2.10^{11}/cm^2
3		1.4 V	$1.8.10^{11}$/cm^2
4		1.6 V	2.10^{11}/cm^2

While measuring in an electrolyte solution /0.1M KCl/ a certain extra charge is induced into the oxide due to a double space-charge layer on electrolyte side. Table II shows the expected charge values in /100/ and /111/ oriented silicon when the measurements in 0.1M KCl are refered to the mercury probe. It has been assumed that the solution-induced extra charge has its centroid localized at the SiO_2 - Si interface.

The boron atoms in the oxide can act as:
- a source of positive charge, if the concentration is small,
- a source of negative charge at higher concentrations,

where the exact concentration corresponding to the change in electrical properties depends upon the amount of HCl added to oxygen in the course of oxidation.

From curves 2-4 in Fig.1 one can calculate charges in implanted wafers presented in Table III.

TABLE III Charges in implanted wafers /after Fig.1/

curve number	boron dose/cm^2	equivalent charge/cm^2	excees equivalent charge/cm^2	remaining inactive boron/cm^2
2 without HCl	$1 \cdot 10^{11}$	$3 \cdot 10^{12}$	$2.9 \cdot 10^{12}$	-
	$1 \cdot 10^{13}$	$-8 \cdot 10^{11}$	-	$6 \cdot 10^{12}$
3 with 5% HCl	$1 \cdot 10^{11}$	$1 \cdot 10^{12}$	$9 \cdot 10^{11}$	-
	$1 \cdot 10^{13}$	$2.5 \cdot 10^{12}$	-	$7.5 \cdot 10^{12}$
	$5 \cdot 10^{15}$	$1 \cdot 10^{10}$	-	$4.9 \cdot 10^{15}$
4 with 7% HCl	$1 \cdot 10^{11}$	$3.5 \cdot 10^{11}$	$2.5 \cdot 10^{11}$	-
	$1 \cdot 10^{13}$	$-3 \cdot 10^{11}$	-	$9.2 \cdot 10^{12}$

The amount of electrically active boron will be different. When chlorine is present in the oxide, the latter gets considerably lowered. It seems reasonable to suggest that, lower concentrations, boron atoms occupy in oxide the sites which refer to interstitial in a perfect crystalline structure. Consequently, they can be considered positive charges able to sweep the adjacent electrons out of their weakly bound positions, which leads to an increase in the total positive charge. When a certain critical concentration is exceeded /dependent on HCl content during oxidation/, boron substitutes the lattice sites as refered to crystalline structure and acts as an acceptor generating the negative charge. From the behaviour of implanted boron one can conclude that the properties observed can be attributed to boron it-

self and have nothing in common with the implantation process.
On the contrary, chlorine performs a gettering-screening action against both positive and negative charges introduced by boron.
As for the number and sign of traps electrochemically detectable in oxide /Table I/, they are dependent on oxidation conditions and implanted boron dose like in the case of equivalent charges.
No changes accompanying the dose changes are observed in the number of traps present in phosphorus-implanted oxide covered samples.

CONCLUSIONS

The boron atoms in the oxide can act as a source of positive or negative charge.

The results of investigations on charges and traps in SiO_2 are generally consistent with those reported by DiMaria[4] /traps/ and Aberg[5] /charges/.

Standard technological methods of MOS structures fabrication can be controlled more accurately if the measuring techniques are understood more unconventionally.

REFERENCES

1. A.Wolkenberg, Surface Science, 50.580./1975/
2. A.Wolkenberg, Applications of Surface Science, 2.502./1979/
3. A.Wolkenberg,B.Wasilewska,. Applications of Surface Science, 3.83./1979/
4. D.J.DiMaria, R.Ghez, D.W.Dong,. Charge trapping studies in SiO_2 using high current injection from Si rich SiO_2 films, IBM Research Report, 11.5./1979/
5. A.T.Aberg,. phys.stat.sol./a/ 42.639./1979/.

CHARGE TRAPPING AND ASSOCIATED LUMINESCENCE IN MOS OXIDE LAYERS

C. Falcony-Guajardo, F. J. Feigl, and S. R. Butler
Sherman Fairchild Laboratory, Lehigh University, Bethlehem, PA 18015

ABSTRACT

Electroluminescence from thermally grown SiO_2 thin films has been studied using pulsed current avalanche injection of electrons. Oxide films 0.06-0.12 μm thick were grown on p-Si at 1000°C in "dry" oxygen. The magnitude of both oxide trapped charge (Q_{ot} and Q_{it}) and luminescence intensity ϕ_λ were measured as a function of the injected electron fluence ϕ_e. The dominant electron traps had effective capture cross section $\sigma_c \sim 10^{-18} cm^2$. An associated luminescence intensity is observed: $\phi_\lambda \propto \dot{Q}_{ot}$, where $\dot{Q}_{ot}$ is the trapping rate for a given electron trapping center (characterized by σ_c). The spectral distribution of luminescence emissions from several trapping centers was determined ($h\nu$ = 2-5 eV). These results are compared with previous studies of cathodoluminescence.

INTRODUCTION

Charge trapping in thin silicon dioxide films is a recurring problem within the silicon planar technology. Ongoing developments in processing, small-dimension device technology, and non-volatile memory have added some immediacy to the long-standing research interest in the phenomenon of electron-hole trapping in SiO_2. Characterization of trapping centers in device films is largely phenomenological, although optical ionization thresholds have been determined.[1,2] Trap characterization by luminescence emissions has been proposed in the past, using cathodoluminescence[3] and high field electroluminescence.[4] In particular, Solomon[4] suggested that electroluminescence might be associated with electron trapping.

We have directly demonstrated correlation of light emission with electron trapping in the MOS oxide. A coarse spectral resolution of the electroluminescence emission associated with specific trapping centers was determined. Finally, the trapping and luminescence phenomena were studied as a function of selected processing variations for the thermal SiO_2 films and/or MOS devices.

EXPERIMENTAL DETAILS

The MOS devices incorporated thermally grown oxide films on (100) p-type polished Si substrates (∼0.07-0.1 Ω-cm). The oxides were grown in "dry" oxygen ambients at 1000°C. The oxide thicknesses d_{ox} were in the range 0.05-0.2 μm. Semi-transparent Al electrodes were vapor deposited on the oxide surface, and a thick

Al layer was similarly deposited on the back of the Si substrate. The active electrode area was in the range 0.01-0.06 cm². A number of samples was given a 400°C post metallization annealing treatment for 30 minutes in dry nitrogen. Other variations in the device fabrication will be discussed later in this paper.

The experimental arrangement used for measurement incorporated a pulsed current avalanche injection circuit designed by Nicollian. This circuit has a feedback network which applies a d.c. correction voltage to the sample in order to maintain a constant (time-averaged) avalanche injection current. The correction voltage V_A was continuously measured throughout the experiment. High frequency C-V characteristic curves were measured intermittently in selected cases. The C-V data was summarized as flat band voltage V_{FB} vs. accumulated time of avalanche injection t_A. The electrical contact to the semitransparent electrode was made with a gold plated fine wire, and the sample was enclosed in a vacuum-tight chamber flushed with dry nitrogen.

The light detecting system for luminescence was a low noise EMI bialkali type photomultiplier tube sensitive in the spectral range 180-650 nm. A silica lens was used to increase optical collection efficiency. The phototube was coupled to a PAR-SSR photon counting system which has a built in integration mode. Thus, we could directly measure either the rate of photon emission or the total number of photons emitted $\int_0^{t_A} \phi dt$ vs. t_A. A rough spectral resolution of the luminescence emission was obtained by inserting wide bandpass filters in front of the photomultiplier tube.

RESULTS

The basic phenomenon studied is a transient light emission intensity under constant current conditions. The measurement programs for this work were either: (A) a sequence of injection steps interrupted by C-V measurement steps, or (B) continuous monitoring of the avalanche circuit voltage. In the first case, $\Delta V_{FB}(t_A) = V_{FB}(t_A) - V_{FB}(0)$ was measured at each step in the sequence. In the latter case, $\Delta V_A(t_A) = V_A(t_A) - V_A(0)$ was measured continuously. Fig. 1 shows the measured voltage shifts for a specimen for which $\Delta V_{FB} = \Delta V_A$. This condition ($\Delta V_{FB} \sim \Delta V_A$) obtains for all the results discussed in the present paper. It is not generally true, as we have previously reported.[6]

For either measurement program, (A) or (B), the luminescence intensity ϕ or the integrated luminescence output $\int \phi dt$ is measured continuously while the constant avalanche injection current is flowing.

Representative luminescence data is exhibited in Fig. 2. The top data set is the total luminescence output, limited by the spectral response of the photomultiplier; the bottom two data sets are obtained with the appropriate band-pass filter in place. Note that the trap filling process is irreversible in these experiments, and that each data set in Fig. 2 was obtained from a different electrode on the oxidized wafer being studied.

ANALYSIS

To calculate the charge trapped in the oxide, we used the relation[5,7]

$$Q_{ot} = e N_{ot} = C_{ox} \Delta V_{FB} = C_{ox} \Delta V_A . \qquad (1)$$

Q_{ot} is the centroid-weighted oxide trapped per unit electrode area

$$Q_{ot} = \frac{1}{d_{ox}} \int_0^{d_{ox}} x\rho(x)\,dx. \tag{2}$$

$\rho(x)$ is the volume density of charge produced by electrons trapped in localized states during transport across the oxide. C_{ox} is the oxide capacitance per unit electrode area: $C_{ox} = \kappa_{SiO_2}\varepsilon_o/d_{ox}$.

A first order kinetic analysis of $Q_{ot}(t_A)$ was used to obtain trapping cross sections and trap densities. This analysis was performed by fitting to data the expression

$$Q_{ot}(t_A) = e \sum_i N_i [1-\exp(-t_A/\tau_i)]. \tag{3}$$

N_i is the density of a particular trap (index i) and $\sigma_{ci} = e(J_A\tau_i)^{-1}$ is the electron capture cross section of that trap. J_A is the avalanche current density in A-cm^{-2} (a time-average dc current). Each distinct trap was measured over a range of currents, and quoted cross section were determined from the slope of τ_i^{-1} vs. J_A/e.

The experimental conditions used permitted study of trapping cross sections from $\sim 10^{-18}$cm^2 to $\sim 10^{-13}$cm^2. The present discussion is restricted to neutral trapping centers,[1] with $\sigma_c = (10^{-18} - 10^{-16})$ cm^2. The solid line in Fig. 1 is a two trap form of Equation (3). The maximum number of traps which could reliably be separated in the present study was three (i = 1,2,3 in Equation (3)).

The luminescence data were fitted with an expression of the form:

$$\Phi(t_A) = \Phi_0 + \sum_i \Phi_i \exp(-t_A/\tau_i)]. \tag{4A}$$

or, as appropriate, the corresponding integrated expression

$$\int_0^{t_A} \Phi(t)dt = \Phi_0 t_A + \sum_i \tau_i \Phi_i [1-\exp(-t_A/\tau_i)] . \tag{4B}$$

The solid lines in Fig. 2 are fitted expressions of the form of Equation (4B), with i = 1,2.

For a given oxide film, this analysis was used to construct the spectral density Φ_λ for electroluminescence associated with a given characteristic time τ_i. Thus, for a time constant τ_i and a given filter (covering wavelength λ and with an effective bandwidth $\Delta\lambda$ and average transmission T)

$$\Phi_{\lambda i} = \frac{\Phi_i}{\eta T \Delta \lambda},$$

where η is the average photomultiplier quantum efficiency for the band pass range of the filter.

Figure 3 shows the spectral distribution of electroluminescence for a nominally dry oxide produced at the IBM Watson Research Center. Figure 4 shows similar results for two different nominally dry oxides produced at Lehigh. Note that each segment of these spectra were determined from complete measurement sequences (B) on several dots on a given experimental wafer. Figures 3 and 4 are composites of approximately ten runs on separate electrodes. This procedure is required because of the irreversible, transient nature of the trapping luminescence, as discussed below.

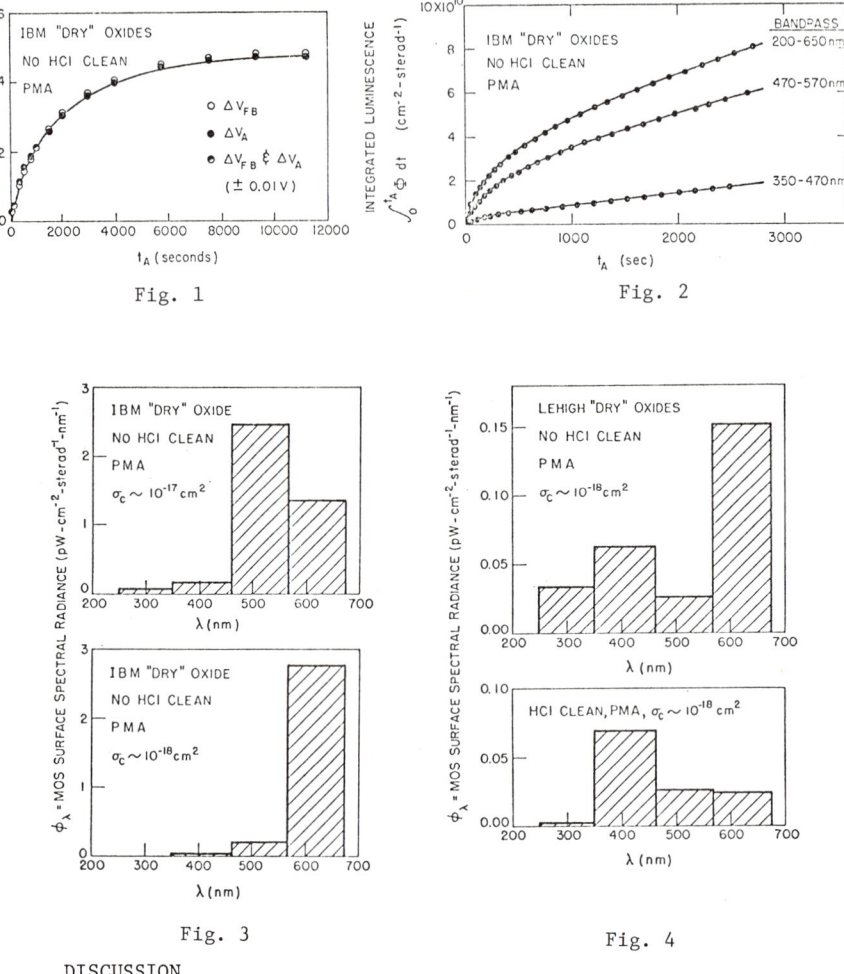

Fig. 1

Fig. 2

Fig. 3

Fig. 4

DISCUSSION

We believe the constant component of the electroluminescence emission -- characterized by the intensity ϕ_0 in Equations (4) -- is generated in the Si depletion layer created during the avalanche injection process. This luminescence is similar to that observed from p-n junctions.[8]

The time dependent component of the electroluminescence emission correlates with charge trapping rates in the oxide. Specifically, the time constants τ_i determined from the luminescence data are the same, within experimental error, as those determined from the charge trapping data. We therefore suggest that the light emissions are directly associated with electron trapping in the oxide film. From Equations (3) and (4), for a trapping center with capture cross section σ_{ci} and areal density N_i in a given oxide layer

$$\phi_i \text{ and } \phi_{\lambda i} \propto \frac{N_i}{\tau_i} \propto J_A . \qquad (5)$$

We have verified the dependence of emission intensity on avalanche current implied by Equation 5.

With this interpretation, the spectral distributions in Figs. 3 and 4 represent light emissions from radiative electron capture into one trapping center or several trapping centers characterized by an effective electron capture cross section σ_c (we shall drop the summing index i in subsequent discussion). Our experience indicates that the experimental and analytical procedures used in this and related investigations can not easily discriminate between cross sections differing by < factor of 2. Thus, the two peaks indicated in Fig. 4 are assumed to arise from two distinct trapping centers, each with $\sigma_c \sim 10^{-18} cm^2$.

The IBM sample (data in Fig. 3) and one of the Lehigh samples (data in Fig. 4A) are nominally identical. Both oxides were prepared in dry oxygen gas, and subjected to a post-metallization annealing treatment. The oxidation furnace had not been subjected to an HCl cleaning treatment. The trapping center with $\sigma_c \sim 10^{-17} cm^2$ and associated luminescence peaked at $\sim$550 nm (IBM sample, Fig. 3A) may be associated with water contamination subsequent to device fabrication.

The Lehigh sample represented in Fig. 4B differs from the others. This oxide was prepared in a furnace previously cleaned with an HCl/O_2 gas mixture. We suggest that the trapping center with $\sigma_c \sim 10^{-18} cm^2$, and emission peaked at $\sim$650 nm, involves non-mobile sodium impurities incorporated during oxidation.[9] HCl pre-cleaning removes such impurities from the oxidizer.[10] Association of a luminescence at $\sim$650 nm with Na impurities has previously been suggested by Jones and Embree.[3] Our interpretation further associates this emission directly with electron capture into an electrically neutral trapping center formed during oxidation or device fabrication.

ACKNOWLEDGEMENTS

This work was supported by the Office of Naval Research, Electronic and Solid State Sciences Program. Mr. Falcony-Guajardo was supported by CONACyT-Mexico.

REFERENCES

1. D. J. DiMaria (1978) in The Physics of SiO$_2$, edited by S. T. Pantelides, Pergamon Press, New York, pp. 160-178.
2. D. D. Rathman, F. J. Feigl, and S. R. Butler (1980) in Insulating Films on Semiconductors 1979, edited by G. G. Roberts and M. J. Morant, The Institute of Physics, London, pp. 48-54. See also D. D. Rathman, F. J. Feigl, S. R. Butler, and W. B. Fowler, this conference.
3. C. E. Jones and D. Embree (1978) in The Physics of SiO$_2$, edited by S. T. Pantelides, Pergamon Press, New York, pp. 289-293.
4. P. Solomon and N. Klein, J. Appl. Phys. 47, 1023 (1976).
5. E. H. Nicollian, C. N. Berglund, P. F. Schmidt, and J. M. Andrews, J. Appl. Phys. 42, 5654 (1971).
6. C. Falcony-Guajardo, F. J. Feigl, S. R. Butler, and J. B. Anthony (1979) in Extended Abstracts Volume 79-1, The Electrochemical Society, Princeton, NJ, pp. 400-402.
7. B. E. Deal, J. Electrochem. Soc. 127, 979 (1980).
8. A. G. Chynoweth and R. G. McKay, Phys. Rev. 102, 369 (1956).
9. S. R. Butler, F. J. Feigl, Y. Ota, and D. J. DiMaria (1976) in Thermal and Photostimulated Currents in Insulators, edited by D. M. Smyth, The Electrochemical Society, Princeton, NJ, pp. 149-161.
10. S. Mayo and W. H. Evans, J. Electrochem. Soc. 124, 780 (1977).

TIME DECAY OF PHOTOLUMINESCENCE FROM AMORPHOUS SiO_2

C.M. Gee and Marc Kastner
Department of Physics and Research Laboratory of Electronics
Massachusetts Institute of Technology, Cambridge, Ma. 02139

ABSTRACT

Photoluminescence (PL) time decay for annealed and neutron-irradiated amorphous (a-) SiO_2 is reported. PL bands at 4.3, 2.7 and 1.9 eV, excited by 7.8 eV radiation from a pulsed F_2 laser, are observed for both annealed and neutron-irradiated Suprasil W. The time decay of the 4.3 eV band is <10ns, but the 2.7 and 1.9 eV PL decay slower. In spite of the complexity of several PL bands, the time decays of the two low energy PL bands are simply composed of a couple of decay rates.

INTRODUCTION

The observation[1,2] of steady state photoluminescence (PL) from SiO_2 has been reported previously. It was concluded from these measurements that the PL is associated with intrinsic defects in amorphous (a-) and crystalline SiO_2. Furthermore, the PL and photoluminescence excitation (PLE) spectra of a-SiO_2 and a typical semiconducting chalcogenide, a-As_2S_3, were found to scale with energy gap, and the temperature dependences of the PL intensities for a-SiO_2, a-As_2S_3 and a-Se were found to scale with glass transition temperature. Therefore, it was suggested that the PL centers in a-SiO_2 are similar to those in narrower gap chalcogenide glasses. This supports the Valence-Alternation Model[3] which has provided a microscopic picture of intrinsic defects in chalcogenides. In the case of a-As_2S_3, the evolution[4] of the PL spectrum with time has been a valuable probe of the properties of the intrinsic defects. We report here measurements of time-resolved PL for annealed and neutron-irradiated a-SiO_2.

EXPERIMENTAL DETAILS

Samples consisted of 3mm polished slabs of pure fused silica (Suprasil W)[5]. The irradiated samples received a dose of $\sim 10^{19}$ cm^{-2} fast neutrons at room temperature. The samples were cooled on the tip of a liquid He cold finger. A Gold (0.7% Fe)-Chromel thermocouple soldered directly to the sample monitored temperature. The sample chamber was evacuated to 10^{-8} torr using a completely oil-free pumping system.

The PL was excited by 1576 Å radiation from a pulsed F_2 laser. The laser, which was operated at 20 Hz repetition rate, supplied $\sim$10ns pulses of $\sim$10µJ energy per pulse at the sample. This corresponds to an average intensity of $\sim$80mW/cm^2 for a spot size of $\sim$0.5mm. Lower intensities were available by aperturing the laser beam. The PL was analyzed by a 0.2 meter grating monochromater and then detected by an S-20 photomultiplier tube. The time-dependent signal from the photomultiplier was processed using a single sweep waveform recorder combined with a multichannel signal averager. This combination yields 2ns resolution of the PL time decay.

RESULTS

Time-integrated PL spectra of annealed and neutron-irradiated Suprasil W for excitation by full laser intensity and by ∿1/100 of full intensity are displayed in Fig. 1 and in Fig. 2. The spectra, although plotted on an arbitrary scale, are correct

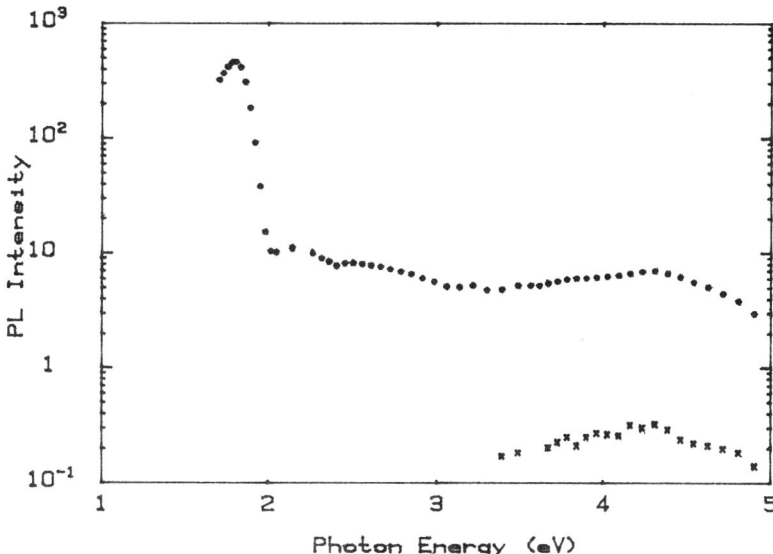

Fig. 1. Time-integrated PL spectra for annealed Suprasil W excited by ∿1/100 of full laser intensity (x) and by full intensity (•).

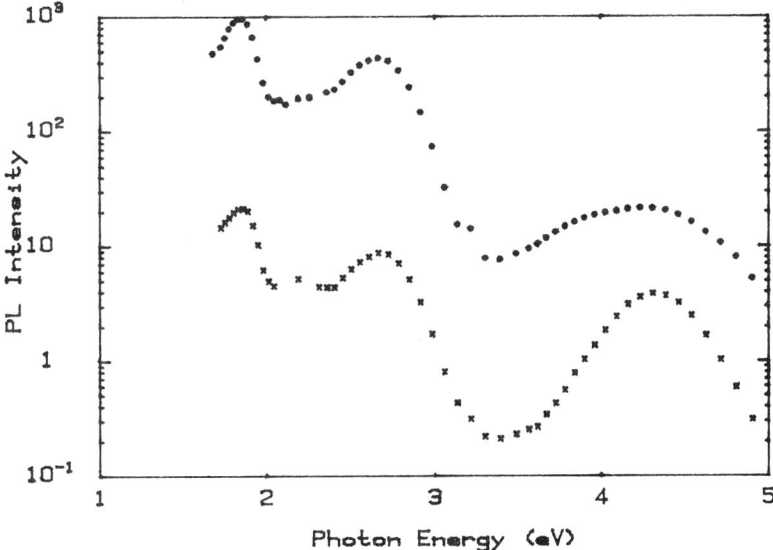

Fig. 2. Time-integrated PL spectra for neutron-irradiated Suprasil W excited by ∿1/100 of full laser intensity (x) and by full intensity (•).

relative to one another. Seen in the spectra are PL bands[6] at 4.3, 2.7 and 1.9 eV. Only the 4.3 eV band is apparent when annealed Suprasil W is excited by ~1/100 of full laser intensity. This PL band is similar to the single PL band observed previously[1] when the PL was continuously excited by light from a hydrogen discharge lamp. That steady state PL consisted of a single band at ~4eV with a full width at half maximum (FWHM) of ~1.5eV for annealed Suprasil W and a band at ~4.3eV with FWHM of ~0.6eV for neutron-irradiated Suprasil W. The difference between the earlier steady state measurements and the present results are not understood. It is possible that the differences may be attributed to the monochromatic radiation or to the high average and peak power of the laser. Furthermore, the increased intensity reveals two additional bands at lower energy. These two bands appear in both neutron-irradiated and annealed Suprasil W, suggesting that the bands are defect-related rather than impurity-induced. The two low energy bands, unlike the 4.3eV PL band, scale linearly with laser intensity. The 4.3eV band of neutron-irradiated Suprasil W, however, appears to be saturated and broadened by full laser excitation. It is not understood why the neutron-irradiated sample should saturate before the annealed sample. However, if saturation is assumed, then

$$n \sim I_s \alpha \tau_s \qquad (1)$$

where n is the density of PL centers, I_s is the incident laser intensity, α is the absorption coefficient at 7.8 eV and τ_s is the decay time of an excited center. Using the measured[2] absorption coefficient at 7.8 eV for neutron-irradiated Suprasil W (~30 cm^{-1}), an estimated upper limit for the density of PL centers is 10^{17} cm^{-3} since the decay of the 4.3 eV PL band is found to be at least as fast as the laser pulse.

However, the decay of the other two PL bands is slower as can be seen from Figs. 3-5. For the 1.9 eV band of annealed Suprasil W, the PL decays at approximately a single rate of time constant $\tau \sim 20 \mu s$. Two distinct decay rates ($\tau \sim 40ns$ and $\sim 10ms$) are observed for the 2.7 eV PL band of neutron-irradiated Suprasil W. The exponential behavior of the PL time decay for the low enery bands are displayed in Fig. 5 and in the insert to Fig. 3.

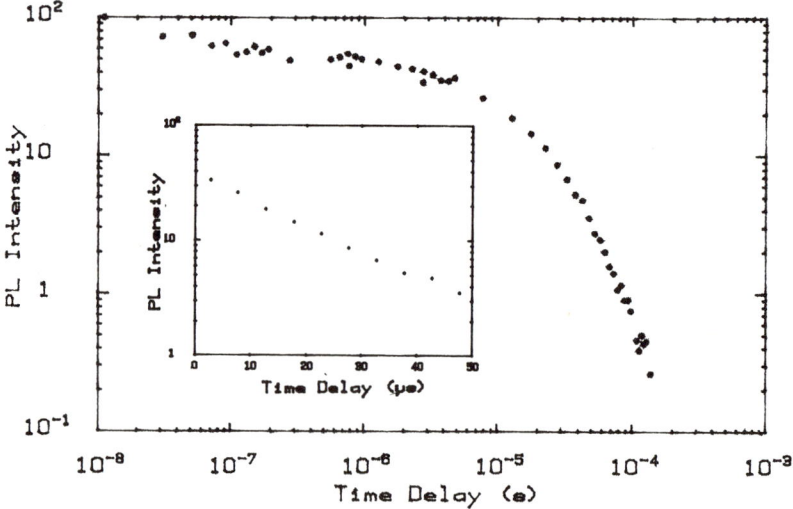

Fig. 3. Time decay of the 1.9 eV PL for annealed Suprasil W excited by full laser intensity.

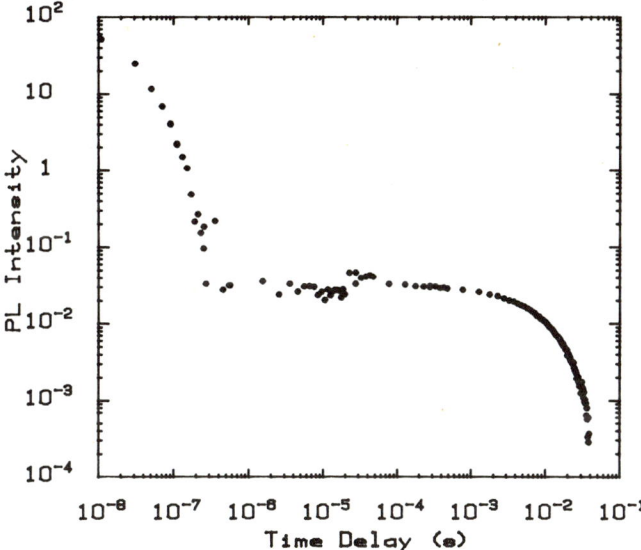

Fig. 4. Time decay of the 2.7 eV PL for neutron-irradiated Suprasil W excited by full laser excitation.

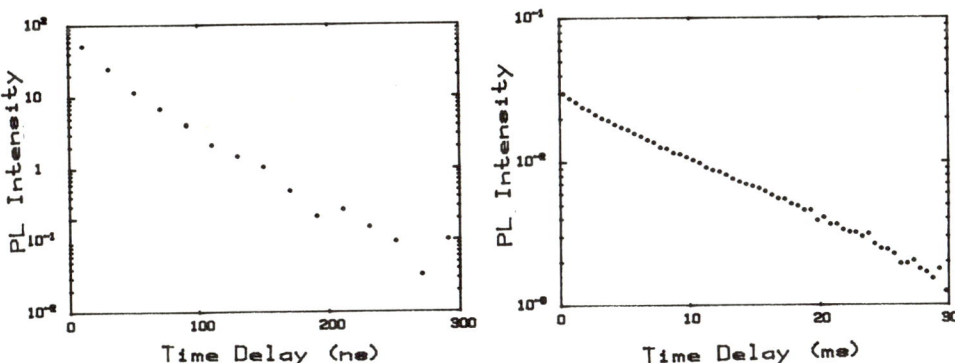

Fig. 5. The time decay of the 2.7 eV PL is composed of two decay rates.

In conclusion, we have measured PL time decay for annealed and neutron-irradiated a-SiO_2. Excitation of the PL by 7.8 eV radiation, unlike electron-beam and gamma-ray excitation, specifically excites a localized defect band (the E absorption band)[7] of SiO_2. The present results indicate that the PL process involves at least three PL bands which can all be excited by 7.8 eV radiation. In spite of the complexity of several PL bands, though, the time decay for two of the bands simply involves a couple of decay rates. A two decay rate process is reminiscent of energy transfer processes between ion species in solids[8], but at the present time, we have no evidence for this in a-SiO_2. Further investigation is necessary to determine the kinetics of the PL process. The PL time-dependence in a-SiO_2, although possibly unlike the PL time-decay of amorphous semiconducting chalcogenides which is characterized by a broad distribution[4] of radiative rates, is in itself interesting and may even be simple.

This work was supported by the Joint Services Electronics Program under Contract DAAG-29-79-C-0020.

REFERENCES

[1] C.M. Gee and M. Kastner, Phys. Rev. Lett. 42, 1765 (1979).

[2] C.M. Gee and M. Kastner, J. Non-Crystalline Sol. 36, 927 (1980).

[3] M. Kastner, D. Adler and H. Fritzsche, Phys. Rev. Lett. 37, 1504 (1976).

[4] G.S. Higashi and M. Kastner, J. Phys. C: Solid State Phys. 12, L821 (1979); G.S. Higashi and M. Kastner, J. Non-Crystalline Sol. 36, 921 (1980).

[5] Heraeus-Amersil trade name for OH-free fused silica.

[6] These bands are also observed by electron-beam excitation of bulk fused silica and SiO_2 films grown thermally on silicon substrates. Refer to S.W. McKnight and E.D. Palik, Proc. of the Fifth University Conference on Glass Science, August 1979.

[7] E.W.J. Mitchell and E.G.S. Paige, Phil. Mag. 1, 1085 (1956); C.M. Nelson and R.A. Weeks, J. Appl. Phys. 32, 883 (1961).

[8] G.F. Imbush, Phys. Rev. 153, 326 (1967).

ELECTRON-BEAM-INDUCED LUMINESCENCE IN SiO_2

S. W. McKnight*, Naval Research Laboratory, Washington, DC

ABSTRACT

Luminescence induced by 100-3000V electron beams (cathodoluminescence) has been studied in thermally grown SiO_2 films and bulk silica. In addition to the commonly observed emission lines at 1.9, 2.7, and 4.4 eV, an additional broad emission was observed in the vacuum ultraviolet. This band is centered at 6.6 eV with a width of approximately 2 eV and a high-energy tail which extends to 10 eV. In contrast to the 4.4 and 2.7 eV lines which fatigue and decrease in intensity on cooling to 80 K, the 6.6 eV line appears insensitive to temperature. Studies of the dependence of luminescence on electron-beam penetration depth indicate a region of about 150 Å at the film-vacuum interface where the 2.7 and 4.4 eV luminescence is strongly inhibited. This effect may indicate the presence of a "dead layer" with a decreased density of luminescence centers near the film surface. The 6.6 eV line does not appear to have a comparable dead layer.

INTRODUCTION

A great deal of luminescence data have been collected on SiO_2 in bulk crystals, bulk fused silica, and thin films.[1] In general, all of these experiments have been carried out either with photon excitation (photoluminescence) or with high energy electron-beam or radiation excitation (cathodoluminescence or radioluminescence). In the visible and UV bands are commonly observed at about 1.9, 2.7, and 4.4 eV that are probably defect-related. There has been relatively little work in the vacuum ultraviolet, and no reported observations extending as far as the reported band gap in SiO_2 (9-10 eV). The present work represents the results of luminescence data in the spectral range from 1.5 to 10.3 eV on bulk and thin film SiO_2 samples excited by low-to-moderate energy (100-3000 eV) electron beams.

Although electron momenta at these beam energies are generally insufficient to directly produce ionic displacements, a wide variety of exposure-dependent effects are observed, including fatiguing effects and the generation of additional luminescence centers. By varying the electron energy the excitation depth can be easily adjusted between 50 and 2000 Å to look for depth dependent effects in thermal oxide films. The first 150 Å of the film from the film-vacuum interface are observed to have a substantially lowered luminescence efficiency for the 2.7 and 4.4 eV emission. Finally, spectra taken in the vacuum ultraviolet on both bulk and film samples indicate a broad emission centered at about 6.6 eV and extending to the band edge at 10 eV. The temperature dependence and lack of a dead layer of this line suggest that the luminescence center is quite different from that responsible for the 2.7 and 4.4 eV lines.

*NRL-NRC Resident Research Associate

EXPERIMENT

Both wet and dry thermal oxides on silicon, and bulk fused silica (Suprasil) and crystal quartz samples were studied. Samples were mounted in an oil-free UHV system (base pressure = 1×10^{-8} Torr) on a cold finger capable of being heated or cooled to temperatures between 77 and 600 K. A Varian grazing-incidence electron gun provided an electron beam about 1 mm square at the sample at voltages from 100-3000 V and currents from 0.1-50 µA. Because the secondary emission coefficient at these voltages is greater than 1, charging even in bulk samples is limited to an insignificant fraction of the electron gun voltage. The low incident power and relatively large beam diameter resulted in negligible sample heating. The predominant energy loss mechanism of the incident electron beam is the production of electron-hole pairs. For typical beam parameters the excitation level is equivalent to 3×10^{22} band-gap excitations per second per cm.^{3} This is to be compared with an estimate of 1×10^{15} photon/sec-cm^3 for photoluminescence.[2] Because of the relative penetration depths, these low voltage cathodoluminescence experiments are in the high excitation limit even compared to most luminescence experiments utilizing electron microscopes at 10-20 kV.

The emitted light was detected with standard photon-counting equipment using either a visible-UV or a vacuum UV monochromator. In both cases the entire optical chain was calibrated by comparison with either a black-body tungsten filament or a standard deuterium lamp. In the vacuum UV the short wavelength cut-off was determined by the MgF_2 photomultiplier window although some data were taken with a wavelength-shifting phosphor. For convenience the corrected spectra are plotted as a function of energy rather than wavelength.

RESULTS AND DISCUSSION

Figure 1 shows the luminescence spectra obtained from a previously undamaged piece of bulk fused silica at room temperature. Each spectra was obtained in about one minute, and the total exposure time preceeding the spectra is indicated on each run. The evolution of the spectra represents the result of changes induced in the sample by beam exposure. Both the emission bands observed and the changes seen on exposure are similar to results obtained in thermal oxide films.[3] The most obvious changes include a threefold increase in the intensity of the line at 2.7 eV and a decrease in the line at 1.9 eV by a factor of 2. The 2.7 line intensity after full exposure was observed to decrease by about 30% on annealing at room temperature for 72 hours. As shown in the figure, annealing to 300C for 20 minutes completely removes the enhanced intensity of this line. The 1.9 eV line on the other hand is not restored by

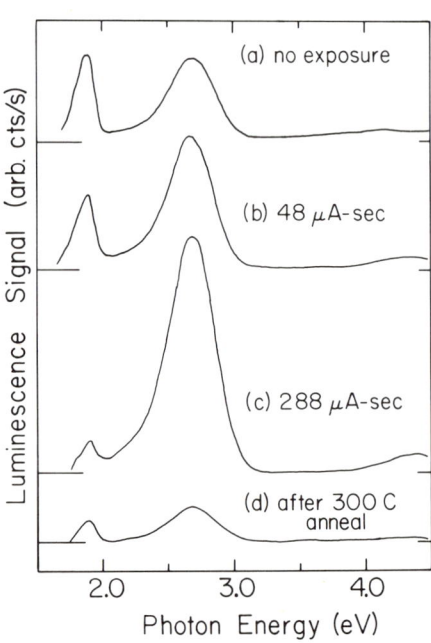

Fig. 1. Cathodoluminescence spectra of fused silica at room temperature with a beam voltage of 1000 V and current of 0.4 µA. The effect of beam exposure as indicated and 300 C anneal is illustrated.

annealing to 300C. The enhancement of the 2.7 eV line under beam exposure has been reported before and attributed to damage of the sample by the electron beam.[3,4] The recovery of this sample after an anneal at 300C shows that the damage center responsible for the 2.7 line must anneal out at this temperature. It is known from EPR work that a 300C anneal is sufficient to remove some radiation-induced paramagnetic centers but not ionic damage such as vacancies.[5] The annealing of the 2.7 eV line at 300C is consistent, therefore, with the suggestions that the line is related to the E' center - an unpaired electron trapped on a silicon dangling bond.[4,5] The 2.0 eV line, on the other hand, has been related to the presence of impurities - especially water[6] - and one explanation for its decrease and non-recovery after the anneal may be electron-beam stimulated desorption of water from the sample. Admitting water vapor to the chamber for short periods of time (1 hour) did not restore this line, however. Exposure at low temperatures resulted in entirely different behavior of the lines. As shown before, (Ref. 3) beam exposure caused the lines at 2.7 and 4.4 eV to fatigue and decrease in intensity dramatically in thermal oxides. Similar behavior was observed in bulk samples. The fatiguing was found to be exponential as a function of exposure (beam current times exposure time) and the lifetime was found to be the same as seen in the thin films when corrected for the different currents. The 2.7 eV line was not found to restore upon annealing at room temperature, and even annealing to 300C resulted in only restoring 75% of the initial intensity.

These exposure dependent effects can obscure the actual temperature dependence of the luminescence. By working at very low currents and short beam pulses we were able to observe that the fatiguing begins to dominate the damage mechanism in the 2.7 eV line at about 140 K and the intrinsic intensity of the luminescence increases by an order of magnitude from room temperature to 77 K.

Studies of luminescence intensity as a function of beam voltage (penetration depth) can give information about the distribution of luminescence centers. Figure 2 shows the results for the intensity of the 4.4 eV line as function of beam voltage at constant power, in a film 1.6 μm thick. Since the electron beam will not penetrate the film at the voltages shown, in the absence of any non-uniform distribution of luminescence centers, the intensity at constant power should be constant. The behavior seen demonstrates that the luminescence efficiency is very much less when the electron beam is confined close to the surface. Dependences of the emission on beam voltage have been observed in cathodoluminescence experiments on semiconductors and attributed to two effects: diffusion of carriers to the surface followed by surface recombination, and a

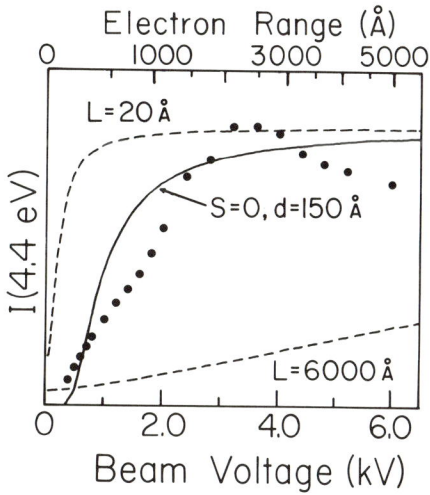

Fig. 2. Luminescence intensity of 4.4 eV line as a function of beam voltage (electron range) at constant power in a 1.6 μ film. Dashed lines are theoretical results for diffusion lengths as shown. The solid line is a calculation with no surface recombination and a 150 Å dead layer.

"dead layer" possibly associated with the surface space charge region near the surface where emission is inhibited. These two effects lead to substantially different effects, as shown in the calculations of Wittry and Kyser.[7] In the case of surface recombination dominated behavior, the shape of the intensity vs. voltage curve is a strong function of the diffusion length for the carriers. The dotted lines in Figure 2 represent the theoretical predictions of the Wittry-Kyser theory for a given reduced surface recombination velocity (S = surface recombination velocity/diffusion velocity = 20) and two limiting diffusion lengths; 6000Å, appropriate for electrons in SiO_2 and 20Å, an upper limit for the holes.

Neither of these curves adequately describes the data, and the fit is not improved by selecting any other value for the surface recombination velocity. The solid line, on the other hand, represents the theoretical curve if the surface recombination is neglected but a dead layer of 150Å is assumed. This curve reproduces the major features of the data. Above 3000V charging effects may be important and interference in the film has not been considered.

These results suggest that the decrease in luminescence efficiency near the front surface is not due to surface recombination with either electron- or hole-limited diffusion. (Exciton transport to the surface with a diffusion length of 400Å can give a reasonable fit to the data, however.) Another explanation is a region of perhaps 150Å near the film-vacuum interface in which luminescence is inhibited. The nature of any such dead layer in SiO_2 is not known. While presumably the surface space charge region of a semiconductor is not a factor in this insulating material, it may well be that the first 150Å of an SiO_2 film is different due to the indiffusion of some species (e.g., water) from the surface.

Such studies are more difficult for the other lines observed because of low intensity or, in the case of the 2.7 eV line, a non-linear behavior of luminescence intensity on current (Ref. 3). It appears, however, that the 2.7 eV line is also characterized by a dead layer of the same thickness as the 4.4 eV line. This is in agreement with other evidence that these two lines are related to the same defect site.

In addition to the previously discussed emission lines which are commonly observed in both cathodoluminescence and photoluminescence, an extension of this work to the vacuum ultraviolet indicates that there is emission extending into the VUV as far as 10 eV. In bulk samples, this emission appears as a shoulder on the 4.4 eV line, reduced in intensity by a factor of 30. The shoulder appears to be centered at about 6.6 eV and extends to the spectrometer cut-off at 10.3 eV. In very thin films where the 4.4 eV line is reduced by the dead layer, the VUV emission becomes comparable to the 4.4 eV line as is shown in Fig. 3. A weak additional feature is suggested at about 5.6 eV. When the sample is cooled to 80 K the line at 4.4 eV fatigues and the emission after the beam-exposure changes have saturated is reduced by one half. The 6.6 eV band, on the other hand, shows very little change in its steady-state intensity on cooling. We have also investigated the beam voltage dependence of this line and find that it shows little or no dead layer in contrast to the other emission lines.

A similar line was observed in radioluminescence measurements by Treadaway, et al.,[8] but their spectrometer cut-off masked the high energy wing of the band. The The extreme width (2.5 eV) of this band and the persistence of its tail to the band band gap, as well as the temperature and dead layer behavior, distinguish it remarkably from the other luminescence lines. These characteristics might be consistent with emission caused by the recombination of a self-trapped exciton. The apparent splitting of the emission line could be explained in this interpretation as recombination of different excitonic states. The 4.4 eV line, for instance, may represent an exciton recombination at the site of an oxygen vacancy as suggested by Griscom.[1] Further study of the exposure dependence and time dependence of this line would be very useful, although experimentally difficult because of the low intensity.

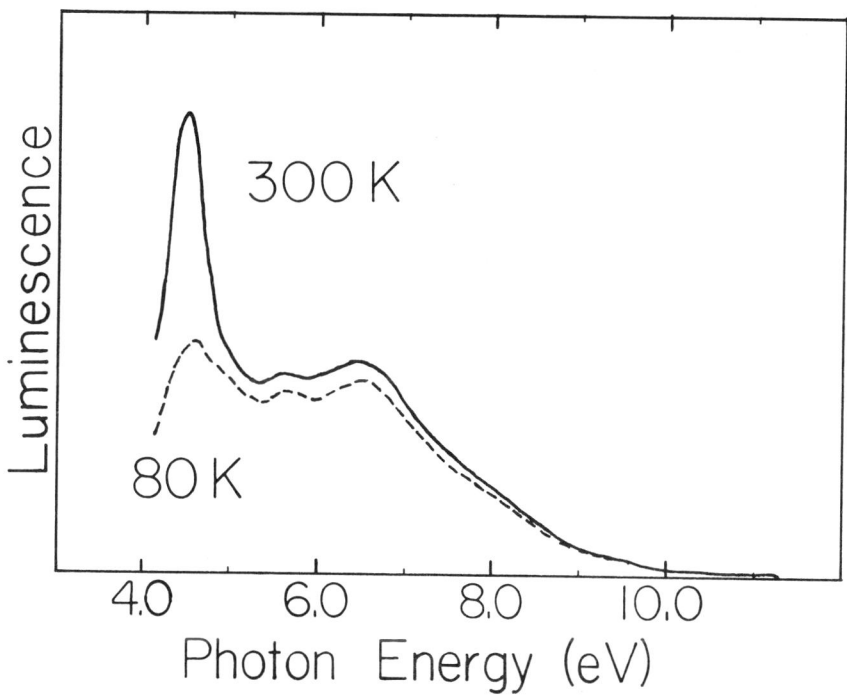

Fig. 3. Luminescence in the vacuum UV with electron beam voltage of 1000 V and current of 32 μA in a 133 Å film at two temperatures.

ACKNOWLEDGMENT:

The author would like to thank E. D. Palik for useful discussions, comments, and suggestions throughout the course of this work. Valuable discussions with R. T. Williams and D. L. Griscom are gratefully acknowledged.

REFERENCES:

1. See review of luminescence work by D. L. Griscom, Proc. 33rd Army Frequency Control Symposium, May 1979, 98.
2. C. M. Gee and M. Kastner, Phys. Rev. Letters 42,1765(1979).
3. S. W. McKnight and E. D. Palik, J. of Non-Cryst. Solids, in press.
4. J. P. Mitchell and D. G. Denure, Solid-State Electronics, 16,825(1973).
5. L. S. Korneienko, A. O. Rybaltovskii, and P. V. Chernov, Fizika i Khimiya Stekla, 2,396(1976).
6. G. H. Sigel, J. of Non-Crys. Solids, 13,372(1973).
7. D. W. Wittry and D. F. Kyser, JAP, 38,375(1967).
8. M. J. Treadaway, B. C. Passenheim and B. D. Ketterer, IEEE Trans. on Nuclear Science, NS-22,2253(1975).

PHOTOIONIZATION CROSS SECTION OF THE 2.5 eV ELECTRON TRAP IN SiO_2

D. D. Rathman, F. J. Feigl, S. R. Butler, and W. B. Fowler
Sherman Fairchild Laboratory, Lehigh University, Bethlehem, PA 18015

ABSTRACT

A modification of the photodepopulation technique has been used to investigate photodetrapping kinetics of a specific electron trapping center in "wet" thermal oxide films. The kinetic data were fitted by a simple first order model. The results are presented as an effective photoionization cross section $\sigma_p(\lambda)$ over the photon energy range 2-7 eV. $\sigma_p(\lambda)$ rises steeply above 2.5 eV, and exhibits a single peak ($\sim 10^{-16} cm^2$) at 3.5 eV. Above 4.0 eV, σ_p decreases slowly with increasing photon energy. With the exception of the peak, this general behavior agrees with calculated photoionization cross sections for coulomb traps in semiconductors, and with similar data for neutral traps in thermal oxide films.

INTRODUCTION

The trapping of electrons during transport across thermally grown silicon dioxide films was first reported by Williams in 1965.[1] Recently, trapped oxide charge effects have become the subject of intensive re-investigation stemming from the development of both 1 μm MOSFET VLSI technology and non-volatile memory technology. A comprehensive review of both electron and hole trapping in SiO_2 films has been presented and published by DiMaria.[2]

Of particular interest to us at Lehigh has been the study of the trapping-detrapping characteristics of a coulombic attractive trapping site with an energy level in the SiO_2 bandgap about 2.5 eV from the bottom of the conduction band.[3] This trapping site is believed to be related to sodium grown in the SiO_2 lattice in an immobile configuration at elevated temperatures (1000-1200 °C).[4] Through the use of photopopulation spectroscopy, the photoionization cross section of this trap has been determined to be approximately $10^{-17} cm^2$ with a threshold energy at approximately 2.5 eV.[3,5] In the present work we have determined the photoionization cross section for this trap to photon energies $h\nu$ well above threshold ($h\nu$ = 2-7 eV).

EXPERIMENTAL PROCEDURE AND DETAILS

Sample Fabrication

MOS devices were prepared on (100), 1-5 Ω-cm, n-type Si substrates. The SiO_2 layer was formed at 1230°C in a wet oxygen ambient, obtained by bubbling

O_2 through a distilled water boiler operated at 95°C. The SiO_2 film thickness d_{ox} was in the range 1.0-5.5 μm. The metal field plate was a semitransparent Al layer, 10-15 nm thick, formed by evaporation from a heated filament through stainless steel masks onto the SiO_2 surface. Thick Al contact pads were deposited over a small part of the field plate, and on the Si back surface. The area of the rectangular uncovered portion of the optically semitransparent Al field plate was 0.2-0.3 cm^2. This was the active device area for the measurements described below.

No post-oxidation or post-metallization annealing steps were performed prior to measurement. The oxidation conditions were designed to maximize incorporation of impurities (notably Na) leached from the fresh glassware of the oxidation system.

Experimental Arrangement

The basic experimental program is a series of electron injection-trapped electron photodepopulation measurements performed with a different light wavelength during the photodepopulation step. The equipment used was quite simple -- a Keithley electrometer operated as either a current meter or current integrator, and a bias supply (Fluke precision voltage source) are connected in series with the MOS device. The light sources used to illuminate the MOS oxide through the semitransparent Al field plate were: (1) a Xe arc lamp and monochromator system which, with appropriate UV grade optics, provides an effective source for the wavelength range 700 nm - 240 nm, (2) a Deuterium lamp with appropriate bandpass filters centered at 225 nm and 205 nm, and (3) a low pressure Hg lamp which, with filters, provided UV radiation at 254 nm and 185 nm. The intensity of the VIS-UV sources within the oxide films was estimated by direct measurement of the UV source intensity plus a correction for the transmittance of the Al film. This intensity is designated hereafter as a photon flux F_λ or $F_{h\nu}$, with units photons-$cm^{-2}-s^{-1}$. λ is the central wavelength for each source and $h\nu$ the corresponding photon energy.

Electron Injection-Photodepopulation

The 2.5 eV trapping states were filled by photoinjection of electrons from the Si substrate into the SiO_2 layer. Under positive bias (V > 0), light of energy $h\nu > E_b$ produces an internal photoemission current I across the oxide. E_b is the $Si-SiO_2$ interfacial potential barrier ($\lesssim$4.2 eV). Individual electrons in transport across the SiO_2 layer are captured into localized states. Not all of the available trapping sites are filled with electrons, however, because of photodetrapping occurring simultaneously with the photoinjection.

What does result in all cases is a steady state level of occupied traps which can vary from 1-95% depending on bias and illumination conditions. This steady state level is highly reproducible provided the illumination and internal field conditions (applied bias plus the field contribution provided by additional trapped charges in the oxide which are not part of the 2.5 eV trap distribution) can be maintained. In both the photodepopulation experiments to be described below, this reproducibility of trap population level for a given sample was critical. The unfiltered Hg lamp (254 nm + 185 nm lines) at an applied positive bias of $5-8 \times 10^5$ V/cm was used to produce injection currents on the order of 10^{-8} A/cm^2. An injection time of at least five minutes produced steady state levels of approximately 25%.

The photodepopulation experiments were performed at zero applied gate bias (V=0) after an injection step as just described. Under these conditions, electron injection from the Si or Al electrodes is suppressed by the large negative oxide space charge. The exciting light can therefore only remove trapped electrons from the oxide film. This will occur for $h\nu > E_T$ where E_T is the trap depth (2.5 eV).

For the present work, two variations of the basic photodepopulation measurement were carried out. For depopulation sources where $h\nu$ was less than 4.0 eV a direct-collected-charge (dcc) technique was used. For $h\nu$ greater than 4.0 eV a residual-collected-charge (rcc) technique was used.

dcc technique ($h\nu$ < 4.0 eV). After the traps were filled to steady state by photoinjection at positive bias, the bias was turned off and the sample illuminated at the designated depopulation wavelength. Using the Keithley electrometer in the coulombmetric mode, the collected charge resulting from integration of the depopulation current was recorded directly as a function of time.

rcc technique ($h\nu$ > 4.0 eV). In this method, the traps were again filled to steady state. This time, however, the zero bias depopulation with the desired illumination conditions continued for a predetermined length of time. This was followed by a 450 nm (2.75 eV) bleaching step in which the total 2.5 eV deep trapped charge remaining after the zero bias UV illumination was removed and measured. The traps were then refilled by photoinjection to the previous steady state level. The zero-bias photodepopulation, and the UV illumination step was then repeated, for an increased duration. The residual charge was again measured. By repeating this process one can determine the residual charge as a function of UV illumination time. This entire sequence was executed for different UV wavelengths.

We should note that the dcc technique was used whenever possible, simply because of the greatly reduced time involved. We did verify that both techniques, when feasible, produced consistent results.

RESULTS AND ANALYSIS

Two typical direct-collected-charge curves are shown in Fig. 1 for depopulation wavelengths of 381 nm (3.25 eV) and 450 nm (2.75 eV). Note that although the rates are different, the total amount of 2.5 eV charge collected is the same. This demonstrates directly the assumption of reproducibility of the steady state level of trapped charge discussed earlier. The points in Fig. 1 represent a model fit to the data. This is described below.

In Fig. 2, a few typical residual-collected-charge results are shown. The data are indicated by discrete points, while the solid lines represent the fitted curves.

The data of Figs. 1 and 2 were interpreted using a single trap model and first order kinetics (no retrapping of the photoexcited electrons). This model was formulated mathematically as follows: the residual charge is

$$\frac{1}{e} Q_{ot} \equiv N_{ot}(t) = N(0) \, e^{-t/\tau_p} \, . \tag{1}$$

More precisely, $N_{ot}(t)$ is the areal density of total centroid weighted[6] oxide trapped charge Q_{ot} associated with the 2.5 eV distribution at time t during

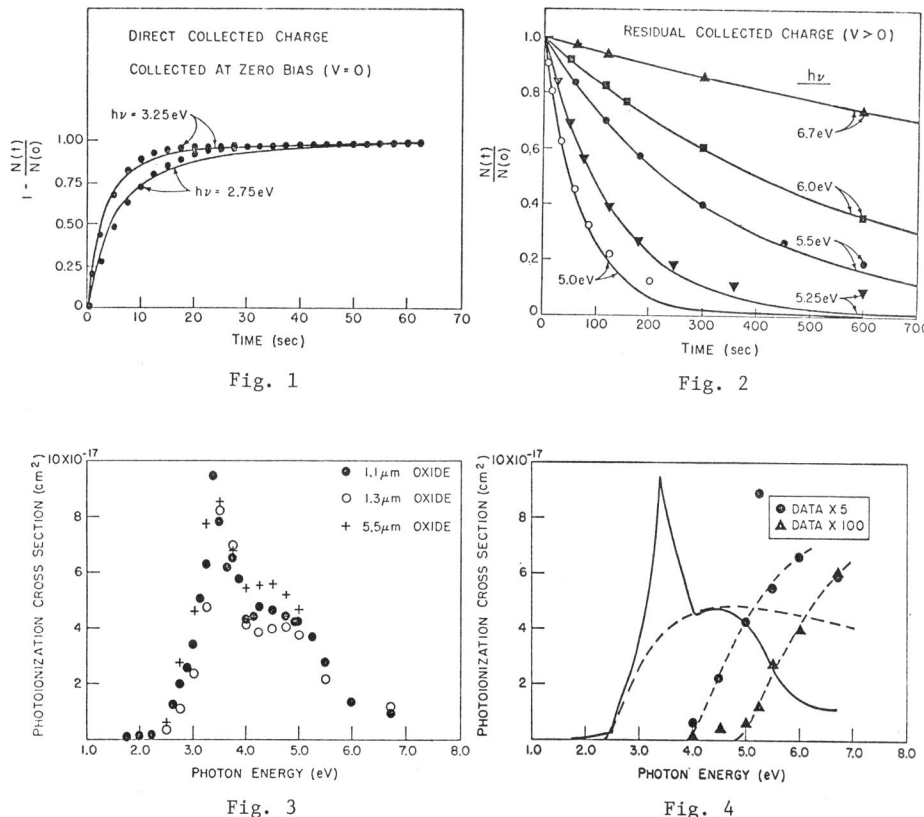

Fig. 1

Fig. 2

Fig. 3

Fig. 4

a given depopulation sequence with fixed illumination (characterized by λ or $h\nu$). t is the total <u>accumulated</u> illumination time under zero bias. N(0) is the areal density of electron-occupied traps at time t = 0 (the steady state trap population discussed above). The trap emptying rate $1/\tau_p$ will, in general, vary with photon energy.

The solid lines in Fig. 2 and the points in Fig. 1 represent statistical best fits of equation (1) to the experimental data. The important result of the fitting program is a determination of the detrapping rates τ_p^{-1}. From these results an effective photoionization cross section σ_p can be defined as follows:

$$\frac{1}{\tau_p} = F_\lambda \sigma_p . \qquad (2)$$

F_λ is the photon flux for exciting light of wavelength λ. The effective photoionization cross section thus obtained is plotted as a function of photon energy $h\nu$ in Fig. 3. Data from several samples are shown.

DISCUSSION

The simple analysis employed above is appropriate to the present experiments. Spatial variations of trapped charge density and/or light intensity within the oxide layer can make analysis of photodepopulation experiments quite complex in general.[7] However, the 2.5 eV trapping center distribution has been shown to be spatially uniform. Also, we have used very thick oxide films (>1 μm) and broadband sources (>20 nm bandpass in the VIS-UV spectral region). These conditions minimize the difficulties associated with optical interference effects within the three layer MOS structure. The results in Fig. 3 show reasonable agreement on the major features of the photoionization curves for all three samples.

The results in Fig. 3 indicate a trap photoionization threshold E_T in the range 2.4 to 2.6 eV, which is consistent with previous results obtained by DiMaria[3] and Kapoor.[5] The important contribution of the present work is the determination of the photoionization cross section for photon energies well in excess of this threshold value. In all samples studied, there is a clearly defined peak approximately 1 eV above threshold, followed by a gradual decline. There may be in addition a broad peak centered approximately 2 eV above E_T. However, this region of the photon energy spectrum (4-5 eV) is complicated by variations in the Si reflectivity.[9] An averaged interpolation of the results in Fig. 3 is presented in Fig. 4 (solid line).

In addition to the 2.5 eV trapping center, the thermal oxide films used in the present investigation contain two additional electron trapping centers which can be photodetrapped with 4-7 eV light. Photoionization cross section results[10] for these two centers are also shown in Fig. 4, as solid circles and solid triangles. The dash lines in Fig. 4 are graphical fits of the cross section model of Lucovsky[11] to these data. We estimate the photoionization thresholds E_T for the three trapping centers to be 2.4, 3.8, and 4.9 eV. We suggest that the cross section peak ∿1 eV above E_T is produced by a conduction band density of states peak, as calculated by Fowler, et al.[12]

ACKNOWLEDGEMENT

This work was supported by the Office of Naval Research, Electronic and Solid State Sciences Program. The permanent address of D. D. Rathman is MIT Lincoln Laboratory, Lexington, MA 02173.

REFERENCES

1. R. Williams, Phys. Rev. 140, A569 (1965).
2. D. J. DiMaria (1978) in The Physics of SiO$_2$, edited by S. K. Pantelides, Pergamon Press, New York, pp. 160-178.
3. D. J. DiMaria, F. J. Feigl, and S. R. Butler, Phys. Rev. B11, 5023 (1975).
4. S. R. Butler, F. J. Feigl, Y. Ota, and D. J. DiMaria (1976) in Thermal and Photostimulated Currents in Insulators, edited by D. M. Smyth, The Electrochemical Society, Princeton, NJ, pp. 149-161.
5. V. J. Kapoor, F. J. Feigl, and S. R. Butler, Phys. Rev. Letters 39, 1219 (1977).
6. B. E. Deal, J. Electrochem. Soc. 127, 979 (1980).
7. R. F. DeKeersmaecker, D. J. DiMaria, and S. T. Pantelides (1978) in The Physics of SiO$_2$, edited by S. T. Pantelides, Pergamon Press, New York, pp. 189-194.
8. V. J. Kapoor, F. J. Feigl, and S. R. Butler, J. Appl. Phys. 48, 739 (1977).
9. H. R. Philipp, J. Appl. Phys. 43, 2835 (1972).
10. D. D. Rathman, F. J. Feigl, and S. R. Butler (1980) in Insulating Films on Semiconductors 1979, edited by G. G. Roberts and M. J. Morant, The Institute of Physics, London, pp. 48-54.
11. G. Lucovsky, Solid State Commun. 3, 299 (1965).
12. W. B. Fowler, P. M. Schneider, and E. Calabrese (1978) in The Physics of SiO$_2$, edited by S. T. Pantelides, Pergamon Press, New York, pp. 70-74.

HYDROGENATION OF AMORPHOUS SILICON NITRIDE*

H. J. Stein, P. S. Peercy and D. S. Ginley
Sandia National Laboratories[†], Albuquerque, New Mexico 87185 USA

ABSTRACT

A relationship between hydrogen and the equilibrium positive charge in silicon nitride films has been confirmed by rehydrogenation of annealed films using a hydrogen plasma. The primary relation between hydrogen and silicon-nitride charge is associated with Si-H formation.

INTRODUCTION

Previous experiments have shown that several atomic percent hydrogen is incorporated in chemical vapor deposited (CVD) silicon nitride (SiN_x) films prepared for MNOS (Metal or Si-gate/nitride/oxide/Si) memory device applications.[1-4] The total hydrogen concentration and the relative concentrations of N-H and Si-H bonds are dependent on the ammonia:silane flow ratio and deposition temperature.[1,4] Subsequent heating of films above the deposition temperature, which is typically 700 to 900°C, results in hydrogen loss by trap-limited diffusion.[4] Depositions of SiN_x for p-channel MNOS are usually performed at temperatures between 700 and 750°C because there is more equilibrium positive nitride charge, Q_{NC}, at these lower temperatures which gives negative equilibrium flat-band and threshold voltages. However, post deposition annealing at either MOS or Si-gate processing temperatures causes degradation of SiN_x properties, particularly for films deposited at these lower temperatures.

Parallels in the dependence of Q_{NC} and hydrogen on deposition and annealing characteristics have been noted previously;[1,5] however, many simultaneous effects occur during deposition and annealing. Methods for introducing hydrogen into SiN_x with minimum disturbance of the structure are thus required to establish the relationship between Q_{NC} and hydrogen. In the present paper we report studies in which hydrogen was introduced into annealed SiN_x films by exposure to a low pressure hydrogen plasma at temperatures between 300 and 700°C. Infrared absorption, nuclear reaction analysis, and capacitance-voltage measurements were used to investigate the reintroduction of hydrogen and to determine the relationship between Q_{NC} and hydrogen in SiN_x.

EXPERIMENTAL DETAILS

Silicon nitride films were deposited onto n-type Si substrates from NH_3 and SiH_4 mixed with N_2 carrier gas in an AMV 1200 reactor. Flow ratios of 100:1 and 200:1 for $NH_3:SiH_4$ and substrate temperatures between 700 and 775°C were used to achieve

---

*This article sponsored by the U. S. Department of Energy under Contract DE-AC04-76-DP00789.

[†]A U. S. Department of Energy facility.

the desired Q_{NC} and chemical bonding of hydrogen. Isochronal (20 min.) annealing was performed in flowing N_2 at temperatures between 750 and 975°C, and radiative heating of samples in a tesla-coil-generated hydrogen plasma at $\sim$ 2 torr was used for rehydrogenation of SiN_x after annealing.

Nuclear reaction analysis (NRA) with the $^1H(^{15}N,(\alpha,\gamma)^{12}C$ reaction (depth resolution of $\sim$ 100 Å) was used to measure the hydrogen depth distribution and concentration before and after plasma hydrogenation. Chemically-bonded hydrogen was determined from Si-H and N-H stretch modes measured by the multiple internal reflection (MIR) technique, and Q_{NC} (assumed proportional to flat-band voltage) was determined from 1 MHz capacitance-voltage (C-V) measurements which were made by using a mercury probe on SiN_x films deposited on Si MIR plates.

RESULTS AND DISCUSSION

Chemically-bonded hydrogen and the C-V characteristics after successively higher annealing temperatures are shown in Fig. 1 for 1000 Å of SiN_x deposited from a 100:1 $NH_3:SiH_4$ ratio at 750°C onto a <100> Si (MIR) plate. The similarity between the reduction of chemically-bound hydrogen and the shift of the C-V characteristic toward zero (reduction of Q_{NC}) is apparent from these data.

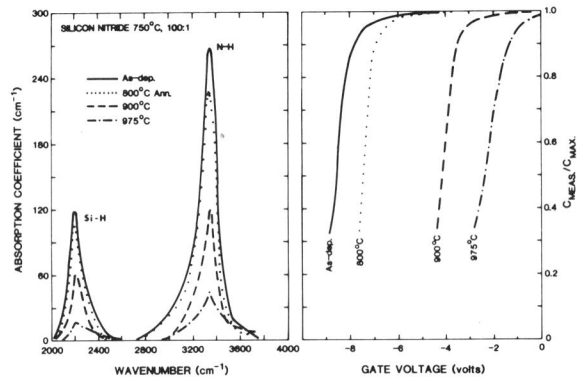

Fig. 1. Si-H and N-H absorption and C-V characteristics for CVD SiN_x/Si after annealing at successively higher temperatures.

After annealing at 975°C, the SiN_x was hydrogenated by exposure to a hydrogen plasma at 650°C for 17 hrs. and then given a 20 min. post-hydrogenation anneal at the deposition temperature of 750°C. Pre- and post-hydrogenation results are compared in Fig. 2. The comparison clearly shows rebonding of hydrogen during plasma exposure. No hydrogenation occurred in an exposure to H_2 at 650°C in the absence of a plasma. The sharpening of the C-V characteristic produced by hydrogenation indicates hydrogen passivation of interface states, and the large negative shift of the C-V characteristic substantiates a relationship between Q_{NC} and hydrogen.

Effects of temperature during plasma treatment are illustrated in Figs. 3 and 4. The data in Fig. 3 show that the intensities of Si-H and N-H absorption bands are largest after 300°C hydrogenation and decrease with increasing temperature. After 750°C annealing, however, the maximum intensities for the bands occur for films hydrogenated at 650°C. These apparently contradictory temperature effects are explained by the hydrogen concentrations and profiles measured by NRA as shown in Fig. 4. The NRA measurements were performed on 800 Å films annealed at 900°C prior

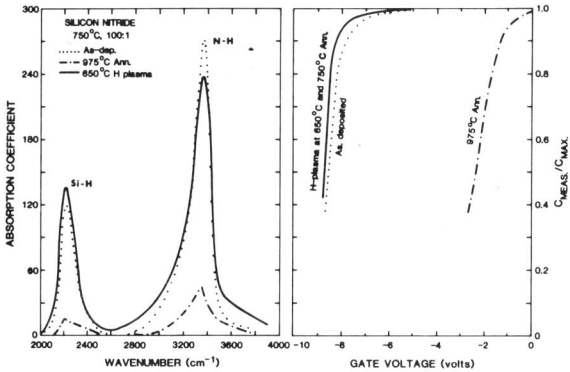

Fig. 2. Chemically bound hydrogen and C-V characteristic after 650°C plasma hydrogenation of annealed SiN_x (solid line). The as-deposited and final anneal data from Fig. 1 are included for comparison.

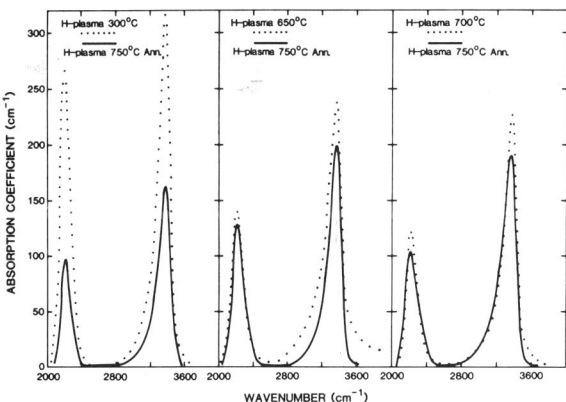

Fig. 3. Plasma treatment temperature effects on hydrogenation of annealed SiN_x as measured by Si-H and N-H absorption before and after 750°C anneal.

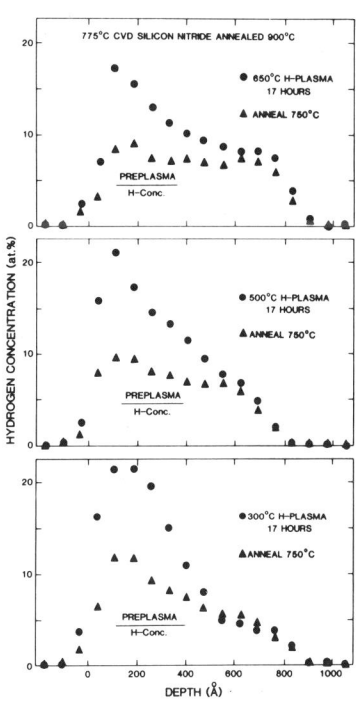

Fig. 4. Effect of plasma treatment temperature on hydrogenation of annealed SiN_x as measured by nuclear reaction analysis before and after 750°C annealing.

to hydrogenation. As can be seen from the data in Fig. 4, hydrogenation at 300°C produces a high concentration of hydrogen near the front surface. Back diffusion at 750°C of this near-surface hydrogen is a major factor in the large decrease in N-H and Si-H intensities produced by annealing. Little or none of the hydrogen introduced in the plasma at 300°C migrates through the film and therefore has little effect on Q_{NC}. On the other hand, hydrogenation at 650°C followed by annealing at 750°C gives a nearly uniform profile with approximately the as-deposited hydrogen concentration and restores the as-deposited C-V characteristic.

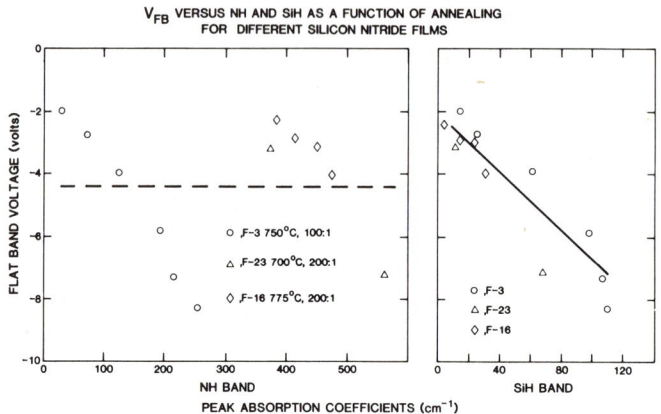

Fig. 5. Flat-band voltage normalized to 1000 Å film thickness plotted versus peak intensities for N-H and Si-H for SiN_x films which have a wide range of N-H and Si-H concentrations.

Although rehydrogenation of SiN_x by exposure to a hydrogen plasma and a dependence of Q_{NC} on hydrogen are well demonstrated by the results presented above, these data do not distinguish between Si-H and N-H centers for altering Q_{NC}. Plotted in Fig. 5 are flat-band voltage V_{FB} (voltage for $C = 0.7\ C_{meas}$) versus N-H and Si-H intensities obtained from annealing SiN_x deposited under conditions which yield a wide range in N-H and Si-H concentrations. These results show the same V_{FB} (and hence the same Q_{NC}) for a factor of four increase in N-H concentration which occurs when increasing $NH_3:SiH_4$ from 100:1 to 200:1. In contrast, the V_{FB} versus Si-H data show a functional similarity for the different SiN_x materials. From these results we conclude that the primary hydrogen-Q_{NC} relation occurs via the Si-H centers.

ACKNOWLEDGEMENT

The authors are indebted to Mary Mitchell and Doug Weaver of the Sandia Microelectronics Laboratory for providing the silicon nitride films.

REFERENCES

1. H. J. Stein, J. of Electron. Mater. 5, 161 (1976); H. J. Stein and H. A. R. Wegener, J. of Electrochem. Soc. 124, 908 (1977); and H. J. Stein and V. A. Wells, Electrochem. Soc. Extended Abstracts 77-1, 303 (1977).

2. F. I. Belyi, F. A. Kuznetsov, T. P. Smirnova, L. V. Chramova and L. K. Kravchenko, Thin Solid Films 37, 139 (1976).

3. G. Stubnya, I. C. Szep, G. Hoffmann, Z. Horvath and P. Tutto, Revue De Physique Appliquee 13, 679 (1978).

4. P. S. Peercy, H. J. Stein, B. L. Doyle and S. T. Picraux, J. Electron. Mater. 8, 11 (1979).

5. R. Hezel, J. Electron. Mater. 8, 459 (1979).

CHAPTER IV
OXIDATION - Si

INITIAL OXIDATION OF ION-SPUTTERED SILICON (100)

D. L. Ellsworth and C. W. Wilmsen
Colorado State University, Fort Collins, Colorado 80523

ABSTRACT

The initial growth of oxide on silicon disordered by ion sputtering has been investigated by AES sputter profiling and XPS oxide/element intensity ratio measurements. Sputtered silicon surfaces were oxidized in-situ at O_2 pressures from 10^{-6} to 10^{-4} Torr for varying times and temperatures in order to observe the resulting growth curves. It was observed that the oxide growth rate decreased with increasing sputter energy. The oxidation rate also decreased with elevated oxidation temperature for O_2 pressures up to 10^{-5} Torr. At a pressure of 10^{-4} Torr the oxidation rate increased with temperature. UPS measures indicated that oxidation reduced the surface state density caused by sputtering.

INTRODUCTION

The importance of inert ion sputtering to both research and manufacturing applications has for several years stimulated considerable interest in the nature of sputter etched surfaces.[1-7] The difficulty of obtaining a clean ion beam has been observed by several workers as an impediment to distinguishing characteristics intrinsic to the sputtered surface.[2,4,5,8] None the less, application of the technique has become quite common e.g. sputter profiling for Auger, XPS, and SIMS surface analysis, fine line pattern etching of semiconductor devices and for surface cleaning during solar cell fabrication.

The present investigation attempts to characterize sputtered silicon surfaces in terms of their oxidation kinetics and photoemission density of states. The oxidation kinetics study requires an accurate method of measuring the effective thickness of the ultra-thin oxides. This is a difficult problem which we have addressed using three different thickness evaluation techniques. The XPS intensity ratio technique[9-11] using the $[Si(2p)]_{Si}$ and $[Si(2p)]_{SiO_2}$ lines was coupled with digital curve resolving and integration procedures. A second means by which in-situ thickness evaluation was performed used AES sputter profiling[3] with pre-determined sputter etch rates. Because surface excitation by electron or ion beams is known to induce profound changes in the initial oxidation of cleaved surfaces[12,13], it was of interest to compare the two procedures. Use of derivative Auger peak-to-peak intensity ratios[5,12] was avoided since the presence of multiple bonding states[23] which tend to broaden the N(E) spectrum have the opposite effect on the derivative peak-to-peak height. Some efforts were made to utilize component resolved N(E) and dN(E)/d(E) Si LVV Auger lines as intensity measures, but the complexity of the line shape and background has not permitted reliable application of the simple models so far attempted.[14,15]

The third thickness evaluation method was ellipsometry of air and furnace grown oxides. As will be discussed, care must be exercised when comparing room-temperature oxide ellipsometric measurements with oxide coverage determined by spectroscopic techniques. However, by application of a combination of measurements to the same oxide, correlations of the various measurements were obtained.

The removal of electronic states from the Si bandgap by O_2 exposure[16,17] is the key to both thick oxide[18] and thin oxide[19] device technologies. To observe electronic behavior of disordered Si, initial UPS measurements were carried out on both the freshly sputtered surface and ones having undergone in-situ re-oxidation.

EXPERIMENTAL

The experimental system consisted of an ion-pumped UHV chamber configured for XPS, AES, UPS and ion sputtering, all without need of re-alignment of the sample. A PHI double pass CMA electron analyzer operated in retarding potential mode was used to obtain the XPS and UPS spectra. The ion gun is a PHI model 04-303 differential rastered ion source. The sample surface is sputtered with ultra-pure Ar at a source internal pressure of 10^{-4} T. During ion source operation, the base chamber pressure of 3 to 5 x 10^{-10} T rose to 2.2 x 10^{-7} T. AES spectra were obtained with electron beam currents of 10 uA at a beam voltage of 3 kV. X-ray radiation of 1486.6 eV was produced by an Al anode (K_α) at 10 kV through an Al filter. For UPS measurements, photons were generated by a differentially pumped He source capable of operation at chamber pressures less than 10^{-9} T. Photons were incident at an oblique angle so that angular emission effects are averaged.

Cleaned Si (100) samples were prepared as oxide thickness standards by both room temperature oxidation and 700°C furnace exposure to dry O_2. Sample wafers were immediately cleaved in two, with one half inserted in the UHV system, and the other ellipsometrically measured for oxide thickness. As has been observed previously,[11] no agreement is obtained between ellipsometric measurements of thin air grown oxides and calculations for XPS $[Si(2p)]_{ox}$ and $[Si(2p)]_{Si}$ ratio vs. thickness. However, it was observed that air grown oxides, while initially exhibiting an extremely low effective oxide thickness as measured by intensity ratio, nevertheless sputter profiled spectroscopically like much thicker oxides. Within a few seconds of application of the ion beam, curve resolved intensity ratios of these oxides increased to values much closer to those expected from ellipsometry. This effect is shown in Fig. 1 for an air grown oxide measured ellipsometrically at 18Å.

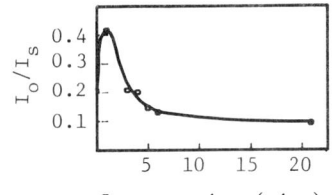

Fig. 1. I_o/I_s vs. sputter time for an air grown oxide measuring 18Å.

The broad oxide Si(2p) peak used in the intensity calculations only exhibited a shift from the elemental peak $\Delta E = 1.5$ eV. As discussed by Bauer et al.[20,21,22] for low oxide coverages at room temperature, the oxide Si(2p) peak is extremely broad on cleaved Si, apparently with multiple components. One component of their higher resolution data is in evidence at approximately the same ΔE observed in the present study. The original air grown oxide of Fig. 1 initially displayed a considerably larger ΔE (~3.1 eV), broadening and shifting to 1.5 eV with sputtering.[7] Within our system resolution, further sputtering is not evidenced by further shifts. With the relatively broad x-ray excitation energy used in our work, these shifts are presumably averages over multiply oxygen-coordinated Si. Continuous-sputtering AES Si LVV profiles were performed on thicker furnace grown oxides, for which XPS intensity ratio measurements agree with ellipsometry values. The etch rate for established ion beam conditions was thus determined. Oxidation of sputtered silicon was carried out in-situ on silicon wafers sputtered continuously for about 10 minutes after all AES evidence of oxide had disappeared. The oxidations were performed with the ion pump valve closed and all instruments off except the cold cathode gauge which was remote from the sample. Heat was supplied by a filament in a gas-filled quartz envelope.

RESULTS

The effect of variation in ion pre-sputter energy on oxide growth is shown for 1 KeV and 5 KeV ions in Fig. 2. It may be noted that the oxide thickness increases approximately logarithmically as has been reported for cleaved surfaces. Oxide thicknesses were obtained by XPS intensity ratios without post-oxidation particle excitation of the surface. Oxygen pressure was held at 1×10^{-5} T for all these exposures. It was determined that the shifted component of the Si(2p) line for the sputtered surface oxidations and for the furnace grown oxides was located at $\Delta E = 1.5$. From the data of Fig. 2, it appears that the surface sputtered at 5 KV oxidizes more slowly than the surface sputtered at 1 KV. At present we have not determined the cause for dependence of oxidation rate on sputter energy, however the separation of the curve can be affected by energy dependence of the atomic dilution of the surface layers by implanted argon,[3,5] and by surface roughening.[12]

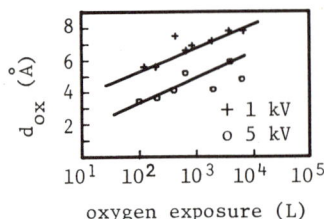

Fig. 2. Oxidation at 10^{-5} T, 22°C.

Figure 3a illustrates that the oxidation rate of a sputtered surface at 375°C is lower than at 22°C for an O_2 exposure pressure of 10^{-5} T. An intermediate rate is observed for similar measurements at 250°C. At an O_2 exposure pressure of 10^{-4} T it was found that the oxidation rate increased with temperature. Several factors may be controlling the initial oxidation of the sputtered surface and causing the reversal of the rate vs. temperature curves. First, the surface damage may be annealing and thus reducing the density of growth sites. While 250°C is a low temperature, thermally stimulated reconstruction of cleaved surfaces has been observed during anneals above 200°C.[24] A second consideration is the desorption of oxygen from the surface. At low pressure and elevated temperature, this effect may reduce the surface coverage of oxygen below that required to support a high oxidation rate, as discussed in the following. The oxidation rate was found to increase if the sputtered surface was exposed to an electron beam. This is seen by comparing the data of Figs. 3a and 3b.

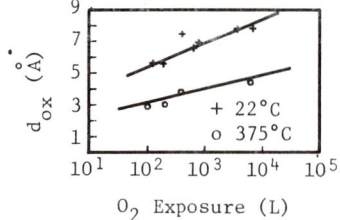

Fig. 3a. 10^{-5} T oxidation, 1 KV pre-sputter.

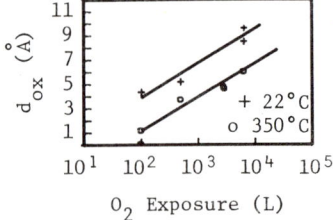

Fig. 3b. 10^{-5} T oxidation rates measured by AES sputter profiling, 1 KV pre-sputter.

A dependence of oxidation rate on pressure has been presented in Figs. 4a,b,c. The time-pressure reciprocity implied by plotting against exposure is not supported by these results at either 22°C or 375°C. Garner et al.[25] have reported that

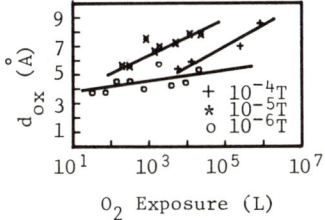

Fig. 4a. Room temperature oxidation, 1 KV pre-sputter.

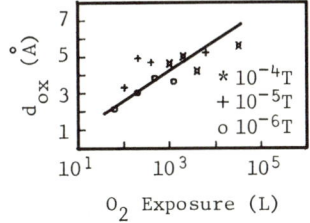

Fig. 4b. Oxidation at 250°C, 1 KV pre-sputter.

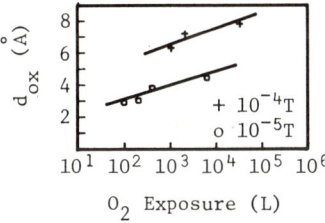

Fig. 4c. 375°C oxidation, 1 KV pre-sputter.

conditions of low pressure exposure exist which do not induce core level energy shifts for Si. Such exposures would therefore not contribute to oxide thickness measurements by this technique. Interestingly, at 250°C, Fig. 4b, reciprocity is observed over two orders of magnitude pressure up from 10^{-6} T. In Fig. 4a, room temperature, it is apparent that the oxidation rate is displaying pressure independence at 10^{-5} T and above, indicating that the surface is saturated with adsorbed oxygen. At 10^{-6} T however, saturation has not been reached, and the rate limiting step in producing Si-O bonds is apparently either the transport of oxygen onto the surface, or the generation of dissociated oxygen atoms.[26,27] In Fig. 4c it can be seen that for higher temperature, the high pressure curve increases and the low pressure curve decreases. This is superficially more consistent with the concept of thermal de-sorption effects mentioned above. The reciprocity of time and pressure demonstrated at intermediate temperature appears to be an artifact of competing processes.

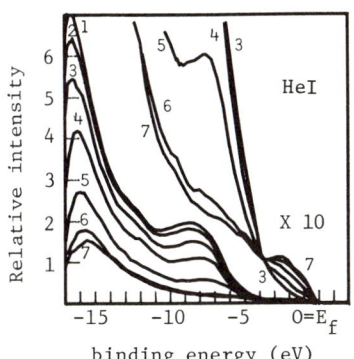

Fig. 5. Photoemission spectra for thin SiO_2 sputtered for varying times at 3 KeV ion energy. $t_1=0$, $t_2=1$ sec, $t_3=3$ sec, $t_4=10$ sec, $t_5=30$ sec, $t_6=2$ min.

Several photoemission spectra (UPS) are shown in Fig. 5 for an air grown oxide as sputtering progresses for 10 minutes at 3 KeV ion energy. Binding energies are referenced to the Fermi level for gold foil. Comparison with similarly obtained spectra which have appeared in the literature for cleaved surfaces[28,29,30] identifies the features at 2-3 eV below the Fermi level as due to dangling bond surface states.[31]

It was of interest to determine the extent to which the observed process of growing dangling bond surface states was reversible upon reoxidation of the surface at room temperature. The UPS data of Fig. 6 shows the reduction of surface state density of the sputtered surface with oxidation. Substantial reduction in the region within 2 eV of the Fermi level occurs at exposures as low as 100 L, exposed at 10^{-6} T. Further work is in progress to determine if device quality oxides can be produced from such a surface.

SUMMARY

The results of this investigation have been applied to the measurement of in-situ low pressure oxidation rates for ion sputtered Si (100). Growth rates were

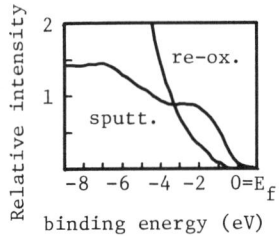

Fig. 6. Photoemission spectra for sputtered and re-oxidized surfaces.

observed to decrease for higher ion bombardment energies and for temperatures elevated to 375°C. Pressure dependence of oxidation rates has been considered from 10^{-6} to 10^{-4} T, indicating rate saturation with pressures above 10^{-5} T at room temperature. Photoelectron emission spectra show a reversibility of the trend toward increased surface state densities resulting from sputtering. Upon re-oxidation of the surface, spectra are obtained which approach those resulting from thick, furnace grown oxides.

ACKNOWLEDGEMENT

This work was supported by SERI and ONR.

REFERENCES

1. P. Sigmund, Phys. Rev. 184, 383 (1969).
2. C. C. Chang, P. Petroff, G. Quintana and J. Sosniak, Surf. Sci. 38, 341 (1973).
3. J. S. Johannessen, W. E. Spicer and Y. E. Strausser, J. Appl. Phys. 47, 3028 (1976).
4. H. R. Deppe, B. Hasler and J. Höpfner, Solid State Electron. 20, 51 (1977).
5. G. E. McGuire, Surf. Sci. 76, 130 (1978).
6. A. R. Neurenther, C. Y. Liu and C. H. Ting, J. Vac. Sci. Tech. 16, 1767 (1979).
7. F. J. Grunthaner, P. J. Grunthaner, R. P. Vasquez, B. F. Lewis and J. Maserjian, J. Vac. Sci. Tech. 16, 1443 (1979).
8. P. Morgen and F. Ryborg, J. Vac. Sci. Tech. 17, 578 (1980).
9. R. Flitsch and S. I. Raider, J. Vac. Sci. Tech. 12, 303 (1975).
10. S. I. Raider and R. Flitsch, IBM J. Res. Develop. 22, 294 (1978).
11. S. I. Raider, R. Flitsch and M. J. Palmer, J. Electrochem. Soc. 122, 413 (1975).
12. H. Ibach, K. Horn, R. Dorn and H. Lüth, Surf. Sci. 38, 433 (1973).
13. H. Ibach and J. E. Rowe, Phys. Rev. B9, 1951 (1974).
14. D. E. Ramaker, J. S. Murday and N. H. Turner, J. Electron. Spec. 17, 45 (1979).
15. S. M. Goodnick, private comm.
16. F. Yndurain and J. Rubio, Phys. Rev. Lett. 26, 138 (1972).
17. T. Iizuka and T. Sugano, Jap. J. Appl. Phys. 12, 73 (1973).
18. R. R. Razouk and B. E. Deal, J. Electrochem. Soc. 123, 1573 (1979).
19. P. V. Dressendorfer, S. Lai, R. C. Barker and T. P. Ma, Appl. Phys. Lett. 36, 15 (1980).
20. R. S. Bauer, J. C. McMenamin, H. Petersen and A. Bianconi, Proc. Intl. Conf. SiO$_2$ and Its Interfaces, ed. by S. T. Pantelides (Pergamon, NY, 1978) p. 401.
21. R. S. Bauer and R. Z. Bachrach, J. Vac. Sci. Tech. 17, 509 (1980).
22. R. S. Bauer, J. C. McMenamin, R. Z. Bachrach, A. Bianconi, L. Johansson, and H. Petersen, Inst. Phys. Conf. Ser. No. 43: Physics of Semiconductors, 1978 (Inst. of Physics, London 1979), p. 797.
23. C. R. Helms, Y. Strausser, and W. E. Spicer, Appl. Phys. Lett. 33, 767 (1978).
24. W. Monch, P. P. Auer and R. Fedder, J. Vac. Sci. Tech. 16, 1286 (1979).
25. C. M. Garner, I. Lindau, C. Y. Su, P. Pianetta, J. N. Miller and W. E. Spicer, Phys. Rev. Lett. 40, 403 (1978).
26. R. Ghez and Y. J. van der Meulen, J. Electrochem. Soc. 119, 1100 (1973).
27. J. Blanc, Appl. Phys. Lett. 33, 424, (1978).
28. J. E. Rowe, H. Ibach, and H. Froitzheim, Surf. Sci. 48, 44 (1975).
29. H. Ibach and J. E. Rowe, Surf. Sci. 43, 481 (1974).
30. M. Chen, I. Batra, C. Brundle, J. Vac. Sci. Tech. 16, 1216 (1979).
31. D. K. Ferry, private comm.

FIXED SURFACE CHARGE DENSITY GENERATION AT THE INTERFACE OF ANODIC SIO$_2$-Si SYSTEMS

J. L. Martínez and E. Gómez
Instituto de Ciencias, Universidad Autónoma de Puebla
Apdo. Postal J-48, Puebla, México

ABSTRACT

Fixed surface charge density (Qss) generation at the interface of anodic SiO$_2$-Si systems has been investigated. Anodic SiO$_2$ layers have been grown on single crystal p-type silicon wafers in several molar concentrations of Potassium Nitrate (KNO$_3$) salt in Ethylene Glycol (C$_2$H$_6$O$_3$). Measurements of Qss in Aluminum-anodic SiO$_2$-Si (MOS) capacitors were made by using the C-V high frequency technique. The results of such measurements indicate that: the Qss centers are positively charged, the amount of Qss depends on the molar concentration of KNO$_3$, the Qss generation depends weakly on the anodization current, and Qss centers are slightly accompanied by fast surface state density generation.

INTRODUCTION

Interface studies of MOS structures have gained continuous interest since Grove and co-workers in 1965, showed the existence of surface state charge (Qss), of positive polarity in n and p-type thermally oxidized silicon, employing Capacitance versus Voltage (C-V) plots.[1-5] Interface studies in anodized silicon are not well documented in the literature. Boroffka[6] reported that anodic oxidation of Si results in a high density of surface states ($\sim 10^{13}$ cm^{-2}) that can be reduced by post-oxidation heat treatment in hydrogen. Revesz[7] using 0.04M of Potassium Nitrate (KNO$_3$) salt in N-Methylacetamide (NMA) as electrolyte measured interface properties of anodized silicon. He showed that low final anodizing current density is required for low surface state charge in the oxide. Beynon and co-workers[8], using KNO$_3$ and Potassium Nitrite (KNO$_2$) salt in ethylene glycol found that films grown using 0.04M KNO$_3$ are easily contaminated although they exhibit no hysteresis in C-V curves for high biasing voltages (representing fields up to 5×10^5 v/cm); on the other hand, films grown using 0.04M KNO$_2$ are less affected by contamination but show large hysteresis for high gate biasing voltage. They subsequently experimented with a mixture of 0.02M KNO$_3$ in ethylene glycol. This mixture produced films having stabilities similar to those grown in KNO$_2$ but which exhibit the good hysteresis qualities of films grown in KNO$_3$. We thought it will be of interest to study the effects of molarity concentration in the electrolyte on the interface properties during anodization of Si. In this work we present the experimental results of interface studies of anodic SiO$_2$ and Si systems using ethylene glycol electrolytes with molarities ranging from 0.02M to 0.08M of KNO$_3$.

EXPERIMENTAL

A set of 29 slices of 4 cm^2 of area, p-type, <111> crystal orientation, Czochralski grown and 10 to 25 Ω.cm resistivity were used. Before anodization, slices were completely degreased and given a DI water rinse, a dip in buffered HF acid, a DI water rinse, and a blow dry with nitrogen gas. The anodization apparatus employed was similar to that described in ref. 8. We used electrolytes of 0.02M, 0.04M, 0.06M, and 0.08M of KNO_3 salt in ethylene glycol. Growth of SiO_2 films was made using a constant current density while forming voltage increased gradually until it reached a predetermined value; at this point forming voltage was kept constant at that value, and current density allowed to decay to about 2% of the original value[7]. Immediately after anodization, wafers were rinsed with DI water and given a complete cleaning procedure before metal deposition. SiO_2 films thickness was measured by a color comparison chart and by ellipsometry using a production L117 Gaertner ellipsometer. Aluminum electrodes of about 6.4×10^{-3} cm^2 of area were vacuum deposited and delineated by standard photolithographic techniques; an 8 by 8 matrix of MOS capacitors was obtained in this way for each Si slice. For C-V measurement a PAR 410 automatic C-V plot system was used. Each sample was subjected to 0, +10, -10 volts bias temperature stress of 250°C for a 5 minutes interval. C-V plots of at least 10 MOS capacitors of each slice were recorded. Calculation of the density of fixed surface charge was carried out by the method given by Terman.[9]

RESULTS

Data to calculate the amount of Qss centers were obtained from the C-V plots of several MOS capacitors of similar anodic oxide growth conditions. These plots were taken under two different conditions: First, with no bias and no thermal stress; these C-V curves always showed a parallel shift to the left of the normalized capacitance axis. Second, with bias and thermal stress; these C-V curves always showed an asymmetric hysteresis behavior with respect to the curves taken under the first condition. The results of these measurements suggest the possibility that although there is some mobile ion contamination in the anodic oxide, also there is always present positive fixed Qss centers. Fig. 1 shows a typical C-V curve of an anodic MOS capacitor.

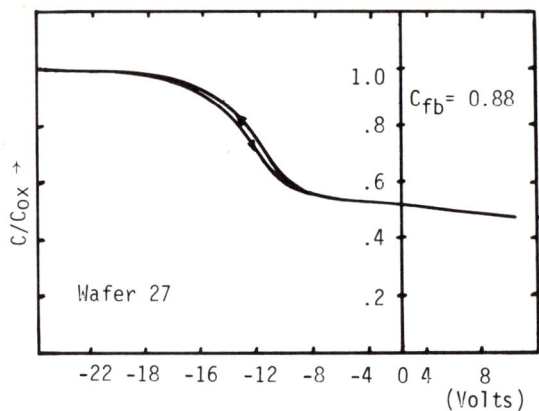

Fig. 1 A typical C-V plot of an anodic MOS capacitor

The effects of the electrolyte molarity in the interface properties of SiO_2 and Si are shown in Fig. 2.

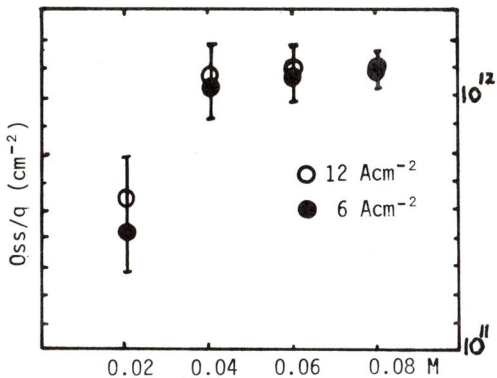

Fig. 2 Average Qss/q vs. Molarity

The figure above shows plots of the average Qss/q vs. Molarity under two constant current density growth conditions. It is easily seen from this figure that the average Qss generation depends on two factors: First, the amount of Qss generated at the SiO_2-Si interface is influenced by the molar concentration of KNO_3. Second, that an increase in Qss generation is produced by an increase in anodic current density. Also we note that, although there exists a density of surface states at the interface of anodic SiO_2-Si, its density does not increase as Qss increases. Calculated values of such states indicate that its density remains constant around 1.2×10^{11} cm^{-2}.

A figure of merit in this experiment is the percentage change Δ in Qss charges using two different anodic current densities. A plot of this figure of merit versus molarity of the electrolyte, Δvs.M, is shown in Fig. 3.

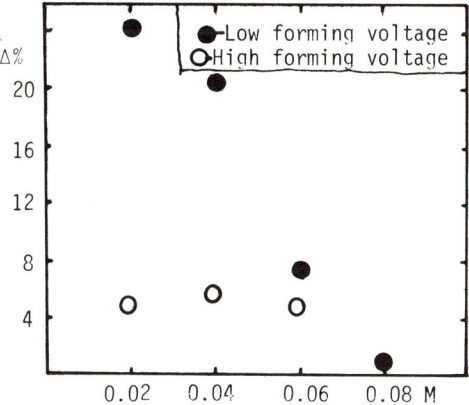

Fig. 3 Percentage change Δ versus Molarity

It can be seen from this figure that the difference in Qss generation (under 2 anodization current densities) is less sensitive to the molarity of the electrolyte at high forming voltage than it is at low forming voltage; on the other hand, given an electrolyte molarity the difference in Qss generation (under two anodization current densities) is more sensitive at low forming voltage than it is at high forming voltage. This is not the case for electrolytes of 0.08M concentrations.

DISCUSSION

In the previous sections we showed that the generation of fixed Qss centers at the interface of anodic SiO_2-Si depends on the growth conditions. Molarity of the electrolyte has an influence on Qss such that a high concentration of KNO_3 salt in the electrolyte produces more Qss centers than a low concentration does. Beckmann and Harrick[10] have reported the presence of hydroxyl radicals (OH^-) in anodic SiO_2 films. These radicals are incorporated into the film during anodization of Si by the applied field and its concentration increases toward the SiO_2-Si interface. At this interface, the OH^- radicals react with Si which results in a further oxidation of Si and a formation of SiOH groups.[10] This situation is in good agreement with the results that only cation movement and electronic conduction through the SiO_2 films are possible.[12] Of particular relevance to our discussion is the fact that OH^- groups are known to be deep electron traps in SiO_2 films.[13] Also of interest is the possibility that during anodization, trapping of electrons by OH^- groups weakens the strength of the SiOH bonds, producing in this way non-bridging Si bonds which can account for the positive fixed Qss centers. This is of interest because an increase in the molarity of the electrolyte (KNO_3 concentration) increases the concentration of OH^- radicals.[11] Anodic oxidation of Si is carried out near room temperature and the oxidation mechanisms are different to those of thermal oxidation.[13] During thermal oxidation fixed Qss centers and surface states are accounted for by the formation of $(SiO)^+$ complexes and the breaking of $Si\equiv$ bonds by the high temperature process. These two chemical entities are known to be proportional to each other. For the case of anodic oxidation of Si, breaking of the $Si\equiv$ bonds could occur less frequently due to the low temperature oxidation process and hence less surface state generation.

Following the discussion on Qss generation, the sensitivity Δ of Qss centers at the interface of anodic SiO_2-Si is of particular interest to us because of their reproducibility. The rate determining step in the reaction of OH^- groups with Si during oxidation can be accounted for of how fast these OH^- radicals are supplied to the SiO_2-electrolyte interface. Increasing current density during anodization could increase the supply rate of hydrated OH^- groups to the SiO_2-Si interface although electrons account for about 95% of the total anodization current density. Thus, the supply rate of OH^- groups to the SiO_2-Si interface will be less sensitive in thicker SiO_2 films.

SUMMARY

Anodic oxide growth conditions have considerable influence on the interface properties of SiO_2-Si. Few reports discuss how their growth conditions affect those properties. In this work we present experimental results on the generation of positive fixed Qss charge centers. These centers were detected in anodic MOS capacitors by means of high frequency C-V technique. Our results show that Qss generation depends on the molarity concentration of KNO_3 salt in ethylene glycol, the forming voltage accounts for the sensitivity of Qss generation under two different anodization current densities, and anodization current density accounts to a less extent to Qss generation.

ACKNOWLEDGMENTS

The authors wish to thank Secretaría de Educación Pública (SEP), México and Universidad Autónoma de Puebla for their financial support.

+Becaria CONACYT

REFERENCES

[1] A. S. Grove, B. E. Deal, E. H. Snow, and C. T. Sah. Solid State Electronics. $\underline{8}$, 145 (1965).
[2] B. E. Deal, M. Sklar, A. S. Grove, and E. H. Snow, J. Electrochem. Soc. $\underline{114}$, 276 (1967).
[3] A. Goetzberger, V. Heine, and E. H. Nicollian, Appl. Phys. Lett. $\underline{12}$, 95 (1968)
[4] B. E. Deal, J. Electrochem. Soc. $\underline{121}$, 198C (1974).
[5] S. I. Raider and A. Berman, J. Electrochem. Soc. $\underline{125}$, 629 (1978).
[6] H. Boroffka, Spring Meeting of the German Physical Soc. Bad Pyrmout (1966)
[7] A. G. Revesz, J. Electrochem. Soc. $\underline{114}$, 629 (1967).
[8] J. D. E. Beynon, G. C. Bloodworth, and I. M. McLeod, Solid State Electronics. $\underline{16}$, 309 (1973).
[9] L. M. Terman, Solid State Electronics. $\underline{5}$, 285 (1962).
[10] K. H. Beckmann and N. J. Harrick, J. Electrochem. Soc. $\underline{118}$, 614 (1971).
[11] E. F. Duffek, E. A. Benjamini, and C. Mylroie, Electrochem. Tech. 3, 75 (1965).
[12] P. F. Schmidt and J. D. Ashner, J. Electrochem. Soc. $\underline{118}$, 325 (1971).
[13] A. G. Revesz, Journal of Non-Crystalline Solids. $\underline{11}$, $\overline{309}$ (1973).

TRACER MEASUREMENTS OF NETWORK OXYGEN EXCHANGE DURING WATER DIFFUSION IN SiO_2 FILMS

Robert Pfeffer
US Army Electronics Technology and Devices Laboratory
(ERADCOM), Fort Monmouth, NJ 07703
Milton Ohring
Stevens Institute of Technology, Hoboken, NJ 07030

INTRODUCTION

As part of the recent development of high-speed integrated circuits with submicron features, new low-temperature processing techniques are being intensively studied. Among these are methods of low-temperature oxide formation on silicon surfaces for preparation of gate oxides in MOS devices. The most fully investigated method to date is thermal oxidation, both in dry oxygen and steam atmospheres. However, previous studies have been largely both device-oriented and restricted to high temperatures. The aim of this work is to clarify the physical mechanisms of the oxidation process in atmospheres containing water vapor, particularly in the low-temperature regime.

It is generally accepted that thermal oxidation of silicon proceeds by the inward transport of oxygen through the oxide layer as a constituent of molecularly dissolved oxygen or water.[1,2] The silicon substrate is converted to new oxide by the incorporation of atomic oxygen at the oxide-silicon interface after molecular dissociation. The layer thickens by formation of new oxide underneath the existing layer. In the interior of the oxide, the bulk diffusion mechanism for oxygen is generally believed to be the interstitial transport of dissolved molecular oxygen. The process occurs without any direct bonding between the diffusing species and the SiO_2 network: it rather involves the transport of nonreacting molecules through the interstices or cavities in the network. The mechanism for bulk diffusion of water is similarly thought to involve the interstitial transport of dissolved molecular water.[3] The transport of the diffusing species, dissolved molecular water, is again postulated to involve no direct reaction with the network. However, the permeation of water in SiO_2 involves a strong reaction with silicon-oxygen bonds in the network, in which the molecularly dissolved water dissociates to form two OH units with the additional O ion being provided by the SiO_2 network. This reversible reaction leads to equilibrium concentrations of immobile reaction products (i.e. SiOH groups) which are large compared to that of molecularly dissolved water, and which depend on the square root of ambient water concentration, in agreement with measurements of water solubility in vitreous silica.[4]

In the present study, it was found that the reaction between dissolved molecular water and the SiO_2 network acts to remove previously immobile implanted ^{18}O from its network site and to convert it instead to a constituent of a mobile species.[5] In the context of this model, that species is molecular water and the exchange proceeds by the reaction

$$H_2^{16}O + Si-^{18}O-Si \longrightarrow Si^{16}OH + Si^{18}OH \longrightarrow H_2^{18}O + Si-^{16}O-Si.$$

The tracer ^{18}O was observed to diffuse through SiO_2 layers with a diffusivity which was linearly proportional to the partial pressure of ambient water vapor, and with an activation energy of diffusion which matched those previously measured for water,[4] even in atmospheres containing far larger concentrations of oxygen. In the absence of ambient water vapor, no diffusion could be detected.

EXPERIMENT

Thermal SiO_2 layers of 2000 Å thickness were formed on low-resistivity (100) silicon wafers in flowing dry oxygen at 1200 C. Isotopically enriched ^{18}O gas (70 atom %) was then implanted into the oxide layers to a fluence of $3(10)^{15}$ cm^{-2} using an accelerating voltage of 40 kV per nucleus. The resulting ^{18}O concentration profile was an Edgeworth distribution[6] centered at 768 Å with a standard deviation of 283 Å and a third-moment ratio of -0.164. The wafers received no post-implant treatment.

The samples, which were prepared by sectioning the wafers, were then given thermal treatments in atmospheres containing various water vapor concentrations. These were carried out in order to determine both the activation energy of tracer ^{18}O diffusion as well as its functional dependence upon ambient vapor phase concentration. The former was determined by observing the temperature dependence of the tracer diffusivity in a particular atmosphere; the latter by observing its dependence on H_2O pressure at a single temperature. All temperature dependence measurements were made in horizontal tubular alloying furnaces lined with fused quartz diffusion tubes. The atmospheres were (a) room air at 21 C and 50% rel hum; temperatures ranged from 413 C to 789 C; durations from 3540 s to 230800 s; (b) flowing dry nitrogen (500 cc/min, 1 atm) obtained from boiloff of liquid nitrogen; temperatures ranged from 543 C to 1160 C; durations from 10980 s to 1799210 s; (c) flowing steam at 1 atm obtained from boiloff of deionized water; temperatures ranged from 263 C to 619 C; durations from 1290 s to 86440 s.

The pressure dependence measurements involved thermal treatments in pure water vapor at reduced pressures; these were performed within the bell jar of a vacuum evaporation station. The furnace was a small quartz-lined tubular resistance furnace powered by a variable autotransformer through feedthroughs in the baseplate. Water vapor was supplied by evaporation of deionized water in a stoppered vacuum flask; a continuous flow was maintained by exhausting the chamber with a mechanical forepump through a choked-down roughing valve. Temperatures ranged from 740 C to 842 C; durations from 7200 s to 609000 s; pressures from 0.062 torr to 8.9 torr. In addition to these treatments, several samples were heated in high vacuum (10^{-6} torr) in order to assess the substitutionality of the implant and the extent of damage to the oxide.

In order to determine the extent of ^{18}O diffusion induced by the thermal treatments, alpha particle yield curves were taken using the method of nuclear resonance profiling.[7] The energy of a beam of protons from a 2 MeV Van de Graaff electrostatic accelerator (Model A, High Voltage Eng. Corp.) was systematically stepped in the neighborhood of the 629 keV resonance in the $^{18}O(p,\alpha)^{15}N$ reaction. The detector was a silicon surface barrier detector at an angle of 150° to the beam direction. Each batch of samples profiled included a reference sample that had been ^{18}O-implanted but not thermally treated; the reference samples provided a continuous self-calibration for the proton beam energy. The alpha particle counts at each beam energy accumulated on a scaler until the proton fluence on the sample had reached a preset value. The range of beam energy settings spanned was typically 13 keV on either side of the energy at which the alpha yield was maximum, corresponding to a distance of 2100 Å on either side of the distribution peak.

The resulting alpha particle yield curves are illustrated in Fig. 1, in which the different symbols denote data recorded for the samples indicated. For the two lower curves the counting time was doubled to obtain sufficient statistical accuracy: the points plotted for those curves are half the recorded counts.

DATA ANALYSIS

By means of thermal treatments conducted in high vacuum, the implantation process was found to effectively replace existing oxygen with immobile ^{18}O as a constituent of the SiO_2 network.[5] According to the model of diffusion described above, the exchange reaction between network oxygen and dissolved molecular water during thermal treatments is solely responsible for any observed change in the distribution of implanted ^{18}O. The value of the ^{18}O tracer diffusivity $\underline{D*}$ which is characteristic of a given thermal treatment is reflected in the change of the implanted ^{18}O distribution resulting from the treatment. Once the alpha particle yield is measured, the extent of change is in principle determinable by inversion of the integral equation for the yield,[5] provided the constituent functions of the integral are previously known. In practice, deconvolution is a complicated procedure entailing large errors whose propagation is difficult to track. However, once the initial ^{18}O distribution is determined, its subsequent evolution is predictable through macroscopic diffusion theory provided the properties of the medium are correctly perceived and the proper boundary conditions are applied. This is treated further in Ref. 5, where the evolved distribution is calculated as a functional of the product $\underline{D*t}$ ($\underline{t}$ being the duration of treatment). Based on this distribution together with the other constituent functions of the yield integral given there, the yield was calculated for each of a number of discrete values of $\underline{D*t}$. The resulting family of yield curves was compared to the experimental yield curve resulting from that thermal treatment. The appropriate value of $\underline{D*t}$ was taken as that which indexed the calculated yield curve most closely matching the experimental yield curve. The goodness of fit can be judged by inspection of Fig. 1, in which the continuous curves are examples of alpha particle yield curves calculated in this manner.

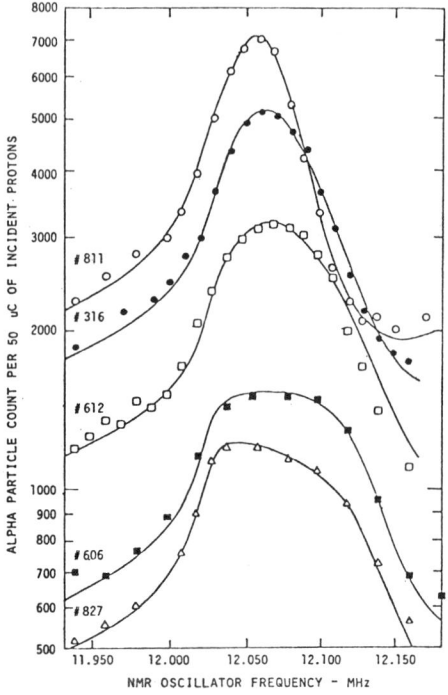

Fig. 1 Observed alpha particle yield curves for selected samples. Continuous curves represent calculated yield curves. The abscissa denotes proton energy.

Based on these calculations, the temperature dependence of $\underline{D*}$ was determined in each of the three atmospheres by fitting to equations of the Arrhenius type with a temperature-independent pre-exponential factor. The results, which appear in Fig. 2 along with the individual values of $\underline{D*}$, are

$$D*_{steam} = 2.7(10)^{-10 \pm 0.4} \exp\left[-\frac{(16.9 \pm 1.3) \text{ kcal/mol}}{RT}\right] \text{ cm}^2/\text{s}$$

$$D*_{room\ air} = 1.3(10)^{-12 \pm 0.3} \exp\left[-\frac{(15.5 \pm 1.4) \text{ kcal/mol}}{RT}\right] \text{ cm}^2/\text{s}$$

$$D*_{dry\ N_2} = 1.8(10)^{-13} \exp\left[-\frac{15.7 \text{ kcal/mol}}{RT}\right] \text{ cm}^2/\text{s}.$$

The three fitted lines shown in Fig. 2 are almost parallel; at 600 C that representing $D^*_{room\ air}$ lies a factor of 93 below D^*_{steam} while $D^*_{dry\ nitrogen}$ lies a factor of 8.1 below that.

The pressure dependence of D^*, which was determined in a similar manner, is shown in Fig. 3. As can be seen, these values are consistent with a purely linear pressure dependence over the entire range of pressures studied.

CONCLUSIONS

These observations of ^{18}O transport in the presence of water vapor are consistent with the conclusion that water diffusion in SiO_2 films proceeds by the transport of molecular water and that it is accompanied by a strong reversible reaction between molecular water and network oxygen. The observed linear dependence of tracer diffusivity upon partial pressure of atmospheric water vapor, for pressures ranging from 1 atm to as little as 0.06 torr, indicates that network oxygen is exchanged with molecular water. The activation energy of tracer diffusion remained about 16 kcal/mol, characteristic of water diffusion, over a range of temperatures from 260 C to 1150 C. Further, the magnitude of the tracer diffusivity was consistent with the presence of free exchange between network oxygen and diffusing water. The extent of tracer diffusion proved to be unaffected by the presence of atmospheric oxygen, indicating the absence of exchange with that species. The results obtained here agreed with corresponding results obtained by most other investigat-

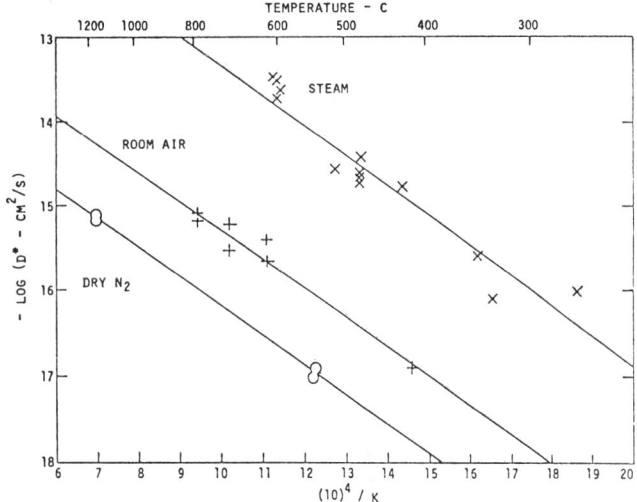

Fig. 2 Observed temperature dependence of ^{18}O tracer diffusivity at 1 atm. Solid lines represent linear least squares fits to the data shown.

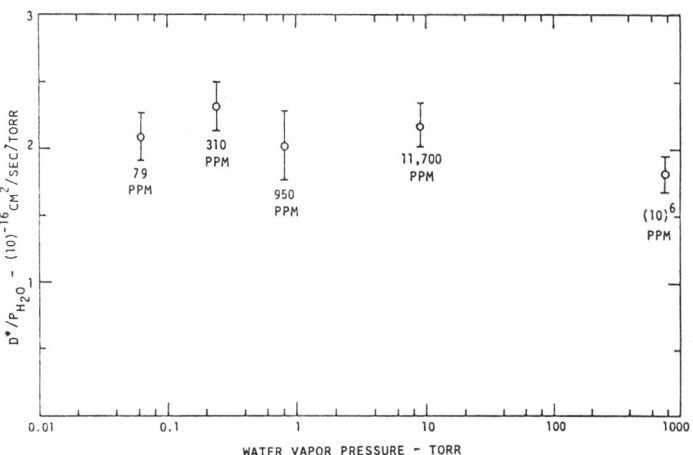

Fig. 3 Observed pressure dependence of ^{18}O tracer diffusivity in water vapor at 820 C. Labels indicate pressure in fractions of 1 atm.

tors using different means. They are in complete agreement with recently proposed models of oxygen and water diffusion in SiO_2 in which the diffusion proceeds by transport of the molecular species through interstices in the network without any direct reaction between the molecular species and the SiO_2 (the strong exchange reaction accompanying water diffusion involved immobile OH groups which are dissociation products of the molecular water).

The process employed for introducing tracer ^{18}O into the network, that of ion implantation, was shown to produce a region within the oxide in which the ^{18}O was fully incorporated in the network: no ^{18}O diffusion was observed in high vacuum, even after 161 hr at 815 C. The properties of the network remained otherwise unaltered, at least as far as diffusion was concerned. This was demonstrated both by the data's close match to theoretical predictions based on a dilute homogeneous medium as well as by the absence of any effect on diffusion of a high-temperature treatment which was sufficient to anneal out any implantation-induced damage. These findings indicate that the thermal treatments performed to induce ^{18}O diffusion were themselves sufficient to anneal out any damage.

The technique of nuclear resonance profiling proved to be a convenient, nondestructive and sensitive method of measuring ^{18}O diffusivities as low as $(10)^{-17}$ cm^2/s at temperatures as low as 260 C.

REFERENCES

(1) B. E. Deal and A. S. Grove, J Appl Phys 36, 3770 (1965).

(2) R. H. Doremus, J Phys Chem 80, 1773 (1976).

(3) R. H. Doremus, in Reactivity of Solids, Mitchell, de Vries, Roberts and Cannon, Eds. (1969) Wiley, New York, p. 667.

(4) J. Moulson and J. P. Roberts, Trans Faraday Soc 57, 1208 (1961).

(5) R. Pfeffer and M. Ohring (unpublished). Submitted to J Appl Phys.

(6) J. F. Gibbons, W. S. Johnson and S. W. Mylroie (1975) Projected Range Statistics, 2nd Edn, Academic, New York.

(7) G. Amsel, J. P. Nadai, E. D'Artemaire, D. David, E. Girard and J. Moulin, Nucl Inst Meth 92, 481 (1971).

AN ^{18}O STUDY OF THE OXYGEN EXCHANGE IN SILICON OXIDE FILMS
DURING THERMAL TREATMENT IN WATER VAPOR[x]

S. Rigo, F. Rochet, A. Straboni[xx], B. Agius
Groupe de Physique des Solides de l'E.N.S., 2 Place Jussieu
Paris V., France.

ABSTRACT

SiO_2 layers (90-300nm) of natural isotopic composition, obtained by anodic or thermal oxidation were treated in ^{18}O enriched water vapor at 13 Torr for 1-3 h at 600 or 930°C. ^{16}O and ^{18}O contents were determined by nuclear microanalysis; ^{18}O profiling was carried out using the 629 keV narrow resonance in the nuclear reaction ^{18}O$(p,\alpha)^{15}$N. Oxygen exchange appears to be the main process even at 930°C. Its kinetics is a function of the type of oxide (anodic or thermal) and increases with temperature. ^{18}O profiles vary drastically with temperature. As a matter of fact, at 930°C it appears that the phenomenon is controled by bulk-diffusion ; in contrast, at 600°C the rate of exchange between network oxygen atoms and mobile species is not infinite any longer and the phenomenon cannot be described by a "common" bulk-diffusion process.

INTRODUCTION

The thermal oxidation of silicon in dry oxygen, as well as in water vapor atmosphere, is commonly described by the theory of Deal and Grove (1). They showed that the oxidation process is controled by the migration, through the oxide, of the oxidizing species which react with the silicon at the Si-SiO2 interface. In the case of thermal oxidation of Si using H2O ambient, the reaction of water with silica induces the formation of hydroxide which may change the nature of the film itself, thereby making the atomic transport phenomena more complicated (2). Doremus (3), in a recent paper, gave an interpretation of the experimental SiOH and SiH profiles, considering the mobility of the water molecules and the immobility of the OH$^-$ species. Spitzer and Ligenza (4), using ^{18}O enriched water have shown that a very important isotopic exchange occurs during growth in high pressure steam (40-160 atm.). As well, very small traces of water vapor might induce exchange during SiO_2 growth in "dry" oxgen (5). The present study is related to a more detailed investigation of the exchange phenomena in a H2O ambient; some preliminary results were published in (12).

[x] This work was supported by C.N.R.S. (R.C.P. N° 157) and D.G.R.S.T.
[xx] Present address : Centre National d'Etudes des Télécommunications, Chemin des Prés, 38240 Grenoble, France.

EXPERIMENTAL PROCEDURES

N-type polished (100) oriented wafers supplied by Chisso, with a 30 mn diameter and a resistivity of 4.5 Ω.cm were used. After cleaning, the wafers were oxidized up to various thicknesses, either anodically or thermally in a medium of natural isotopic composition in oxygen (0.2 % of ^{18}O). Thermal oxidation treatments were performed during various times and under different temperature conditions in a dry natural oxygen flow with a water content less than 10 ppm at $\sim$ 1 atm (Table 1). Anodic oxidations were performed in a diethylene-glycol solution containing 0.4 % by weight of KNO_3 and $H_2^{16}O$. Anodization was carried out at a constant current density of 1 mA/cm^2 up to a predetermined voltage V_f followed by a stage at constant voltage. The final current was allowed to decrease down to about 10 µA/cm^2 for all samples. The oxidation duration was about 24 h. Under these conditions the thickness of SiO_2 film is about 0.6 V_f.nm. The films were then annealed in water vapor at 13 Torr and at 700° C during five hours in order to stabilize their exchange properties (6). Each wafer was cut in several pieces, one being kept as a reference sample. The other pieces were simultaneously submitted to exchange treatments in 92 % ^{18}O enriched water vapor[x]. The label gas-treatment was performed in a quartz furnace connected to a turbomolecular pump. After pumping down the set-up to the limit vacuum, the furnace was isolated from the pumping system and connected to a flask containing the previously degased water. The water pressure in the furnace was 13 Torr : the temperature of the flask was maintained at 15°C. Then the samples were introduced into the high-temperature section, either at 600°C or 930°C. The nuclear reactions $^{16}O(d,p)^{17}O^{x}$ and $^{18}O(p,\alpha)^{15}N$, induced by deuteron or proton beams from a 2-MeV Van de Graaff accelerator, were used to measure the number of $^{16}O(N_{16_O})$ and $^{18}O(N_{18_O})$ atoms/cm^2 in the oxide layers (7). Absolute values of N_{16_O} and N_{18_O} were determined by comparison to standard references known to within $\pm$ 3 % (8) ; their relative precision was better than $\pm$ 1 % for ^{16}O and up to 3 % for ^{18}O. Oxide thicknesses were deduced from their oxygen content determined by nuclear microanalysis (7), assuming a silica density of 2.2. The ^{18}O concentration profiles were deduced from the analysis of the excitation curves of the $^{18}O(p,\alpha)^{15}N$ reaction near the 2-keV wide 629-keV resonance (7,9).

RESULTS

The results of ^{16}O and ^{18}O content measurements on various oxide films before and after the $H_2^{18}O$ treatments are summarized (respectively) in tables 1 and 2.

Wafer	T (°c)	t (mn)	V_f (V)	N_{16_O}	X_O (nm)
A_1	-	-	150	395	89.9
A_2	-	-	200	530	120.2
A_3	-	-	250	669	151.5
T_1	1000	110	-	598	135.5
T_2	1000	210	-	841	190.4
T_3	1100	120	-	1111	252
T_4	1100	180	-	1358	307.5

Table 1 : Results of ^{16}O content (N_{16_O}) measurements on the various oxide films after the ^{16}O oxidation. N_{16_O} is given in 10^{15} atoms/cm^2 units. X_O is the equivalent oxide film thickness (0.2 % of natural ^{18}O being taken into account).

[x] this labeled gas was produced by the Weizmann Institute, Rehovot, Israël.

Wafer	T (°C)	t (min)	N_{16_O}	N_{18_O}	X_{16+18} (nm)
A_1	600	60	360	21.6	86.3
	600	180	324	77.3	90.7
A_2	600	60	468	29.3	112.4
	600	180	397	124	117.7
A_3	600	60	581	31.7	138.5
	600	180	546	97	145.4
T_1	930	60	467	186	147.5
	600	60	592	14.1	137
	600	180	585	41	141.5
T_2	600	60	842	14.5	193.6
	600	180	822	59	199.1
T_3	930	60	967	200	264
	600	60	1114	20	256.3
	600	180	1094	57	260
T_4	930	60	1230	204	324.1

<u>Table 2</u> : Results of ^{16}O and ^{18}O contents (N_{16_O} and N_{18_O} respectively) measurements on the various oxide films after the exchange treatments in $H_2^{18}O$. N_{16_O} and N_{18_O} are given in 10^{15} atoms/cm^2 units ; X_{16+18} is the equivalent oxide film thickness.

The relative ^{16}O amounts decrease for all treatments while ^{18}O is incorporated in the film, evidencing an exchange phenomenon. The increase in thickness cannot be estimated because of imprecision on absolute values of oxygen amounts. Let us note that using kinetic constants published by Deal and Grove (1) the thickness increase would be of about 3.0 to 5.0 nm at 930°C and less than 0.1 nm at 600°C. The exchanged quantities increase with time and, at 600°C, seem to depend strongly on the nature of the oxides, the anodic oxides exchanging much more than the thermal ones. At 930°C where a slight growth occurs, there is a very important ^{18}O exchange which is about ten times greater than the corresponding growth.

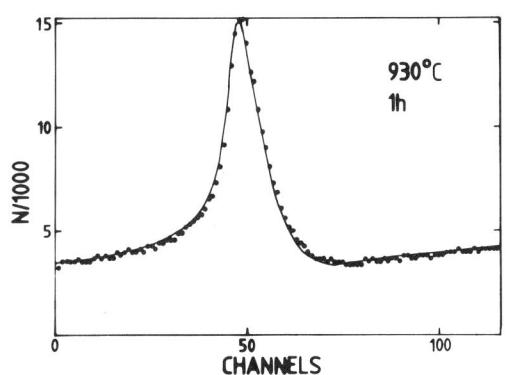

<u>Fig.1</u> - ^{18}O concentration profile measurement using the 629 keV resonance of the $^{18}O(p,\alpha)^{15}N$ reaction on a thermal oxide (wafer T3) treated during 1h at 930°C in $H_2^{18}O$. The solid line is the fit calculated for an erfc profile with an ^{18}O surface concentration of 92% and $\sqrt{D^{18}t}$ = 43 nm. For this figure and the others the energy step is equal to 384 eV/channel.

Fig. 1 shows the excitation curve of the $^{18}O(p,\alpha)^{15}N$ resonant reaction for sample T_3 after $H_2^{18}O$ treatment at (930°C, 60 mn). It is fitted by a curve calculated assuming an erfc profile with an ^{18}O surface isotopic concentration equal to the isotopic concentration of water vapor (i.e. $\sim$ 92%) and a value of "$\sqrt{D^{18}t}$" equal to 43 nm ; the latter value is equal to the one deduced from Table 2 assuming a diffusion limited process with an infinite value of surface exchange current (i.e. constant surface concentration). These results indicate that this oxygen exchange phenomenon is controled by a bulk-diffusion process and gives a value of D^{18} equal to $5 \ 10^{-15}$ cm^2/sec.

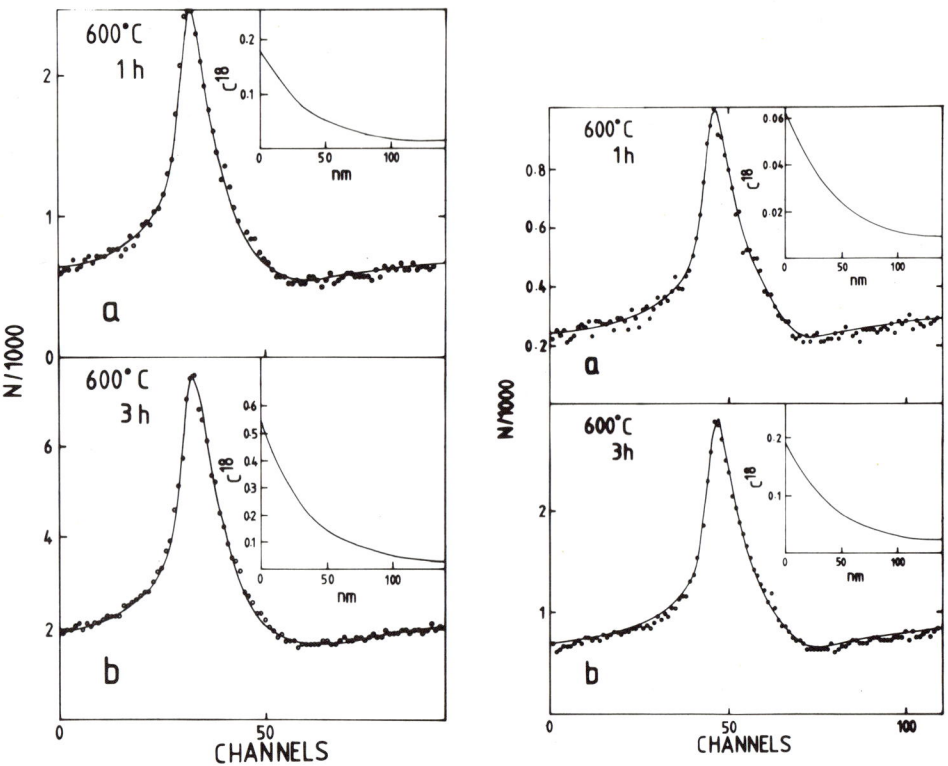

Fig. 2 - ^{18}O concentration profile measurements on an anodic oxide (wafer A_3) treated during 1h (a) and 3h (b) at 600°C in $H_2^{18}O$. The solid lines are the fits calculated for profiles represented in insets.

Fig. 3 - ^{18}O concentration profile measurements on a thermal oxide (wafer T_1) treated during 1h (a) and 3h (b) at 600°C in $H_2^{18}O$. The solid lines are the best fits calculated for profiles represented in insets.

Fig. 2 and Fig. 3 show the excitation curves for sample A_3 and T_1 respectively, after two treatment times (60 mn and 180 mn) at 600°C. Each of them is fitted by a theoretical curve calculated from a concentration profile represented in the insets. Isotopic surface concentration appears markedly lower than water vapor isotopic concentration ; this surface concentration $C_S^{\cdot}$ increases with time :
- for sample A_3, $C_S^{\cdot}$ is equal to 18% (60 mn of treatment) and 54% (180 mn).

- for sample T_1, C_S is equal to 6% (60 mn of treatment) and 19% (180 mn). Thus, in contrast with what happens at 930°C, the surface exchange current is not infinite. The general shape of these concentration profiles C^{18} has been obtained by a linear combination of exponential functions, verifying grad C^{18} equal to zero at the SiO_2/Si interface. For each of the two oxides the concentration curves obtained after two different time treatments are nearly homothetic. This suggests that exchange process is not a simple and common "bulk-diffusion-phenomenon". Although the shapes of depth profiles are similar for an anodic and a thermal film, for the latter the curve is somewhat less steep with a surface concentration which is about three times smaller. The bulk and surface exchange in the thermal film is hence significantly lower than in the anodic film.

DISCUSSION

An attempt of explanation will be done considering that the oxygen exchange is due to the reaction of water molecules with the network giving rise to the formation of two Si-OH groups per interacting water molecule (10).

At 930°C the practically-constant-with-depth value of the self-diffusion coefficient could be explained by the fact that the concerned thicknesses are greatly smaller than X_c ($\sim$ 250 nm); X_c being the thickness up to which the oxide growth remains in the linear domain (1), where H_2O concentration is very close to the SiO_2/gas interface water concentration C^*. Thus ^{18}O self-diffusion coefficient D^{18} should be equal to $D_{H_2O} \times C^* \times N_o^{-1}$ where D_{H_2O} is the diffusion coefficient of water in silica and N_o the number of network oxygen atoms per volume unit. Now, this expression is equal to B/2, B being the parabolic coefficient of oxide growth in Deal and Grove (1). The extrapolated value of B at 13 Torr is 6×10^{-15} cm^2/sec which is in good agreement with the D^{18} value that we find experimentally. As regards the bulk limited diffusion process, we could explain it by a fast reaction of water with the network at this temperature.

At 600°C in contrast with the former case, the process is not dominated by water diffusion any longer, probably because of a slower rate of reaction between water and the network. Water molecules would hence have a longer migration range without exchanging the labeled oxygen. This assumption is supported by the fact that the general shape of the ^{18}O concentration profile seems to be nearly time-independent. In the case of anodic films a more open structure might increase the number of mobile species and the network reactivity. This might explain more steep ^{18}O concentration profiles and greater exchange values than in the case of thermal films. The history of the sample has hence a great influence on the exchange phenomenon; note that this was also the case for water diffusion in silica at low temperature (11).

REFERENCES

(1) B.L. Deal and A.S. Grove, J.Appl.Phys., 36, 3770 (1965).
(2) A.G. Revesz, J.Electrochem.Soc., 126, 122 (1979).
(3) R.H. Doremus, J.Phys.Chem., 80, 1773 (1976).
(4) W.G. Spitzer and J.R. Ligenza, J.Phys.Chem.Solids, 17, 196 (1961).
(5) E. Rosencher, A. Straboni, S. Rigo and G. Amsel, Appl.Phys.Lett., 34, 254(1979).
(6) S. Rigo, Thesis (Université de Paris, 1977).
(7) G. Amsel et al., Nucl.Instr. and Meth., 92, 481 (1971).
(8) G. Amsel et al., Nucl.Instr. and Meth., 49, 705 (1978).
(9) B. Maurel, Thesis (Université de Paris, 1980).
(10) G.J. Roberts and J.P. Roberts, Phys.Chem.Glasses, 7, 82 (1966).
(11) G.J. Roberts and J.P. Roberts, Phys.Chem.Glasses, 5, 26 (1964).
(12) M. Croset, S. Rigo and G. Amsel, International Conf. on MIS Structures, Grenoble, June 1969, Edit. C.E.N. Grenoble - LETI, 1969, p. 259.

X-RAY PHOTOELECTRON SPECTROSCOPY OF SILOXENE: A MODEL COMPOUND REPRESENTING INTERMEDIATE OXIDATION STATES OF SILICON AND INTERFACE DEFECT SITES

J. A. Wurzbach, Jet Propulsion Laboratory, California Institute of Technology, Pasadena, California 91103

ABSTRACT

Siloxene is a stable Si(+1) compound with a molecular formula of $H_6Si_6O_3$, which has been prepared in 50% purity. An O1s-Si2p separation of 432.9 eV was found for Si(+1) in this system. The impurities represent a distribution of silicon oxidation states from -1 to +4. Controlled oxidation of this system generates high concentrations of Si(+2) and Si(+3). Electron irradiation converted Si(+3) to Si(+4) similar to thermal SiO_2, which provided a convenient reference for approximate comparison of the lower oxidation states with each other, with crystalline Si(0), and with the interface of a thin, thermal SiO_2 layer on a crystalline silicon substrate.

INTRODUCTION

Recent investigations by x-ray photoelectron spectroscopy (XPS) on silicon/oxide interfaces have shown consistent evidence of silicon oxidation states intermediate between zero and +4.[1-3] In particular, the Si(+1) state has been difficult to characterize because of its close proximity to the silicon element signal, which is usually much more intense. Siloxene is an unusual silicon oxide in which all the silicon is found in the Si(+1) state. It has been prepared in reasonably pure form to help characterize the Si(+1) state and aid our understanding of silicon interfaces. Siloxene was first studied in pure form by Kautsky and Hengge.[4] It is a planar molecule composed of six-membered Si rings and has a molecular formula of $H_6Si_6O_3$, or H_2Si_2O (see Fig. 1). The rings, which resemble the Si(111) crystal face, are interconnected by oxygen bridges, thus forming a two-dimensional polymer. Siloxene shows EPR activity attributed to a silicon dangling bond defect.[5]

EXPERIMENTAL

Siloxene was prepared by hydrolysis of $CaSi_2$ in alcoholic HCl.[4] The product, a powder, was mounted for XPS by two methods. An optically flat, gold-plated silicon wafer was covered with a thin hydrocarbon polymer film by solution evaporation. This film was found to be reasonably conductive in the photon beam. The powder was evenly distributed over this film and run at 250 K. Another sample of powder was pressed into a graphite pellet and run at room temperature. Spectra were obtained on a modified Hewlett-Packard 5950A spectrometer using monochromatic

Al x-rays. Stoichiometries were determined by correcting integrated intensities for relative cross-section[6] and escape depth.[7]

RESOLUTION ENHANCEMENT/DECONVOLUTION

A powerful method of spectral reconstruction has been under development at JPL for several years; it has been described elsewhere.[2,3,8] Application to experimental data requires spectra with reasonable signal-to-noise ratio, knowledge of a broadening function, and the ability to accurately reconstruct the unbroadened signal. Selection of the broadening function depends on mathematical criteria relatively free of user prejudice. Reconstruction involves Fourier division and linear prediction based on a maximum-entropy criterion. Test cases based on trial data generated by computer from known functions have shown the power (Fig. 7, Ref. 3) and the limitations of the method. Major peaks can be reconstructed within ± 0.2 eV and ± 10% in intensity. Moderate peaks are within ± 0.3 eV and ± 30% in intensity. Minor features may be subject to averaging.

Given siloxene's reactivity and the absence of a crystal structure, it could not be assumed that the main Si2p peak was due to H_2Si_2O. To confirm the assignment of the main peak, deconvolution will be used to separate components due to differential charging and higher oxide impurities from the intensity of the main peak and calculate a mass balance between silicon and oxygen. It will not be necessary to make unique and positive identification of the higher oxides; these determinations will be made in subsequent oxidation experiments without deconvolution. It will be sufficient to estimate intensities and oxygen stoichiometries (i.e., oxidation states) from approximate positions.

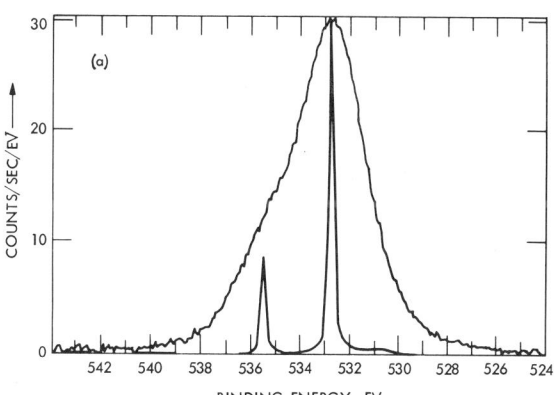

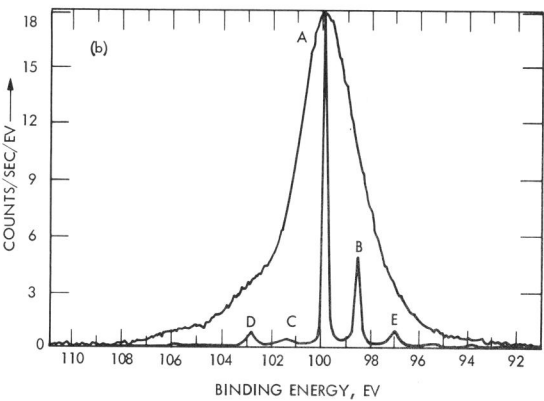

Fig. 1. Siloxene structure proposed by Kautsky and Hengge based on optical spectra, composition, substitutional chemistry, and crystallite morphology. No crystals large enough for x-ray diffraction have been obtained.

Fig. 2. Siloxene XPS spectra: (a) O1s, (b) Si2p.

RESULTS AND DISCUSSION

O1s and Si2p raw data with background removed are shown in Fig. 2 for siloxene mounted on the hydrocarbon film. The results are a combination of chemistry and charge broadening. The integrated raw data show an overall stoichiometry of $Si_{1.4}O$ (cf. Si_2O theo.). This ratio, plus the high binding energy (HBE) shoulders on O1s and Si2p, indicate the presence of oxidized impurities.

Deconvolutions are also plotted in Fig. 2. For clarity, they have been scaled to match the height of the raw data. The O1s data show a moderate peak at 536 eV and a main peak. The Si2p data show a main peak, A, a moderate contribution, B, and three minor signals: C, D, and E. Still lower intensity projects out toward 106 eV.

The Si/O stoichiometry of the 106 eV and 536 eV signals is reasonably close to SiO_2. The O1s-Si2p separation is also similar to that observed for amorphous SiO_2. These HBE signals probably represent SiO_2 impurities translated toward HBE by differential charging.

The main Si2p peak in Fig. 2b may be attributed to Si(+1) as H_2Si_2O, as shown by the mass balance below. Peak B, found at 1.3 eV toward low binding energy (LBE), can be attributed to a 23% impurity of Si(0) which, unlike crystalline Si(0), is in the form of a dihydride. This assignment is consistent with combining the 0.3 eV group shift observed for H on Si[9] with a 1 eV shift corresponding to the formation of a single Si-O bond.[2,3,5] Peak E may correspond to a $CaSi_2$ impurity.

The higher oxides indicated by the Si2p shoulder at 103 eV correspond to peaks C and D. Peak D is best interpreted as representing the combined intensity of both Si(+3) and Si(+4), having a position near the average between these two states.

Peak C shows many characteristics of Si(+2). Its 1.5 eV separation from the H_2Si_2O Si(+1) peak is consistent with previous work[2,3] and consistent with combining the shifts due to loss of an Si-H[9] bond and formation of an Si-O bond.[2,3,5] Its separation from crystalline Si(0) (Table I) agrees with data obtained on SiO films.[5] Its O1s-Si2p separation of 431.4 eV is similar to the value (431.0 eV) derived from SiO.[5] This striking correspondence with data from many sources prompts the entry of peak C as Si(+2) in Table I. However, given the limitations of the deconvolution algorithm, it has been appropriately marked with an approximation sign.

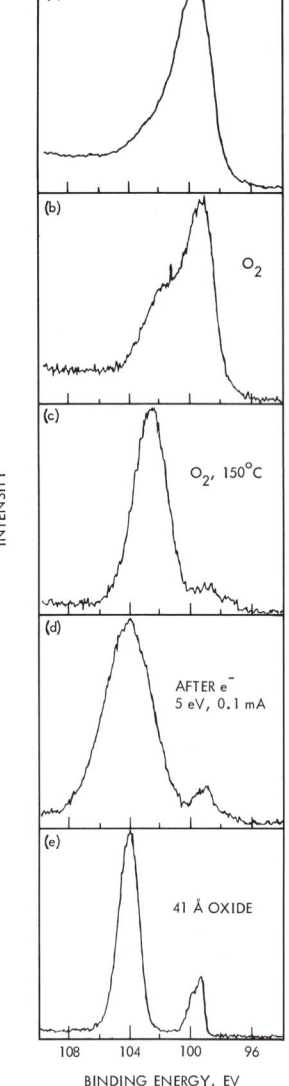

Fig. 3. Siloxene Si2p spectra (a) before and (b,c) after exposure to oxygen and (d) electrons; (e) 41 Å SiO_2 on Si for comparison.

It is conspicuous that Fig. 2 shows only two O1s peaks compared to a manifold of Si2p peaks. To a first approximation, the main O1s peak represents oxygen atoms in the same oxidation state, -2. The corresponding silicon oxidation states range from -1 to +4. In reality, the O1s signals do not coincide; Table I implies a shift of 0.9 eV between O1s from H_2Si_2O and SiO_2. The corresponding Si2p shift is four times greater. Further, there is four times more oxygen per silicon atom in SiO_2 than Si_2O. Thus, in the Si2p spectrum, an SiO_2 peak may be small compared to the main

Si_2O peak, but the corresponding O1s peaks will be nearly comparable in intensity. Therefore, in Fig. 2, the main O1s peak represents closely spaced peaks of comparable height whereas the Si2p spectrum is composed of a main peak with widely space secondaries.

Since discrete O1s signals could not be resolved for the separate oxides, one cannot calculate stoichiometries directly. One can, however, obtain a mass balance between the main O1s signal and Si2p peaks A, C, and D. Indeed, if one estimates the amount of O1s intensity expected from Si2p peak A based on H_2Si_2O, and adds similarly calculated estimates for peaks C and D based on average stoichiometries for Si(+2) and Si(+3, +4), the sum is equal within 3% to the observed intensity of the main O1s peak.

Thus, the O1s and Si2p spectra in Fig. 2 are consistent with an assignment of Si2p peak A as Si(+1) in the form of H_2Si_2O, as expected. The product is 50% H_2Si_2O with 15% higher oxide impurities—apparently as Si(+2) and Si(+3, +4). An O1s-Si2p separation of 432.9 eV is observed for Si(+1) as H_2Si_2O.

It is significant that the mass balance shows close consistency between O1s and Si2p. The lineshapes and parameters were selected independently; the agreement between the two spectra was not built into the deconvolution. In general, it is essential to compare two different core levels from the same sample to reach reliable conclusions.

To obtain high concentrations of the other sub-oxides and to obtain a suitable Si(+4) reference signal for comparing siloxene to a real silicon interface, the graphite sample was subjected to controlled oxidation. The initial sample is shown in Fig. 3a. Exposure to an atmosphere of O_2 produced a modest result. The unexpected stability against O_2 is shown in Fig. 3b, where a shoulder roughly in the Si(+2) region was formed. Exposure to O_2 at 150 C generated nearly complete oxidation and left a low-level signal which probably represents Si(0) at various levels of hydrogenation (Fig. 3c). The Si2p position is consistent with Si(+3) (see Table I); the O1s spectrum showed stoichiometry of Si_2O_3. It seems, therefore, that this method generates a signal which is probably a mixture of states, predominantly Si(+3), with an average atomic ratio of Si_2O_3.

It was found that this signal could be converted, at least in part, to Si(+4) as SiO_2. A 0.1 mA beam of 5 eV electrons deposited sufficient energy to drive the conversion. The resulting peak shifted toward HBE and showed an O1s-Si2p separation of 429.4 eV, in agreement with the separation of 429.5 eV observed for thermal SiO_2 on silicon.[3] Thus, it appears that it is possible to prepare an oxide on siloxene which is very similar to thermal SiO_2. On this assumption, the siloxene spectra of Figs. 2 and 3a-d were shifted to be consistent with this reference point, and compared to a 41 Å thermal oxide (Fig. 3e). Siloxene peak positions with respect to crystalline Si(0) were derived from this comparison. The results are summarized in Table I along with similar data from Si/SiO_2 interfaces[2,3] and SiO[5] films. Any such reference is, to some extent, arbitrary due to differential charging and relaxation effects, but the absence of abnormal shifts in O1s plus a secondary reference of C1s from adsorbed hydrocarbons both suggest that the comparison is accurate to within 0.5 eV and probably much better.

CONCLUSIONS

It has been possible to prepare, with 50% purity, A Si(+1) compound which has structural relevance to the Si/SiO_2 interface. Preliminary results on controlled oxidation showed an unexpected resistance toward molecular oxygen at room temperature; only partial conversion to Si(+2) was achieved. Moderate temperatures converted the system to an average stoichiometry of Si_2O_3, i.e., Si(+3), in high concentration. This form of Si_2O_3 was found to be unstable with respect to conversion to Si(+4), SiO_2, under 5 eV electron irradiation. Signals attributable to Si in various stages of hydrogenation have also been derived. Stoichiometries, O1s-Si2p separations, and Si2p separations from crystalline Si(0) have been established. These are useful parameters for characterizing low-intensity signals of intermediate oxidation states at the Si/SiO_2 interface. Despite possible

TABLE I. Binding Energy Separations (eV).

	O1s-Si2p Separation	Separation from silicon element peak		
		Siloxene	Si/SiO$_2$[3,7]	SiO[5]
Si(0) (dihydride)		-0.7b		
Si(+1) (hydride, H$_2$Si$_2$O)c	432.9	0.5		
Si(+1) (Si$_2$O)d		0.8	0.6	
Si(+2)	~431.4	~2.1	1.7-2.2	2.2
Si(+3) (Si$_2$O$_3$)	430.0	3.4	2.9-3.3	
Si(+4) (SiO$_2$)	429.5	4.8	4.5	4.4

a) Approximate. From comparing oxidized siloxene to 41 Å SiO$_2$ film.
b) Minus sign means toward lower binding energy.
c) This is the hydrogenated form of Si(+1) found in siloxene.
d) An estimate of where Si$_2$O might be found at a silicon interface. Corrects (c) by silicon group shift of 0.3 eV (Grey et al.).

differential charging [-0.2 - 0.4 eV for Si(+3) and Si(+4)], the results from the model systems show close to excellent agreement with ranges of results expected and observed at a real Si/SiO$_2$ interface. Thus, the question of intermediate oxidation states has been approached from complementary points of view: the silicon interface and a system of model compounds; similar results are reached in both cases. This agreement lends support to conclusions reached previously on the distribution of oxidation states at the Si/SiO$_2$ interface.[2,3]

ACKNOWLEDGMENTS

The author gratefully acknowledges helpful discussions with F. J. Grunthaner, A. Madhukar, and R. P. Vasquez, sample preparation by M. Altobelli and C. Zachman, data reduction by C. Akers, and manuscript preparation by M. Brandenberg, T. Pipes, and J. Justice. This work was performed at the Jet Propulsion Laboratory, California Institute of Technology, under NASA Contract NAS7-100.

REFERENCES

1. C. R. Helms, J. Vac. Sci. Technol. 16, 608 (1979). See therein: Raider, Ref. 17, 22; Grunthaner, Ref. 20; Bauer, Ref. 33; Garner, Ref. 33.
2. F. J. Grunthaner and J. Maserjian in The Physics of SiO$_2$ and Its Interfaces, S. Pantelides, Ed., Pergamon Press, New York, 1978, p. 389.
3. F. J. Grunthaner, et al., J. Vac. Sci. Technol. 16 (5), 1443 (1979).
4. E. Hengge, Silicon Chemistry II, Topics in Current Chemistry, Vol. 51, Springer-Verlag, New York, 1974; Chem. Ber. 95, 645 (1962); and H. Kautsky, Z. Anorg. u. Allg. Chemie 117, 209 (1921).
5. Y. Ono, Y. Sendoda, T. Keii, J. Amer. Chem. Soc. 97, 5284 (1975).
6. J. Scofield, Lawrence Livermore Laboratory Report #UCRL-51326 (1973).
7. C. J. Powell, Surf. Sci. 44, 29 (1974); and C. D. Wagner et al., Handbook of X-ray Photoelectron Spectroscopy, Perkin-Elmer Corp., 1979, p. 5.
8. R. P. Vasquez, J. D. Klein, J. J. Barton, and F. J. Grunthaner, submitted to J. Electron Spectrosc. and Relat. Phenomena.
9. R. C. Grey et al., J. Electron Spectrosc. and Relat. Phenomena 8, 343 (1976).

EFFECT OF ANNEALING IN O_2/N_2 MIXTURE ON THE MOS CHARACTERISTICS

A. K. AboulSeqoud and S. Masoud
Faculty of Engineering, University of Alexandria, Egypt

ABSTRACT

The inerface charge density in an MOS was found to be senstive to annealing after oxidation. It is affected by annealing ambient, temperature and duration. The presence of small amounts of oxygen in the annealing ambient was reported to decrease the interface states charge density since it saturates the oxygen deficient bonds. It was also reported that the oxidation in quartz tubes of thermally heated furnaces develops hydride and hydroxyl bonds. Both bonds are unstable and may decompose during annealing. The process of saturating the oxide bonds takes place in a slower rate than the decomposition of the hydride bonds in an ambient containing oxygen. The saturation rate as well as the hydride oxidation rates depend on the oxygen ratio.

INTRODUCTION

The annealing temperature, duration and ambient are parameters of the process outcome. The interface charge density as well as the interface states density were attributed to the oxygen deficient bonds and to the dengling bonds respectively(1). Hess and Deal investigated the effect of annealing in a mixture of oxygen and nitrogen(2). They reported that the interface charge density decreased to a steady state after several hours of annealing. This improvement was attributed to the saturation of the oxygen deficient bonds during annealing in the mixture. Annealing in hydrogen was reported to decrease the interface states density due to the saturation of the silicon dangling bonds(3). Beckmann and Harrick(4) found out evidences of the presence of hydride and hydroxyl bonds in the silicon/silicon oxide interface using infrared spectroscopy. This was attributed to the diffusion of water from the silica walls of the thermally heated furnace to the oxidation ambient(5). Both bonds are metastable and may decompose at high temperature. The hydride bonds decompose to silicon while the hydroxyl bonds decompse to oxygen deficient oxide(6). Both bonds should decompose contributing to the MOS characteristics, and both are affected by the annealing ambient.

In this work it is intended to study the contribution of the hydride and hydroxyl bonds formed during dry oxidation of silicon in a thermally heated furnace with a silica tube. It is also intended to show the interface states density senstivity to the annealing conditions both in inert and in oxygen blended ambients. The study will also cover the effect of annealing on the interface charge density.

PROCEDURE

Chisso p-type silicon wafers of resistivity of about 4 ohm-cm and (1,1,1) orientation were used to prepare the specimens. After standard cleaning procudure, specimens were oxidized in a thermally heated furnace. The oxidation continued for 70 min. in dry oxygen. The oxidation temperature was fixed at 1100+ 0.5°C. The chamber was utilized to anneal the oxidized specimens under different conditions. The ambient of mainly nitrogen was blended with oxygen with a ratio varying between 0 and 10% at different runs. Six specimens were drawn out of the furnace in each run at intervals of 10 min. Some specimens were left without annealing to be taken as references,as will be explained later on. Oxide was stripped off one side of each specimen, and then the face was coated with aluminum. One m.m. diameter electrodes were deposited by evaporation of aluminum through metallic masks on the oxidized surfaces. Finally specimens were annealed for two minutes at 600° C. in nitrogen. The C-V and G-V curves for all specimens were determined using a G.R. digital bridge tuned to 10 KHz. The ideal C-V curve was also determined after measuring the exact wafer resistivity. The interface charge density for each specimen was then determined. The interface states density was only determined for specimens annealed in pure nitrogen,using the C-G method(7). In this measurement the frequency spectrum varied between 1 and 100 KHz, and the gate voltage was -3 volts.

RESULTS

The data collected from 28 sets of specimens were analysed to determine the interface charge density. Figure 1 shows the effect of annealing time and ambient on the interface charge density in reference to the value taken for specimens without annealing. It is noted that the interface charge density decreases slowly with time to a steady state when annealing in pure nitrogen. It reaches a steady state value within one hour. Similar results were previously reported(8),where it was noticed that the interface charge density never increases when specimens were annealed in an inert ambient. When specimens were annealed in a mixture of oxygen and nitrogen, a sharp increase in the interface charge density was noted in the first few minutes then a gradual decrease was noted with time. It should be mentioned here that the data in this period is not precise when determining the value and the location of the peak. It is also noted that the interface charge density decreases sharply with the increase in the oxygen ratio in the annealing ambient. A steady state value is observed for the interface charge density after 60 min. when specimens were annealed in an ambient containing 10% of oxygen. The decay region in fig. 1 fits the relation:

$$Q_{ss}(t) / Q_{ss}(0) = A(x) \exp(-\alpha(x)t)$$

as shown in fig. 2. Both the value of A and α were plotted as function of the oxygen ratio x. This relation was plotted in fig. 3 and it is noticed that both A and α are linear functions of x. The value of the interface states density was evaluated for specimens annealed in pure nitrogen only. The density seems to increase with the increase in the annealing time, as shown in fig. 4.

DISCUSSIONS AND CONCLUSIONS

It was previously reported that both hydride and hydroxyl bonds are formed during oxidation in fused silica thermally heated furnaces due to the diffusion of water from the silica tube(9). Both bonds are unstable(6). When specimens are annealed in nitrogen at 1100° C, the hydride bonds decompose to silicon atoms increasing the interface states density as shown in fig. 4. At the same time the hydroxyl bonds decompose to oxide deficient bonds decreasing the interface charge density in magnetude. But since the hydroxyl bonds are much less than the hydride bonds density (5), then the decrease in the interface charge density is small. When specimens are annealed in an ambient containing small amounts of oxygen, the hydride bonds are oxidized to oxygen deficient bonds raising the interface charge density to a peak. The rate of oxidation of the hydride bonds is proportional to the oxygen ratio in the ambient. This explains the sharp increase in the interface charge density when the

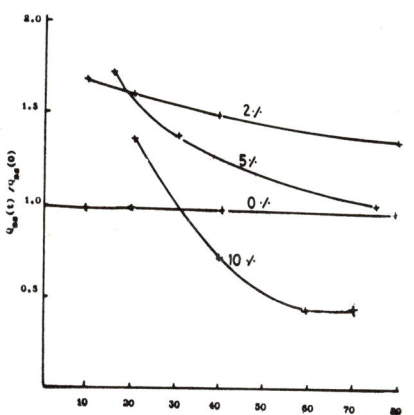

Fig. 1. Dependence of $Q_{ss}(t)$ on the annealing time and oxygen ratio

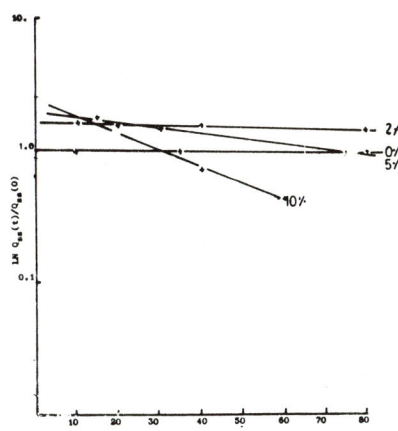

Fig. 2. $\ln(Q_{ss}(t)/Q_{ss}(0))$ versus annealing time and oxygen ratio.

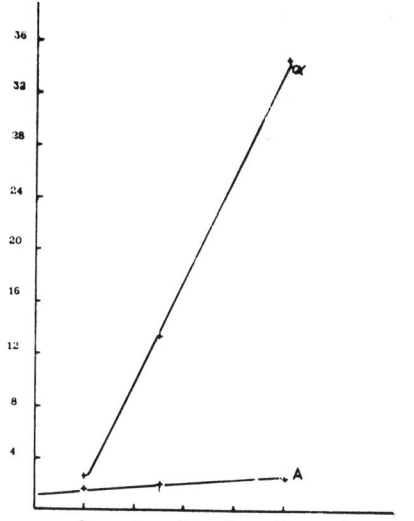

Fig. 3. Dependence of A and α on the oxygen ratio.

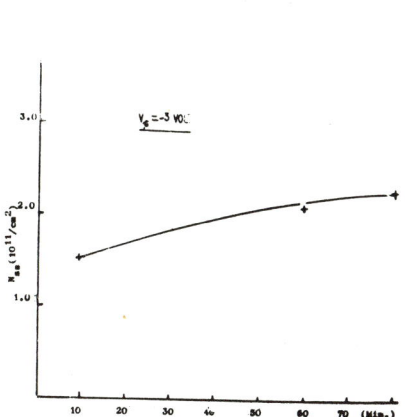

Fig. 4. Surface states density versus annealing time in pure nitrogen under -3 volt bias.

oxygen ratio is 10% compared to the increase when the ratio is 2% as shown in fig. 1. In all cases, the decay in the interface charge density is due to the saturation of the oxygen deficient oxide bonds. Hess et al.(2) suggested this to justify their data which differed from ours in the early few minutes of annealing.

It is of interest to show here that the change of the interface charge density fits the relation

$$Q_{ss}(t) / Q_{ss}(0) = A(x) \exp(-\alpha(x)t),$$

where A is a linear function of the oxygen ratio x. It was determined by extrapolating the semilogarithmic plot of $Q_{ss}(t)$ to t=0. Therefore, it depends on the maximum value of the interface charge density attained during annealing. The linear relation is in agreement with the explanation that the increase in the interface charge density depends on the ratio of oxygen in the annealing ambient. The coefficient $\alpha(x)$ acts to decrease the value of the interface charge density with time. This is physically done by the saturation of the oxygen deficient oxide bonds. Therefore, the coefficient should increase with the increase in the oxygen ratio in the ambient.

REFERENCES

(1) B. E. Deal, J. E. Chem. Soc., 121, 198C (1974).
(2) B. E. Deal, ibid, 122, 1123 (1975).
(3) A. K. Sinha etal., Solid St. Electron., 2=, 531 (1978).
(4) K. H. Beckmann and N. J. Harrick, J. E. Chem. Soc., 118, 614(1971).
(5) A. G. Revesz, ibid, 124, 1811(1977).
(6) A. G. Revesz, ibid, 126, 122(1979).
(7) E. H. Nicollian and A. Goetzberger, Bell Syst. Tech. J., 46, 1055(1967).
(8) F. Montillo and P. Balk, J. E. Chem. Soc., 118, 1463(1971).
(9) A. G. Rvesz, J. Non-cryst. Solids, 11, 309(1975).

CHAPTER V
OXIDATION - COMPOUND SEMICONDUCTORS

CHEMICAL REACTIONS IN NATIVE OXIDE
FILMS FORMED ON III-V SEMICONDUCTORS

G. P. Schwartz
Bell Laboratories, Murray Hill, New Jersey 07974

ABSTRACT

Electrochemically formed native oxide films on III-V substrates are subject to two distinct chemical reaction processes upon thermal annealing. For anodic films on GaAs and GaSb, oxide-substrate reactions have been observed which yield interfacial deposits of As and Sb via $As_2O_3 + 2GaAs \rightarrow Ga_2O_3 + 4As$ and $Sb_2O_3 + 2GaSb \rightarrow Ga_2O_3 + 4Sb$. The determination of relevant sections of the Ga-As-O and Ga-Sb-O condensed phase ternary diagrams has provided a thermodynamic basis by which these reactions and others can be understood and predicted. Thermally annealed anodic films on GaP and InP do not appear to be subject to an oxide-substrate reaction. Instead, oxide-oxide reactions are predicted to occur involving P_2O_5 and either Ga_2O_3 or In_2O_3. These reactions can yield ortho or metaphosphates according to $P_2O_5 + M_2O_3 \rightarrow 2MPO_4$ and $3P_2O_5 + M_2O_3 \rightarrow 2M(PO_3)_3$, where M is Ga or In.

INTRODUCTION

At the present time neither native oxide films nor deposited insulators can be produced on III-V substrates with bulk dielectric and interface properties comparable to the $Si-SiO_2$ system. The current absence of compatible dielectric-substrate pairs has generally limited the utilization of III-V materials in many device-related applications. Despite the considerable experimental effort aimed at forming a viable native oxide by thermal, plasma, or electrochemical oxidation techniques, little data is available concerning either the thermal or long term stability of such films.

The present paper briefly summarizes the efforts of my colleagues and myself in identifying and understanding thermally induced solid state reactions in native oxides on the III-V materials GaSb, GaAs, GaP, and InP. Although most of this work has emphasized electrochemically anodized films, some data are also available on rf and dc plasma oxides on GaAs. These reactions are in fact not specific to the electrochemical anodization process, but will occur whenever certain oxide components are present in the films. Reflection Raman scattering has been the primary tool in identifying these reactions. Portions of the relevant ternary phase diagrams have been estimated from thermodynamic calculations and binary-mixture phase stability experiments. The phase diagrams provide the basis by which the observed reactions can be understood and general reaction schemes predicted.

EXPERIMENTAL

Sample Preparation

N-type substrates of GaSb (2×10^{17} cm^{-3}, (111)), GaAs ($2-5 \times 10^{18}$ cm^{-3}, (100)), GaP (6×10^{17} cm^{-3}, (111)), and InP (2×10^{18} cm^{-3}, (100)) were polished to a mirror finish and then anodized at room temperature in an ethylene glycol based electrolyte. This electrolyte consists of 3% H_3PO_4 adjusted to a pH of 6.2 with NH_4OH, which is then diluted in a 1:2 volume ratio with ethylene glycol.[4] The wafers were anodized at constant current (1 mA/cm^2) up to 50 volts. The samples were sealed into evacuated (5×10^{-7} torr) quartz tubes prior to the annealing step.

Raman Scattering

Surface reflection Raman spectra were obtained by coupling the incident laser light (5145 Å) into the films at Brewster's angle and detecting the light scattered normal to the sample surface. An Instruments S.A. RAMANOR monochromator (Model HG-2S) equipped with holographic gratings and f/1.8 collection optics was used to analyze the scattered light. The analyzed light was detected with a cooled Hammamatsu R-928P photomultiplier coupled to conventional photon counting electronics. The electric vector of this incident light was adjusted with a polarization rotator to lie in the scattering plane of the sample (H). For the work reported here, the polarization of the Raman scattered light was unanalyzed (U). This combination of incident and scattered polarization properties is designated as HU.

Phase Diagrams

In addition to performing thermodynamic calculations whenever data were available, binary mixture experiments were also performed in order to check critical sections of the ternary phase diagrams. Powdered mixtures of the III-V materials and the respective group V oxide (Sb_2O_3, As_2O_3, or P_2O_5) were sealed in evacuated quartz tubes and reacted at elevated temperatures. The reaction products were then analyzed by x-ray powder diffraction and Raman scattering and compared with the phase field predictions resulting from the thermodynamic calculations. Details of these experiments and calculations can be found in Ref. 1 for the Ga-As-O system and in Ref. 2 for the Ga-Sb-O and Ga-P-O systems.

RESULTS AND DISCUSSION

General

Two general comments are in order prior to discussing detailed results of thermally induced native oxide reactions with III-V materials. The first point to consider concerns the bulk composition of electrochemically anodized films on GaSb, GaAs, GaP, and InP. Currently available data suggest that for the Ga based materials the primary oxide components consist of Ga_2O_3 and Sb_2O_3, As_2O_3, or P_2O_5 respectively.[3] For plasma oxidized GaAs, the major components also are Ga_2O_3 and As_2O_3[4] with elemental As typically present as a minor constituent. Anodic films on InP appear to consist primarily of In_2O_3 and P_2O_5[5]; the actual ratios of these components in the films are sensitive to the pH of the electrolyte.[6] As will be discussed later, the film constituents resulting from anodic or plasma growth are not those anticipated from thermodynamic considerations in which film formation is assumed to occur under "near equilibrium" conditions.

The second point concerns the detection sensitivity of the Raman scattering technique in terms of which chemical species can actually be detected in thin (1000 Å) films in the present scattering configuration. The topic has been examined in detail elsewhere;[2,7] the oxide components being optically transparent in the visible are not detected at the excitation wavelength (5145 Å). In contrast, the metalloids Sb, As, and P all absorb at 5145 Å and are subject to resonance enhancement of their Raman cross sections. The substrate materials also absorb and in addition are present in bulk quantities exceeding 1000 Å.

GaSb and GaAs

Surface reflection Raman spectra from as-anodized (A) and thermally annealed (B) anodic films taken from Ref. 2 are shown in Fig. 1. S_1 and S_2 denote the positions of the A_{1g} and E_g modes of crystalline Sb in its A7 structure. Amorphous antimony displays a broad, featureless band between 120-170 cm^{-1}. The numbered peaks (n=1,16) can all be assigned to one and two-phonon scattering from the substrate.[2] The most prominent features in the as-anodized spectrum are the longitudinal and transverse (LO, TO) optic modes of the substrate. Within the peak-to-peak noise ($\sim$ 30 c/sec) in the region between 120-170 cm^{-1}, no differences were detected between the bare substrate (not shown) and the as-anodized GaSb wafer. The estimated detection limit for elemental Sb is an equivalent thickness of 5-10 Å.[2] By contrast, the annealed (450°C/1 hour) sample spectrum is dominated by the A_{1g} and E_g modes of crystalline Sb.

Chemical etching experiments were performed in order to locate the spatial position of the Sb. No reduction in the Raman signal associated with Sb was observed down to a remaining oxide film thickness of $\sim$ 300 Å; below that point interference colors could no longer be used to gauge the film thickness. The Sb formed during annealing is thus localized within 300 Å of the oxide-GaSb interface, consistent with binary mixture experiment data which indicated a reaction between Sb_2O_3 and GaSb according to

$$Sb_2O_3 + 2GaSb \rightarrow Ga_2O_3 + 4Sb \tag{1}$$

It is worthwhile pointing out that any growth technique (electrochemical, plasma, etc.) which produces Sb_2O_3 in a film on GaSb

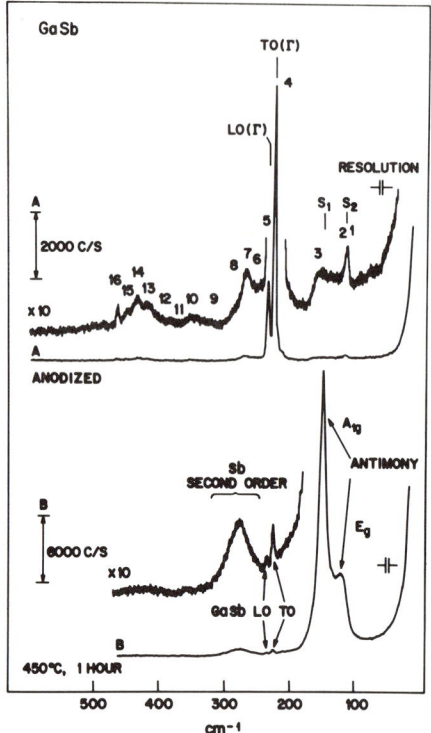

Fig. 1. Raman scattering from as-anodized and thermally annealed anodic films on GaSb. Data from Ref. 2.

leaves a dielectric layer which is subject to eventual chemical reactions. This point is illustrated in Figs. 2 and 3 for anodized and plasma oxidized films on GaAs. Details of the experiments can be found in Ref. 4.

One notes in Fig. 2 that some elemental As (estimated at ~ 40 Å), can be detected in the as-grown plasma films (1000 Å) whereas any residual As in the electrochemical film is below the detection limits. The LO and TO modes and two-phonon scattering from the bare substrate dominate all the spectra in the as-grown condition. Thermal annealing however generates considerable arsenic deposits in all cases (see Fig. 3). Although the reaction rates may vary somewhat, all three films are incipiently unstable due to the presence of As_2O_3 in these films. A reaction equation similar to Eqn. 1 is operative, i.e.

$$As_2O_3 + 2GaAs \rightarrow Ga_2O_3 + 4As \qquad (2)$$

Chemical etching experiments once again verify that the arsenic deposits lie close to the oxide-substrate interface[7] as is implicitly

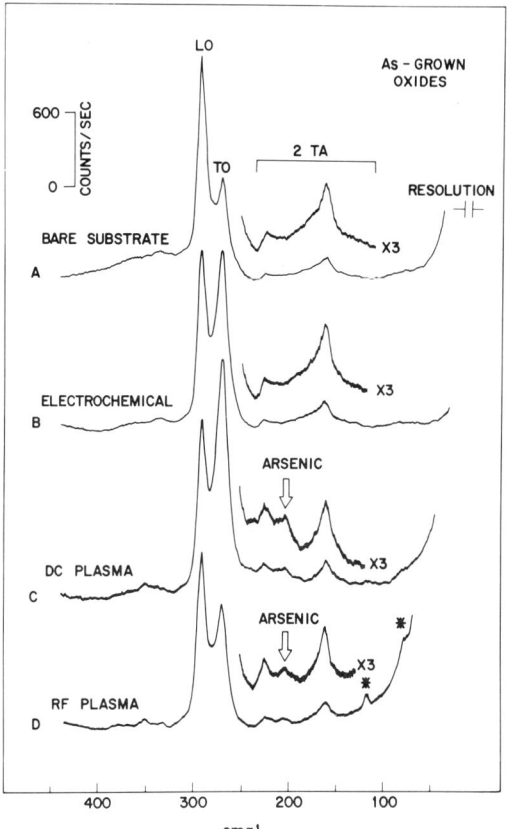

Fig. 2. Raman scattering from anodically and plasma oxidized films on GaAs prior to thermal annealing. Data from Ref. 4.

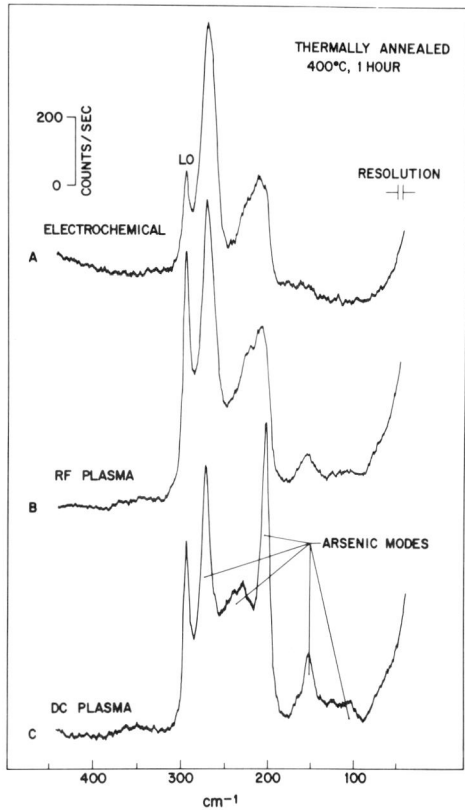

Fig. 3. Raman scattering from the wafers in Fig. 3 after thermal annealing. Data from Ref. 4.

suggested in Eqn. 2. Estimates of the Ga-Sb-0 and Ga-As-0 ternary phase diagrams provide a uniform basis on which to understand and predict the observed chemical reactions. Details of the phase diagram constructions can be found in Refs. 1 and 2. Figure 4 shows the condensed phase portion of these ternary diagrams in which mutual solid solubility has been neglected. Only the lower section of the Ga-Sb-0 diagram has been estimated; it suffices however to explain the observed interfacial reaction on GaSb.

Consider the Ga-As-0 diagram with the stable tie lines as shown. Materials which are directly connected to GaAs by a tie line (Ga, Ga_2O_3, As) will not react with it. The other condensed phases ($GaAsO_4$, As_2O_5, As_2O_3) will react with GaAs however (including O_2, which is not a condensed phase). For thin native oxide films composed primarily of Ga_2O_3 and As_2O_3, thermal annealing will drive

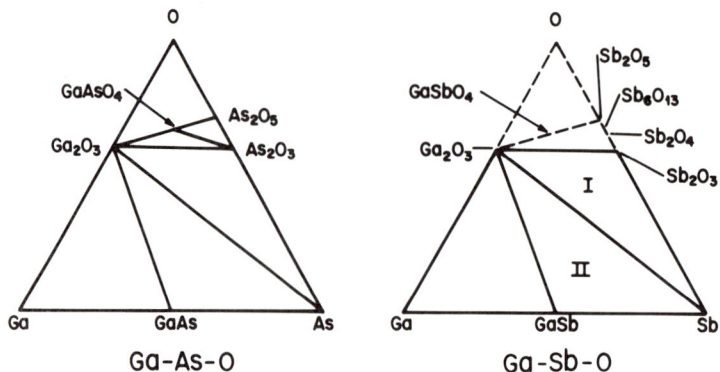

Fig. 4. Estimated condensed phase ternary diagrams for Ga-As-O and Ga-Sb-O. Data from Refs. 1 and 2.

the system toward the thermodynamically favored configuration. Equation 2 is nothing more than a mass balance expression for the crossing point of two lines which connect the pairs Ga_2O_3/As and As_2O_3/GaAs. Assuming that all the As_2O_3 is reacted in the annealing process, the final equilibrium phase field is that bordered by GaAs, Ga_2O_3, and As. The thermodynamically driven reaction thus results in the conversion of a Ga_2O_3/As_2O_3 film into one containing Ga_2O_3/As. Similar considerations obtain in the Ga-Sb-O system and suffice to explain the observed reaction between Sb_2O_3 and GaSb.

GaP and InP

Raman scattering from as-grown and thermally annealed anodic films on GaP are shown in Fig. 5. The data are from Ref. 2. All of the numbered structures shown in the top panel can be rigorously assigned to the substrate; P_1 and P_2 denote the anticipated positions of the strongest modes in crystalline red phosphorous. As careful comparison of panels A and B show, there is no evidence for elemental phosphorous in either the as-grown or annealed films. Similar results have also been obtained for anodic films in InP.[9] Higher annealing temperatures (up to 700°C on GaP) have also failed to yield evidence for a reaction between P_2O_5 and either substrate. In a separate binary mixture experiment,[2] P_2O_5 and powdered GaP failed to react conclusively after heating at 850°C for 1 day.

At the present time there are insufficient data available for a complete calculation of either the Ga-P-O or In-P-O ternary diagrams. Of the 24 possible diagrams associated with a ternary system composed of 8 condensed phases, 18 of the diagrams can be eliminated in the case of Ga-P-O using currently available data. The remaining possibilities are shown in Fig. 6. Only one of these diagrams

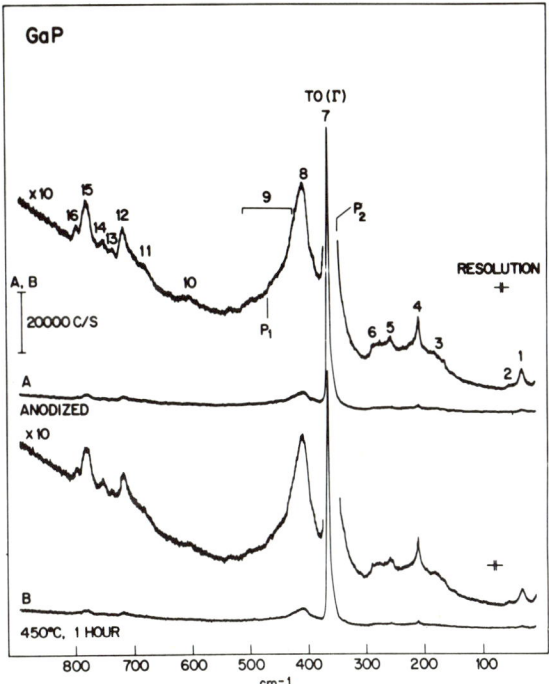

Fig. 5. Raman scattering from as-anodized and annealed anodic films on GaP. Data from Ref. 2.

contains a tie line connecting P_2O_5 and GaP; it is the diagram which we currently favor. It should be noted however that not all of the remaining 5 diagrams imply the formation of elemental phosphorus as a reaction product between P_2O_5 and GaP. The top most three diagrams would not yield phosphorus, and it is conceivable that if these reaction products did form they might not have been detected by Raman scattering in such thin films because they are not optically absorbing. In continuation of this point, one notes that two stable compounds intersect the Ga_2O_3-P_2O_5 tie line; these are $GaPO_4$ and $Ga(PO_3)_3$. For an anodic film containing Ga_2O_3 and P_2O_5, reaction Eqns. 3 and 4 listed below may compete to tie up the oxide film components.

$$Ga_2O_3 + P_2O_5 \rightarrow 2GaPO_4 \qquad (4)$$

$$Ga_2O_3 + 3P_2O_5 \rightarrow 2Ga(PO_3)_3 \qquad (5)$$

We have observed that powdered mixtures of Ga_2O_3 and P_2O_5 react readily at 650°C and suspect that this product is indeed being formed although we have not detected it yet in thin annealed films using Raman scattering.

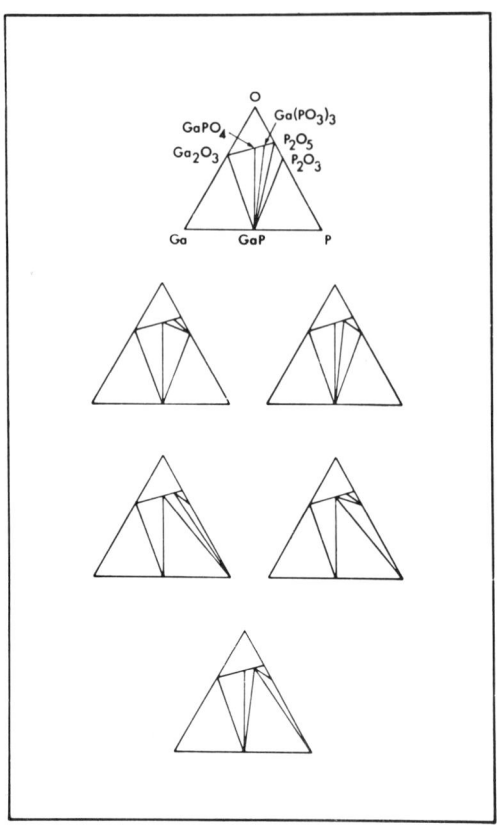

Fig. 6. Six possible choices for the Ga-P-O phase diagram. Data from Ref. 2.

The situation with InP is even more uncertain since even less thermodynamic data is available. Preliminary experiments on the direct thermal oxidation of InP[9] suggest that the In-P-O system may be similar to the Ga-P-O diagram. All of the diagrams shown in Fig. 6 would predict that the "near equilibrium" oxidation of GaP using O_2 should yield a single phase film of $GaPO_4$. X-ray diffraction[10] and photoemission[11] experiments verify this expectation for both high and low oxidation temperatures. Figures 7 and 8 indicate that the situation on InP may be complicated by kinetic factors.

For oxidation temperatures below 550°C, we have observed that the oxide film contains elemental crystalline red phosphorus,[9] as can be observed in Fig. 7c by comparison with Raman spectra from the substrate and red P. For oxidation temperatures exceeding 650°C, our preliminary results obtained in thick films suggest that the native oxide is primarily composed of $InPO_4$ (see Fig. 8b) with

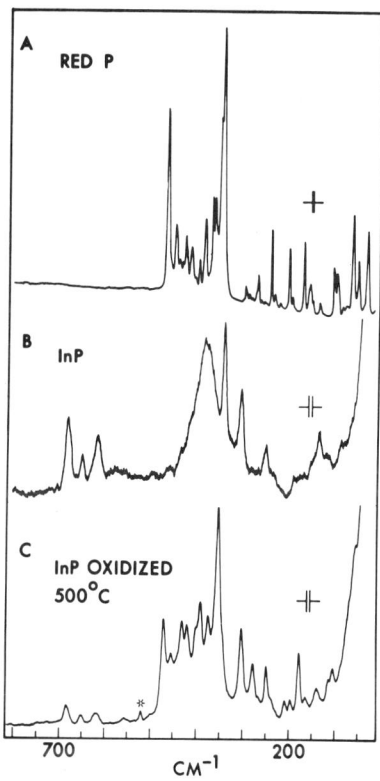

Fig. 7. Raman scattering from crystalline red P (A), bare InP (B), and air oxidized InP (C). Data from Ref. 9.

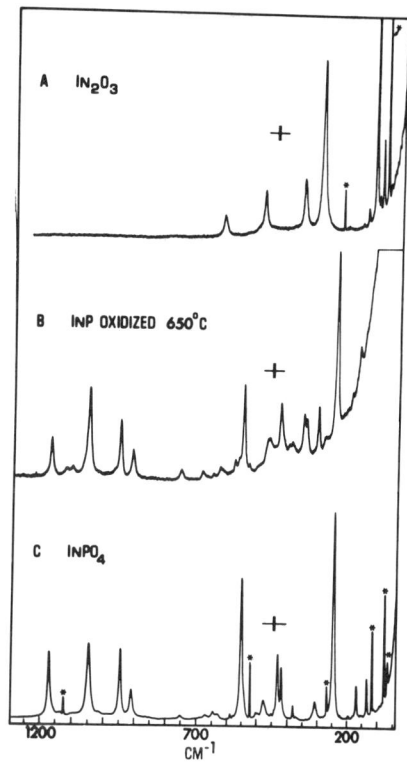

Fig. 8. Raman scattering from In_2O_3 (A), air oxidized InP (B), and $InPO_4$ (C). Asterisks denote laser plasma lines. Data from Ref. 9.

some In_2O_3 present as a minor constituent. Our current interpretation of this data is that kinetic factors dominate product formation below 550°C, but above this temperature thermodynamic considerations obtain. In the latter case, the phase diagram for the In-P-O system appears to bear some similarity to those suggested for the Ga-P-O diagrams. Considerable work clearly remains to be accomplished before a final version of either ternary diagram can be put forth with any degree of confidence.

SUMMARY

Observations of thermally induced interfacial oxide-substrate reactions have been presented for anodic films on GaSb and both anodic and plasma films on GaAs. Similar reactions have not been observed in anodic films on GaP and InP; instead it is suggested that oxide-oxide reactions may occur which can yield either ortho or metaphosphates.

A general understanding of these reactions is possible based on thermodynamic considerations implicit in the appropriate forms of the respective ternary phase diagrams.

Acknowledgements

It is a pleasure to acknowledge the many contributions of my colleagues and coauthors J. E. Griffiths, G. J. Gualtieri, C. D. Thurmond, B. Schwartz, and W. A. Sunder at Bell Laboratories and Dr. Takuo Sugano of the University of Tokyo.

References

1. C. D. Thurmond, G. P. Schwartz, G. W. Kammlott, and B. Schwartz, J. Electrochem. Soc. 127, 1366 (1980).

2. G. P. Schwartz, G. J. Gualtieri, J. E. Griffiths, C. D. Thurmond, and B. Schwartz, J. Electrochem. Soc., in press.

3. For a review of the literature up to 1977, see C. W. Wilmsen and S. Szpak, Thin Solid Films 46, 17 (1977). Reference 2 contains selected references for data after 1977.

4. G. P. Schwartz, B. Schwartz, J. E. Griffiths, and T. Sugano, J. Electrochem. Soc., in press.

5. C. W. Wilmsen and R. W. Kee, J. Vac. Sci. Technol. 15, 1513 (1978).

6. C. W. Wilmsen, presented at the Workshop for Dielectrics on III-V's (San Diego, 1980).

7. G. P. Schwartz, J. E. Griffiths, and B. Schwartz, J. Vac. Sci. Technol. 16, 1383 (1979).

8. J. S. Lannin, Phys. Rev. B. 15, 3863 (1977).

9. G. P. Schwartz, J. E. Griffiths, and W. A. Sunder, unpublished.

10. M. Rubenstein, J. Electrochem. Soc. 113, 540 (1966).

11. R. Nishitani, H. Iwasaki, Y. Mizokawa, and S. Nakamura, Japan J. Appl. Phys. 17, 321 (1978).

ANODIC OXIDE INSULATORS ON InP AND InAs

D. A. Baglee, D. H. Laughlin, C. W. Wilmsen and D. K. Ferry
Colorado State University, Fort Collins, Colorado 80523

ABSTRACT

The insulating properties of anodic oxides grown on InP and InAs have been investigated and the composition of these oxides were qualitatively correlated to the resistivity. The composition of the InP anodic oxide layers were found to be strongly dependent on the electrolyte and the pH. The resistivity of oxide layers containing a low concentration of P_2O_5 was $\approx 10^{10}$ ohm-cm at 1 volt while those with high P_2O_5 concentration had a resistivity of $\approx 10^{15}$ ohm-cm at 10 volts. The current through the high resistivity InP oxides appears to be controlled by a bulk mechanism. The composition of the InAs oxides did not change significantly with electrolyte and pH, however there appeared to be variation in the interface width. The current for InAs oxides appears to be electrode limited and the resistivity ranges from 10^{13} to 10^{15} ohm-cm at 10 volts. All of the oxides are sensitive to exposure to air.

INTRODUCTION

The insulating characteristics of the anodic oxides of III-V compound semiconductors have been investigated only superficially, considering the many different ways of growing anodic oxides, e.g. varying the electrolyte, pH and current density. The insulating properties of GaAs anodic oxides grown by the AWG method have been reported. Weimann[1] has carefully examined the current-voltage (I-V) curves for one growth condition, Weiss et al.[2] reported the effects of annealing and Zeisse et al.[3] investigated the affects of exposure to a humid environment. Kohn and Hartnagel[4] reported on the charge injection into these oxides. The insulating properties of GaSb[5] and GaP[6] oxides have been investigated far less. I-V curves for anodic oxides of InP[7-10], InAs[7,11] and InSb[12-14] have been reported for oxides grown in a number of electrolytes. Of these, the insulating characteristics of the InSb oxides have been the most completely investigated. While current-volt characteristics have been investigated, very little has been reported on any possible correlation between the I-V characteristics of III-V oxide layer and its chemical composition (only InSb[14] and InAs[9] have been examined). Thus, there is presently little understanding of the relationship between the composition of an oxide and it's electrical insulating properties. In this paper we report the I-V curves and chemical compositional profiles of anodic oxides grown on InP and InAs in a variety of electrolytes at pH values from 2 to 7.5. From these data, it is postulated that the band gaps of In_2O_3, As_2O_3 and P_2O_5 and the amount of water in the oxide layer are important parameters which affect the insulating quality.

EXPERIMENTAL TECHNIQUE

Anodic oxides were grown on n-type InP wafers oriented <100> with a net carrier concentration of ~2 x 10^{17} cm^{-3}. The InAs wafers were p-type, <111> oriented zinc doped to a concentration of 2 x 10^{17} cm^{-3}. The surface of each sample was prepared by first soaking the wafer in a 10% HCl solution for 10-15 min. in order to remove any oxide. Then the surface was chemomechanically polished in 1% bromine-methanol, rinsed in methanol, dipped in Chemsol Z and soaked in ultra-pure water. The anodic oxides were grown with constant current; the current density ranged from 1 to 5 mA/cm^2 depending on the electrolyte. A number of electrolytes were used as listed in Table 1. The electrolytes for InP anodization were mixed 1:1

TABLE 1 Electrolytes used to anodize InP and InAs

InP	InAs
3% arsenic acid	25% arsenic acid
10% citric acid	3% tartaric acid
3% phosphoric acid	0.1N KOH

with propylene glycol and for InAs, 1:2. The pH of the acidic electrolytes were adjusted with NH$_4$OH. HCl was used to adjust the KOH solution. The composition and bonding of the grown oxide layers was determined by ESCA profiling using the 2P$_{3/2}$ line of As and P and the 3d$_{3/2}$ line of In. Discussion of our profiling technique is reported elsewhere.[15,16] The metal contacts were deposited either immediately after oxide growth or after storage in a desiccator.

DISCUSSION OF RESULTS

InP

The current density vs. (applied voltage)$^{\frac{1}{2}}$, J-V$^{\frac{1}{2}}$, for anodic oxides of InP grown in the three above electrolytes at a pH of 6.5 and 2.0 are given in Figs. 1 and 2.

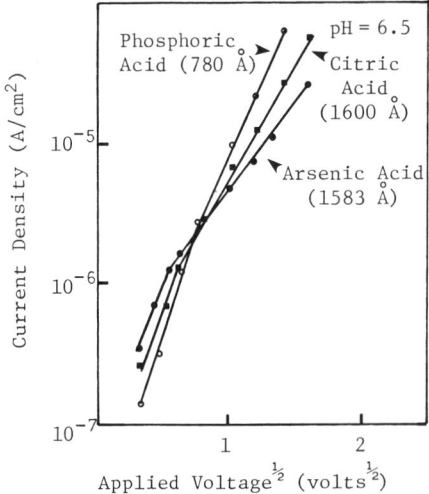

Fig. 1. A comparison of the insulating characteristics of anodic oxides of InP grown in 3 different electrolytes at pH = 6.5.

The curves were obtained from freshly grown oxides which were stored and measured in a dry environment at room temperature. The curves for various samples were, within experimental error, independent of the applied voltage polarity. Occasionally the current through an oxide sample had a definite polarity dependence. In some cases this could be attributed to applying too large a gate voltage but for other sampes there was no apparent reason for the polarity dependence.

Figure 1 shows that the J-V$^{\frac{1}{2}}$ curves for oxides grown in an electrolyte at a pH of 6.5 are essentially the same and that the resistivity of these oxides is low, ≈10^{10} ohm-cm at 1 volt. In these same electrolytes at a pH of 2.0, the J-V$^{\frac{1}{2}}$ curves vary dramatically with the electrolyte used to grow the oxides, as seen in Fig. 2. The breakdown strength of the oxide grown in phosphoric acid at a pH = 2 was ≈2 x 10^6 V/cm. ESCA profiles

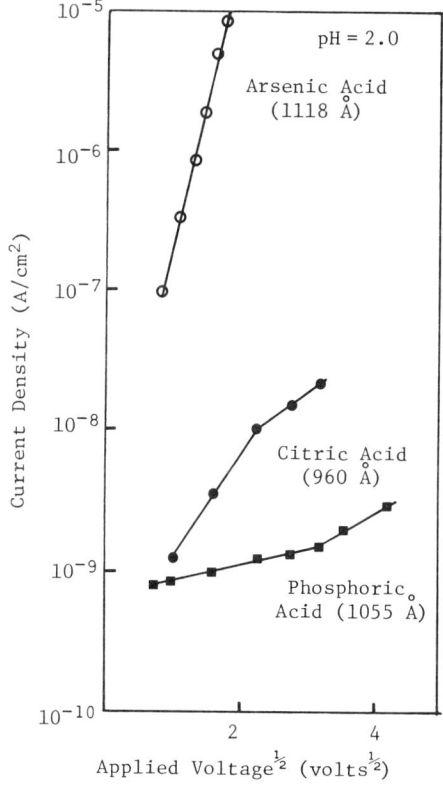
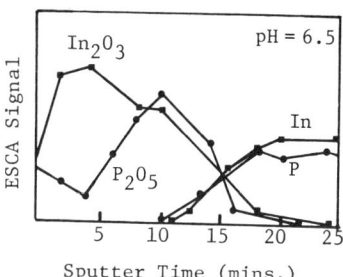
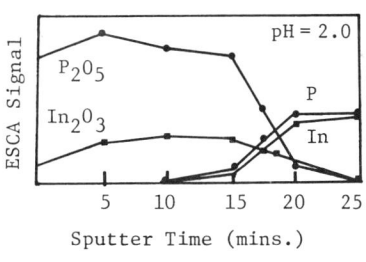

Fig. 2. A comparison of the insulating characteristics of anodic oxides of InP grown in 3 different electrolytes at pH = 2.0.

Fig. 3. ESCA profiles of anodic oxides of InP grown in 3% phosphoric acid.

of these oxides, such as those shown in Fig. 3 for oxides grown in 3% phosphoric acid, were used to determine the composition of the oxide layers. The large increase in the P_2O_5 concentration[17] of the film grown at a pH of 2 appears to be due to the rapid dissolution of In from the growing oxide film[18], although both P and In are thought to be dissolving. From the ESCA profiles of Fig. 3 and profiles of the oxides grown in the other electrolytes (not shown here) it is observed that the oxide layer resistivity increases as the P_2O_5/In_2O_3 ratio increases. This is modified somewhat by the uniformity of the layer composition.

The increase in resistivity with increasing P_2O_5/In_2O_3 ratio may be related to the band gaps of the P_2O_5 and In_2O_3. The band gap of In_2O_3 has been reported[19,20] to be between 3.1 and 3.75 eV while P_2O_5 should have a band gap of between 6 and 10 eV based on the position of P in the periodic table. The large band gap of P_2O_5 may yield a higher resistivity insulating layer. Other factors, such as water content, impurities and oxide imperfection, must be considered however, since these will influence the location of the Fermi energy in the oxide band gap. For example, In_2O_3 can have oxygen vacancies[18,21] which introduces a donor site near the conduction band edge. This can greatly reduce the resistivity of the In_2O_3 and in fact can cause the Fermi energy to move into the conduction band.

The as-grown oxide appears to be relatively free of water since high resistivity oxides can be grown and annealing in dry N_2 does not increase the resistivity very much. However, the oxide layer readily adsorbs water from the atmosphere and the resistivity decreases rapidly when exposed to room air.

InAs

The room temperature J vs. $V^{\frac{1}{2}}$ characteristics of freshly prepared InAs anodic oxides grown in different electrolytes and with different electrode metals are illustrated in Figs. 4 and 5. Unlike the InP oxides, these characteristics are

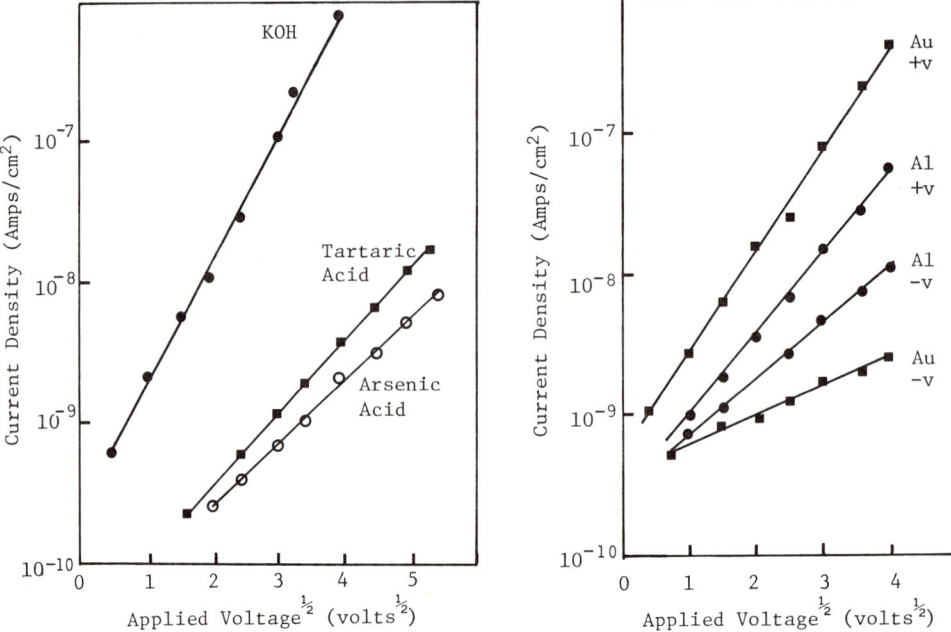

Fig. 4. A comparison of the insulating characteristics of anodic oxides of InAs grown in 3 different electrolytes.

Fig. 5. Electrode dependence of the InAs anodic oxide J vs. $V^{\frac{1}{2}}$ curves.

polarity dependent and do not change with the pH of the electrolyte. ESCA profiles of all the oxides are similar to that shown in Fig. 6. Note that the composition of the oxide is uniform and composed of approximately equal parts of In_2O_3 and As_2O_3, however there are some apparent differences in the interfacial regions for oxides grown in different electrolytes, as listed in Table 2. The fact that the current is polarity dependent, the bulk oxide composition is the same for all layers, the interface widths of the oxides change with electrolyte and the linear lnJ vs. $V^{\frac{1}{2}}$ plots all suggest that the current is electrode limited (possibly Schottky emission). An experimental barrier height of ≈1 eV is obtained from the lnJ vs. V curves of Fig. 4 by assuming a simple Schottky emission model. This is below the maximum value estimated from the band gaps of the oxides by the

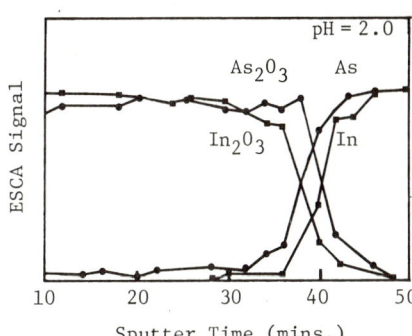

TABLE 2 Characteristics of InAs anodic oxides

Electrolyte	Apparent Interface width	Dielectric breakdown field
KOH	120Å	1.0×10^6 V/cm
Tartaric Acid	55	3.2
Arsenic Acid	50	3.7

Fig. 6. ESCA profile of an anodic oxide on InAs grown in arsenic acid at a pH of 2.

expression ½ oxide band gap – band gap of InAs. The band gaps of the oxides are $In_2O_3 \doteq 3.1$ to 3.5 eV (ref. 19,20) and $As_2O_3 \doteq 4$eV (ref. 22).

As seen from Table 2, the breakdown field for the best InAs oxides is almost twice that for the best InP oxide, even though the band gap is much larger than that of As_2O_3. This requires further investigation.

Conclusions

It appears that the insulating properties of freshly prepared anodic oxide layers of InP and InAs are determined, largely, by the band gap and concentration of the oxides in the layers, and to some extent the uniformity of the layers. The concentration of water can greatly reduce the resistivity of the layers but this does not appear to be significant unless the oxides are exposed to air for several hours.

Acknowledgements

The authors wish to thank Rich Henry of NRL and Art Clawson of NOSC for providing the InP and InAs wafers used in this research. This work was supported by the Army Research Office.

References

1. G. Wiemann, Thin Solid Films, 47, 127 (1977).
2. B. Weiss, E. Kohn, B. Bayraktaroglu and H. L. Hartnagel, Inst. Phys. Conf. Ser., No. 33a, 168 (1977)
3. C. R. Zeisse, L. J. Messick and D. L. Lile, J. Vac. Sci. Tech., 14, 957 (1977).
4. E. Kohn and H. L. Hartnagel, Solid State Electron., 21, 409 (1978).
5. C. W. Fischer, N. Leslie and A. Etchells, J. Vac. Sci. Tech., 13, 59 (1976).
6. J. M. Poate, P. J. Silverman and J. Yahalom, J. Phys. Chem. Solids, 34, 1847 (1973).
7. C. W. Wilmsen, Crit. Rev. Solid State Sci., 5, 313 (1975).
8. D. L. Lile and D. A. Collins, Appl. Phys. Lett., 28, 554 (1976).
9. C. W. Wilmsen, Thin Solid Films, 39, 105 (1976).
10. K. P. Pande and G. G. Roberts, J. Vac. Sci. Tech., 16, 1470 (1979).
11. D. A. Baglee, D. K. Ferry, C. W. Wilmsen and H. H. Wieder, J. Vac. Sci. Tech., 17, Sept./Oct. issue (1980).
12. T. Sakurai, T. Suzuki and Y. Noguchi, Jap. J. Appl. Phys., 7, 1491 (1968).

13. C. W. Wilmsen, G. C. Vasbinder and Y. K. Chan, J. Vac. Sci. Tech., 12, 56 (1975).
14. C. W. Wilmsen, J. Vac. Sci. Tech., 13, 64 (1976).
15. C. W. Wilmsen and R. W. Kee, J. Vac. Sci. Tech., 15, 1513 (1978).
16. C. W. Wilmsen, R. W. Kee and K. M. Geib, J. Vac. Sci. Tech., 16, 1434 (1979).
17. In this paper we refer to the phosphorus oxide as P_2O_5 since the ESCA spectra for P indicates that P is in the P^{+5} oxidation state. The P bonding may, however, be more complex.
18. Atlas of Electrochemical Equilibrium in Aqueous Solution, M. Pourbaix, Pergamon Press, 1976, New York.
19. R. L. Weiher, J. Appl. Phys., 33, 2834 (1962).
20. R. L. Weiher and R. P. Ley, J. Appl. Phys., 37, 299 (1966).
21. Physics of Thin Films, Vol. 9, Ed. G. Hass, M. H. Francombe and R. W. Hoffman, 1977, p. 1-71, J. L. Vossen, Academic Press, New York.
22. W. P. Doyle, J. Phys. Chem. Solids, 4, 144 (1958).

OPTICAL PROPERTIES AND INTERFACE ANALYSIS OF THE GaAs-ANODIC OXIDE SYSTEM

D. E. Aspnes, G. J. Gualtieri, B. Schwartz,
G. P. Schwartz, and A. A. Studna
Bell Laboratories, Murray Hill, New Jersey 07974

ABSTRACT

The optical properties and interface widths of a highly uniform electrochemically grown anodic oxide on GaAs were determined by spectroscopic ellipsometry. Modeled as amorphous As, the GaAs-oxide interface is less than 3Å wide. The oxide-ambient interface shows a total distributed density deficit equivalent to 10Å lost material, probably due to leaching of As_2O_3 in the surface region.

INTRODUCTION

Analysis of recent Raman[1,2] and ion scattering spectroscopy[3] (ISS) data involving the thermal, plasma, and electrochemical oxidation of GaAs indicates that the relative elemental arsenic content of as-grown, electrochemically formed films can be suppressed for low current density growth. A second feature associated with lowering the current density appears to be a concomitant decrease in the oxide-substrate interface width.[3] Both of these features allow a more accurate determination of the intrinsic dielectric response of anodic films on GaAs. The latter data are not only useful in their own right, but also as input for determining composition and interface parameters of oxides formed by other techniques.

The work reported here is part of a project whose objective is to determine the dielectric properties of electrochemically formed oxides and to establish new limits on widths and possible compositions of interface regions in these films.

EXPERIMENTAL

The data analyzed here for interface parameters were obtained on a <111>B single-crystal GaAs substrate with a carrier concentration of 2×10^{16} cm^{-3}. The initial surface was prepared by etch-polishing in bromine-methanol. The sample was anodized in an AGW electrolyte[4] at a constant current of 0.1ma cm^{-2} to a limiting potential of 60V. The oxide, consistent with those grown on other samples, was visually perfect.

Optical measurements were made from 1.5 to 6.0 eV with an automatic spectroscopic ellipsometer described elsewhere.[5] The sample was maintained in a dry N_2 ambient in a windowless cell to minimize surface oxidation and adsorbed contamination effects. The anodized sample was cleaned with methanol and after the tan ψ, cos Δ spectra were measured, the oxide was stripped with 1:1::HCl:methanol to obtain the dielectric function of the GaAs substrate. All cleaning and chemical processing steps were performed with the sample already aligned in the ellipsometer. Peak substrate values of ε_2 (at 4.75 eV) increased ∼0.5 after anodizing and stripping, indicating that surface quality was improved by this procedure.

RESULTS AND DISCUSSION

Figure 1 shows the dielectric function spectrum that represents the best approximation to the intrinsic response of the substrate used in this work. This spectrum differs slightly from that actually measured in that 2Å of amorphous GaAs was removed mathematically to account for etching-induced surface disorder. This is done in accordance with results obtained from similar measurements on single crystals of Ge whose surfaces were fully ordered in uhv.[6] Also shown for comparison are the previous best spectra for GaAs, obtained from a Kramers-Kronig transform of reflectance data measured over the range 1-27 eV.[7] The differences in energy positions of the structures are probably due to the smaller energy separation (17 meV) between the ellipsometric data points, which consequently were able to follow the structure more faithfully. The substantial discrepancies in absolute values are almost certainly due to the presence of inadvertent oxide or contamination layers on the reflectance samples, which in contrast to the ellipsometric samples could not be stripped in situ.

Figure 2 shows the approximate dielectric function spectrum of the 60V anodic oxide. This was calculated without regard to interface effects in the three-phase model from measured tan ψ, cos Δ spectra and the substrate data of Fig. 1. The oxide thickness was chosen to yield a value of "zero" for ε_2 in the region of transparency below 4.5 eV. In fact, ε_2 is not identically zero here but rather shows small oscillations of amplitude ~0.01 due to interfaces.[8] Nevertheless, the results show that the three-phase model is already a very good approximation, indicating that the oxide is uniform and the interfaces are narrow.

To independently determine both oxide and interface parameters, it is necessary to separate the small oscillations from the slowly varying intrinsic response of the oxide. This can be done by setting $\varepsilon_2 = 0$ in the region of transparency below 4.5 eV and by fitting in the same region a four-term Sellmeier expression,

$$\varepsilon_1(E) = 1 + \sum_{i=1}^{4} E_{gi} E_{0i}/(E_{gi}^2 - E^2), \qquad (1)$$

to ε_1. Choosing $E_{g1}...E_{g4}$ as 0.075, 5.5, 7.5, and 16 eV, respectively, corresponding approximately to the centroids of the major optical transitions, and setting $E_{01} \cong 0.35$ eV to account for the difference between infrared and dc dielectric constants, we find E_{02}, E_{03}, and E_{04} to be 1.62 ± 0.27 eV, 5.5 ± 1.9 eV, and 18.8 ± 3.7 eV, respectively.

With the above expression as a starting point, we now minimize the mean square deviation between measured tan ψ, cos Δ spectra and those computed by models of increasing complexity using the techniques previously developed for interface analysis.[8,9] For example, we suppose first the Eq. (1) exactly describes the oxide-interface system and use the oxide thickness as the single free parameter in a three-phase model. The discrepancies between measured and calculated tan ψ, cos Δ spectra are shown in Fig. 3 and the free parameter, 90% confidence limit, and mean-square deviation values given in Table 1. The one-parameter model is too simple, but it does establish a scale of reference by which to judge more elaborate models.

Logical extensions allow for a density change (void fraction) and an absorbing component in the oxide, and an interface between GaAs and the oxide overlayer. These introduce three more free parameters, which are the thickness of the interface region and the respective constituent fractions of the oxide. Results obtained by assuming that amorphous arsenic (a-As) represents both interface material and absorbing fraction are also shown in Fig. 3 and Table 1. Surprisingly, even with three additional free parameters there is very little improvement, showing that the major discrepancy is due to yet another factor.

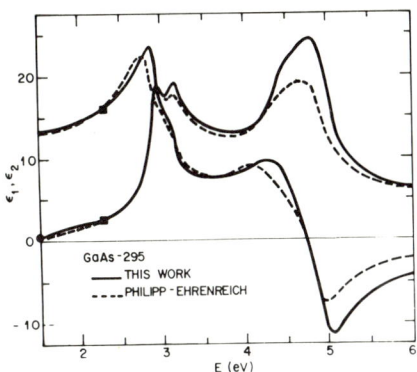

Fig. 1 Dielectric function spectra of GaAs as determined from ellipsometric (this work) and reflectance (ref. 7) measurements.

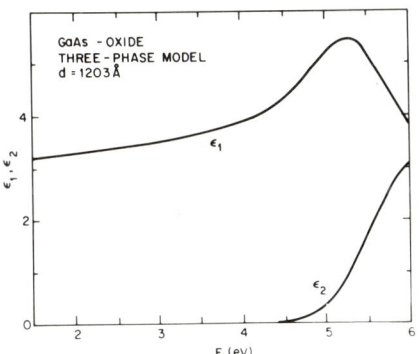

Fig. 2 Approximate dielectric function of AGW oxide on GaAs calculated in the three-phase model.

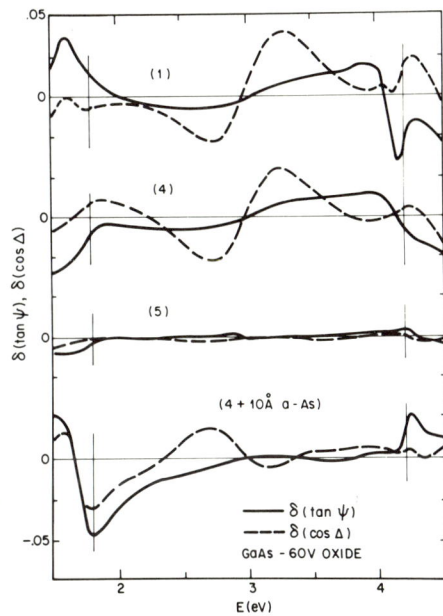

Fig. 3 Differences between measured and calculated tan ψ, cos Δ spectra for various models.

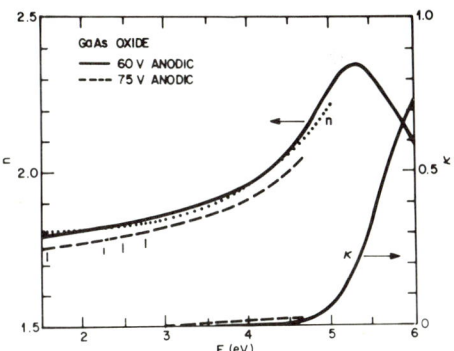

Fig. 4 Refractive index spectra of oxides on GaAs. Other data sources: (....) Palik et. al., ref. 15; (||||) Barnes and Schinke, ref. 16.

TABLE 1 Values of free parameters, 90% confidence limits, and mean square deviations of various models for a 60V AGW anodic oxide on GaAs

		1-param	4-param	5-param
GaAs-oxide interface	Composition: thickness(Å):	– –	a-As −0.9±1.9	a-As 0.3±0.3
Oxide	Void frac.: a-As frac.: thickness(Å):	– – 1197.5±0.5	0.019±0.008 0.0014±0.0008 1212±6	0.015±0.001 0.0021±0.0001 1158±2
Oxide-ambient interface	Void frac.: thickness(Å):	– –	– –	0.15(fixed) 68±2
Mean-square deviation		0.020	0.015	0.002

Noting that radioactive tracer,[10] photoemission,[11,12] and ISS[3] studies have all shown As_2O_3 depletion at the oxide-electrolyte interface for anodizations in glycol-based electrolytes, we are led to suppose a density-deficient outer layer at the oxide surface. Figure 3 and Table 1 show that this extension accomplishes what the preceding three could not -- good agreement between measurement and calculation. Thus the major discrepancy between this GaAs-oxide-ambient system and the ideal three-phase model occurs at the oxide-ambient interface, not at the substrate-oxide interface.

Of particular interest is the value of 0.3 ± 0.3Å obtained for the width of the substrate-oxide interface. Although a-As was assumed to be the interface material, clearly the region is so thin that the choice of material is essentially irrelevant as has been verified by additional calculations. To emphasize that the sensitivity to this interface width is indeed real, we have also calculated the result of minimizing the remaining four free parameters while holding the substrate-oxide thickness fixed at 10Å a-As. We find that δ = 0.020 has increased by an order of magnitude and as shown at the bottom of Fig. 3 the good agreement between measurement and calculation has vanished. But because the confidence limits do not include possible systematic errors, an upper limit of 3Å to the substrate-oxide interface width is probably more realistic. Our main point is that this interface can be very abrupt -- narrower in fact than those previously determined for thermally grown oxides on Si.[13] This suggests that the high density of interface states typically observed in CV studies of GaAs-oxide systems may have its origin in parameters unrelated to, or at least not dominated by, interfacial sharpness.

The final results for the optical response of the AGW oxide are given in Fig. 4, along with similar spectra for other electrochemically grown anodic oxides on GaAs. Our previous results,[14] obtained on films grown in an H_3PO_4-water solution under constant-voltage conditions and then heat treated at 200C for 1h, agree almost exactly with our 60V data. Spectra obtained by Palik et. al.[15] for films grown in AGW and measured without heat treating also agree almost exactly with the 60V results. Barnes and Schinke[16] obtained n values in the range 1.770 - 1.807 at λ = 6328Å for a number of samples anodized in a phosphoric acid solution and observed a drop of the order of 0.03 upon heat treating at 250C for 1h. The latter data are those given in Fig. 4. We also show in Fig. 4 a spectrum obtained by us on a film also grown in AGW but on a <100> sample with $N_D \sim 10^{18}$ cm^{-3} and to a limiting potential of 75V. This oxide has somewhat different properties from that grown to 60V.

The scatter, of the order of ±0.03 in n, may reflect differences in density as modeled here, but it could also represent small differences in the bulk film stoichiometry that result from selective dissolution of one component during film growth, or from changes that occur upon heat treatment. One should note that stoichiometry is both current-density and electrolyte dependent, making it difficult to compare these oxides directly. Anneals above 300C subject the film constituents to a variety of chemical reactions dependent on the nature of the ambient.

In our opinion, the question of elemental As content in these films is better addressed with nondestructive optical techniques such as Raman scattering and spectroscopic ellipsometry than with either photoemission or Auger profiling. One does not have to contend with electron- or ion-beam chemical reduction artifacts of As_2O_3, nor with background interference from the substrate as the interface is approached. On the other hand, spectroscopic ellipsometry does require a relatively uniform film or else simple model representations for the system will be inadequate.

REFERENCES

1. R. L. Farrow, R. K. Chang, S. Mroczkowski, and F. H. Pollak, Appl. Phys. Lett. 31, 768 (1977).
2. G. P. Schwartz, B. Schwartz, J. E. Griffiths, and T. Sugano, J. Electrochem. Soc. (in press).
3. M. Croset, J. Diaz, D. Dieumegard, and L. M. Mercandalli, J. Electrochem. Soc. 126, 1543 (1979).
4. H. Hasegawa, K. E. Forward, and H. L. Hartnagel, Appl. Phys. Lett. 26, 567 (1975).
5. D. E. Aspnes and A. A. Studna, Appl. Opt. 14, 220 (1975); Rev. Sci. Instrum. 49, 291 (1978).
6. D. E. Aspnes and A. A. Studna, Surface Sci. (in press).
7. H. R. Philipp and H. Ehrenreich, Phys. Rev. 129, 1550 (1963).
8. D. E. Aspnes, J. B. Theeten, and R. P. H. Chang, J. Vac. Sci. Technol. 16, 1374 (1979).
9. D. E. Aspnes, J. B. Theeten, and F. Hottier, Phys. Rev. B20, 3292 (1979).
10. J. C. Verplanke and R. P. Tijberg, J. Electrochem. Soc. 124, 802 (1977).
11. P. A. Breeze, H. L. Hartnagel, and P. M. A. Sherwood, J. Electrochem. Soc. 127, 454 (1980).
12. G. P. Schwartz and G. J. Gualtieri (to be published).
13. D. E. Aspnes and J. B. Theeten, Phys. Rev. Lett. 43, 1046 (1979).
14. D. E. Aspnes, B. Schwartz, A. A. Studna, L. Derick, and L. A. Koszi, J. Appl. Phys. 48, 3510 (1977).
15. E. D. Palik, N. Ginsberg, R. T. Holm, and J. W. Gibson, J. Vac. Sci. Technol. 15, 1488 (1978).
16. P. A. Barnes and D. P. Schinke, Appl. Phys. Lett. 30, 26 (1977).

XPS STUDY OF GaAs(100) SURFACE OXIDE CHEMISTRY AND INTERFACE POTENTIAL

R. W. Grant, S. P. Kowalczyk, J. R. Waldrop, and W. A. Hill
Rockwell International Electronics Research Center
Thousand Oaks, California, 91360

Considerable research effort has been expended in recent years searching for a successful GaAs MOS technology. Although promising capcitance-voltage results have been recently reported (1-3), the methods used have, for the most part, relied on technologically complex processing procedures (e.g., molecular beam epitaxy) to obtain the insulator-GaAs interface. It is also noteworthy that the results to date for GaAs MOSFET structures have not been very encouraging. We report experiments in which x-ray photoemission spectroscopy (XPS) has been used to correlate chemistry, electrical potential, and charge at the native-oxide/GaAs interface (n-type material) as a function of various processing treatments. Our results indicate that the lowest interface charge is obtained for the Ga_2O_3/GaAs interface and that the prospect of a MOS structure involving native oxides should not yet be ruled out.

It is well known that x-ray photoemission spectroscopy (XPS) can be used to determine the surface chemistry of a solid sample (4). A much less utilized aspect of the XPS technique is the measurement of surface (or band-bending) potential, V_{BB}, of a semiconductor sample (5). This measurement is illustrated in Fig. 1; a flat-band condition has $V_{BB} = 0$. All binding energies are referenced to the Fermi level, E_F. The conduction-band minimum is E_c, the valence-band maximum is E_v, the core-level binding energy is E_{c1}, $\Delta = E_F - E_v$, W is the depletion width and q is the electronic charge.

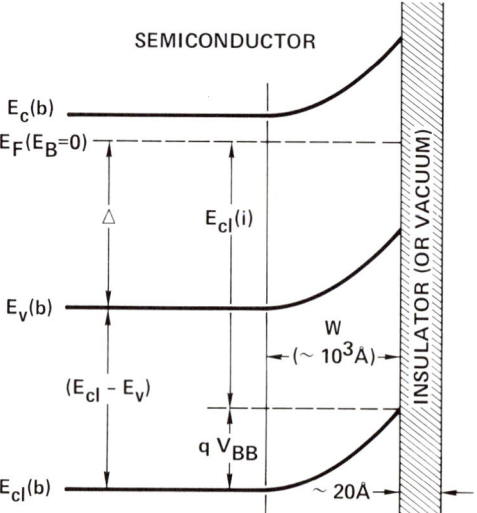

Fig. 1. Schematic energy-band diagram which illustrates the XPS measurement of band bending.

The (b) and (i) notations refer to bulk- and interface-quantities, respectively. A HP5950A XPS spectrometer which utilizes AlKα radiation (1486.6 eV) was used for the measurements. The most energetic photoelectrons analyzed in these measurements have escape depths of about 25Å. Thus, if only a thin insulating layer of ≈20Å is present on a semiconductor surface, unscattered photoelectrons from the underlying semiconductor will be observed and $E_{cl}(i)$ can be determined. For moderately doped semiconductors ($\lesssim 10^{17} cm^{-3}$), W≈10^3Å and thus band bending will not complicate the $E_{cl}(i)$ measurement. From Fig. 1 it is seen that

$$qV_{BB} = \Delta + (E_{cl} - E_v) - E_{cl}(i). \tag{1}$$

A knowledge of the semiconductor bulk doping characteristics determines Δ, ($E_{cl} - E_v$) can be obtained from XPS measurements. For GaAs, a precise value of

$$E_{As3d}^{GaAs} - E_v^{GaAs} = 40.73 \pm 0.02 \text{ eV}$$

has been recently reported (6).

Changes in surface chemistry during thermal treatment of a chemically etched GaAs (100) surface were monitored by XPS. The "initial" surface was prepared by etching in a (4:1:1) $H_2SO_4:H_2O_2:H_2O$ solution. The sample (n-type, ≈$1 \times 10^{17} cm^{-3}$) was attached to a Mo plate by using In (this involved heating in air to ≈160°C) and was inserted into the XPS spectrometer within a few minutes. Within the XPS system the relative sample temperature was measured by a thermocouple attached to the sample heater (in all cases the stated temperature is, therefore, considerably higher than the actual sample surface temperature, by as much as ≈20%). The sample was heated for five minutes at a specific temperature followed by cooling to ≈30°C for the XPS analysis. Each successive anneal was to a higher temperature; vacuum during annealing was ≈10^{-9} torr. The small amount of initial C surface contamination became undetectable above ≈300°C and O was removed from the surface at ≈677°C. The initial surface exhibited oxidized forms of both As and Ga (upper part of Fig. 2). The chemical shift of the As3d level in As oxide relative to GaAs ($\Delta E_B \approx 3.2$ eV) and of the Ga3d level in Ga oxide relative to GaAs ($\Delta E_B \approx 0.95$ eV) suggests that the oxides are primarily As_2O_3 and Ga_2O_3 (7). The lower part of Fig. 2 shows the same spectral region after the sample was heated to 650°C. The As_2O_3 has clearly been removed from the sample surface, however the shoulder on the high binding-energy side of Ga3d line (which corresponds to Ga_2O_3) has clearly increased in intensity.

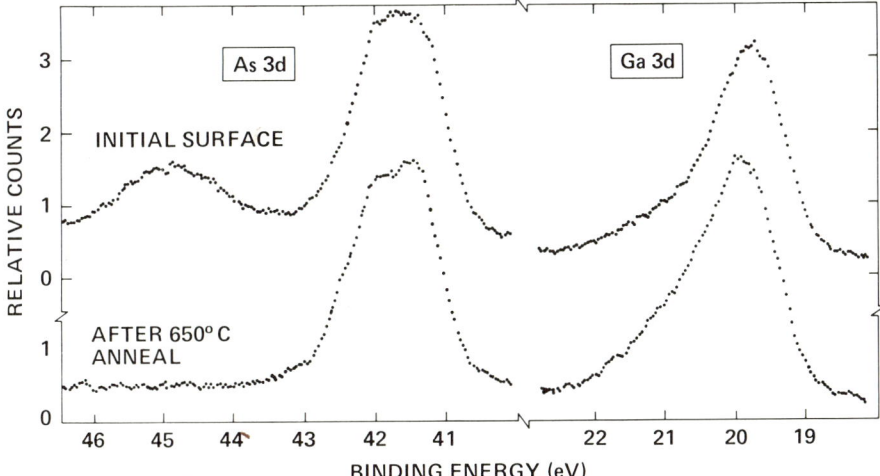

Fig. 2. XPS spectra in the region of the As3d and Ga3d levels. Sample is described in text.

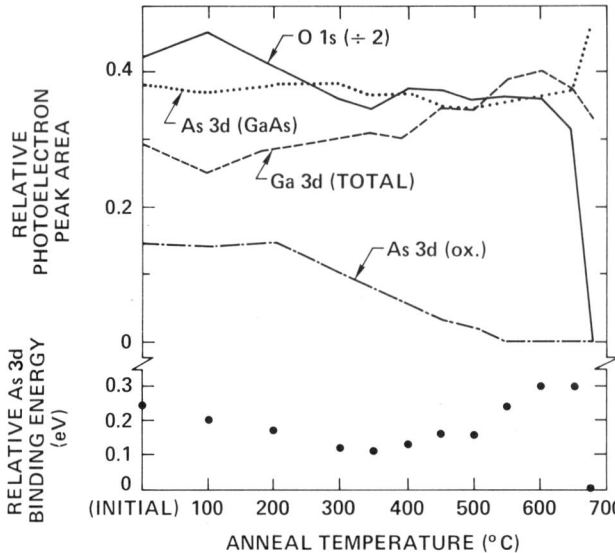

Fig. 3. Relative photoelectron peak areas and As 3d binding energy vs. annealing temperature. Sample described in text.

In Fig. 3 the relative XPS peak areas above background are plotted as a function of annealing temperature. The oxidized As, labeled As 3d (ox.), decreases slowly to zero between ≈300 and 550°C. The O 1s line intensity decreases until ≈300°C; however, in the temperature range where the oxidized As is removed from the surface, the O 1s line intensity is essentially constant. Also shown in Fig. 3 are the peak areas associated with As 3d in GaAs and the total Ga 3d peak area (oxidized Ga + Ga in GaAs). The As 3d(GaAs) remains relatively constant up to ≈650°C while Ga 3d (total) is observed to increase significantly in the 300 to 550°C temperature range. These results suggest a new view of the surface chemistry associated with thermally cleaning a GaAs surface. It is usually assumed (see e.g., ref. 8) that As oxides are initially removed from the surface due to the high volatility of these compounds. The XPS results of Figs. 2 and 3 show that As is lost from the surface as elemental As (presumably As_2 or As_4) and in the process additional Ga_2O_3 is formed. We conclude that a surface chemical reaction

$$As_2O_3 + 2GaAs \rightarrow Ga_2O_3 + \frac{4}{x} As_x \uparrow \qquad (2)$$

is associated with the removal of As_2O_3 from the GaAs(100) in the 300-550°C temperature range. Additional GaAs is consumed in this process. This same thermally induced solid-state reaction has been proposed (9) to explain the formation of elemental As at interfaces between GaAs and relatively thick (~10Å^3) anodic oxides. Interfaces prepared at room temperature which do not initially exhibit elemental As are observed to form elemental As rapidly at 450°C (9), which, due to the thickness of the oxide, remains trapped at the interface. Recent condensed phase diagram calculations (10) have also shown that only elemental As and Ga_2O_3 can exist in thermodynamic equilibrium with GaAs, consistent with Eq. (2).

There has been considerable recent interest in the electrical effect of the elemental As detected at the thermally annealed native-oxide/GaAs interface (11, 12). These elemental As atoms may be a major source of interface traps and thus play a significant role in determining the GaAs MOS characteristics (13). For GaAs MOS applications, it is important to reduce the interface charge and interface state density. As mentioned above, XPS can be used to determine V_{BB} and thus the net interface charge, Q_S. In the lower part of Fig. 3, the E_B for the As 3d core level relative to the thermally cleaned surface, δE_B, is shown (we also define δV_{BB} as the change in V_{BB} relative to the thermally cleaned surface, $\delta E_B = -\delta V_{BB}$). As the surface contamination associated with C and O is removed at temperatures below ≈300°C, δE_B is observed to decrease by ≈0.15 eV, which indicates an increase in V_{BB} and a corresponding larger net negative Q_S. However, in the temperature range where As is lost from the surface, which is associated with additional Ga_2O_3

formation and consumption of GaAs, δE_B increases by ≈ 0.2 eV (V_{BB} decreases by ≈ 0.2 eV) and the net negative Q_S decreases. This behavior is caused by a change in the interface-state density with the possibility that acceptor states are being removed from the interface during the Ga_2O_3 formation process. A marked decrease in δE_B is again observed when the sample temperature is raised high enough to clean the surface (Fig. 3).

In the Table, surface-potential variations for several surface treatments are presented for two additional series of experiments. Both samples were from the same GaAs boule and the initial surface preparation was the same as for the sample discussed previously. The E_{As3d}^{GaAs} scale for sample #1 was established by the evaporation of a thick Au layer onto the sample at the conclusion of the experiment and indexing the observed XPS peak position of the $Au4f_{7/2}$ line to 84.00 ± 0.01 eV (14).

GaAs(100) Surface Potential Variation With Surface Treatment

Sample #	Treatment	E_{As3d}^{GaAs} (eV)	δV_{BB}* (eV)	Remarks
1	500°C; Vacuum 2 days at 25°C	41.95	-0.50	Ga_2O_3
	688°C; 6×10^4 L O_2 at 610°C	41.71	-0.26	Ga_2O_3
	695°C	41.45	0.00	Clean, LEED
2	700°C	-----	0.00	Clean, LEED
	6×10^3 L H_2O at 25°C	-----	0.00	
	680°C	-----	-0.02	Clean
	6×10^4 L O_2 at 600°C	-----	-0.23	Ga_2O_3
	10^5 L H_2O at 38°C	-----	-0.44	
	600°C	-----	-0.22	
	10^4 L H_2O at 39°C	-----	-0.41	

*$\delta V_{BB} = -\delta E_B$

For sample #1, the initial chemically etched surface was heated to 500°C followed by storage in vacuum for two days at 25°C; this procedure produced a very high E_{As3d}^{GaAs}, and low V_{BB}. Subsequent thermal surface cleaning to obtain a surface with no detectable O or C as determined by XPS and which exhibited a characteristic LEED (low energy electron diffraction) pattern, followed by reoxidation at 610°C to form Ga_2O_3, did not immediately produce a low surface potential. In several experiments, we have observed that room temperature vacuum storage ($\sim 10^{-9}$ torr) for a period of 1-2 days markedly reduced V_{BB} on n-type GaAs(100). A possible explanation is provided by the experiments reported on sample #2 in the Table. An initially clean surface was exposed to H_2O vapor at 25°C; no change in δE_B was observed. This surface was thermally recleaned and oxidized to form Ga_2O_3. A change in δE_B similar to that observed for sample #1 was observed. Subsequent exposure of this surface to small amounts of H_2O vapor for ~ 1 hour resulted in large values of δE_B similar to that observed in sample #1. Any alteration of the surface chemistry as a result of the H_2O vapor exposure was below detectable limits of the XPS analysis. Because H_2O vapor is a residual gas component of the ion pumped XPS system, exposure to H_2O vapor may offer an explanation for the variation in E_B with the room temperature vacuum storage noted above.

The net interface charge can be calculated by using Eqs. 1 and 3. From the Table, the surface of sample #1 (which was annealed at 500°C and stored in vacuum for 2 days) exhibited E_{As3d}^{GaAs} = 41.95 eV. XPS analysis of this surface showed only Ga_2O_3 (in addition to the GaAs), no oxidized As was observed. For a donor density of $N_D \approx 10^{17} cm^{-3}$, $\Delta \approx 1.39$ eV and thus V_{BB} = 0.17 eV. Q_S can be calculated from (15)

$$Q_s = (2 q \varepsilon N_D |V_{BB}|)^{\frac{1}{2}} \qquad (3)$$

where ε is the permitivity of GaAs. Q_S is found to be $0.8 \times 10^{-7} coul/cm^2$ ($\approx 5 \times 10^{11}$ electrons/cm^2).

The results presented here suggest the possibility of forming a useful GaAs/oxide interface by heat treating a native oxide of moderate thickness (≤ 100Å). If this oxide is annealed in the 500-600°C range, it may be possible to form pure Ga_2O_3, evaporate the elemental As which is formed, and produce a new GaAs interface by consuming additional GaAs during the process. Our observations show that this process reduces Q_S (and hopefully the interface-state density). A MOS structure would be formed by subsequent deposition of a good quality dielectric.

This work was supported by WPAFB Contract No. F33615-78-C-1532.

References

1. W. T. Tsang, M. Olmstead and R. P. H. Chang, Appl. Phys. Lett. 34, 408 (1979).
2. R. P. H. Chang and J. J. Coleman, Appl. Phys. Lett. 32, 332 (1978).
3. D. W. Langer, F. L. Schuermeyer, R. L. Johnson, H. P. Singh, C. W. Litton, and H. L. Hartnagel, Physics of Compound Semiconductor Interfaces Conference, Estes Park, Colorado, Jan. 29-31, 1980.
4. K. Siegbahn, et al., ESCA: Atomic, Molecular and Solid State Structure Studied by Means of Electron Spectroscopy, Almquist and Wiksells, Uppsala, 1967.
5. J. Auleytner and O. Hörnfeldt, Arkiv för Fysik 23, 165 (1963).
6. E. A. Kraut, R. W. Grant, J. R. Waldrop, and S. P. Kowalczyk, Phys. Rev. Lett., June, 1980 (In Press).
7. P. W. Chye, P. Pianetta, I. Lindau, and W. E. Spicer, J. Vac. Sci. Technol. 14, 917 (1977).
8. C. W. Wilmsen, R. W. Kee, and K. M. Geib, J. Vac. Sci. Technol. 16, 1434 (1979).
9. G. P. Schwartz, B. Schwartz, D. DiStefano, G. J. Gualtieri, and J. E. Griffiths, Appl. Phys. Lett. 34, 205 (1979).
10. G. P. Schwartz, C. D. Thurmond, G. W. Kammlott, and B. Schwartz, Physics of Compound Semiconductor Interfaces Conference, Estes Park, Colorado, Jan. 29-31, 1980.
11. J. A. Cape, W. E. Tennant, and L. G. Hale, J. Vac. Sci. Technol. 14, 921 (1977).
12. R. L. Farrow, R. K. Chang, S. Mroczkowski, and F. H. Pollack, Appl. Phys. Lett. 31, 768 (1977).
13. R. P. H. Chang, T. T. Sheng, C. C. Chang, and J. J. Coleman, Appl. Phys. Lett. 33, 341 (1978).
14. F. R. McFeely, S. P. Kowalczyk, L. Ley, R. A. Pollak, and D. A. Shirley, Phys. Rev. B, 7, 5228 (1973).
15. A. S. Grove, Physics and Technology of Semiconductor Devices, (John Wiley and Sons, Inc., New York, 1967), p. 266.

GERMANIUM (OXY)NITRIDE BASED SURFACE PASSIVATION
TECHNIQUE AS APPLIED TO GaAs & InP

B. Bayraktaroglu, Universal Energy Systems, 3195 Plainfield Road,
Dayton, OH 45432
R. L. Johnson, D. W. Langer and M. G. Mier, AFWAL Avionics Lab,
W.P.A.F.B., OH 45433

Introduction

Germanium nitride has a low heat of formation and therefore during its formation on GaAs it is likely to produce less damage on the GaAs surface than compounds with a high heat of formation like Si_3N_4. In this study germanium nitride was deposited by reactive evaporation. Most films contained varying degrees of oxygen originating from the residual gasses in the reactor. A general formula for the deposited insulator should therefore be in the form $Ge_xO_yN_z$.

The major part of this work was performed on GaAs samples. However, some InP samples prepared in the same manner as the GaAs samples were also coated with $Ge_xO_yN_z$. InP MIS devices were analyzed alongside with the GaAs sample. The interface properties produced on both semiconductors with the present passivation technique were then compared.

Experimental

The reactor used in this study was a 1" diameter quartz tube, sealed in one end and connected to vacuum pump on the other. The sample was placed on a quartz carrier facing the source which consisted of a quartz cup filled with 99.999% pure Ge_3N_4 powder. The portion of the reactor containing the source and the sample was placed inside a resistively heated furnace. During the deposition the source temperature was kept at 700-750°C whereas the sample temperature was 300°C. Prior to the deposition the reactor was evacuated and then backfilled with hydrazine vapor or ammonia gas to a pressure of 0.2-0.5 torr. The presence of hydrazine or ammonia serves two purposes. It reduces the decomposition of the germanium nitride source material and the deposited layers on GaAs at the deposition temperatures. It also reduces the thickness of the natural oxide on the GaAs surface prior to the deposition.

The samples studied here were of n-type, bulk grown, (100) orientation and had carrier concentrations of $5-7 \times 10^{15} cm^{-3}$ for GaAs, and 3×10^{15} for InP. One side of each sample was polished to a mirror finish. Au-Ge ohmic contacts were evaporated onto the other side and annealed at 450° for 2 minutes in H_2 ambient. Minutes before loading the samples into the reactor at least 1.5 μm of the surface was removed in an etchant consisting of $NaOH: H_2O_2:H_2O$ in 1:2: 100 volume ratio. The sample was then rinsed in D.I. water and blown dry with a jet of N_2 gas. A 30 second dip in 48% HF as a final surface treatment was found to be useful in obtaining better interfaces. Typical deposition rates of 50-100Å/min were obtained when the source temperature was kept in the 700-750°C range. Faster deposition rates were possible by increasing the source temperature. This, however, produced Ge rich films which had lower electrical breakdown strengths and larger d.c. leakage currents.

The thickness of the deposited layers were measured with a Gaertner 7122-47 automatic scan ellipsometer. For electrical measurements $2 \times 10^{-3} cm^2$ area Au field plates were deposited on the insulator by evaporation through a metal mask.

The set-up used for electrical characterization of the interfaces contained two capacitance meters, together covering a frequency range of 50 Hz-10 MHz. As a first step capacitance and conductance of the MIS structures were measured at different frequency and ramp speeds. These capacitance and conductance measurements were automatically fed into a HP 9825A calculator for C-V, G-V, $G_p/\omega C_I$-log ω, $G_p/\omega C_I$-C_p/C_I plots. Information regarding the interface states was extracted from these plots with the use of a.c. conductance method[1].

Results and Discussion

The choice of germanium (oxy)nitride films for the passivation of GaAs originates from the low heat of formation of germanium nitride. As shown in Table 1, the heats of formation of most dielectrics used on GaAs for surface passivation have larger values than that of GaAs. From a simple point of view one would expect that during the formation of these dielectrics on GaAs, some Ga-As bonds are likely to be broken giving rise to states at the interface. For the same reason the formation of germanium (oxy)nitride films on GaAs is expected to cause the least damage among the dielectrics shown in Table 1.

However, the practical surfaces of GaAs are usually covered with a thin layer of natural oxides due to oxidation in air. This itself is enough to create large number of interface states[4]. For successful surface passivation, this thin layer of natural oxides and possibly some of the damaged GaAs surfaces must be removed prior to the deposition of the dielectric. In the passivation technique described here the removal of the natural oxides was achieved by heating GaAs in hydrazine vapor or ammonia gas. AES depth profiles taken on germanium (oxy)nitride coated GaAs samples showed that the interface between the insulator and GaAs was very sharp ($\leq$50Å) and no oxygen peak was observed at the interface. Instead, a distinct and sharp nitrogen peak was present near the GaAs surface. On all samples analyzed this pattern was repeated. The bulk concentration ratios of oxygen to nitrogen varied widely with no apparent effect on the interface properties.

TABLE 1 Heats of formation of some insulators used on GaAs.
*Ref. (2), †Ref. (3)

Insulator/Semiconductor	ΔH_f° (kcal/mol)
SiO_2	-205.0 *
Si_3N_4	-497.3 *
Al_2O_3	-399.09 *
Ga_2O_3	-258 *
$As_2O_3 \cdot As_2O_5$	-351.1 *
Ge_3N_4	- 14.8 *
GaAs	- 21.7 †

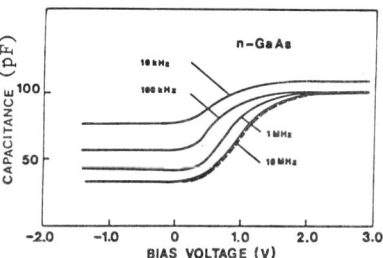

Fig. 1 C-V curves of GaAs MIS device

Figure 1 shows the C-V curves of MIS devices produced on n-type GaAs. The variation of the insulator capacitance (C_I) between 10 kHz and 100 kHz is due to the frequency dispersion of the dielectric constant of the insulator. Depending on the deposition conditions, the degree of dispersion was varied. At all frequencies, the C-V curves in Fig. 1 reached the insulator capacitance. The capacitance level in inversion, on the other hand, had a frequency dependence. As the frequency was increased the capacitance approached the theoretically expected high frequency capacitance level.

Another important feature of Fig. 1 is the practically nonexistence of hysteresis effects. With ramp speeds between 1 mV/sec. and 500 mV/sec., the hysteresis width remained below 100 mV. The hysteresis is only shown on the 10 MHz curve in Fig. 1, but the C-V curves at other frequencies also had a similar hysteresis behavior. The direction of the hysteresis is clockwise on n-type GaAs. Since the hysteresis is very small the injection of charges into the traps in the insulator away from the GaAs surface must be negligible. This is in contrast to the results obtained when anodic native oxides[5] and CVD Si_3N_4[6,7] were produced on GaAs.

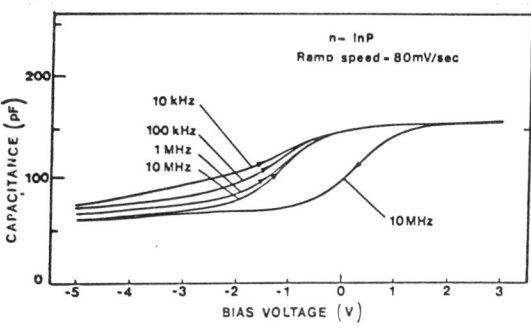

Fig. 2 C-V curves of InP MIS device

C-V curves obtained with MIS devices on n-type InP (see Fig. 2) showed strong similarities with those of Fig. 1. The basic differences are the flat-band voltage values and the hysteresis effects. In Fig. 2, the hysteresis is shown only on the 10 MHz curve but all other C-V curves had similar hysteresis behavior. The direction of the hysteresis is consistent with the charge injection mechanism as was the case with the GaAs devices. No threshold voltage was observed for the start of the hysteresis. Its magnitude gets larger as the bias is increased from zero volt towards both negative and positive voltages. With a bias range of -5 to +3 V, the hysteresis width was 1.4 V. Other dielectrics produced on InP such as Al_2O_3[8], SiO_2[9], Si_3N_4[10], anodic oxides[11] all showed C-V curves with hysteresis similar to those of Fig. 2. Even prolonged hydrazine or ammonia pre-treatment of the InP surfaces was not effective in reducing this large hysteresis.

Information regarding the fast surface states on both GaAs and InP was obtained with the use of a.c. conductance method[1,12]. A set of typical $G_p/\omega C_I$-log ω curves for GaAs MIS devices at selected biases is shown in Fig. 3. In the bias range 0 to 4 V, the device is in depletion. The experimental curves were fitted to the theoretical curves when a variance factor[12] $\sigma g = 0.5$ was used. As the bias is increased to larger positive values, the $G_p/\omega C_I$-log ω curves retain their shape but shift to the right due to the change in the time constants of the surface states. Also, as the bias is increased towards negative values the curves shift to the left, but this time their shapes started to alter, becoming much steeper. At $V_b \geq -1.5$ V, the $G_p/\omega C_I$-log ω curves were characterized by a single time constant behavior. This is an indication that the weak inversion conditions are reached[1]. In depletion all $G_p/\omega C_I$-log ω curves showed good symmetry about the peak (see Fig. 3). This indicates that the fast surface states are located at or very close (≤ 10Å) to the GaAs surface. This result together with the small hysteresis of the C-V curves and the AES depth profiles point out that the germanium (oxy)nitride-GaAs interface is indeed sharp.

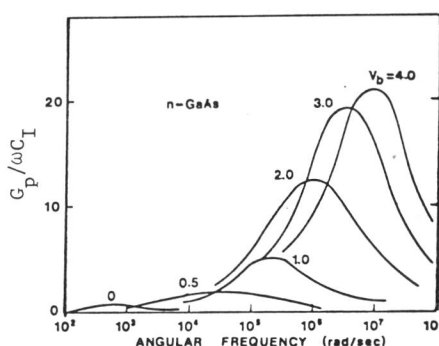

Fig. 3 $G_p/\omega C_I$-log ω curves at different biases

A few InP MIS devices were also analyzed with the a.c. conductance method. $G_p/\omega C_I$-log ω curves showing a sharp peak could only be obtained when the bias was zero (near flat-band conditions) or positive. As the bias was increased towards more positive values, $G_p\omega C_I$-log ω

curves shifted to the left indicating that the time constants of the states increased. This behavior is opposite to that observed with GaAs devices. An explanation of this difference can be given assuming that the surface states are located in the insulator causing longer time constants. Those states closer to the conductance band must be further away from the interface. Since most of the $G_p/\omega C_I$-log ω curves were symmetrical about the peak, these states are not likely to be spacially distributed in the insulator, but are rather localized.

The surface state density N_{ss}, and the time constant of the states τ, were calculated from the $G_p/\omega C_I$-log ω curves by taking into account the variations of σ_g. The surface potential ψ_s, as a function of the applied bias was estimated from the 10 MHz C-V and the $G_p/\omega C_I$-C_p/C_I (Cole - Cole plots[13]) curves. The results are shown in Figs. 4 and 5 for GaAs. In Fig. 4 it can be seen that the surface state density is in the lower $10^{11} cm^{-2} eV^{-1}$ range over the large portion of the upper half of the band-gap. At 0.9-1.1 eV from the valance band, a small but distinct shoulder was seen which was present with all the MIS devices tested. A similar feature was also observed when PED silicon nitride films were used for surface passivation[14]. Recently, with the use of deep-level transient spectroscopy Yamasaki and Sugano[15] found a trap level at $E_c - E_T = 0.36$ eV when anodic oxides were grown on n-type GaAs surfaces. The agreement between the energy levels found in these three different cases is very good. The surface state density of InP MIS devices was 3-4 x $10^{11} cm^{-2} eV^{-1}$ near midgap, 4 x $10^{12} cm^{-2} eV^{-1}$ at flat-band increasing rapidly towards the conduction band-edge.

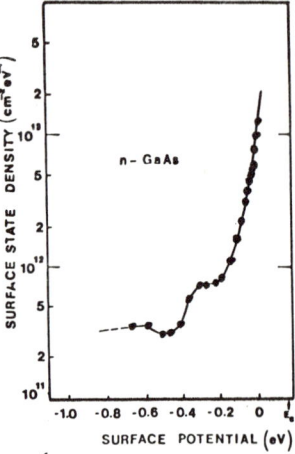

Fig. 4 The surface states in the upper half of the band-gap

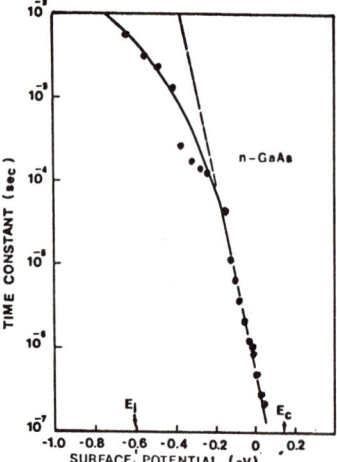

Fig. 5 The time constants of the surface states

As seen in Fig. 5, the time constants of the surface states on GaAs are exponentially dependent on the energy position of the states. This relationship persists over a large portion of the upper half of the band-gap. Nearer the midgap τ varies at a slower rate with energy. The capture cross section of the traps was calculated using the expression

$$\sigma_n = (\bar{v}\tau N_D)^{-1} \exp(-\psi_s/kT) \qquad (1)$$

where $\bar{v}$ is the average thermal velocity, N_D is the bulk carrier concentration. The results are shown in Fig. 6. Near the conduction band edge σ_n is the 10^{-16} - $10^{-17} cm^2$ range slowly increasing towards the mid-gap. Between $\psi_s = -0.35$ and -0.6 eV, σ_n increases at a much faster rate reaching unrealistically large values e.g. $10^{-8} cm^2$. These large values of σ_n can not be explained by the interface state band (ISB) model[16] since the interface state density in the upper half of the band-gap is not large enough to form a band (see Fig. 4).

Conclusions

It was shown that the passivation of the GaAs and InP surfaces was possible by the deposition of germanium (oxy)nitride films. The surface state densities of low $10^{11} cm^{-2} eV^{-1}$ could be achieved on both semiconductors. $N_{ss}-\psi_s$ curves of GaAs had a shoulder at 0.9-1.1 eV from the valance band similar to those curves obtained with PED silicon nitride films[14]. C-V curves of GaAs MIS devices showed small hysteresis whereas InP devices had bias dependent hysteresis. The electron capture sections of the surface states on GaAs were in the $10^{-16} - 10^{-17} cm^2$ range near the flat-band conditions.

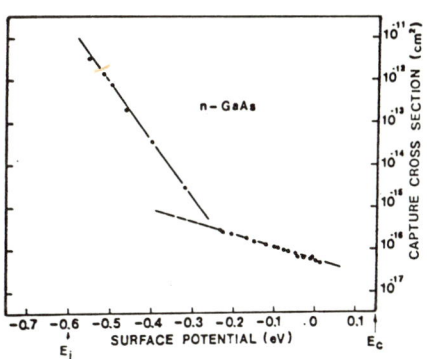

Fig. 6 The electron capture cross-sections of the surface states

References

(1) E. H. Nicollion and A. Goetzberger, Bell Syst. Techn. J. **46**, 1005 (1966).

(2) CRC Handbook of chemistry and physics, 57th edition, Edited by R. C. Weast, (1976), CRC Press, Cleveland.

(3) C. Pupp, J. J. Murray and R. F. Pottie, J. Chem. Thermodynamics, **6**, 123 (1974).

(4) W. E. Spicer, P. Pianetta, I. Lindau and P. W. Chye, J. Vac. Sci. Technol. **14**, 885 (1977).

(5) E. Kohn and H. L. Hartnagel, Solid State Elect. **21**, 409 (1978).

(6) J. A. Copper, Jr., E. R. Ward and R. J. Schwartz, Solid State Elect. **15**, 1219 (1972).

(7) J. E. Foster and J. M. Swartz, J. Electrochem. Soc. **117**, 1410 (1970).

(8) P. N. Favennec, M. LeContellec, H. L'Haridon, G. P. Pelous and J. Richard, App. Phys. Lett. **34**, 807 (1979).

(9) L. Messick, J. App. Phys. **47**, 4949 (1976).

(10) A. J. Grant, D. C. Cameron, L. D. Irving, C. E. Greenholgh and P. R. Norton, Inst. Phys. Conf. Ser. No: **50**, 266 (1980).

(11) D. L. Lile and D. A. Collins, Appl. Phys: Lett. **28**, 554 (1976).

(12) A. Goetzberger, E. Klausmann and M. J. Schulz, CRC Crit. Rev. in Solid St. Sci. **6**, 1 (1976).

(13) P. C. Malmin, Phys. Stat. Sol (a), **8**, 597 (1971).

(14) B. Bayraktaroglu and R. L. Johnson, Workshop on the Dielectric Systems on III-V Compounds, San Diego, May 19-20, 1980.

(15) K. Yamasaki and T. Sugano, Appl. Phys. Lett. **35**, 932 (1979).

(16) H. Hasegawa and T. Swada, J. Vac. Sci. Technol. **16**, 1478 (1979).

KrF--LASER ANNEALING OF NATIVE OXIDES ON GaAs

R. K. Ahrenkiel, G. Anderson, D. Dunlavy, C. Maggiore,
R. B. Hammond, and S. Stotlar
Los Alamos Scientific Laboratory, Los Alamos, NM 87545

ABSTRACT

Annealing of native oxides grown on GaAs has been performed using a pulsed KrF laser. This process allows the oxides to be heated to temperatures well above 350°C without arsenic loss from the GaAs substrate. The physical, chemical, and electronic properties of the oxide are markedly changed by laser processing.

INTRODUCTION

Laser annealing of semiconductors has been used for some time for the removal of ion implantation damage (1). Here the annealing of a native oxide by an ultraviolet (KrF) laser is demonstrated for the first time. This type of processing may be useful in applications to a number of III-V semiconductor MOS structures.

Native oxides have been grown on GaAs by thermal oxidation (2), plasma (3) and electrochemical (4) anodization. MOS device applications require a low density of interface states ($<10^{11}/cm^2 \cdot eV$) and oxides that are highly insulating and relatively trap-free.

Most GaAs native oxides contain high densities of electronic traps, which produce hysteresis in the capacitance-voltage (C-V) characteristic. Prior work has shown that these traps are primarily slow traps; that is, electronic states that are located deep in the oxide (5). Thermal annealing in inert gas or vacuum partially removes these traps, and one finds that the trap density decreases with increasing annealing temperature. However, the maximum annealing temperature is limited to about 350°C. Higher temperatures produce As loss from the oxide-semiconductor interface, and the electrical properties of the interface are degraded. By monitoring the 1-MHz conductance-voltage (G-V) signal (6), we can see the onset of interface deterioration.

EXPERIMENTAL RESULTS

Here we demonstrate for the first time the annealing of native oxides on GaAs, InP, and GaP with a KrF laser (τ_p = 15 ns, λ = 2484 Å). We will primarily discuss the results on GaAs/native oxides, which are anodically grown.

The absorption edge of such oxides is about 2600 Å, and the absorption coefficient at 2484 Å is about 3×10^4 cm^{-1} (7). Hence there is a fairly uniform deposition of laser energy in oxides of 1000 to 2000 Å thickness. Experimentally, laser fluxes greater than 60 mJ/cm^2 produced color changes in the oxide, indicating a change in oxide thickness and/or index of refraction.

Calculations of the energy density deposited in the oxide as a function of distance from the front surface are shown in Fig. 1 for a 2000 Å oxide. Here the input laser flux Φ_0 is a parameter, and the reflection from the interface is

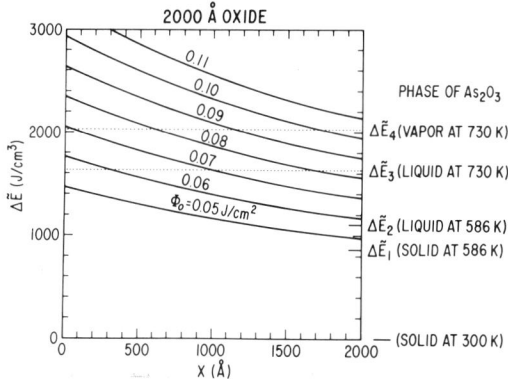

Fig. 1. The calculated energy density in a 2000 Å thick oxide for various laser fluxes.

incorporated in the calculations. The horizontal dotted line represents phase changes in the As_2O_3 component of the oxide corresponding to the appropriate energy density. A linear model (constant oxide composition as a function of laser flux) was assumed throughout. The model verifies the onset of color change at 50-60 mJ/cm^2 corresponding to the melting of As_2O_3. These initial color changes are probably related to densification of the somewhat porous as-grown oxide.

Vaporization of As_2O_3 should occur at fluxes greater than about 110 mJ/cm^2. Our work has shown that evaporation of the As_2O_3 component occurs at fluxes from 200 mJ/cm^2 to 1.0 J/cm^2.

The "crater" produced by laser annealing has been examined by several techniques. A micro-ellipsometer was used to obtain the data of Fig. 2. Here a focused HeNe laser beam was polarization modulated, and the technique of Jasperson et al. was used to determine the ellipsometric parameters ψ and Δ. From these parameters, the index of refraction n and the oxide thickness t could be determined. A profile of the crater is shown in Fig. 2. Here the oxide thickness has been reduced by about 34% by a 1-J/cm^2 laser pulse. The roughness is believed to be produced by intensity structure in the laser beam. The Los Alamos Scientific Laboratory nuclear microprobe (9) using 2-MeV deuterons produced the oxygen concentration data of Fig. 3. The nuclear reaction $^{16}O(d,p)^{17}O$ was used to obtain quantitative oxygen data. The oxygen content here is about 60% of the initial value and scales with the thickness shown in Fig. 2. The primary oxygen loss mechanism is probably the vaporization of As_2O_3 as predicted by our calculations. The relative thickness of annealed and unannealed areas was also verified by the deposition of 20-mil diameter gates followed by the measurement of oxide capacitance. As shown in Fig. 2, the oxide thickness as determined from $\varepsilon A/d_{OX}$ is also about 66% of the initial value. Here A is the cross sectional area and d_{OX} is the oxide thickness.

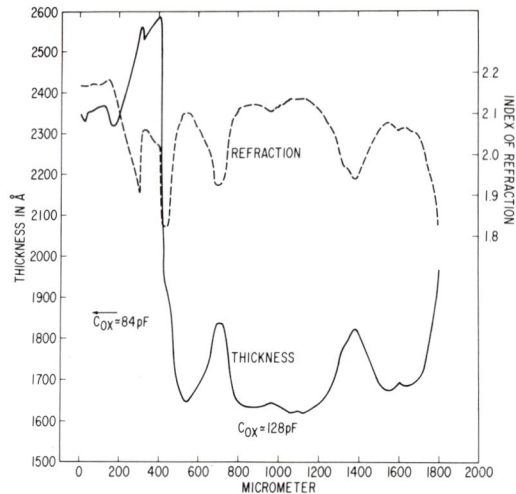

Fig. 2. The thickness and index of refraction of a laser-annealed "crater" with an input flux Φ_0 of about 1.0 J/cm^2.

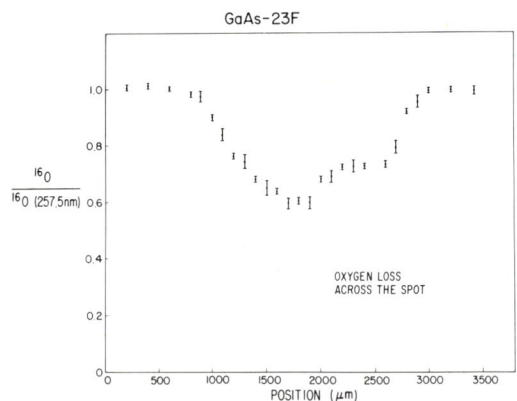

Fig. 3.. The oxygen concentration across the laser-annealed crater of Fig. 2 as determined by ion backscattering.

G-V studies of MOS structures both inside and outside the crater show no significant difference in interface properties even though the local temperature may have exceeded 450°C during laser annealing. By comparison, thermal annealing of oxides at 400°C produces significant interface damage as seen by enhanced G-V loss peaks. Hence laser annealing of oxides appears to be a means of processing native oxides without damaging the interface of the III-V material.

Pulsed 1-MHz capacitance measurements were used to evaluate the slow trap density. Here the MOS structure was pulsed to -20 V for about 10 ms. At the termination of the voltage pulse, the capacitance is given by:

$$C(t) = \frac{C_{OX}}{\sqrt{1-K(V_{FB} + \Delta V_{FB})}} \qquad (1)$$

$$\text{where } K = \frac{2C_{OX}^2}{N_A \varepsilon A^2} \qquad (2)$$

Here ε is the oxide dielectric constant, A is the area of the MOS structure, C_{OX} is the oxide capacitance, and N_A is the density of free holes.

The pulsed capacitance recovery is shown in Fig. 4 for three different oxides. Curve A is the recovery of an as-grown oxide. Curve B is the response of a furnace-annealed oxide for an annealing temperature of 350°C--the maximum permissible steady-state temperature that avoids arsenic loss. Curve C results from an oxide that was thermally annealed at 350°C and laser annealed at a flux of 230 mJ/cm^2.

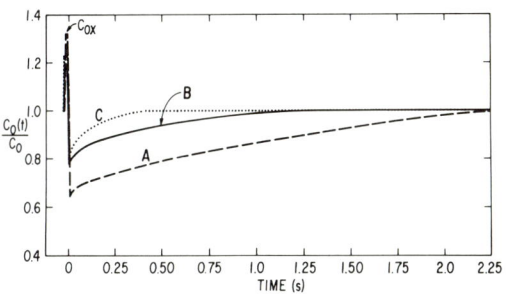

Fig. 4. Pulsed-capacitance recovery for three oxides. A) unannealed; B) thermal-annealed at 350°C; C) thermal-annealed at 350°C plus laser-annealed at 230 mJ/cm^2.

The charge in oxide traps is:

$$N_{OX}(t) = \frac{C_{OX}|V_{FB}|}{Ae} \qquad (3)$$

From the pulsed capacitance recovery, N_{OX} may be calculated as a function of time. Figure 5 shows N_{OX} for the three MOS structures. The reduction of slow oxide traps with the various treatments is obvious here. Multiple laser pulses seem to produce more desirable changes in the oxide without producing radiation damage. Such studies are still in progress.

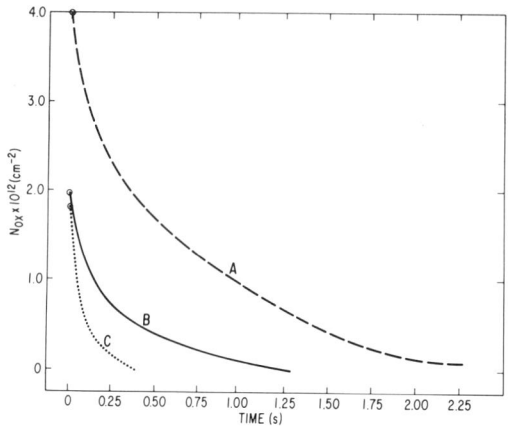

Fig. 5. Concentration of trapped oxide charge for the three sample of Fig. 4.

Initial studies have shown that the chemical etch rates for laser-annealed areas are different than for unannealed areas. This effect could be useful in developing new photolithographic processes.

In conclusion, we have demonstrated a new method of changing the physical, chemical, and electronic properties of native oxides grown on GaAs. This method suggests a variety of uses in semiconductor technology.

REFERENCES

(1) For a review see, A. E. Bell, RCA Review 40, 295 (1979).

(2) S. P. Murarka, Appl. Phys. Lett. 26, 180 (1975).

(3) R. P. H. Chang and J. J. Coleman, Appl. Phys. Lett. 32, 332 (1978).

(4) H. Hasegawa, K. E. Forward, H. L. Hartnagel, Appl. Phys. Lett. 26, 567 (1975).

(5) E. Kohn and H. L. Hartnagel, Solid-State Electron 21, 409 (1978).

(6) E. H. Nicollian and A. Goetzberger, Appl. Phys. Lett. 7, 216 (1965).

(7) E. D. Palik, N. Ginsburg, R. T. Holm, and J. W. Gibson, J. Vac. Sci. Technol. 15, 1488, (1978).

(8) S. N. Jasperson, D. K. Burge, and R. C. O'Handley, Surface Science 37, 548 (1973).

(9) C. J. Maggiore, "Materials Analysis with a Nuclear Microprobe", Scanning Electron Microscopy/1980, edited by Om Johari.

ANODIC OXIDATION OF $Hg_{0.68}Cd_{0.32}Te$

Bruce K. Janousek, Michael J. Daugherty, and Richard B. Schoolar
The Aerospace Corporation, P.O. Box 92957, Los Angeles, California 90009

INTRODUCTION

The compound semiconductor, $(Hg_{1-x}Cd_x)Te$, in which x denotes the mole fraction of CdTe, has received considerable attention in recent years as a potentially useful infrared sensing material. Single crystal $(Hg_{1-x}Cd_x)Te$ not only has a high sensitivity to infrared radiation, it also allows for bandgap tunability by the variation of crystal alloy composition, x.[1] However, the electro-optical properties of $(Hg_{1-x}Cd_x)Te$ are a strong function of the semi-conductor surface treatment. For example, it has been determined that $(Hg_{1-x}Cd_x)Te$ performance can be surface dominated when used as a photoconductor. Likewise, both charge transfer efficiency and storage time will depend critically on surface treatment when $(Hg_{1-x}Cd_x)Te$ is employed in the charge-coupled device configuration.[3,4] Although many laboratories have been active in the area of $(Hg_{1-x}Cd_x)Te$ surface passivation, there is at present only a limited amount of published work in this area, particularly for compositions with peak sensitivity in the 3-5 μ range. In this paper, we report the results of an investigation of the anodic oxidation passivation technique on $Hg_{0.68}Cd_{0.32}Te$. The electronic properties of the anodic oxide-$Hg_{0.68}Cd_{0.32}Te$ interface have been determined using capacitance-voltage and conductance measurements.

EXPERIMENTAL

$Hg_{0.68}Cd_{0.32}Te$ wafers (15 x 15 x 0.5 mm) of random orientation were supplied by Cominco American, Inc. For this composition, the material has a bandgap of 0.293 eV at 77°K which corresponds to a threshold wavelength of 4.2 μ. The samples investigated were n-type with a carrier concentration of 1.5×10^{14} cm^{-3} and mobility of 1.4×10^4 cm^2/V-sec at 77°K. The wafers were cut into slices of approximately 2 x 6 mm with a diamond impregnated stainless steel wire saw. The individual slices were etched in a 5% by volume bromine in methanol solution for approximtely one minute immediately prior to anodization.

The anodization was carried out in a solution of 0.1 M KOH in 90% ethylene glycol/10% water at a constant current density of 0.3 ma/cm^2.[5] After the desired film thickness was achieved (as monitored by the voltage drop across the oxide film and the film color), the anodization was allowed to continue at a constant voltage ~ 60% that of the final voltage. The current density decreased during the constant voltage anodization and resulted in films of at least two orders of magnitude greater resistivity.[6] Films were grown to thicknesses ranging from 500 Å to 2000 Å.

MIS structures were constructed by e-beam evaporation of 10,000 Å of indium onto the anodic oxide through a mask to produce 0.50 mm diameter gates. Electrical contact to a particular gate is made with a 0.05 diameter (2 mil) Cu-Be wire which is attached to a positioning stage. In some samples, where additional electrical insulation was desired, 3500 Å of ZnS was evaporated over the anodic oxide prior to metal gate deposition.

Capacitance-voltage and conductance data were obtained frm 100 Hz to 210 KHz with a PAR Model 126 lock-in amplifier. A PAR Model 410 C-V Plotter was employed to obtain capacitance data at 1 MHz. Quasi-static C-V curves were generated using a Model 616 Keithley Electrometer.

RESULTS

Film Properties

Anodic oxides on $Hg_{0.68}Cd_{0.32}Te$ exhibit a resistivity of 10^{10}–10^{13} Ω-cm at 77°K. Film breakdown occurs at field strengths of 2×10^5 - 2×10^6 V/cm at 77°K. For samples on which ZnS has been evaporated over the anodic oxide, the resistivity is greater than 10^{16} Ω-cm.

The anodic oxide dielectric constant, ϵ_{ox}, was determined by measuring the MIS capacitance in accumulation. A plot of (capacitance)$^{-1}$ vs. oxide thickness is linear and gives a value for ϵ_{ox} of $14.8\epsilon_o$ at 77°K where ϵ_o is the permittivity of free space.

The anodic oxide has been examined using x-ray photoelectron spectroscopy (ESCA) in combination with Ar-ion etching. The film is composed of TeO_2, oxidized Cd (CdO and/or $Cd(OH)_2$), and HgO.

MIS Characteristics

Fixed charge density. Anodic oxides on $Hg_{0.68}Cd_{0.32}Te$ contain a fixed positive charge as evidenced by the negative flat-band voltages observed in the C-V curves. Flat-band voltages of -0.5 to -1.0 V were typically observed for oxide thicknesses of 1000-1500 Å. For these small negative flat band voltages, the oxide fixed charge density[7] was calculated to be $6.4 \pm 0.7 \times 10^{11}$ cm^{-2}. This value was calculated assuming all the charge is at the semiconductor-oxide interface and the In-(Hg,Cd)Te work function difference is zero. On samples which have been overcoated with ZnS, a positive fixed charge density of $5.8 \pm 2.5 \times 10^{11}$ cm^{-2} is observed. When ZnS is evaporated directly onto the (Hg,Cd)Te, the insulator exhibits a fixed <u>negative</u> charge density of approximtely 10^{11} cm^{-2}.

Hysteresis. The capacitance-voltage curves exhibit a small amount (< 0.2 V) of horizontal hysteresis at 77°K indicating the presence of charge trapping at the interface[8] (commonly referred to as slow surface states). The magnitude of the hysteresis is dependent upon the voltage extremes to which the MIS structure has been stressed and increases as a function of increasing temperature. The effect appears to be more significant for samples with oxides thicker than 1200 Å.

Interface State Density. The interface state density, N_{ss}, was determined through the comparison of a high-frequency C-V curve (1 MHz) and quasi-static C-V curve.[9] The high-frequency and quasi-static curves at 77°K for a ZnS-anodic oxide MIS structure are compared in Fig. 1. The capacitance of the interface states with time constants longer than 1 μsec results in a greater total capacitance in the quasi-static curve when compared to the portion of the 1 MHz curve from accumulaton to the onset of inversion. This allows a determination of N_{ss} as a functon of position in the semiconductor bandgap. By this method, N_{ss} was determined to vary continuously from 1×10^{13} cm^{-2} (eV)$^{-1}$ at the conduction band edge to 4×10^{10} cm^{-2} (eV)$^{-1}$ at mid gap.

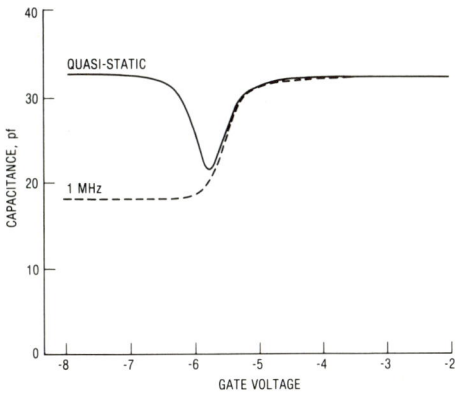

Fig. 1

Interface state time constants. The interface state time constant distribution was found by measuring the equivalent parallel conductance of the MIS structure as a function of frequency at 77°K.[10] The time constants were determined with the sample biased in accumulation and depletion but not in inversion where the fast minority carrier response time would interfere with the measurement (vide infra). The equivalent parallel conductance in accumulation/depletion as a function of frequency is shown in Fig. 2. The results indicate the presence of interface states of both long time constant (> 0.01 sec) and short time constant (< 5 μsec).

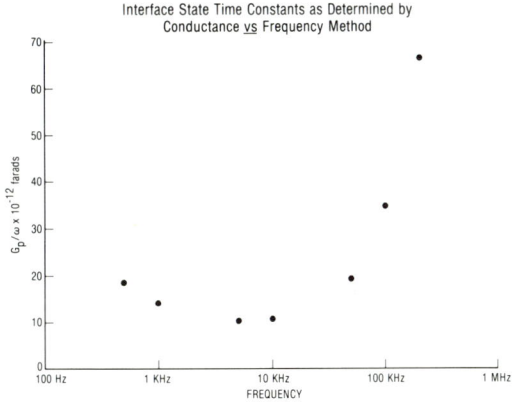

Fig. 2

Storage time. The minority carrier storage time which is determined by the dark current density is of particular importance in $(Hg_{1-x}Cd_x)Te$ MIS structures due to potential applications as charge-coupled devices. Capacitance-voltage curves of our initial MIS structures at 77°K and low infrared backgrounds exhibit low frequency behavior up to approximately 10 KHz which is indicative of minority carrier dark currents of 1×10^{-5} amp/cm^2. This current is anomalously large in these samples in view of reported dark current densities at least two orders of magnitude smaller for $Hg_{0.7}Cd_{0.3}Te$ MIS structures.[4]

We have investigated the temperature dependence of the dark current from $77^{\circ}K$ to $195^{\circ}K$ in an attempt to determine the mechanism of minority carrier generation. The generation mechanisms of depletion region generaton, bulk diffusion, and surface generation are expected to exhibit strong temperature dependences over this temperature regime.[11,12] However, the dark current is observed to be temperature independent from $77^{\circ}K$ to $153^{\circ}K$. Therefore, we conclude that neither depletion region generation, bulk diffusion, nor surface generation is giving rise to the observed dark current at $77^{\circ}K$. Likewise, tunneling currents are negligibly small for material of this bandgap and doping density.[13] The most reasonable explanation for the observed dark current density is the presence of p-type regions in the MIS depletion region which would serve as sources of holes. Inhomogeneities such as this have previously been observed in $(Hg_{1-x}Cd_x)Te$ samples which are p-type as grown and annealed in mercury to achieve p to n conversion.[14] This is the processing procedure used to produce the material currently under study. More recent MIS structures made on a different slice from the same supplier have exhibited the expected low dark current densities and storage times of 0.1-1.0 sec. This provides further evidence that the large dark currents observed in the initial samples were likely due to inhomogeneities in the semiconductor slices.

REFERENCES

1. D. Long and J. L. Schmit, in "Semiconductors and Semimetals," R. K. Willardson and A. C. Beer, eds., Vol. 5, Academic Press, New York, 1970, p. 175.
2. R. B. Schoolar, R. L. Alt, and M. J. Daugherty, "Proceedings of the 157th Meeting of the Electrochemical Society," St. Louis, May 1980, p. 414.
3. R. A. Chapman, M. A. Kinch, A. Simmons, S. R. Borrello, H. B. Morris, J. S. Wrobel, and D. D. Buss, Appl. Phys. Lett., 32, 434 (1978).
4. M. A. Kinch, R. A. Chapman, A. Simmons, D. D. Buss, and S. R. Borrello, Infrared Physics, 20, 1 (1980).
5. P. C. Catagnus and C. T. Baker, U.S. Patent No. 3,977,018 (24 August 1976).
6. Y. Nemirovsky and E. Finkman, J. Electrochem. Soc., 126, 768 (1979).
7. S. M. Sze, "Physics of Semiconductor Devices," Wiley-Interscience, New York, 1969, pp. 425-504.
8. S. R. Hofstein, Solid-State Electron., 10, 657 (1967).
9. M. Kuhn, Solid-State Electron, 13, 873 (1970).
10. E. H. Nicollian and A. Goetzberger, Bell Syst. Tech. Journal, 46, 1055 (1967).
11. A. Goetzberger and E. H. Nicollian, Bell Syst. Tech. Journal, 46, 514 (1967).
12. P. H. Zimmermann, M. E. Mathews, and D. E. Joslin, J. Appl. Phys. 50, 5815 (1979).
13. W. W. Anderson, Infrared Physics, 17, 147 (1977).
14. W. F. H. Micklethwaite and R. F. Redden, Appl. Phys. Lett., 36, 379 (1980).

CHAPTER VI

INTERFACES

CHEMICAL BONDING AT METAL/SiO_2/Si(111) INTERFACES

R. S. Bauer and R. Z. Bachrach
Xerox Palo Alto Research Centers, Palo Alto, CA 94304
and
L. J. Brillson
Xerox Webster Research Center, Webster, NY 14580

ABSTRACT The composition and extent of Si(111) MOS interface regions have been probed following *in situ* growth using electron spectroscopies excited with soft X-ray synchrotron radiation. Our 2p core electron photoemission shows that Si_2O-type bonds bridge Si(111) to SiO_2. For non-abrupt interfaces (>3.85 Å wide), a unique SiO layer is present which contains negligible SiO_4 tetrahedra as characterized by Si(2p) CFS-photoyield. For Al contacts to SiO_2, Al_2O_3 grows at the interface by reducing SiO_x (x<2). The liberated oxygen first bonds within the Al overlayer in an intermediate aluminum-oxide configuration of three Al atoms bound to each surface O. In contrast, Au forms islands on SiO_2 with evidence of Au-Si bonding.

INTRODUCTION The interfaces connecting SiO_2 to adjoining metal and Si layers can control the electronic properties of MOS devices. In this paper we summarize our latest understanding of the extent and composition of the transition region between bulk silicon and the silicon dioxide thermally grown onto its (111) face. The chemical bonding and reaction products are then examined for Al and Au deposited onto these characterized SiO_2-Si(111) structures. Particular emphasis is given to microscopic properties which can be determined using soft x-ray photoemission energy distribution curves and electron yield, utilizing synchrotron radiation excitation.

EXPERIMENT In studying the Si-SiO_2 interface with surface senstive probes, one either can begin with an oxidized silicon wafer and etch the surface by chemical (Ref. 1) or plasma (Ref. 2) techniques to expose the interfacial region, or can oxidize a silicon surface in the measuring system. We have employed the later by controlling the oxide formation in a manner as similar to oxidation furnaces as possible. The silicon was oxidized as described elsewhere (Ref. 3) to form the UHV saturated oxide layer which is independent of preparation technique (Ref. 4). The major limitation is that the UHV environment limits the effective oxygen arrival rate at the silicon surface to some three orders of magnitude below that normally occuring during atmospheric pressure oxidation of silicon devices. Investigation of such interfaces should provide very representative results since many basic properties are the same for both systems; for example, the transition from bulk silicon to SiO_2 occurs over atomic dimensions, identical to those reported for controllably etched (111) and (100) Si (Refs. 1,2,5).

The initial stages of metal-SiO_2 interface formation were investigated for room temperature deposition. Gold was evaporated from a tungsten filament, while aluminum was evaporated from a water-cooled MBE-type shuttered evaporation cell with a boron nitride crucible. The evaporation rates were calibrated using a Sloan quartz crystal monitor and could be controlled to fractions of an Angstrom/sec. The pressure was maintained below 1×10^{-9} Torr during the evaporation. The measurements presented were made using the variable photon energy excitation available on the 4° beam line at the Stanford Synchrotron Radiation Laboratory (Ref. 6). Photoemission and photoyield (Ref. 7) were measured with a PHI double-pass CMA and a computer-aided data-acquisition system.

THE Si(111) - SiO$_2$ INTERFACE With the thermally controlled *in situ* oxidation technique described above, we are able to produce samples which exhibit the same characteristic Si(2p) partial yield and photoelectron energy-distribution in Fig. 1 over a wide range of conditions. Reproducible interfaces are obtained by cleavage in an atmospheric pressure of O$_2$, by repeated dosing with an excited oxygen plasma, and by high temperature dosing with ground-state molecular dioxygen. Detailed discussion of the evolution of Si(111) + O$_2$ to an SiO$_2$ layer has been presented elsewhere (Ref. 4). The major result to be emphasized here is that the transition from Si to SiO$_2$ is observed within one to two atomic layers. From the core photoemission, the interface oxide is characterized by a 2.6eV Si(2p) binding energy shift from that of bulk Si as shown in Fig. 2. Importantly, when this oxidation state occurs at the interface, it exhibits a Si(2p) absorption resonance identical to that of evaporated SiO (Ref. 8). The only other explanation of this interface oxide would be for its absorption fingerprint to correspond to that of Si$_2$O$_3$. For this to be consistent with our data, the "standard" oxide formed by UHV depositon of SiO (Ref. 8) would have to form as such a Si^{+3} state. However, the slightest addition of O$_2$ to the vacuum causes SiO$_4$ quantum-well resonances characteristic of SiO$_2$ to appear imediately in the CFS-yield (Ref. 8). As SiO$_2$ is the most thermodynamically stable oxidation state of Si, we conclude that the product of the purest evaporation corresponds to the stoichiometry of the starting SiO. Thus, the absence of the SiO$_4$ tetrahedral signature in the partial-yield of the *interfacial* oxide allows identification of a single connective layer of unique SiO stoichiometry just beneath the SiO$_2$. This may be thought of as a disordered GeS-type structure, rather than Hollinger's mixture model (Ref. 9). Such an SiO layer is suggested in Tiller's growth model (Ref. 10) and allows for the large mismatch between Si and SiO$_2$.

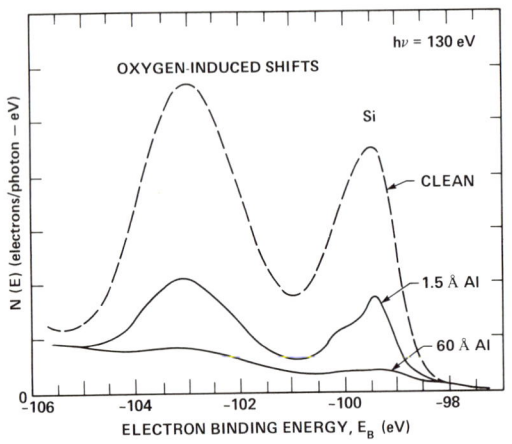

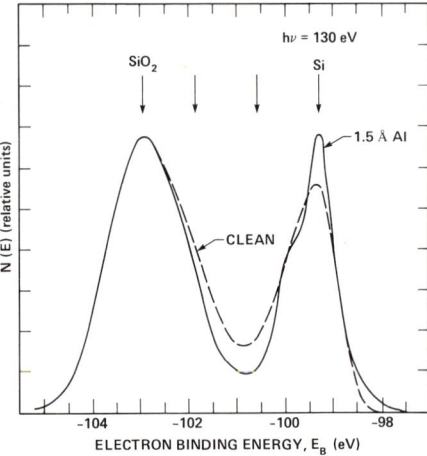

Fig. 1 Energy distributions normalized to constant photon flux for Si(2p) core electrons photoemitted from oxidized Si onto which Al was evaporated by MBE to 1.5 A and 60 A.

Fig. 2 Energy distributions normalized to constant total Si(2p) core electron yield for Si oxidized with 10 doses of 5x10^3 L unexcited O$_2$ at 625°C and aluminized (data same as in Fig. 1).

The connection of this intervening SiO layer to the underlying Si(111) is accompanied by a chemically shifted peak in the Si(2p) photoemission at 1.3±.3 eV as shown in Fig. 2. The variations in this energy depending on analysis and preparation conditions is the subject of another investigation. Suffice it to say, that since these Si atoms are bound with an energy between that of bulk Si and the 2.6 eV intermediate state identified as SiO above, we conclude that Si_2O is the appropriate stoichiometry describing the Si interface transition on the Si side of the MOS junction. This means that the first Si which is not 4-coordinated with Si retains three Si nearest neighbors and has a single O-bond which replaces the inter double-Si-layer bond. Since this interface bonding is directed toward the (111) surface, incomplete oxidation in this first non-Si plane would lead to the P_b defect, consistent with Poindexter's work (Ref. 11). From the strength of this emission within our very short photoelectron escape depth (7A for Si, ~14A for SiO_2), this O is one of the atoms forming the Si^{+2} bonds of the transition layer. We conclude that the net distance from a Si atom in bulk Si to one in SiO_2 is ~5.3A. There is no evidence in our data for the Si_2O_3-type interface bonding state reported in Ref. 1.

THE Al - SiO_2 INTERFACE Al was deposited at room temperature onto the terminating SiO_2 surface layers studied above. The attenuation of the normalized Si(2p) emission in Fig. 1 is consistent with uniform coverage of the oxidized Si surface by Al. Most of the aluminum at this interface is reacted to form aluminum oxide as characterized by the Al(2p) core level chemical shift for approximately a monolayer coverage shown in Fig. 3. We observe in the top of Fig. 4 that this formation takes hours to occur at room temperature. Further in the bottom of Fig. 4 we see that a thick aluminum film is not significantly oxidized in a 5-hour period by the vacuum system ambient of 5×10^{-10} Torr total pressure. Thus we can observe the kinetic evolution of the true interface oxidation-reduction reaction at room temperature. After some 40 minutes, the Al at the interface has formed an intermediate state characterized by a 1.8 eV chemical shift from bulk Al (see arrow in Fig. 4) and negligible Al_2O_3.

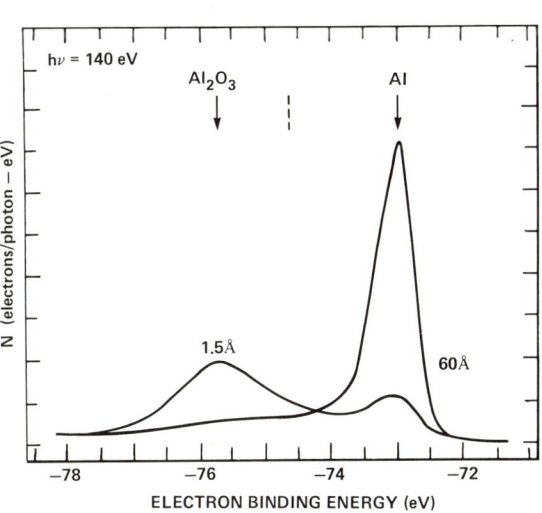

Fig. 3 Energy distributions normalized to constant photon flux for Al(2p) core electrons photoemitted from two overlayers of Al on oxidized Si.

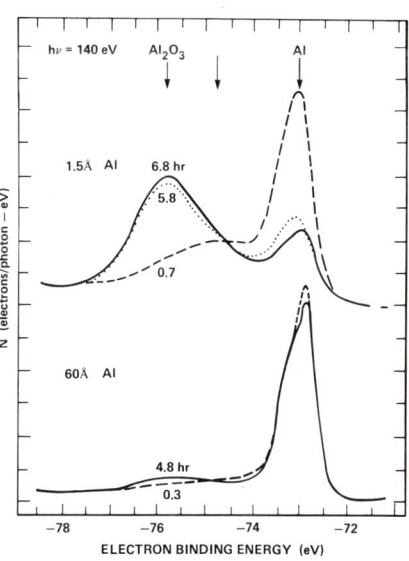

Fig. 4 Time evolution of the Al(2p) core levels of Fig. 3. The lower half shows the minimal oxidation of the Al film due to the vacuum system ambient.

This oxidation state of Al corresponds to a geometry of oxygen within or above an Al(111) close-packed layer (Ref. 13). Detailed studies of this O chemisorption state observed on the Al(111) surface identifies the geometry as an oxygen occupying the hollow site between three close-packed Al atoms spaced 2.86 A apart (Ref. 13, 14). Thus for the Al MOS contact, we can understand the formation of the actual interface as three Al atoms clustering around each of the terminating O atoms at 2.1 A (Ref. 13) or 1.79A (Ref. 14). The Al oxidation studies of Ref. 13 suggest that this intermediate oxidation state will be converted to Al_2O_3-type bonding before a fourth Al atom fills the hollow above this Al plane. Since it is more favorable for the Al to have a higher oxidation state, eventually O is leached from the SiO_2 to form the thermodynamically more stable Al_2O_3 layer in the Al-SiO_2 interface region.

As the Al_2O_3 grows, the Si remains while the oxygen diffuses into the Al overlayer. From measurements of the Si(2p) core electrons shown in Fig. 2, we find that the oxygen consummed has come from the surface of the oxidized Si leaving the SiO_2 and intermediate SiO largely intact below the top layer. A layer of excess Si then resides between the oxidized Al and oxidized Si regions as seen by the relative increase in the unshifted Si(2p) part of the core photoemission in Fig. 2. We cannot discern reactions of this excess interfacial Si and the Al contact since we measure identical Si(2p) core photoemission and absorption for clean Si(111) 7x7 and the Si(111) $\sqrt{3}\times\sqrt{3}$: Al overlayer (Ref. 15).

THE Au - SiO_2 INTERFACE The depostion of Au on the free surface of SiO_2 is distinctively different from the interface formed with Al. One of the most striking features is monitored by the Au(5d) photoemission for a monolayer coverage.

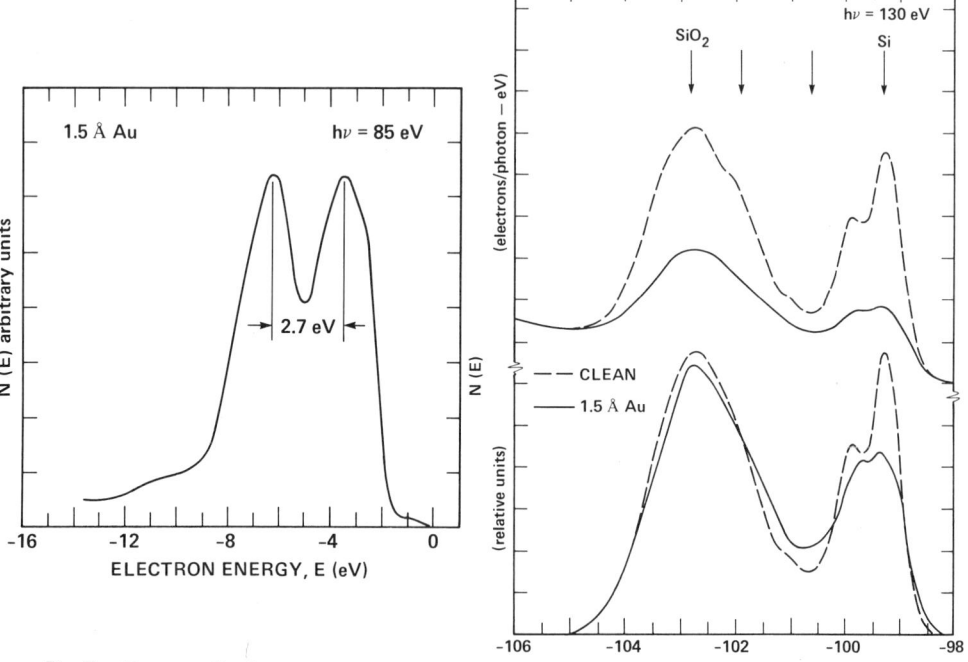

Fig. 5 Energy distribution of Au(5d) electrons for 1.5A of Au on oxidized Si in the valence band emission region. The 2.7 eV spin-orbit splitting establishes the clustering of Au atoms on the surface.

Fig. 6 Si(2p) photoemission, normalized as in Figs. 1 and 2, for clean and Au-covered oxidized Si.

As seen in Fig. 5, the spin-orbit splitting of the Au(5d) levels is $\Delta = 2.7$ eV for an equivalent uniform layer of 1.5Å Au on the oxidized Si surface. This compares to $\Delta = 2.5 \pm .2$ eV for metallic Au and 1.6 ± 0.1 eV for isolated Au atoms. Thus, the Au aglomeragtes in metallic islands on SiO_2 at the lowest coverages measurable. This is to be compared to the uniform deposition of Al and formation of Al-O bonds discussed above. It is also distinquished from Au deposition on *clean* semiconductor surfaces, where the Au diffuses into the material as dispersed atoms [e.g., $\Delta = 1.4$-1.8 eV for Si (Ref. 16), 1.7eV for GaSb (Ref. 17) and 1.6eV for InP (Ref. 17)].

At the interface of the Au islands and the SiO_2, the bonding interactions are opposite to those observed for Al. As seen in Fig. 6, the normalized Si(2p) emission decreases at the binding energies for both bulk Si and SiO_2. The relative increase in the intermediate energy region is indicative of two possible occurances. The Au deposition could have induced a reduction of SiO_2 to intermediate oxidation states at the SiO_2-Au interface; the disposition of the oxygen thus liberated is not obvious from our data. More likely, the observed shift in binding energy for SiO_2 to lower binding could be the opposite of the effect seen as SiO_2 builds up on the Si surface (Ref. 4); this would suggest the formation of a strained SiO_2 region at the Au interface [or perhaps metallic screening of the insulator from which the electrons are photoemitted (Ref. 18)]. This needs to be considered in light of the reduction in bulk Si emission in Fig. 6, with a relative increase in the spin-orbit split region some 0.5 eV more tightly bound. This latter data could suggest an increase in Si_2O oxidation species, as was indicated by similar photoemission changes for room temperature, unexcited oxygen dosing of Si(111) (Ref. 3). It seems more reasonable to us that this near bulk-Si increase results from a bonding of the Au to Si atoms at the interface, since comparable shifts are observed when Au diffuses into Si in large amounts (Ref. 16). This would imply the breaking of some Si-O bonds in the process and obviously cause strain in the near SiO_2 surface region, consistent with the discussion above. Recent reports of dipole layers at Au MOS contacts are consistent with such metal-SiO_2 interface reactions (Ref. 19).

CONCLUSIONS The composition and extent of MOS interface regions have been probed during *in situ* growth. One of our purposes is to illustrate the chemical identification of interfacial reaction products and spatial distribution that can be delineated using soft x-ray, synchrotron-radiation excited core-photoelectron spectroscopy. Simply relying on the binding energy of intermediate oxidation states, as measured by XPS, can lead to spurious conclusions for the local coordination of the excited atom when large variations in structure and ionicity occur. Correlating local atomic transition resonances in the corresponding core level absorption, as measured by the partial photoyield in the same experiment, provide a more reliable indication of microscopic bonding.

SiO_2 was formed by low-pressure, controlled O_2 dosing of the *in situ* cleaved Si(111) surface held at 625°C; the width of the substrate Si transition and the composition of evaporated Al interfaces on this insulator are identical to those reported for MOS devices. Our 2p core electron photoemission shows that Si_2O-type bonds bridge Si(111) to SiO_2. No evidence is found for the previously reported Si_2O_3 interface oxidation state. A unique SiO layer is identified at most interfaces which contains negligible SiO_4 tetrahedra. Thermal annealing generally causes greater coordination of interfacial SiO_x (x<2).

For Al contacts to SiO_2, Al_2O_3 grows at the interface and frees excess Si by the reduction of SiO_x (x<2). The reaction kinetics are slow enough at room temperature that one can follow the conversion of oxidized Si to Al_2O_3 through an intermediate oxidation state identical to the chemisorption geometry of oxygen on Al (111). The liberated oxygen first bonds within the Al overlayer in a characteristic intermediate aluminum-oxide configuration of three close-packed Al atoms bound to each surface O. In contrast, Au forms islands on SiO_2 with the Au bonding to Si atoms in the now strained metal-SiO_2 interface region. These primary chemical aspects of MOS interfaces need to be understood in order to attempt detailed correlations and electronic calculations of realistic device properties (Ref. 20).

ACKNOWLEDGEMENTS We greatfully acknowledge Dr. J. C. McMenamin for his collaboration during the course of this work. The measurements were performned at the Stanford Synchrotron Radiation Laboratory which is supported by the NSF (contract No. DMR7727489) in co-operation with the DOE.

REFERENCES

1. F. J. Grunthaner, P. J. Grunthaner, R. P. Vasquez, B. F. Lewis, J. Maserjian, and A. Madhukar, Phys. Rev. Letters 43, 1683 (1979); R. P. Vasquez and F. J. Grunthaner, Surface Science (in press).

2. C. R. Helms, W. E. Spicer, and N. M. Johnson, Solid State Comm. 25, 673 (1978).

3. R. S. Bauer, J. C. McMenamin, H.J. Petersen and A. Bianconi, *Proc. Int'l Conf. SiO_2 and Its Interfaces*, edited by S. T. Pantelides (Pergamon, New York 1978) p. 401 and cited references.

4. R. Bauer, J. McMenamin, R. Bachrach, A. Bianconi, L. Johansson and H. Petersen, *Inst. Phys. Conf. Ser. No. 43: Phys. of Semicon., 1978*, B. Wilson,ed. (Inst. of Physics, London 1979) p. 797.

5. D. E. Aspnes and J. B. Theeten, Phys. Rev. Letters 43, 1046 (1979); N. W. Cheung, L. C. Feldman, P. J. Silverman, and I. Stensgaard, Appl. Phys. Letters 35, 859 (1979).

6. F. C. Brown, R. Z. Bachrach and N. Lien, *Nuclear Ins. and Methods*, 152, 73, (1978).

7. R. S. Bauer, R. Z. Bachrach, S. Flodstrom, and J. McMenamin, J. Vac. Sci. Technol. 14, 378 (1977).

8. A. Bianconi and R. S. Bauer, Surface Science (in press); R. S. Bauer and R. Z. Bachrach, J. Vac. Sci. Technol. 17, 509 (1980) and cited references.

9. G. Hollinger (these Proceedings) and cited references.

10. W. A. Tiller, Materials Research Society Annual Meeting, Boston, 1979 (unpublished).

11. P. J. Caplan, E. H. Poindexter, B. E. Deal, and R. R. Razouk, J. Appl. Phys. 50, 5847 (1979).

12. R. Z. Bachrach and R. S. Bauer, J. Vac. Sci. Technol. 16, 1149 (1979).

13. S. A. Flodstrom, R. Bachrach, R. Bauer, and S. Hagstrom, Phys. Rev. Lett. 37, 1282 (1976); S. A. Flodstrom, C. Martinson, R. Bachrach, S. Hagstrom and R. Bauer, Phys. Rev. Lett. 40, 907 (1978).

14. J. Harris, and G. S. Painter, Phys. Rev. Lett. 36, 151 (1976); L. I. Johnsson and J. Stohr, Phys. Rev. Lett. 43, 1882 (1979) and Phys. Rev. B (to be published).

15. R. S. Bauer, R. Z. Bachrach, G. V. Hansson, R. H. Williams, D. J. Chadi, and P. Chiaradia, Proc. XV Intern. Conf. on Physics of Semiconductors-Kyoto, 1980.

16. L. Braicovich, C. Garner, P. Skeath, C. Su, I. Lindau, and W. Spicer, Phys. Rev. B 20, 5131 (1979).

17. P. W. Chye, I. Lindau, P. Pianetta, C. Garner, C. Su. and W. Spicer, Phys. Rev. B 18, 5545 (1978).

18. G. Hollinger, Y. Jugnet, P. Pertosa, and T. M. Duc, Chem. Phys. Letters 36, 441 (1975).

19. T. W. Hickmott (these Proceedings).

20. N. M. Johnson, D. K. Biegelsen, and M. D. Moyer (these Proceedings).

DIPOLE LAYERS AT THE GOLD–SiO$_2$ INTERFACE

T. W. Hickmott
IBM Thomas J. Watson Research Center
Yorktown Heights, New York 10598

INTRODUCTION

The conventional analysis of metal–SiO$_2$–semiconductor (MOS) structures is extended to include the effect of dipoles at the metal–insulator interface or in the bulk of the insulator on capacitance–voltage (C–V) characteristics. The occurrence of both dipole layers and bulk charge can be established by measuring the flat–band voltage of MOS capacitors as a function of oxide thickness. C–V measurements of Au–SiO$_2$–Si capacitors are combined with thermally–stimulated ionic conductivity (TSIC) measurements of Na$^+$ in SiO$_2$ to show that annealing of the Au–SiO$_2$ interface between 150 °C and 250 °C produces a positive dipole at the Au–SiO$_2$ contact while annealing between 250 °C and 400 °C results in a negative dipole at the Au–SiO$_2$ interface as well as introducing negative charge into the insulator. C–V measurements give the magnitude of work function changes while TSIC measurements show that the changes occur at the Au–SiO$_2$ interface. The work function at the metal–insulator interface is not a well-defined constant but depends on metal–insulator interactions that depend, in turn, on processing of the interface.

THEORY

The gate voltage, V_G, of an MOS capacitor is obtained by inspection of electron energy-band diagrams such as in Fig. 1a.[1] For p–silicon, if there is no charge in surface states, Q_{ss}, nor bulk charge in the oxide, $\rho(x)$,

$$qV_G = \phi_m - (\chi + E_G/2) + q\psi_S - q\phi_F + qV_{ox} \tag{1}$$

where ϕ_m is the metal–insulator work function, χ is the electron affinity of the silicon–SiO$_2$ interface, V_{ox} is the potential drop across the insulator layer, E_G is the band gap of silicon, ψ_S is the surface potential of the silicon, ϕ_F is the Fermi potential of the silicon measured from the intrinsic Fermi level, E_I, and q is the magnitude of the electron charge. The units of ϕ_m, χ, and E_G are electron volts while ψ_S, ϕ_F, V_G, and V_{ox} are in volts. Similar equations are valid for n–silicon.

The work function difference, ϕ_{ms}, with respect to the intrinsic energy level E_I, can be defined as $q\phi_{ms} \equiv \phi_m - (\chi + E_G/2)$. At the flat-band voltage, V_{FB}, there is no net charge in the silicon, and $\psi_S = 0$.

The voltage appearing across the oxide, V_{ox}, depends on the charge in the insulator. For carefully prepared MOS capacitors, the bulk charge, $\rho(x)$, is zero but there is charge at the Si–SiO$_2$ interface, Q_{ss}/cm^2. Q_{ss} includes both fixed charge and charge in surface states. If $\rho(x)$ is zero,

$$V_{FB} = \phi_{ms} - \phi_F - (Q_{ss}/C_{ox}) = \phi_{ms} - \phi_F - Q_{ss}(L/\varepsilon_{ox}) \tag{2}$$

where L is the oxide thickness, ε_{ox} is the dielectric constant of SiO$_2$, and C_{ox} is the capacitance due to the oxide. A standard method to measure ϕ_{ms} for MOS capacitors is to plot V_{FB} for capacitors of different oxide thicknesses as a function of thickness.[2] The slope gives Q_{ss}, the intercept gives ϕ_{ms}.

Both space charge and dipole layers in the insulator can affect the apparent value of ϕ_{ms}. If the integral of the moment of $\rho(x)$ can be approximated by a centroid of charge, Q_I, at a distance d from the metal–SiO$_2$ interface, as in Fig. 1c, $\Delta V_{FB} = -(d/L)(Q_I/C_{ox})$.

Dipole layers at the metal-insulator interface can change ϕ_{ms}, just as they can at the metal-vacuum interface. In addition the dielectric can support a dipole layer which a vacuum cannot. An ideal dipole layer in a dielectric is a sheet of positive charge of density σ /cm^2 separated by a distance t from a sheet of negative charge of equal magnitude, as shown schematically in Fig. 1b. Such a configuration has no net charge but adds a term to V_{FB},[3] $\Delta V_{FB} = - (\sigma t/\varepsilon_{ox})$, which is independent of oxide thickness. In principle, a dipole layer can exist at either interface or at any position in the dielectric.

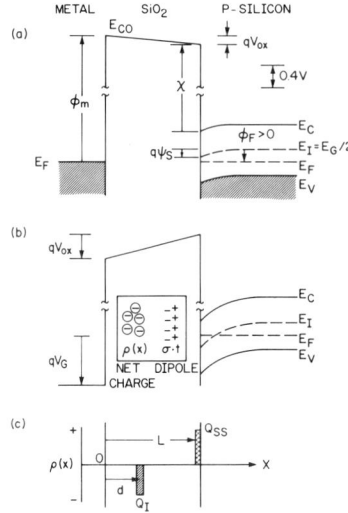

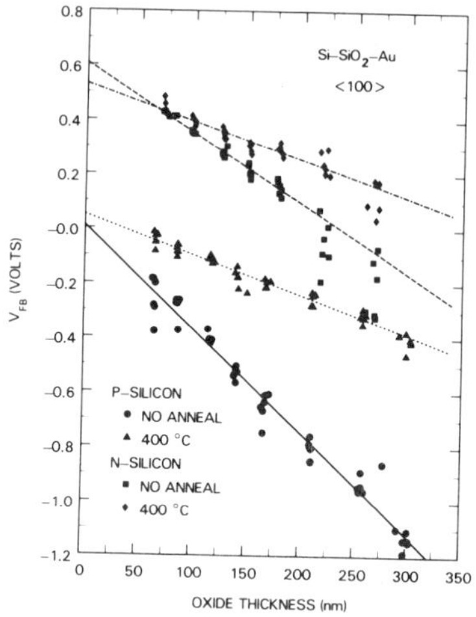

Fig. 1. (a) Schematic MOS energy-band diagram. (b) With applied bias, and with schematic dipole layer and charge distribution in the oxide. (c) Schematic charge distribution.

Fig. 2. Flat-band voltage versus oxide thickness for Si–SiO$_2$–Au capacitors, as-deposited and after 400 °C anneal in N$_2$.

The generalized expression for V_{FB} of an MOS capacitor which results is

$$V_{FB} = \phi_{ms} - \frac{\sigma t}{\varepsilon_{ox}} - \frac{d}{L}\frac{Q_I}{C_{ox}} - \frac{Q_{ss}}{C_{ox}} + \phi_F. \tag{3}$$

If the centroid of net charge is at a constant distance d from the metal–SiO$_2$ interface, independent of oxide thickness, as might occur due to metal–SiO$_2$ reaction or ion implantation through the outer surface, the third term of Eq. 3 is independent of oxide thickness, and Eq. 3 can be put in the form

$$V_{FB} = (\phi_{ms} - \frac{\sigma t}{\varepsilon_{ox}} - \frac{dQ_I}{\varepsilon_{ox}}) - \frac{Q_{ss}L}{\varepsilon_{ox}} + \phi_F = \phi_{eff} - \frac{Q_{ss}L}{\varepsilon_{ox}} + \phi_F \tag{4}$$

Thus, plotting V_{FB} as a function of oxide thickness does not give ϕ_{ms} alone. It reflects the presence of dipole layers anywhere in the insulator of MOS structures, regardless of whether there is net charge in the insulator. If there is a centroid of charge at a constant distance from the metal–insulator interface, it appears in ϕ_{eff}, but does not change Q_{ss}. If d in Eq. 3 is a constant fraction of L, the apparent value of Q_{ss} changes but the intercept is unchanged. Other distributions of Q_I can change both ϕ_{eff} and Q_{ss}. If bulk charge, $\rho(x)$ is distributed throughout the insulator, V_{FB} versus L can be non-linear and determination of either Q_{ss} or ϕ_{eff} is not possible. We will show in the present work that reaction of SiO$_2$ and Au can introduce charge in the insulator, and that annealing of the Au–SiO$_2$ interface changes ϕ_{eff}.

EXPERIMENTAL

To determine ϕ_{ms} and Q_{ss} from Eq. 4, it is essential to have samples of different oxide thickness with the same values of Q_{ss}. To insure constancy of Q_{ss}, <100>-oriented silicon wafers were steam-oxidized at 1000 °C to form oxide films about 4000 Å thick. After oxidation, 6 mm wide stripes of varying oxide thickness were formed by step-etching in dilute HF. After etching, wafers were cleaned and were annealed in forming gas (10% H_2, 90% N_2) for 20 minutes at 400 °C to reduce surface states and surface charge to low levels. Gold capacitors were formed by evaporating 99.999% pure gold through a metal mask to form 0.075 mm circular dots. C–V curves were measured at 1 MHz at an average sweep rate of 10 mV/sec. Oxide thicknesses were determined from the measured oxide capacitance in strong accumulation, C_{ox} and the measured sample area.

Detailed procedures for measuring TSIC curves due to Na^+ motion in SiO_2 have been published.[4] TSIC curves in $Si-SiO_2$–metal structures depend on the observation that current at room temperature and above is primarily carried by alkali ions such as Na^+. To measure a TSIC curve, an MOS sample is heated from room temperature to high temperature (~420 °C) using a controlled heating rate and a constant voltage on the capacitor. As the sample temperature increases, Na^+ in the sample is released from traps and moves from one interface to the other. For metal positive, positive charge moves to the $Si-SiO_2$ interface; for metal negative, the charge returns to the metal–SiO_2 interface.

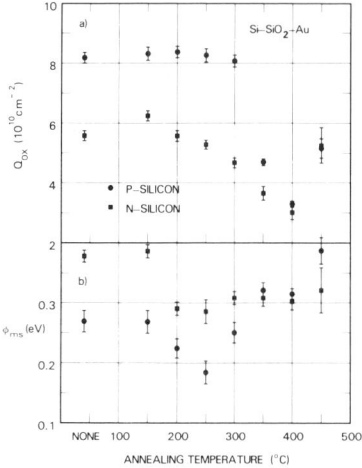

Fig. 3. (a) Effective oxide charge in Au–SiO_2–Si capacitors after annealing in N_2. (b) ϕ_{ms} of capacitors after annealing in N_2.

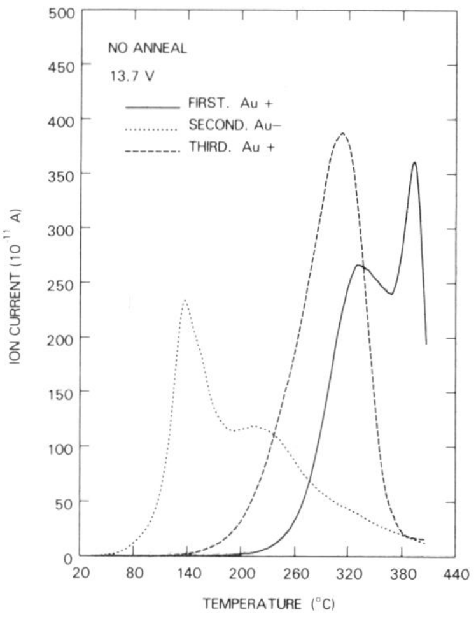

Fig. 4. Successive TSIC curves for Au–SiO_2–Si samples with 9×10^{13} Na^+/cm^2. No sample heating before TSIC curves.

For TSIC samples, oxide films of 4000 or 5000 Å thickness were steam-grown on <100> oriented p–silicon wafers at 1100 °C. Na^+ was deposited on the SiO_2 surface by evaporation of NaCl through a metal mask; Au electrodes were deposited through the same mask immediately after NaCl evaporation. Samples were not heated after metal deposition.

TSIC curves were measured in a dry N_2 atmosphere. For TSIC annealing measurements, samples were annealed *in situ*, with no contact made to the capacitor during annealing. The total amount of charge moved, n_o, is obtained by integrating experimental TSIC curves;

$$n_o = \frac{1}{Aq} \int I dt \qquad (5)$$

where A is the sample area, I is the current, and t is time.

RESULTS

Figure 2 shows V_{FB} versus oxide thickness for Si–SiO$_2$–Au samples with both n–Si and p–Si, before and after annealing at 400 °C in N$_2$. Samples were first measured after Au electrodes were deposited. The samples were then annealed for seven minutes in N$_2$ at 50 °C temperature increments and V_{FB} versus oxide thickness was remeasured. The points labeled 400 °C were measured after 400 °C anneal. In all cases V_{FB} is linearly dependent on oxide thickness. The most striking effect is a reduction in the slope of the curve after 400 °C anneal, i.e. there is a significant reduction in the net positive charge by annealing. Figure 3 shows Q_{ox} and ϕ_{ms} for both n–Si and p–Si samples as a function of annealing temperature. The error bars show the standard deviation of the data. Q_{ox} is defined as the effective charge/cm^2 at the Si–SiO$_2$ interface derived from plots of V_{FB} versus oxide thickness. The initial fixed charge level is higher in p–Si samples than in n–Si samples. Kar[2] reported this, though the charge level is about a third of the value he found. For p–Si, Q_{ox} is constant for anneal through 300 °C; it drops abruptly after anneal at 350 °C and 400 °C. For n–Si, the decrease in Q_{ox} is more gradual but the same level is observed after anneal at 400 °C. Annealing at 450 °C increases Q_{ox}, though there is a substantially greater spread in the data than for annealing at lower temperatures.

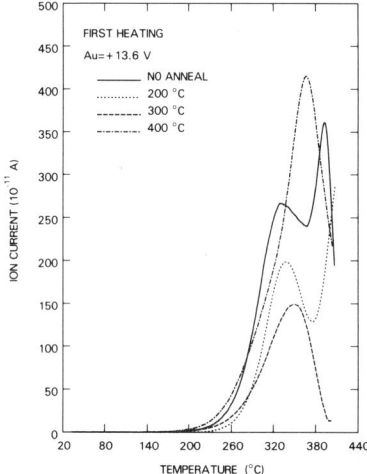

Fig. 5. Effect of annealing without bias on first TSIC curve of Au–SiO$_2$–Si samples with 9x10^{13} Na$^+$/cm^2. Au=+13.6 V.

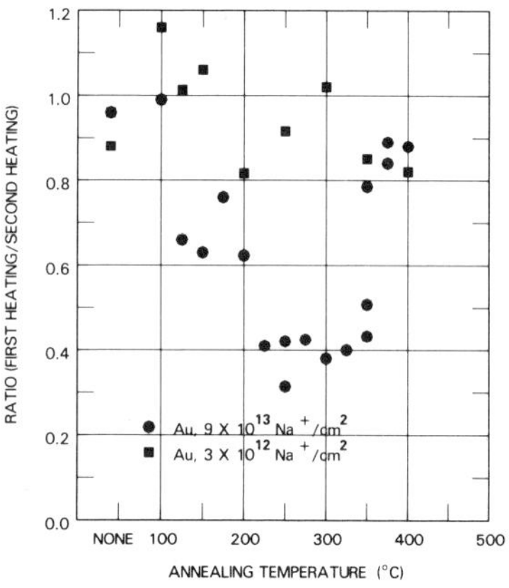

Fig. 6. The ratio, (Na$^+$ moved in first TSIC curve)/(Na$^+$ moved in second TSIC curve) for different annealings of Au–SiO$_2$–Si samples without bias.

ϕ_{ms} for Si–SiO$_2$–Au is shown in Fig. 3b. The initial values of ϕ_{ms} differ by ~0.1 eV for p–Si and n–Si. However, both show a drop of ~0.1 eV after annealing at 250 °C, followed by an increase on annealing at 300 °C. The values of ϕ_{ms} are identical, 0.31 eV, after annealing at 350 °C and 400 °C. The decrease in ϕ_{ms} at 250 °C corresponds to a positive dipole forming at the Au–SiO$_2$ interface. The subsequent increase can be due to a negative dipole forming after anneal at 300 °C and higher. Thus, ϕ_{ms} is not necessarily a constant depending only on metal and semiconductor; it also depends on annealing of the metal–insulator interface.

The decrease in Q_{ox} can be due either to a reduction of positive charge at the Si–SiO$_2$ interface or to an increase of negative charge by reaction of Au with SiO$_2$. TSIC measurements of Na$^+$ motion in Si–SiO$_2$–Au capacitors confirm that negative charge is generated by reaction of Au and SiO$_2$ in the same temperature range as Q_{ox} decreases and ϕ_{ms} increases.

Na$^+$ motion from the metal–SiO$_2$ interface depends on the metal and on the thermal history of the sample. Typical TSIC curves for Au–SiO$_2$–Si samples are shown in Fig. 4. The first heating with Au +13.7 V gives a peak in the TSIC curve at T_m=300 °C and at 400 °C. For Au–, Na$^+$ moves from the Si–SiO$_2$ interface at much lower temperatures; the peak in the TSIC curve is at 140 °C. For the third heating, T_m is ~310 °C and only one peak is observed. Subsequent curves for Au– are similar to the second heating; those for Au+ are similar to the third heating. The actual values of T_m for Au capacitors depend on the applied voltage; however, the general features of the changes of TSIC curves of Fig. 4 are observed for a wide range of bias voltages during TSIC measurements. Annealing of the metal–SiO$_2$ interface profoundly affects the kinetics of Na$^+$ motion under bias.

The effect of annealing at intermediate temperatures on TSIC curves for Si–SiO$_2$–Au samples is shown in Fig. 5 for Au positive. For each curve, a new sample was heated for five minutes at temperature in the TSIC equipment, without contact to the Au dot. After heating, TSIC curves were measured, as in Fig. 4. For Au=+13.6 V, annealing at 200 °C decreases the magnitude of the peak but the TSIC curve retains vestiges of two peaks. Annealing at 300 °C gives a single smaller peak on the first TSIC curve, while annealing at 400 °C gives a substantially larger single peak. By contrast, Na$^+$ motion from the Si–SiO$_2$ interface gives TSIC curves of approximately constant shape and magnitude, independent of annealing.

By Eq. 5, the total charge moved is obtained by integrating an experimental TSIC curve. In Fig. 6, the ratio of the amount of Na$^+$ moved in the first heating of a sample (i.e. Na$^+$ moving from the Au–SiO$_2$ to the Si–SiO$_2$ interface) to the amount moved in the second heating (i.e. Na$^+$ moving from the Si–SiO$_2$ to the Au–SiO$_2$ interface) is plotted as a function of annealing temperature. For samples with 9x10^{13} Na$^+$/cm^2, the ratio is ~1.0 for annealing below 100 °C. Annealing between 125 °C and 300 °C gives a ratio decreasing to about 0.4. The decrease is due to the reduced amount in the first heating, as shown in Fig. 5; the total amount in the second heating is roughly constant for all samples. Annealing above 325 °C, however, *reverses* the trend of the data. The ratio of the amount in the first heating to the amount in the second heating again reaches values of 0.9 for annealing at 375 °C and 400 °C. The decrease in the ratio for annealing below 300 °C is due to the migration of Na$^+$ from the Au–SiO$_2$ interface to the Si–SiO$_2$ interface during the five minute annealing without bias. Since Na$^+$ moves readily from the interface when positive bias is applied above 300 °C, some reaction between Au and SiO$_2$ occurs above 300 °C that changes the zero-bias field and helps to trap Na$^+$ close to the Au–SiO$_2$ interface. This is consistent with the introduction of negative charge, as found by C–V measurements.

The extent of reaction may depend on the amount of Na$^+$ at the Au–SiO$_2$ interface. Figure 6 shows the ratio of Na$^+$ in the first heating to that in the second heating for samples with 3x10^{12} Na$^+$/cm^2. The data spread more and the decrease in ratio followed by an increase is not observed with 3x10^{13} Na$^+$/cm^2.

DISCUSSION

The most notable results of the present work are the observation of the injection of negative charge at the Au–SiO$_2$ interface, and the low temperature range (250–400 °C) required for charge generation. From Fig. 3, ϕ_{ms} decreases up to 250 °C, Q_{ox} remains constant. The change in ϕ_{ms} is a change in the dipole at the Au–SiO$_2$ interface. Above 250 °C, both ϕ_{ms} and Q_{ox} reflect the injection of negative charge. According to Eq. 4, if charge in the oxide were generated uniformly at a constant distance from the Au–SiO$_2$ interface, only ϕ_{ms} would change. Therefore the centroid of negative charge must penetrate deeper for thicker oxides. Above 300 °C, the depth of penetration of negative charge is sufficient to change the internal field at the Au–SiO$_2$ interface and trap Na$^+$ that could otherwise move during anneal under zero bias. A more extensive account of this work will appear elsewhere.[5]

REFERENCES

1) A. Goetzberger and S. M. Sze, *Applied Solid State Science*, Vol. 1, ed. R. Wolfe, (Academic Press, New York, 1969), p. 153.
2) S. Kar, Solid-State Elec. *18*, 169 (1975).
3) J. A. Stratton, *Electromagnetic Theory*, (McGraw-Hill, New York, 1941), p. 190.
4) T. W. Hickmott, J. Appl. Phys. *46*, 2583 (1975).
5) T. W. Hickmott, J. Appl. Phys., In press.

MEASUREMENT OF TUNNELING INTO INTERFACE STATES

Walter E. Dahlke and David W. Greve**
Sherman Fairchild Laboratory
Lehigh University, Bethlehem, PA, USA

INTRODUCTION

In the past, tunneling to interface states has been invoked to explain Schottky barrier heights (1), excess currents in MOS tunnel diodes (2), and leakage in MNOS devices (3). But there is little detailed understanding of the tunneling process due to a lack of methods for studying the process directly. In this paper, we adapt our technique of optical studies of states (4) to investigate tunneling into interface states.

EXPERIMENT

We observed tunneling to interface states in Cr-SiO$_2$-nSi tunnel diodes. Oxidation time t_{ox} in a batch was varied between 2 and 4 minutes at 894°C in dry oxygen. The devices exhibited in reverse bias linear $1/C^2$-V curves characteristic for MOS tunnel diodes in deep depletion. Both the capacitance and the photocurrent transients were measured at low temperature T, photon energy $h\nu < E_G$, the capacitance at 100 kHz, and the photocurrent with light chopped at 338 Hz (5).

Measurements of the diode with t_{ox} = 2 min. at 77 K and $h\nu$ = 0.65 eV are presented in Fig. 1. The tunnel diode was biased in accumulation to fill the states, and then switched to deep depletion at t = 0. Without light, the measured capacitance increased logarithmically with t due to thermal electron emission from the states. With light on at delay time t_d, optical carrier emission began. Photocapacitance and current rose immediately and relaxed slowly to values at t_1 when states were practically emptied by optical release of charge Q_R. Then with light on at t_1, interface states were partially refilled by metal electrons tunneling through the oxide. This tunneling charge Q_T increased while the capacitance decreased logarithmically with time $\Delta t = t-t_1$. Finally, with light on at t_2, the refilled states were optically discharged. Photocapacitance and current resumed their values at t_1, after the charge Q_T was completely released.

ANALYSIS

As illustrated in Fig. 2a,b, filled states emit thermal electrons during delay time t_d. Therefore, the trap electron Fermi level (6),

$$E_{tn} = E_c - kT \ln \nu_n t \qquad (1)$$

---

*Supported by Grant ENG-78-05917-01 from the National Science Foundation.

**Present address: Philips Research Laboratories, Sunnyvale, Calif. 94086.

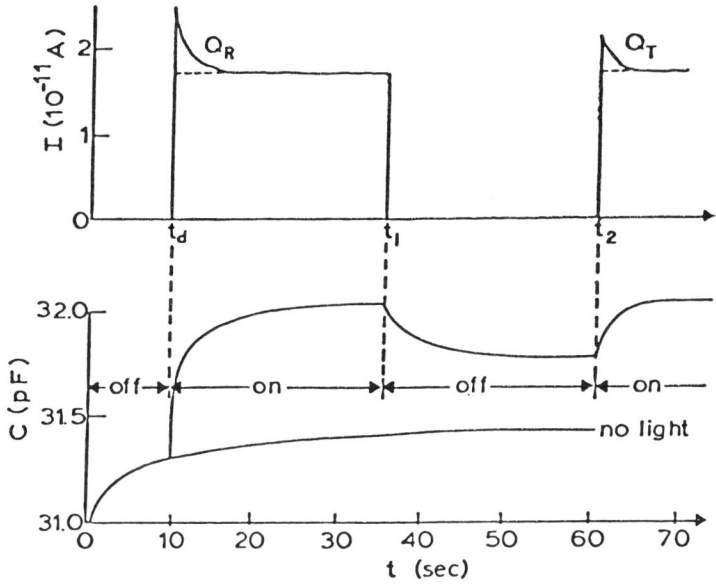

Fig. 1 Photocurrent I and photocapacitance C of MOS tunnel diode measured at 77 K and $h\nu = 0.65$ eV as function of time t.

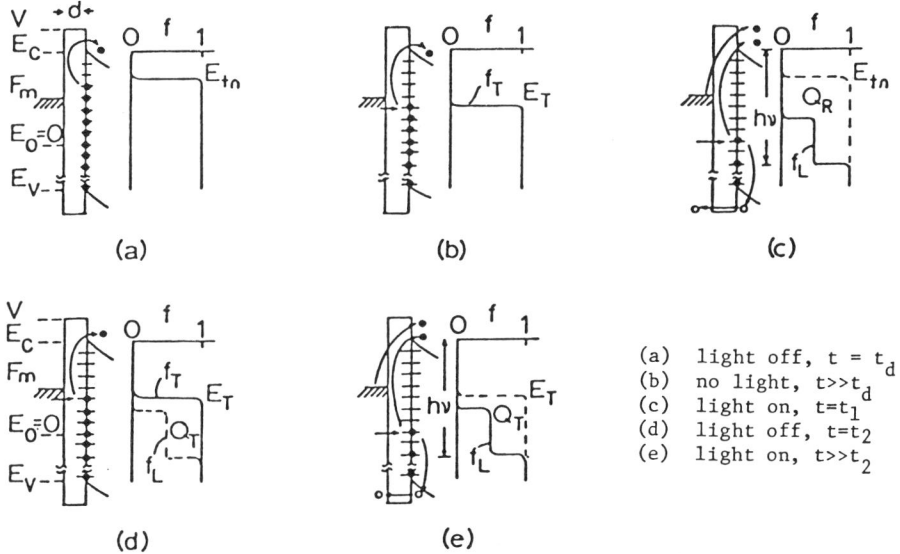

(a) light off, $t = t_d$
(b) no light, $t \gg t_d$
(c) light on, $t = t_1$
(d) light off, $t = t_2$
(e) light on, $t \gg t_2$

Fig. 2 Band diagrams and trap occupancy f

with occupancy 1/e and escape frequency ν, drops with time below E_c causing the increase of C without light in Fig. 1. For sufficiently large temperature, E_{tn} falls below the metal Fermi level F_m, reaching steady state occupancy f_T and steady state trap Fermi level E_T. Traps $E_t > E_T$ are thermally controlled and empty, while traps $E_t < E_T$ are tunneling controlled and full. We note that the tunneling region shrinks to zero with increasing oxide thickness since $E_T \to 0$. An analytic expression for E_T is found by equating for $E_t = E_T$ the relaxation times of tunneling electrons (4),

$$\tau_T = \tau_o \exp(V-E_t)^{1/2}d \simeq \tau_T(0)\exp(-E_t d/2V^{1/2}) \quad \text{for } E_t \ll V, \tag{2}$$

with oxide barrier energy [eV], effective barrier thickness d [Å], midgap tunneling time $\tau_T(0)$, and of thermally emitted electrons

$$\tau_n = \nu_n^{-1} \exp(E_c - E_t)kT. \tag{3}$$

With light on at t_d, optical electron emission begins and then slowly decays, as reflected by current I, charge Q_R, and capacitance C in Fig. 1. The dashed thermal occupancy in Fig. 2c relaxes to the optical occupancy $f_L = \sigma_p^o / (\sigma_n^o + \sigma_p^o)$ with photoionization cross sections σ_n^o, σ_p^o. Since Q_R depends strongly on $E_{tn}(T)$, the current I measured as function of T has been used to determine density $N_{ss}(E_t)$ and cross section $\sigma_n^o(h\nu, E_t)$ of the optically active states (7). The time independent photocurrent in Fig. 1 has two components: Electron emission of metal electrons through the oxide over the silicon barrier, and optical pair generation of states near midgap, as indicated in Fig. 2c. With light off at t_1, metal electrons tunneling through the oxide refill the empty states with charge $Q_T < Q_R$, and the trap occupancy relaxes from f_L to f_T. The dependence of the occupancy on $\Delta t = t-t_1$ as illustrated in Fig. 3, can best be characterized by a hole trap Fermi level

$$E_{tp} = 2V - (2V^{1/2}/d)\ln(\Delta t/\tau_o) \tag{4}$$

obtained from $\Delta t = \tau_T(E_{tp})$ and eq. (2). This level defines the trap energy, at which the hole occupancy $1-f_L$ is reduced to $(1-f_L)/e$ shown as dashed line in Fig. 3. States are full above E_{tp}, but empty below E_{tp} and above E_T. With growing Δt, tunneling fills up first the level E_T. Then as the occupancy of E_T reaches 2/3 at $\Delta t = \tau_T(E_T)$, the hole Fermi level starts dropping logarithmically with Δt. It reaches midgap at $\tau_T(0)$ filling all states between E_o and E_T. In steady state finally, capture of metal electrons is equal to thermal emission from E_T, but this component of photocurrent is unchopped and not recorded. An always present variation of d changes only the speed $2V^{1/2}/d$ with which E_{tp} is falling below E_T, but does not disturb the (observed) logarithmic time dependence of eq. (4).

With light on at t_2 in Fig. 2e, the occupancy f_T without light relaxes to the optical occupancy f_L. We note that the light controlled region (4) is governed by $\tau_L = \phi_o(\sigma_n^o + \sigma_p^o) < \tau_T$ with optical trap lifetime τ_L, photon-flux ϕ_o, and tunneling time τ_T, eq. (2). Since the released charge Q_T depends strongly on $E_T(T)$, similar to Q_R in Fig. 2c on $E_{tn}(T)$, density N_{ss} and optical cross section σ_n^o of states refilled by tunneling can be obtained by differentiating the measured photocurrent I(T) against T.

CONCLUSIONS

We have directly observed tunneling to interface states from the metal. The same states can also be filled from or emptied to the silicon bands. We obtained a density $N_{ss} = 10^{11} \text{cm}^{-2}\text{eV}^{-1}$, and a photoionization cross section $\sigma_n^o = 10^{-18} \text{cm}^2$ in

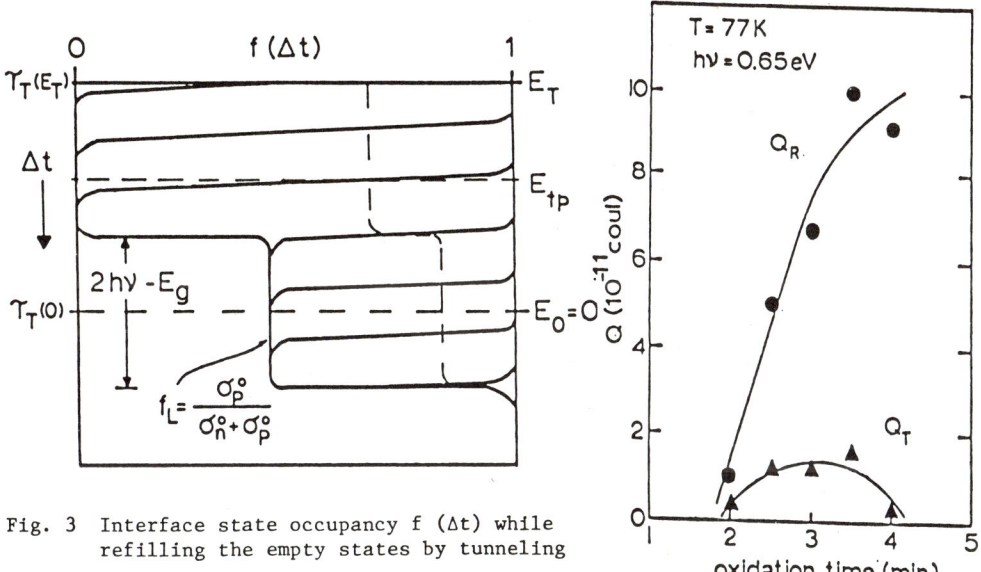

Fig. 3 Interface state occupancy f (Δt) while refilling the empty states by tunneling

Fig. 4 Interface charges Q_R and Q_T

reasonable agreement with earlier data for tunnel diodes with d ≃ 30Å (7). Measurements of the released charges Q_R and Q_T versus t_{ox} in Fig. 4 show that the ratio Q_T/Q_R approaches one monotonically when d decreases. This occurs when $E_T = E_{tn}$ in Figs. 2c and d. For $t_{ox} \simeq 3$ min, Q_T reaches a maximum when the tunneling time $\tau_T \gtrsim \tau_L \simeq 10$ sec. It falls with increasing t_{ox} since the tunneling region shrinks to zero with $E_T < 0$ in Fig. 2d. Both charges Q_R and Q_T decrease with decreasing t_{ox}, since the occupancy of tunneling controlled states cannot be changed by light if $\tau_T < \tau_L$. Additional information on the tunneling process, such as transmission through and voltage drop across the interface, can be obtained (8) by studying the component of the photocurrent emitted over the silicon barrier shown in Figs. 2c and e.

REFERENCES

(1) A. M. Cowley and S. M. Sze, J. Appl. Phys. <u>36</u>, 3212 (1965).
(2) W. E. Dahlke and S. M. Sze, Solid St. Electron <u>10</u>, 865 (1967).
(3) M. H. White, J. W. Dzimianski, and M. C. Pecherar, IEEE Trans. ED-24, 577 (1977).
(4) W. E. Dahlke and D. W. Greve, Solid St.-Electron. <u>22</u>, 893 (1979).
(5) D. W. Greve and W. E. Dahlke, in <u>Insulating Films on Semiconductors</u>, Durham England, Inst. Phys. Conf. <u>50</u>, 107 (1979).
(6) J. G. Simmons and L. S. Wei, Solid St. Electron. <u>17</u>, 117 (1974).
(7) D. W. Greve and W. E. Dahlke, Appl. Phys. Lett. <u>36</u>, 1003 (1980).
(8) J. A. Shimer, PhD thesis in progress at Lehigh University.

IMPROVED EXPERIMENTAL CHARACTERIZATION OF THE Si/SiO$_2$ INTERFACE

A. Sher and Y.H. Tsuo
SRI International, Menlo Park, California 94025

Pin Su
College of William and Mary, Williamsburg, Virginia 23185

W.E. Miller
NASA Langley Research Center, Hampton, Virginia 23665

ABSTRACT

The well-established quasistatic and conductance methods are improved by using (i) effectively thin composite insulators, (ii) low carrier concentration substrates, and [most importantly] (iii) low-level illumination at a wavelength that creates electron-hole pairs. Items (i) and (ii) extend the dynamic range of the interface-state density measurement to four decades. Item (iii) decreases the response time of the slow states (for n-type samples, those in the lower part of the band gap), so the optically driven electronic system follows a slow voltage ramp "quasistatically," and the conductance method can be used on all the states throughout the band gap. The sample investigated had a 250-Å, thermally grown oxide prepared in dry oxygen. It was never exposed to H$_2$ or H$_2$O at an elevated temperature. The composite gate insulator was completed by having an e-gun-deposited, 250-Å layer of LaF$_3$. The resulting interface, subjected to the improved experimental model at 77°K and 298°K, yields a wealth of distinctive structure, rather than the often-reported featureless, U-shaped interface-state density. Sharp interface peaks resulting from 20 ppm of fluorine on the interface are also seen.

We have recently suggested[1] several improvements of the quasistatic capacitance[2] (QSC) and conductance[3,4] methods. In this paper, we present extensions of the previously reported characterization of a (100)-oriented Si/SiO$_2$ interface. Several samples have been examined that all display the same qualitative behavior. However, because the collection and reduction of the experimental data is a highly labor-intensive activity, only three (n-type) samples have been completely characterized. Only one of these data sets is presented in this paper; details on the preparation and physical characteristics of this MIS structure (Si-58) are given in Ref. 4. Of special importance here is the composition of the insulating layer: a 250-Å, thermally grown (in dry O$_2$ at 1150 °C with no exposure to hydrogen) SiO$_2$ film on a (100) surface of silicon, followed by a 250-Å e-gun-deposited LaF$_3$ film. The LaF$_3$ has such a large capacitance ($\gtrsim 1$ μF/cm^2 for all frequencies $< 10^5$ Hz)[4] which provides a good blocking contact for the oxide, that the net insulator capacitance, $C_0 = 139$ nF/cm^2, results entirely from the thin SiO$_2$ layer. Since it is possible to resolve changes in the device capacitance to better than 1%, interface-state densities as high as 100 $C_0/e \approx 10^{14}$ states/eV–cm^2 can be accurately measured. At the other extreme, our low carrier concentration, $N_b = 2.4 \times 10^{14}$ cm^{-3}, reduces the minimum depletion-layer capacitance, C_d^{min}, to a measured value in strong inversion of 5.79 nF/cm^2. The smallest capacitance variation we can accurately detect is about 5% of this value, so that the smallest interface-state density we can resolve is $C_d^{min}/20e \approx 2 \times 10^9$ states/eV–cm^2. Thus, the effective range of our measurements extends at least over four decades ($10^{10} - 10^{14}$ states/eV–cm^2).

We have measured the QSC characteristics of our sample as a function of light intensity, with the radiation normally incident on the (transparent) top metal electrode of the MIS structure. Results in the dark and for six low levels of light intensity are illustrated in Fig. 1. All of the data were obtained with a slow, positive ramp rate (9.04 mV/s), and were unchanged when either the rate was lowered or a negative ramp was used. There are several noteworthy features about these results.

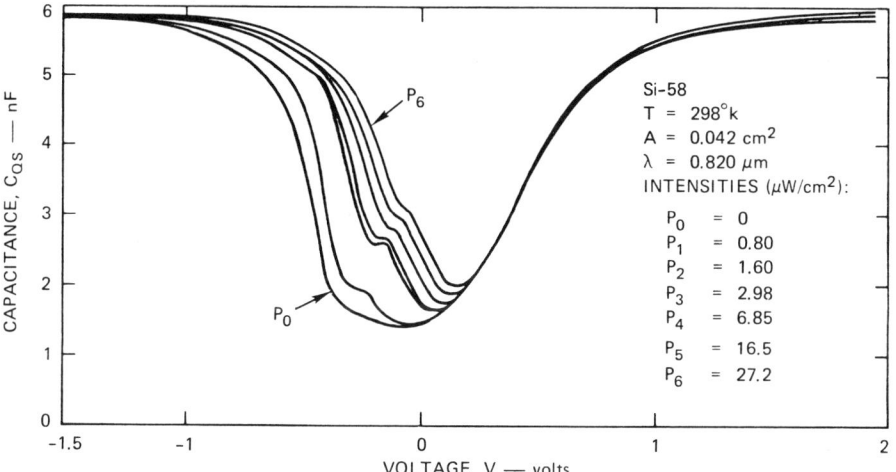

Figure 1 Quasistatic capacitance vs. voltage for six light intensities.

1) Even for the slow ramp speed used, the normal quasistatic condition,[2]

$$\beta\tau\frac{d\psi_s}{dt} < 1, \tag{1}$$

does not appear to be well-satisfied in the dark. Here, ψ_s is the surface potential at the interface, τ is the effective response time of the system, and $\beta \equiv e/kT = 38.9 V^{-1}$ at room temperature. From the conductance measurements discussed below, we have determined that $\tau > 3s$ for the "slow" valence-band side states in the dark, so that $\beta\tau d\psi_s/dt > 1$. In the presence of low-intensity light ($\approx 1.6\mu W/cm^2$), however, τ is reduced to the millisecond range and Eq. (1) is easily satisfied.

2) The shapes of the QSC curves change rapidly with light intensity between 0.80 and 1.60 $\mu W/cm^2$, as the time constants are reduced to a point where Eq. (1) is satisfied; they then change slowly from 1.60 to 2.98 $\mu W/cm^2$. Finally, the curves change once more at still higher light intensities where the depletion layer thickness and surface potential are significantly modified by the light. For low light intensities, the response time of the interface-states in the light τ is related to those in the dark τ_{dark} by

$$\tau = \tau_{dark} e^{-\beta\delta\psi_s}, \tag{2}$$

where the light-induced change in the surface potential $\delta\psi_s$ is

$$\delta\psi_s = \frac{eA}{C_{d/a}}\tau_r\Phi, \tag{3}$$

A is the area of the capacitor, $C_{d/a}$ is the depletion/accumulation layer space charge capacitance, τ_r is the recombination time for electrons and holes initially separated by the space charge layer, and Φ is the photon flux. It is evident from Eq. (2) that τ changes more rapidly than ψ_s (or C_d) does with light intensity. The functional dependence of τ predicted by Eqs. (2) and (3) on light intensity, temperature, and space charge capacitance have been verified by separate experiments (not reported in this brief account).

3) Two small peaks in C(V) near -0.5 and +0.2 V are revealed by the application of the light. These peaks correspond to energies in the Si/SiO$_2$ system, where ion-implanted fluorine is known to produce discrete levels.[5] We therefore ascribe these features to fluorine ions that have diffused from the LaF$_3$ to the interface.

We have also measured the frequency dependence of the parallel conductance and capacitance of our sample as a function of applied voltage and light intensity.[5] The net parallel conductance divided by the frequency, G_p/ω, after the effect of C_0 has been removed from the data, is shown in

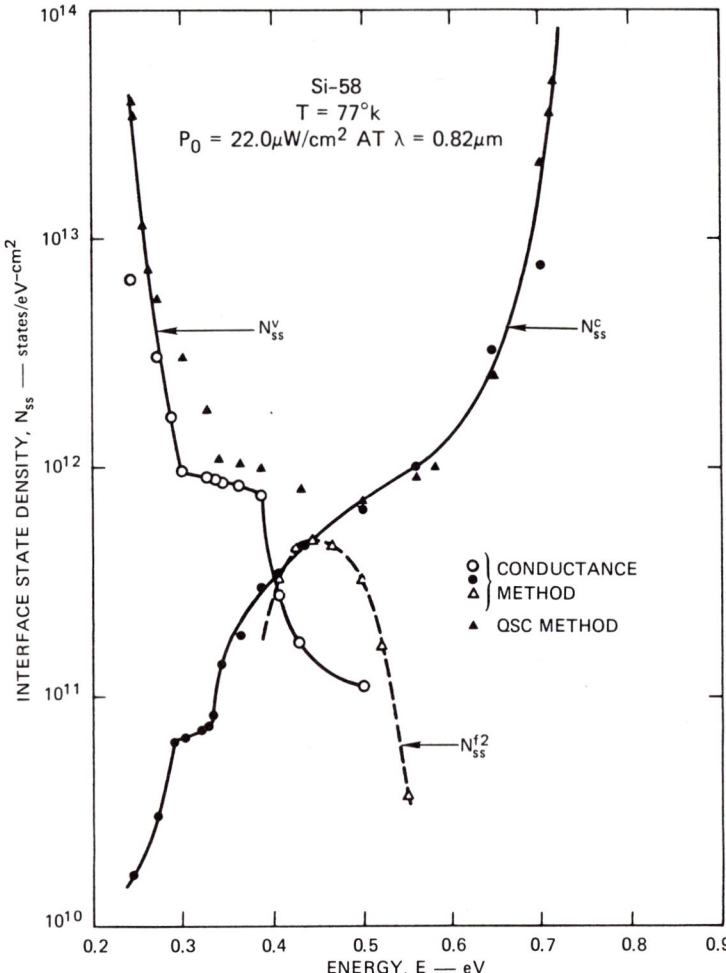

Figure 2   Interface state density vs. energy.

Fig. 2 of Ref. 1. The two principal peaks correspond to "slow" valence-band and "fast" conduction-band side states. Unlike many previous results,[3,5] however, no statistical broadening had to be introduced to explain the observed width of the peak in the depletion region. In addition, the high-frequency G_p/ω peaks were found to be nearly independent of light intensity, as expected. The fluorine-related states near $E = 0.46$ eV contribute to a discernible third peak to G_p/ω.

From a combination of our quasistatic and conductance measurements, we have thus far been able to identify four separate contributions to the interface density of states in our sample: "slow" valence-band-side and "fast" conduction-band-side densities N_{ss}^v and N_{ss}^c, respectively, and fluorine-related densities N_{ss}^{f1} and N_{ss}^{f2}. These four components at 298°K are plotted as a function of energy in Fig. 3 of Ref. 1, and N_{ss}^v, N_{ss}^c, and N_{ss}^{f2} taken at 77°K are plotted in Fig. 2. For N_{ss}^v and N_{ss}^c, good agreement is found between the results obtained from the quasistatic and conductance methods, respectively, so long as the continuum nature of these distributions is taken into account in interpreting the conductance data.[3] In contrast to the often-reported, featureless U-shaped total density of states,[5] our well-resolved components show a great deal of structure with a number of prominent features: The densities N_{ss}^v and N_{ss}^c are rapidly varying in energy and seem to have near-infinite slopes at energies of 0.23 eV (0.23 eV) and 0.90 eV (0.72 eV), respectively, at 298°K (77°K), which are denoted as E_{vs} and E_{cs} in Fig. 3 of Ref. 1 and in Fig 2. The magnitudes of N_{ss}^v near E_{vs} and N_{ss}^c near E_{cs} are considerably larger than obtained in most previous measurements,[5] while the sum of N_{ss}^v and N_{ss}^c in the midgap region (0.3 - 0.8 ev) is somewhat more typical in magnitude. The areas under N_{ss}^v and N_{ss}^c correspond to 7.9×10^{-4} and 1.0×10^{-3} states/atom, respectively.

The principal interface state densities evidently lie below E_{vs} and above E_{cs}, where they are too large for us to be able to observe them. The features of N_{ss}^v and N_{ss}^c that can be seen resemble band tails[6] of the principal interface state densities. If this is the case, then the state densities ordinarily observed are not broadened local states caused directly by interfacial imperfections (e.g., dangling bonds, bond angle distortions), but rather the result of their indirect effect on the majority species. Given this picture, the observed scattering-induced states $N_{ss}^v(E)$ and $N_{ss}^c(E)$ can be related to the "ideal" surface interface states $N_{ss0}^v(E)$ and $N_{ss0}^c(E)$ that would exist if there were no scattering by the exact relation[7]

$$N_{ss}^j(E) = \int d\epsilon \frac{N_{ss0}^j(\epsilon)\Delta_j(E)/\pi}{[\epsilon - E - \Lambda_j(E)]^2 + \Delta_j^2(E)}, \quad j = v,c \quad , \tag{4}$$

where Λ_j and Δ_j are the real and imaginary parts of the scattering-induced self-energies. If one examines this expression in energy-intervals E far removed from the band edges (i.e., $|E - E_{js}| \gg \Delta_j, \Lambda_j$), then Eq. (4) simplifies to

$$N_{ss}^j(E) = \frac{\Delta_j(E)}{\pi} \frac{1}{(E - E_{js})^2} \int d\epsilon N_{ss0}^j(\epsilon)$$

$$= \frac{\sigma_s \Delta_j(E)}{\pi} \frac{1}{(E - E_{js})^2} \quad . \tag{5}$$

For definiteness, we have identified the remaining integral in Eq. (5) with σ_s, the density of Si surface atoms, where $\sigma_s = 6.8 \times 10^{14}$ atoms/cm² for the (100) surface. The imaginary parts of the self-energies $\Delta_v(E)$ and $\Delta_c(E)$ calculated from Eq.(5) at 77°K and 298°K are plotted in Fig. 3. If the scattering causing the band tails all arose from states, with energies well away from the band edges (e.g., charged centers), then the Δ_js would decrease monotonically from the band edges. The structure in Δ_v and Δ_c is a fingerprint of the scattering mechanisms contributing to the measured interface-state density.

There is a wealth of information in these curves and controlled experiments on a number of samples will be needed to sort through it all. However, several general trends are easily extracted:

- At low temperature, the lines narrow, their peak height increases, and they tend to shift in energy toward the valence band edge.

- There is a distinct (if small) peak in $\Delta_c(E)$ at room temperature at exactly the same energy (0.46 eV) as the peak of the line designated f-2. Presumably this is a scattering resonance in the continuum conduction band tail states caused by the localized fluorine states.

- There is a great deal of structure in Δ_v and Δ_c near the energy where the valence and conduction band tails overlap. This structure may reflect effects caused by hybridization between states in the two band tails.

- The large peak in $\Delta_c(E)$ at high energy looks like a composite feature produced by several resonances. It may prove possible to indentify some of the states causing the resonances with those predicted by Laughlin et al.[8] to arise from Si-O-Si bond angle distortions.

In conclusion, we have shown that our refined quasistatic and conductance methods can lead to an improved experimental characterization of the Si/SiO$_2$ interface. Not only have we been able to measure accurately the total density of interface states across the bulk band gap on a single sample, but we have also devised a new way to interpret it that resolves the density of states into consider-

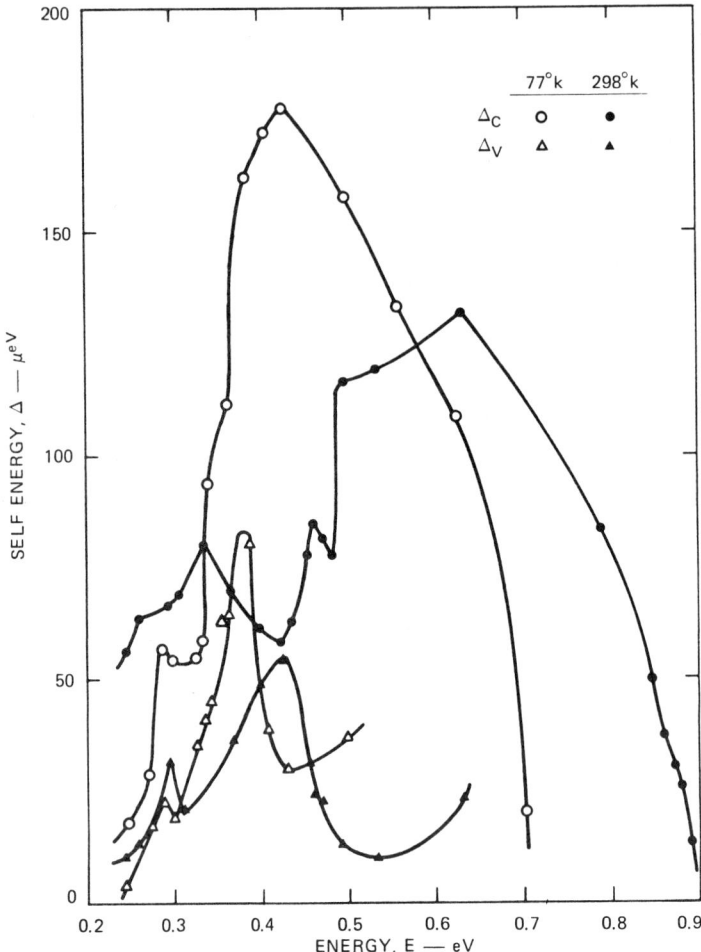

Figure 3 Scattering self-energy vs. energy.

able fine structure. Such a multicomponent interface-state density is consistent with current understanding and we expect that a closer correspondence with theory will be possible in the future through carefully planned experiments and theoretical calculations.

This work was supported in part by NASA Grant NSG-1385. We wish to thank W.R. Feltner, who grew the SiO_2 layers, and J.A. Moriarty, who contributed to the initial phase of this work.

REFERENCES

1. P. Su, A. Sher, Y.H. Tsuo, J.A. Moriarty, and W.E. Miller, Appl. Phys. Lett. (15 June 1980).
2. R. Castagne and A. Vapaille, Surf. Sci. *28*, 157 (1971).
   M. Kuhn, Sol. State Electron. *13*, 873 (1970).
3. E. H. Nicollian and A. Goetzberger, Bell Sys. Tech. J. *45*, 1055 (1967).
4. A. Sher, Y.H. Tsuo, J.A. Moriarty, W.E. Miller, and R.K. Crouch, J. Appl. Phys. *51*, 2137 (1980).
5. A. Goetzberger, E. Klausman, and M.J. Schultz, CRC Crit. Rev. in Sol. State Sci. *6*, 1 (1976).
6. N.F. Mott and E.A. Davis, *Electronic Processes in Non-Crystalline Materials*, Clarendon Press, Oxford, 1979.
7. A.B. Chen, G. Weisz, and A. Sher, Phys. Rev. B *5*, 2897 [Eq. (130)] (1972).
8. R.B. Laughlin, J.D. Joannopoulos, and D.J. Chadi, *Proc. Int. Topical Conf. on the Phys. of SiO_2 and Its Interfaces*, p. 321, Pergamon Press, New York 1978.

GAP STATES OF CRYSTALLINE SILICON AND AMORPHOUS SiO_2 SYSTEM

Takayasu Sakurai and Takuo Sugano
Department of Electronic Engineering,
The University of Tokyo, Bunkyo-ku, Tokyo 113, Japan

ABSTRACT

Various energy levels caused by micro-structural defects such as dangling bonds of silicon or oxygen atoms, oxygen vacancies and impurity atoms (H, OH, Cl and F) at the interface between amorphous SiO_2 and the crystalline Si substrate, and in the SiO_2 film are calculated based on Green's function formulation and the parametrized Hamiltonians for Si and O. The extended Hückel theory in which overlap integrals are not included is used to calculate Hamiltonians for H, Cl, and F.

The major results are (1) the perfect interface and the interface with oxygen dangling bonds do not have a energy level in the Si bandgap, whereas $Si_3 \equiv Si$-dangling bond and oxygen vacancy at the interface have, (2) this dangling bond level moves out from the Si bandgap with bonding any of H, OH, Cl and F, (3) Si-Si bond and Si-O weak bond at the interface give rise to localized states in the Si bandgap whose energy varies by changing the geometrical configuration of the chemical bonding such as the bond length and the bond angles, these are thought to be possible origins of interface states continuously distributed in energy, (4) amorphous SiO_2 without Si dangling bonds or oxygen vacancies has no localized level in the SiO_2 bandgap, even if the Si-O-Si bond angle is varied in a wide range, and (5) the Si dangling bond level disappears from the gap when any of H, OH, O, Cl and F is bonded.

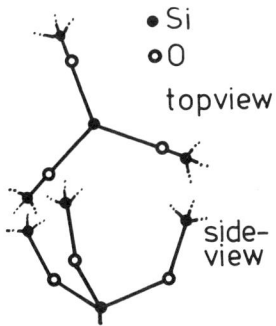

Fig. 1. Bethe-lattice representing amorphous SiO_2.

INTRODUCTION

The physical and chemical origins of the trap states at Si and SiO_2 interface and in SiO_2 have not been fully understood although some attempts were made to establish the theoretical background to these problems (1, 2). This work is an extention of Chadi, Joannopoulos and Laughlin's approach. Assuming a crystalline Si substrate and various types of defects, we calculated the energy level of trap states at Si-SiO_2 interface and in SiO_2 film. Here we propose possible model for the interface trap states continuously distributed in energy and the annealing behaviors. Our model basically consists of amorphous SiO_2 represented by Bethe-lattice as shown in Fig. 1 and crystalline Si substrate with (111) orientation.

FORMULATION OF HAMILTONIAN

A semiempirical formulation of parametrized Hamiltonian in which various interaction parameters are determined so as to fit the calculated band structure to the experimental results and the band structure calculated by pseudopotential method was used for SiO_2, whose parameters were given by Chadi et al (2). For crystalline Si, the agreement of the band structure calculated by using previously published parameters (3, 4) to the band structure which were experimentally known and calculated by the pseudopotential method is not necessarily satisfactory, so that the parameters up to 2nd nearest neighbor interactions have been determined to improve the result. The parameters are shown in Table 1 in the standard notation.

TABLE 1 Interaction Parameters of Si in eV

E_s	E_p	V^1_{ss}	V^1_{sp}	$V^1_{pp\sigma}$	$V^1_{pp\pi}$	V^2_{ss}	V^2_{sp}	$V^2_{pp\sigma}$	$V^2_{pp\pi}$
-10.44	-4.101	-2.144	2.090	2.346	-0.588	0.123	-0.366	0.435	-0.154

the superscripts 1 and 2 refer to 1st and 2nd nearest neighbors

In varying the bond length, we assume that each interaction parameter is altered according to

$$V_{ij}(R_{kl}) = V_{ij}(R^o_{kl}) \cdot S_{ij}(R_{kl})/S_{ij}(R^o_{kl}) \qquad (1)$$

within the approximation of the extended Hückel theory. In Eq. 1 the interaction parameter $V_{ij}(R_{kl})$, and the overlap integral $S_{ij}(R_{kl})$ are those at the bond length R_{kl} respectively and the superscript o denotes the bulk bond length.

The second method is based on the extended Hückel theory (5) and applied when H or Cl or F is concerned. The ij element of the Hamiltonian is calculated as

$$H_{ij} = -K\, S_{ij}(VOIP_i + VOIP_j)/2$$
$$H_{ii} = -VOIP_i \qquad (2)$$

, where S_{ij} is an overlap integral between i-th and j-th orbitals, $VOIP_i$ a valence orbital ionization potential of the i-th orbital whose value was given by Basch et al (6), and K a proportional constant. K is varied between 1.0 and 2.0 since this range is empirically believed to be probable (5, 8). To calculate S_{ij}, the atomic orbitals were approximated by Slater-type orbitals with Clementi's orbital exponents (7). The nearest neighbor interactions among s and p orbitals for valence electrons were taken into consideration.

MODEL AND CALCULATION PROCEDURE

Models of the atomic configurations used for the calculation are shown schematically in Figs. 2(a)-(g). The following is the procedure of the calculation.

(1) Fix one certain k-vector.
(2) Calculate the Green's function for Si (111) free surface (Fig. 2(a)).
(3) Connect Bethe-lattice and one oxygen to this surface (Fig. 2(b)).
(4) Sum up the Green's functions over various k-vectors.
(5.1) Separate Bethe-lattice and one oxygen to form a Si≡Si- dangling bond (Fig. 2(c)).
(5.2) Bring Bethe-lattice closer to Si dangling bond to simulate Si-Si bonding and O-vacancy (Fig. 2(d)).
(5.3) Bond any of H, O, OH, Cl, and F to Si dangling bond to represent the bonding of impurity atom (Fig.2(e)).
(5.4) Connect two Bethe-lattices with the variation of Si-O-Si angle and Bethe-lattice and other atomic groups to represent various atomic configurations in the SiO_2 (Figs. 2(f)-(g)).

The normal bond lengths between O and H, Si and O, Si and H, Si and Cl, and Si and F are chosen to be 0.97, 1.61, 1.50, 1.50, and 1.50 Å respectively.

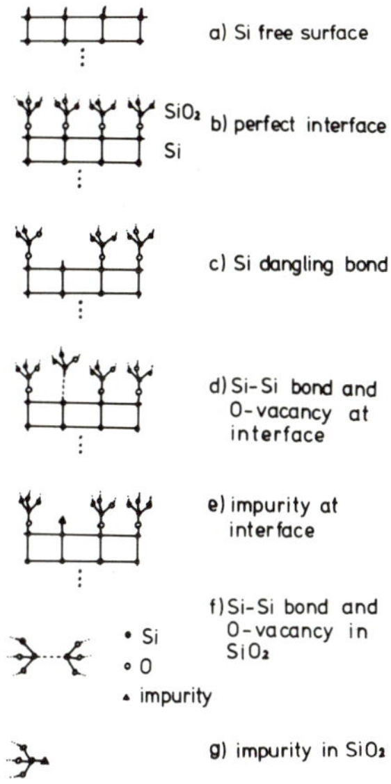

a) Si free surface
b) perfect interface
c) Si dangling bond
d) Si-Si bond and O-vacancy at interface
e) impurity at interface
f) Si-Si bond and O-vacancy in SiO_2
g) impurity in SiO_2

Fig. 2 Models for calculation.

RESULTS

Both the perfect interface with the Si-O-Si bond angle ranged from 120° to 180° and the interface with oxygen dangling bond do not have a gap state but Si_3≡Si- dangling bond gives rise to a gap state at about the midgap of bandgap, as Laughlin et al (2) have indicated. But it should be noted that Si bandgap in their calculation was about 2.5 (eV) due to the nearest neighbor approximation and a Bethe-lattice approximation for the crystalline Si, whereas in our model Si bandgap is calculated to be exactly 1.1 (eV). The Si-Si bond and Si-O weak bond at the interface produce trap levels in the Si bandgap, whose energy varies by changing the bond lengths (Figs. 3, 4) and the bond angles. The bonding parameters at the actual Si-SiO_2 interface can be assumed to vary because of the amorphous structure and the large internal stress included in this system. Therefore, trap states are thought to be the possible origins of the interface trap states continuously distributed in energy. These levels move out of the energy range between 0.5 eV below the top of Si valence band and 0.5 eV above the bottom of Si conduction band when any of H, OH, Cl and F. is bonded to the Si atom at the interface (Fig. 2(e)). Since a energy level outside the Si bandgap can not work as a trap state at the interface under normal operating conditions, this result explains the reduction of the interface trap density by H_2 annealing, trichrolo-ethylene annealing, and HCl oxidation and further suggests the possibility of F annealing.

Amorphous SiO_2 without Si dangling bonds or oxygen vacancies has no localized level in the SiO_2 bandgap, even if the Si-O-Si bond angle is varied from 120° to 180°. Si dangling bonds, oxygen vacancies, and Si-Si bond in SiO_2 produce levels in the SiO_2 bandgap (Fig. 5). This dangling bond level changes about 1 eV and the top of SiO_2 valence band changes about 0.4 eV by the Si-O-Si bond angle variation, that

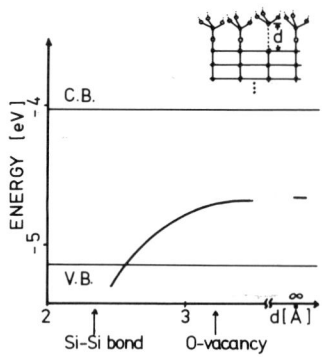

Fig. 3 Energy level of Si-Si bond at interface and its dependence on Si-Si bond length.

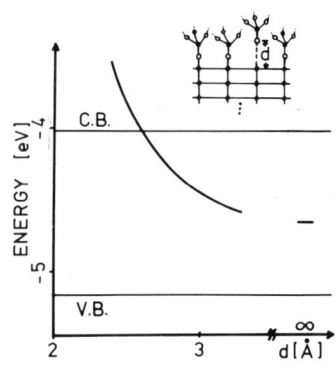

Fig. 4 Energy level of Si-O weak bond at interface and its dependence on Si-O bond length.

is, by amorphous effect. The Si dangling bond level and antibonding level of oxygen vacancy can be possible origins of neutral traps in the SiO_2 film (9) and the calculated result that these levels disappear from the SiO_2 bandgap with bonding any of H, OH, O, Cl and F indicates the annealing effect on the neutral traps. If the Si dangling bond level exists near the Si-SiO_2 interface, the electron in this level can go into the Si conduction band by tunneling and leaves the positive fixed charges, which can be a physical origin of the positive fixed charges commonly observed in the Si-SiO_2 system (10).

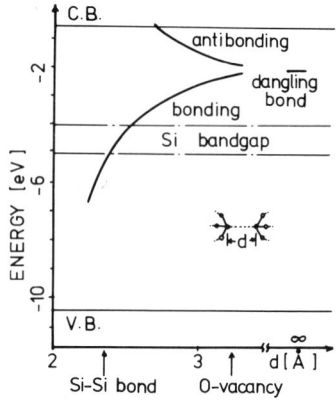

Fig. 5 Energy level of Si-Si bond in SiO_2 and its dependence on Si-Si bond length.

CONCLUSIONS

Empirical tight-binding Hamiltonians together with the Green's function formulation have enabled the energy level calculation of crystalline Si and amorphous SiO_2 system with and without defects. Possible origins of interface trap states which are distributed continuously in Si bandgap are suggested to be $Si_3\equiv Si-$ dangling bond, Si-Si bond, and Si-O weak bond at the interface. The reduction of the gap state density by H_2 annealing, trichrolo-ethylene annealing, or HCl oxidation is understood using this model. Some chemical origins of neutral traps in the thermally grown oxide might be the Si dangling bond level and the anti-bonding level of the O-vacancy in the oxide.

ACKNOWLEDGMENT

We express our sincere thanks to Dr. Y. Okabe at the Univ. of Tokyo for helpful suggestions in the course of this study.

REFERENCES

1. D. J. Chadi, R. B. Laughlin and J. D. Joannopoulos, Electronic structure of crystalline and amorphous SiO_2, The physics of SiO_2 and its interfaces, Pergamon Press, pp.55-59 (1978).
2. R.B. Laughlin, J. D. Joannopoulos and D.J. Chadi, Electronic states of Si-SiO_2 interfaces, ibid., pp.321-327 (1978).
3. K. C. Pandey and J. C. Phillips, Atomic densities of states near Si(111) surfaces, Phys. Rev. B13, 750 (1976).
4. G. Dresselhaus and M. S. Dresselhaus, Fourier expansion for the electronic energy bands in silicon and germanium, Phys. Rev. 160, 649 (1967).
5. N. P. Il'in and V. F. Masterov, Electron structure of deep centers in gallium arsenide doped with iron-group transition elements, Soviet Phys. Semicond. 11, 874 (1977).
6. H. Basch, A. Viste, and H. B. Gray, Valence orbital ionization potantials from atomic spectral data, Theoret. Chem. Acta 3, 458 (1965).
7. E. Clementi, and D. L. Raimondi, Atomic screening constants from SCF functions, J. Chem. Phys. 38, 2686 (1963).
8. T. Shimizu et al, Energy bands and their pressure dependence of diamond-and zincblende-type crystals by molecular orbital method, Report of Eng. Dept. of Kanazawa Univ. 11, 11 (1977).
9. T. H. Ning, Thermal reemission of trapped electrons in SiO_2, J. Appl. Phys. 49, 5997 (1979).
10. C. M. Svensson, The defect structure of the Si-SiO_2 interfaces, a model based on trivalent silicon and its hydrogen "compounds". The physics of SiO_2 and its interfaces, Pergamon Press, 328 (1978).

INTERFACE WIDTH AND STRUCTURE OF THE SiO_2 LAYER ON
OXIDIZED Si

H. Frenzel[+] and P. Balk, Institute of Semiconductor
Electronics/SFB 56 "Festkörpertechnologie", Technical
University Aachen, 5100 Aachen, Federal Republic of
Germany.
[+]present address: Atomika Technische Physik GmbH.,
Munich, Federal Republic of Germany

ABSTRACT

AES depth profiling was performed to study the structure of the SiO_2/Si interface. The upper limit for the inhomogeneity of the SiO_2 layer thickness was determined to be 0.7 nm on a confidence level of 95 %. The interface width, represented by the SiO_x bonding state, is 0.6 ± 0.4 nm. Previously published results from AES depth profiles giving 2 nm for the observed transition region in the signal were reproduced. However, this result is an artifact of the ion beam sputter profiling technique caused by the random removal of surface atoms.

INTRODUCTION

The defect structure of the SiO_2/Si interface has been the topic of considerable discussion in recent years. Depending on the applied method for physical analysis values for the width of the transition region ranging from zero (abrupt) to 2 nm were[1] reported. As a unified interpretation of these data it has been proposed[1,2,3] that the SiO_2/Si transition region per se is virtually abrupt (< 0.8 nm), but that the film exhibits thickness undulations of approximately 2 nm. This model would explain the abrupt interface observed in TEM studies[1]. In the opinion of these authors the broad 2 nm transition determined from AES or ISS signals would be caused by sputtering through a film of non-uniform thickness, since these techniques sample a large surface area.

The fact that other techniques (NBS[4], ESCA[5]), which also show poor lateral resolution yield very small values for the interface width, raises doubt regarding this model. It is worthwhile noticing, that the large interface width was concluded from measurements which utilize sputter profiling. Based on this observation we have made a critical examination of the sputtering process. It will be shown that indeed in cases where sputter removal was used the roughness generated by the technique accounts for the observed width of the transition region.

GENERATION OF ROUGHNESS BY ION SPUTTERING

The depth resolution of AES is limited by the escape depth (0.5-2 nm) of the Auger electrons. The use of ion sputtering opens the possibility to measure concentration profiles, but also changes the bonding and composition near the surface by knock-on and preferential sputtering. These perturbations, which yield an additional smear-out of the profiles, can be reduced by the choice of a small range for the primary ions. However, it will be shown in the following, that there is a further contribution to the broadening which is related to the very nature of the sputtering process: The randomness of the removal of surface atoms, which is independent of the parameters of the ion beam, leads to a roughness of some monolayers and determines the minimum width of the transition region for the signal.

When first impinging on the virgin surface at different locations, the ion beam removes atoms from the top monolayer. Subsequently not only atoms from the upper surface layer, but also atoms from the exposed second layer are sputtered. Next, the third and following layers become exposed to the ion beam. The microroughness generated by this process will continuously increase with sputter depth, if it would not be limited by differences in the chemical bonding strength of the exposed atoms. A single atom on a surface has a larger probability to be sputtered than one imbedded in the surface layer, because the bonding energy is in first approximation proportional to the number of neighbours[6]. The generation of roughness is determined by these differences in bonding energy.

To estimate the resulting roughness, we performed a Monte Carlo study using the above model for layers containing identical atoms. We assumed that the probability to remove an atom is inversely proportional to its total bonding energy, as has been proposed by Sigmund[6]. For simplicity a simple cubic lattice was considered. The calculation shows, that the roughness develops rapidly. A steady state smear-out of four interatomic distances (10 % - 90 % level) is already reached after the average sputtering front has proceeded by 3 monolayers. Since we deal in this case with statistical effects, there will be a certain probability to generate holes with a given depth. Fig. 1 shows the distribution of distances between the centers of 5 monolayer deep holes after removal of 30 atomic layers. It may be seen that the average distance amounts to approx. 50 atomic distances. Our calculation reproduces the result of a TEM study[7] on Ar^+ sputtered SiO_2, which reports an average distance of approx. 25 nm for approx. 2.5 nm deep holes, when we make the assumption, that one monoatomic layer amounts to 0.5 nm and that one atom in our model represents a SiO_2 unit. This finding leads us to conclude, that in sputtering SiO_2 one has to take into account a smear out of 4 interatomic distances, i.e. 2 nm, in profile measurements.

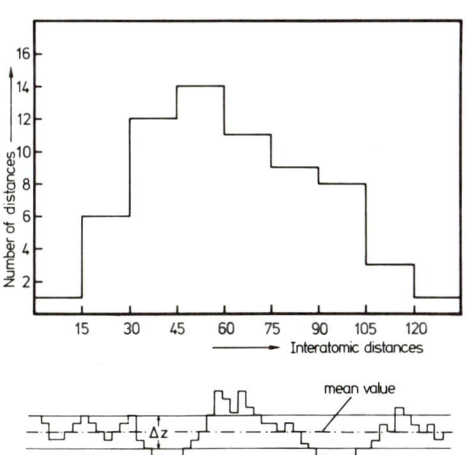

FIG. 1: Distribution of distances between 5 monolayer deep holes generated by statistical sputtering

EXPERIMENTAL

SiO_2 layers 20 to 100 nm thick were obtained by thermal oxidation at 1050°C in dry oxygen on (100) Si substrates. Films with thicknesses from 2.4 to 5 nm were grown at 900°C with 10 % O_2/N_2 on (100) or (111) Si substrates.

AES depth profiles were measured using primary electrons with energy of 5 keV. The primary beam (current: 1 µ A) was scanned over an area of 5 x $10^{-5} cm^2$ to prevent electron stimulated desorption of oxygen. The sputtering was carried out with an Ar^+ ion beam (energy: 3 keV) at an angle of 70° relative to the surface normal. These parameters led to an Ar^+ ion range in SiO_2 of 1.1 nm[8].

Corrections for knock on and information depth were carried out in the same manner as done by Helms et al.[3]. The depth calibration for the measurements on SiO_2 films in the range of 20 to 100 nm thickness was obtained from elipsometrical measurements. The thickness of thinner layers was determined with three independent methods: First, it was derived from a previously determined plot of thickness versus oxidation time. In the second technique we used the rise of the implanted argon signal (primary ion) in the Si for obtaining a depth calibration. The depth to reach a steady state is known and proportional to the ion range[8]. As a third way to get a depth scale we utilized the known different escape depth[3] of O_{KLL} and Si_{LMM} Auger electrons (1.0 resp. 0.5 nm) and the corresponding widths of the signal transitions at the interface. Values obtained by these three methods were averaged and resulted in an error for the mean value of at most 10 %.

RESULTS AND DISCUSSION

The width of the transition regions derived from the Auger signals (Si at 76 eV as in SiO_2, Si at 92 eV as in Si, O at 502 eV) and corrected as indicated above, is 2 ± 0.4 nm. It is independent of layer thickness in the range 20 to 100 nm. In addition to the measurements made in previous studies[2,3] we sampled the SiO_x signal, which is visible as a negative peak in the differentiated spectrum at 84 eV (see fig. 3 of our earlier paper[9]). This line was observed only over a distance (full width at half maximum) of 0.6 ± 0.4 nm, in contrast to the value of 2 nm mentioned above. This difference suggests, that sputter artifacts have to be accounted for to explain these observations.

We will first consider the Si (76 eV), Si (92 eV) and O (502 eV) signals, which all showed a corrected transition width of 2 nm. Special tests were made on Helms' hypothesis that this smear-out is caused by an undulation in the thickness of the SiO_2 film: A check on the concept of thickness inhomogeneity was performed by comparing the data on oxidized (111) samples with the above mentioned results for (100) substrates. Since the growth kinetics of thin layers depends strongly on surface orientation, one would not expect the same undulation in layer thickness on (100) and (111) samples. However, we measured in both cases the same transition width of 2 nm from the three Auger profiles. Furthermore, in thin layers of approx. 2 nm a thickness variation of 2 nm can hardly be expected. Nevertheless, experiments on the interfaces of samples with 2.4 to 5 nm thick SiO_2 films showed the same transition width as observed in thick layers (Fig. 2).

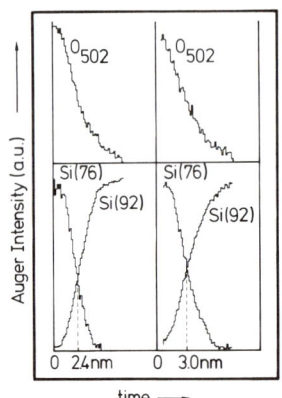

FIG. 2: AES depth profile of thin SiO_2/Si layer systems. Si (76 eV) for silicon bonded like SiO_2; Si (92 eV) for elemental silicon; O (502 eV)

The assumption of an undulating oxide thickness was also used by Wager and Wilmsen[2] to explain their results on samples with 1.8 to 4.5 nm thick SiO_2 layers. However, if a thickness inhomogeneity of 2 nm occurs, one would expect from the known information depth that the elemental Si (92 eV) signal would yield a concentration of 7 % at the surface of 2 nm oxide layers. Such a signal was neither observed in the data of Wager and Wilmsen[2] nor in ours. We determined from the missing Si (92 eV) signal at the surface of the thinnest SiO_2 layer an upper limit of 0.7 nm (95 % confidence level) for the thickness inhomogeneity.

Our results are clearly incompatible with the existence of 2 nm large variations in the thickness of the grown SiO_2 film. We propose, that the 2 nm broad transition region is related to roughness generated by the sputtering process, which has reached a steady state value already before removal of the thinnest (2.4 nm) SiO_2

film investigated. As discussed in an earlier section, a rapidly developing roughening of the observed magnitude (2 nm) is to be expected from statistical sputtering.

We also invoke this effect to explain the difference between the interface width obtained from the SiO_x peak (0.6 ± 0.4 nm) and that determined from the earlier mentioned lines (2 nm). This discrepancy can be explained by making the following assumptions: The SiO_x peak emerges from a chemical bonding state, which only occurs near the interface. This bonding structure is disturbed by the atomic sputtering cascade. However, the Auger signal remains weakly visible because of the finite information depth. Our model will only describe the data, if the lateral extend of the sputtering cascade (3.3 nm at 70°C incidence, 3 keV Ar^+ ions) is at least of the same magnitude as the period of the roughness. Therefore, this argument does not apply to the model of Helms[3] where the assumed periodicity is in the micrometer range.

The only way to obtain reliable information about the "true" interface structure is to utilize methods that do not require sputter profiling. This may be simply realized by measuring the Auger signal on films with thicknesses of the same order of magnitude as the information depth. The before mentioned experiment of samples with 2.4 nm thick SiO_2 layer allowed to limit thickness inhomogeneities to 0.7 nm. A broad interface would also cause a SiO_x (84 eV) or Si (92 eV) signal on the SiO_2 surface. However, such signals were not observed. From this finding it follows not only that the nonuniformity is less than 0.7 nm, but it implies also that the extend of the interfacial region has the same value as an upper limit. A lower limit is one bond length, since the SiO_x peak is observable.

CONCLUSIONS

In studies using ion sputter techniques a microroughness is generated, leading to a surface roughening of approximately 4 monolayers on a 10 % to 90 % level. An earlier interpretation of this broadening as caused by inhomogeneity in the thickness of the SiO_2 layer is incorrect.

The thickness of thermally grown SiO_2 films (thickness smaller 100 nm) is uniform within 0.7 nm. It appears likely that the region of gross non-stoichiometry is only one bond length wide, as would be concluded from a simple model of the interface.

REFERENCES

1. J. Blanc, in Semiconductor Characterisation Techniques, E.C.S. Electron Div. 78-3, 139 (1978)
2. J.F. Wager and C.W. Wilmsen, J. Appl. Phys. 50, 847 (1979)
3. C.R. Helms, N.M. Johnson, S.A. Schwarz and W.E. Spicer in "The Physics of SiO_2 and its Interfaces" edited by S.T. Pantelides (Pergamon Press, New York), 1978
4. N.W. Cheung, L.C. Feldman, P.J. Silverman and I. Stensgaard, Appl. Phys. Lett. 35, 859 (1979)
5. F.J. Grunthaner, P.J. Grunthaner, R.P. Vasques, B.F. Lewis and J. Maserjian, J. Vac. Sci. Techn. 16, 1443 (1979)
6. P. Sigmund, Phys. Rev. 184, 383 (1969)
7. C.F. Cook Jr., C.R. Helms and D.C. Fox, J. Vac. Sci. Technol. 17, 4 (1980)
8. V.V. Yudin, Appl. Phys. 15, 223 (1978)
9. H. Frenzel and P. Balk, J. Vac. Sci. and Technol. 16, 1454 (1979)

CHEMICAL COMPOSITION AND KINETIC LAW OF THE SiO_2/Si INTERFACE

Amelia Lora-Tamayo, Enrique Dominguez, Emilio Lora-Tamayo, Andrés Payo
Lab. de Microelectrónica, I.E.C. Serrano 144, Madrid, Spain
Francisco Ferrer
Lab. de Criminalística, D.G.G.C., Madrid, Spain
and Juan Llabrés
Inst. Física del Estado Sólido, C.S.I.C., Canto Blanco, Madrid, Spain

ABSTRACT

In this study the kinetic law of the thermal growth of silicon dioxide and the chemical composition of the transitional region between SiO_2 and Si have been investigated by SEM-EDAX techniques.

The quantitative analysis shows the variation of the silicon atom concentration at the interface. The observed composition in the transition region is SiO_x (0.5 < x < 2) where the x parameter depends on the interface depth. It was also found that the interface width is a function of the silicon dioxide thickness.

INTRODUCTION

The physico-chemical properties of the interface between Si and SiO_2 is a research subject that has received considerable attention. A knowledge of these properties is necessary for the manufacture of MOS devices [1].

Several and varied problems are faced in this study. Firstly the reported experimental data [2,3] on the silicon and silicon dioxide interface thickness give differing values. In fact, with samples prepared under identical conditions widely varying interface thicknesses have been measured using one of the approved methods (He-back-scattering, ESCA, Auger spectroscopy, Photo-emission, high resolution electrode microscopy). In general terms, it can be said that some methods give information on the relative location of the atoms in the solid compound, whilst with others the electron behaviour of the atomic shell is observed [4]. Hence, the diversity in results cannot come as a surprise. Another problem connected to the above mentioned is based on the definition of the transition region [5]. A strict definition could be given in terms of the variation in the chemical composition that changes from one chemical and structural component to another. The object of the present work was to study the variation in chemical composition in the transition region. This was carried out using a non-destructive method, the Energy Dispersive Analysis of X-rays, which gives a description of the chemical topography of a given material. In thermally grown SiO_2 samples (thickness >0.5μm) measurements were taken of the relative variation of the silicon atom concentration in a perpendicular direction to the interface.

A further physico-chemical problem arises in the determination of the kinetic law of growth of the silicon dioxide. It has been shown that the oxidizing agent can be present in three distinct ionic forms (O^-, O_2^- and $O_2^=$). Some authors have proposed alternative models, but the given arguments do not seem, in our opinion, to be sufficiently convincing. A scanning electron microscope technique (SEM) enables one to measure the thickness of the SiO_2 layer whilst carrying out X-ray fluorescence analysis. The range of silicon dioxide thicknesses we have prepared is larger than that previously studied by us using the Secondary Ion Mass Spectroscopy (SIMS) technique aided by "Talystep" (5).

EXPERIMENTAL PROCEDURE

Single-crystalline silicon wafers (resistivity 2-5 Ω.cm) orientated in the (100) and (111) crystallographic directions were oxidized at temperatures of 970 and 1110°C. The wafers were chemically polished in a FH, HNO_3, CH_3-COOH, H_2O mixture and dried, for three minutes, in a N_2 jet. Finally, they were oxidized in a quartz tube furnace. Dry oxygen (water content less than 1 ppm) was introduced into the quartz tube.

The investigation of the chemical composition at the interface and of the silicon dioxide thickness was carried out using a Philipps scanning electron microscope and an Energy Dispersive Analyzer of X-ray EDAX 711. An EDAX System module, the EDAX 352 Line Scan Rate-meter provides for line scan analysis which is representative of the relative concentration of a selected element, in this case silicon. The analyzer contains a standard solid state detector with a surface area of 10 mm^2. The acceleration voltage was 15 KV. The X-ray take-off angle was 38° in all measurements. The obtained counts correspond to the silicon K_α fluorescence peak. The time of analysis was 32 seconds.

The analysis of a cross-section of each sample in several points permits us a general view of the silicon wafer, interface and silicon dioxide layer. A cathode ray tube monitors the thickness of the SiO_2 layer in contrast with the other regions. To avoid doping or changes in the state of the observed surface, all oxidized wafers underwent mechanical cleavage, through easy-cleavage planes, without a further mechanical or chemical polish. The cut samples were fixed with silver paint to an aluminium sample holder. The quantitative analyzer was calibrated with the aid of α-quartz and silicon standards.

The SiO_2 layer thickness was systematically evaluated by "Talystep" and ellipsometry. The former giving a rough determination of the SiO_2 + interface thickness.

RESULTS AND DISCUSSION

The grown silicon dioxide thicknesses ranged between 5000 Å and 3 μm. Figure 1 shows a micrograph of a (111) sample with 2.0 μoxide (treatment temperature = 970°C). The oscillating trace is a representation of the relative silicon concentration through the scan line. There appear, from left to right, single crystalline silicon, interface and silicon dioxide. The determination error in the horizontal scale is < 200 Å (the electron beam spot diameter), while the vertical scale error due to the ripple can be estimated at $\pm$ 12%. The broad white linear bands on the left are fracture lines.

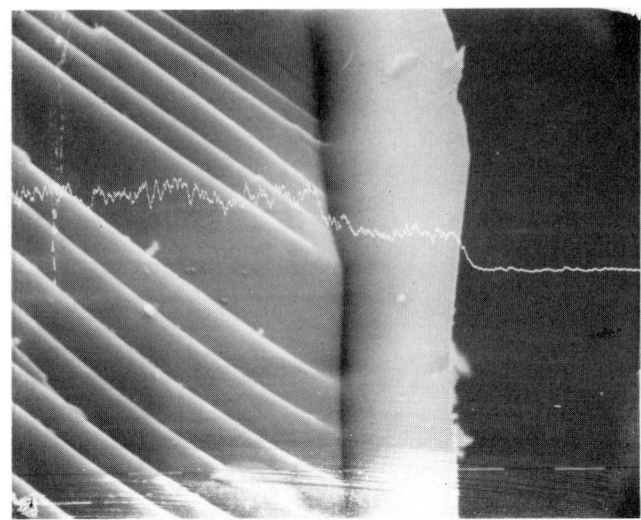

Fig. 1. SEM-EDAX micrograph of the cross-section of a thermally grown SiO_2 layer on (111) silicon. The white segment is 1 µm.

Some authors [7] using Transmission Electron Microscopy (TEM) on ultra-thin silicon dioxide samples, have found an undulating interface. In the above micrograph there appears a certain undulation of about 2 µm period . Nevertheless, this effect does not appear systematically in all samples. This leads us to assume that the above mentioned undulation could be due to distortion during cleavage and has not connection with the growth process. The interface width is calculated from an average of the scan parallel lines which show, in the microscope screen, the change in composition.

The plateau (a constant Si concentration region) shown at the interface corresponds to a chemical formula - SiO. The silicon concentration per unit area in this region is approximately 1.1×10^{15} atoms/cm^2. This result is in agreement with X-ray photoelectron data [8] on non-bonded silicon concentration in ultra-thin SiO_2. The deduced composition in the whole transition region is SiO_x ($0.5 < x < 2$) which is in agreement with ref. (9).

The kinetic laws determined by the described method are reported for temperatures of 970 and 1110°C (Fig. 2).

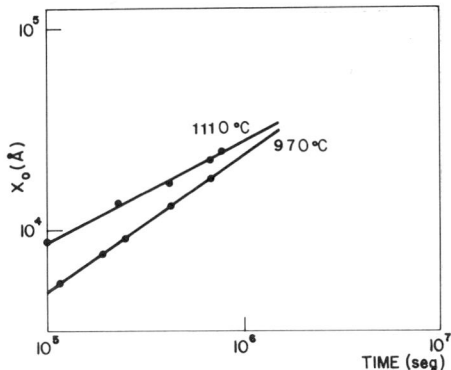

Fig. 2. Silicon dioxide thickness vs. oxidation time plot for 970 and 1110°C.

At 970°C the kinetic law obeys the expression $x_o = A \cdot t^{2/3}$ (t = oxidation time, x_o = thickness). At the higher temperature, however, the expression becomes $x_o = B \cdot t^{1/2}$. These growth laws coincide with those we determined for thinner oxide samples [5]. The kinetic law was not found to depend on the crystallographic orientation. The method described allows the investigation of the interface growth law. Table 1 shows the values of oxide and interface thickness (SEM-EDAX measurements) in five silicon (111) samples.

TABLE 1 SiO_2 thickness values determined by "Talystep", ellipsometry and SEM-EDAX

x_o (Å) SEM-EDAX	Interface width (Å)	x_o (Å) Talystep	x_o (Å) Ellipsometry
8,400	1,900	10,600	9,620
12,500	3,000	15,300	13,890
17,800	4,800	23,700	19,760
20,100	5,100	25,200	22,620
28,400	7,200	35,100	31,640

The interface widths have been determined by the Line Scan EDAX technique. Samples were oxidized at 970°C.

It is deduced that the kinetic law of the interface coincides with that found for the grown oxide at the same temperature. Therefore, the interface thickness clearly depends on the silicon dioxide thickness.

CONCLUSIONS

We have shown the application of a scanning electron microscope technique and Energy Dispersive Analysis of X-rays in the determination of thick SiO_2 samples of a) chemical topography of the SiO_2/Si interface and b) thickness of the silicon dioxide layer.

The kinetic law for the SiO_2 growth has been found to depend on the growth temperature. These evaluated laws coincide with those found in samples grown under identical conditions but with thinner silicon dioxide layers. The kinetic laws are not significantly dependent on the crystallographic orientation of the silicon wafer. The interface follows the same growth law as the oxide layer grown at the same temperature. The transition region is a reduced oxide SiO_x ($0.5 < x < 2$).

REFERENCES

1. B.E. Deal, Jour. Electrochem. Soc. 121, 198C (1974).
2. R. Williams, Jour. Vac. Sci. Technol. 14, n° 5, 1106 (1977).
3. I. Barsony and J. Giber, Applications of Surface Science 4, 1 (1980).
4. S.T. Pantelides and M. Long, Proc. of the Int. Conf. The Physics of SiO_2 and Its Interfaces, New York. Pergamon Press 1978, p. 339.
5. A. Lora-Tamayo, E. Domínguez, E. Lora-Tamayo and J. Llabrés, Appl. Phys. 17, 79 (1978).
6. P.J. Jorgensen, Jour. Chem. Phys. 37, 874 (1962).
7. J. Blanc, C.J. Buiocchi, M.S. Abrans and M.E. Ham, Appl. Phys. Lett. 30, 120 (1977).
8. S.I. Raider and R. Flitsch, Jour. Vac. Sci. Technol. 13, 58 (1976).
9. F.J. Grunthamer and J. Maserjiam, see the Proceedings quoted in ref. 4, p.389.

AUGER ANALYSIS COUPLED WITH CAPACITANCE STUDIES OF THE Si-SiO$_2$ INTERFACE

M. Hirose, S. Sakano, and Y. Osaka
Department of Electrical Engineering, Hiroshima
University, Hiroshima 730, Japan

T. Hattori
Department of Electrical Engineering, Musashi
Institute of Technology, Tokyo 158, Japan

ABSTRACT

An intermediate chemical state of Si in the Si(111)-SiO$_2$ interface, SiO$_x$, has been confirmed as a clear dip of the derivative Auger spectrum around 84 eV, and the in-depth profile of the SiO$_x$ state has been determined by comparison of the measured and synthesized SiLMM Auger spectra. The amount of SiO$_x$ in the SiO$_2$ near the interface is found to be increased by irradiation of 20 keV electron-beam (dose: 1x10^{-4} C/cm^2) through an Al film. From an intimate correlation between the interface state density and signal intensity of the SiO$_x$ appearing in the SiO$_2$ near the interface, it is concluded that the amount of the SiO$_x$ in the interface region is also increased by electron-beam irradiation.

INTRODUCTION

The width of the transition layer in the Si-SiO$_2$ interface and a corresponding intermediate chemical state of Si have been studied by Auger electron spectroscopy and X-ray photoelectron spectroscopy[1,2]. It is also reported that the SiO$_x$ state can be observed as a dip of the derivative Auger spectra at energies around 84 eV[3,4].

This paper describes the SiO$_x$ state near and in the Si-SiO$_2$ interface, which is affected by electron-beam irradiation as well as subsequent hydrogen annealing, based upon Auger analysis and capacitance measurement of thin SiO$_2$-Si(111) structures.

EXPERIMENTAL

Oxides on nSi(111) (10-20 ohm·cm) and pSi(111)(1.0-1.7 ohm·cm) substrates were grown at 1050°C in dry oxygen diluted by nitrogen gas [5]. MOS diodes fabricated with 48- and 105-Å-thick SiO$_2$ exhibited the saturation of capacitance in the accumulation bias[5]. Note that nonstoichiometric oxides result in the lack of capacitance saturation on account of large leakage current through the oxides.
The oxide thickness of an as grown specimen obtained from the measured accumulation capacitance and static dielectric constant of the bulk SiO$_2$ is in good agreement with the thickness determined by

the ellipsometry. Interface state density calculated from the low frequency capacitance or frequency dispersion of conductance of as fabricated MOS structures was low 10^{11} cm^{-2}eV^{-1} at midgap for Si(111) substrates and low 10^{10} cm^{-2}eV^{-1} for Si(100).
Electron-beam irradiation of oxides was performed through a 5000-Å-thick Al film in the dose range $1 \times 10^{-5} - 1 \times 10^{-4}$ C/cm^2. Hydrogen annealing of irradiated specimens was made at 450°C for 90 min before gate metallization.
Auger spectra of specimens sputtered by a rastered 2 keV ion beam in 4×10^{-5} Torr of Ar were measured using an analyser with a 3 keV and 1 μA electron beam.

RESULTS AND DISCUSSION

SiLMM Auger spectra at different depths into the Si-SiO$_2$ interface are shown in Fig.1, where each synthesized line was obtained as a weighted average of the Si spectra from pure Si and SiO$_2$ surfaces. The synthesized curves are well fitted to the measured ones except for energies around 84 eV. Discrepancy between the measured and synthesized amplitudes at 84 eV increases with sputtering time. In order to justify the synthesized SiLMM spectra, we prepared a specially designed Si-SiO$_2$ structure composed of an angle-polished Si substrate covered with an 800-Å-thick SiO$_2$ (Fig. 2). Ar ion etching from the back side of the Si substrate provides appropriate surface areas of pure Si and SiO$_2$, which determine an weighted average of the Si signal from the corresponding surfaces. Thus obtained SiLMM Auger lines showed no significant dip at 84 eV, in agreement with the synthesized curves in Fig. 1. In the case of Fig. 2, Auger signal from the transition layer is negligible since it comes only from a very small area of S$_2$. Also, Ar ion sputtering and bombardment of the primary electron beam introduce no appreciable dip in the spectrum near 84 eV. Therefore, the 84-eV signal in the derivative Auger spectrum is attributable to a feature of the SiO$_x$ state.

Figure 3 represents in-depth profiles of the SiO$_x$ as indicated by the intensity difference between the measured and synthesized SiLMM signal and full width at half-maximum (FWHM) of the SiO$_x$ region is estimated to be about 30 Å. Helms et al. (1) have obtained a FWHM of 20 Å for a SiO$_2$-pSi(100) system, and interpreted the apparent transition layer width in terms of topographical and beam-induced artifacts, because actual interface width is considered to be of the order of one monolayer. This is also the case of a SiO$_2$-Si(111) system and the larger value of FWHM for pSi(111) substrates is consistent with result of XPS (2).

Effect of electron-beam irradiation on a thin SiO$_2$-Si(111) system was studied in conjunction with the SiO$_x$ state not only in the bulk SiO$_2$ matrix but also in the interface. Direct bombardment of oxides by 20 keV electron-beam causes no appreciable change in the signal intensity of the SiO$_x$ state, while indirect irradiation through a 5000-Å-thick Al film leads to a significant increase in the SiO$_x$ signal even in the SiO$_2$ layer near the interface, as shown in Fig. 4. In this figure, three types of specimens are compared; i.e., a pSi(111) wafer with a 48-Å-thick SiO$_2$ was divided into three groups refered to as grown specimens, irradiated, and hydrogen annealed. Specimens of each group are metalized for C-V measurements to obtain interface state density (Fig. 5), and then ungated areas of the specimens were analyzed by AES. The Auger spectra for the three types of specimens

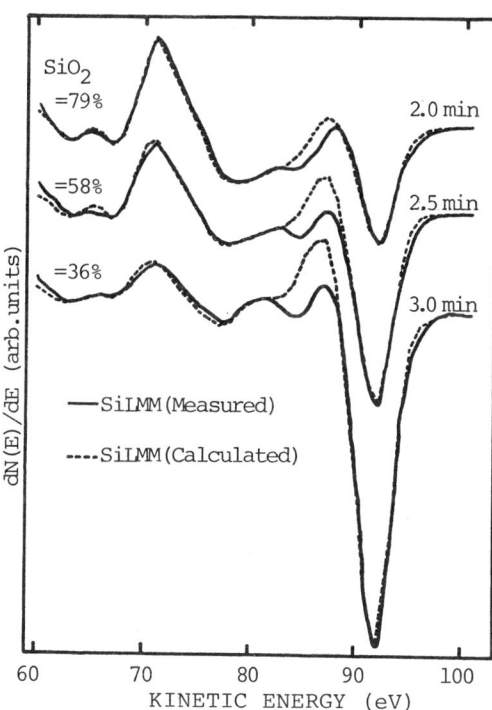

Fig. 1 Comparison of the measured and synthesized SiLMM Auger spectra at different depths into the interface for a 105-Å-thick SiO_2/nSi (111) specimen.

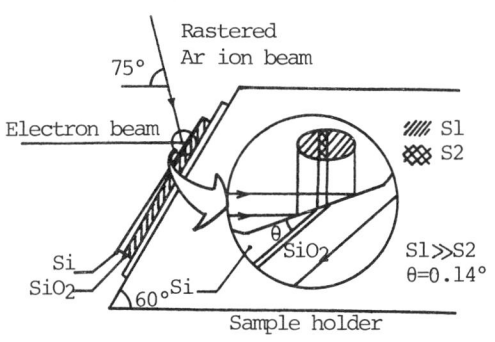

Fig. 2 A Si–SiO_2 structure with an angle-polished Si substrate for synthesizing the Si signal from the surfaces of pure Si and SiO_2.

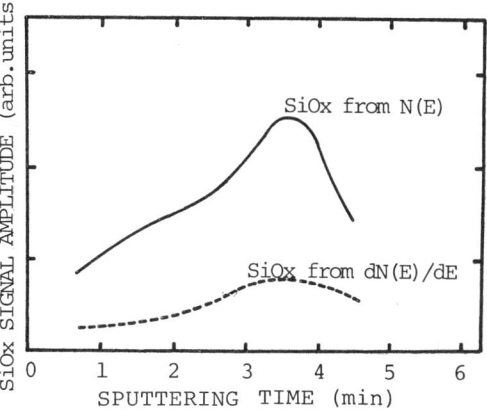

Fig. 3 Auger in-depth profiles of the SiO_x state obtained from the difference between the measured and synthesized Auger LMM lines at 84 eV for a 105-Å-thick SiO_2/nSi(111) system.

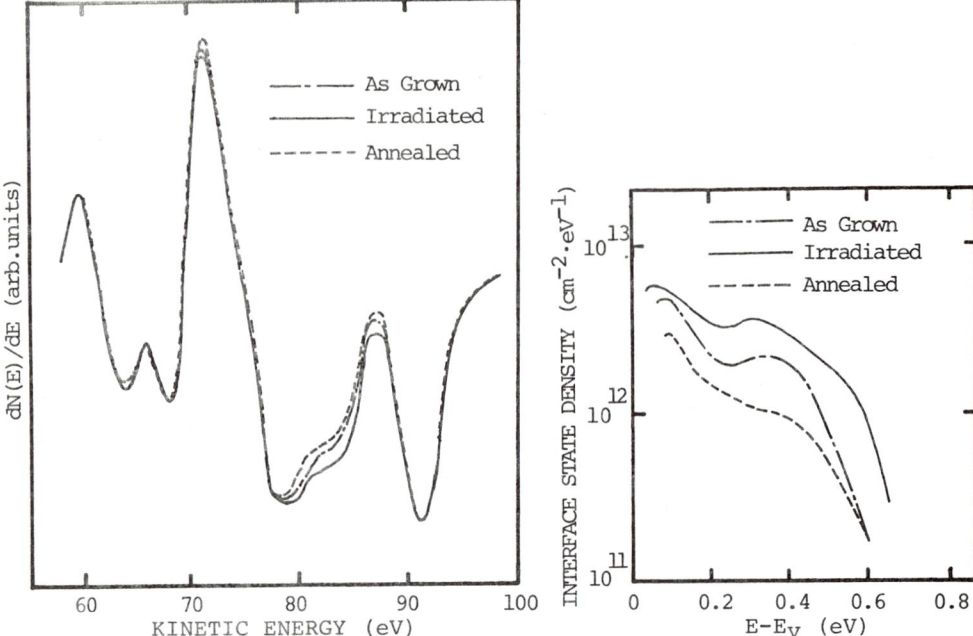

Fig. 4 Effect of electron irradiation (dose: 1x10^{-4} C/cm^2) and subsequent hydrogen annealing (at 450 °C) on an intermediate chemical state of Si in the SiO$_2$ near the Si-SiO$_2$ interface.

Fig. 5 Interface state distributions for the specimens obtained from the same samples as those in Fig. 4.

at the same depth have been obtained by noting that the ratio of the Si-92 eV signal intensity to 78 eV, F, could be used as a measure of the depth near and in the interface region. The value of F for an as grown sample sputtered for 2 min is taken as a standard for comparing spectral lines of three types of specimens. In order to obtain the spectrum for an irradiated sample with the same F-value, two measured Auger spectra, one of which has a slightly large value of F and the other a small, were synthesized to produce the desired F-value. Thus obtained Auger line for the irradiated sample is shown in Fig. 4. The same analysis was made for an annealed sample. It is clear from Fig. 4 that the SiO$_x$ state is introduced even in the SiO$_2$ near the interface by indirect irradiation of electron-beam and that subsequent hydrogen annealing results in a lower amount of SiO$_x$ than that of an as grown sample.

The effect of irradiation on the SiO$_x$ state in the interface region is apparently less pronounced. This is explained as follows: Increase in the Si-92 eV peak height due to bombardment of the primary electron-beam is more appreciable in the interface than in the SiO$_2$, so that the enhanced Si-92 eV signal tends to mask a change in the amplitude of the SiO$_x$ signal in the interface. If one takes into account this effect due to the primary electron-beam as well as an intimate correlation between the measured interface state density dependent on irradiation (Fig. 5) and the behavior of the signal intensity of the SiO_x state in the SiO_2 near the interface (Fig. 4), it is likely that a significant increase in the SiO_x signal by irradiation should occur also in the interface region. Indirect irradiation of electron-beam with doses less than 3×10^{-5} C/cm^2 has less influence on the density of interface states and therefore on the SiO_x state.

Reduction of the SiO_x signal by hydrogen annealing might be interpreted in terms of a silicon dangling bond (6) or of trivalent silicon (7) which are considered to be responsible for the interface states and passivated by forming a SiH bond during hydrogen annealing. Creation of the SiO_x state by irradiation could be related to radiation induced holes in the oxides which will produce silicon dangling bonds or trivalent silicon.

In summary, it is demonstrated that an intermediate chemical state of Si is introduced in the SiO_2 near the Si-SiO_2 interface and in the interface as well through indirect irradiation of electron-beam and that the SiO_x state is reduced by hydrogen annealing.

REFERENCES

1. C.R.Helms, Y.E.Strausser, and W.E.Spicer, Appl. Phys. Lett. 33,767 (1978).
2. S.I.Raider, and R.Flitsch, in Proceedings of the International Topical Conference on the Physics of SiO_2 and Its Interfaces (Pergamon, New York, 1978) p.384.
3. C.R.Helms, N.M.Johnson, S.A.Schwarz, and W.E.Spicer, J. Appl. Phys. 50, 7007(1979).
4. S.Sakano, M.Hirose, and Y.Osaka, 1979 Fall Meeting of Japan Society of Applied Physics(Sept. 1978) p.478 (unpublished).
5. M.Hirose, S.Hiraki, T.Nakashita, and Y.Osaka, Jpn.J.Appl.Phys.14, 999(1975).
6. R.B.Laughlin, J.D.Joannopoulos, and D.J.Chadi, In Ref.2, p.321.
7. C.M.Svensson, In Ref.2, p.328.

SOME METAL - SILICON DIOXIDE INTERFACE PHENOMENA

Christer M Svensson
Dept of physics and measurement technology, Linköping
University, 581 83 Linköping, Sweden

ABSTRACT

I will discuss a few phenomena related to the metal-silicon dioxide interface. I will review the effect of hydrogen on the metal work function and adhesion in the $Pd-SiO_2$ system, and the effect of sodium on the metal work function in the $Hg-SiO_2$ system. Then I will discuss some recent results on hydrogen induced drift of the flat band voltage of Pd-gate MOS capacitors. The latter effect indicates interactions of atomic hydrogen with the silicon dioxide or with sodium therein.

INTRODUCTION

In spite of the very large amount of work done on the silicon dioxide - silicon interface, little have been done on the other interface in an MOS structure. I would therefore like to discuss some metal-silicon dioxide effects observed by us. Of special interest are phenomena which may cause changes in MOS device characteristics after fabrication. An extreme example of such a phenomenon is the interaction of hydrogen with metals like palladium and platinum. A similar effect was observed in the system mercury with small amounts of sodium. These effects are very fast (seconds) making them useful for special devices like chemical sensors. We have also observed slow phenomena in the system palladium with small amounts of hydrogen. The slow phenomena are of great importance for device stability, for example in connection with the sensor devices mentioned above.

THE EFFECT OF HYDROGEN ON PALLADIUM WORK FUNCTION

The effect of hydrogen gas on the effective work function of palladium was discovered 1975[1]. We observed that an MOS transistor with palladium gate could be used as a hydrogen gas sensor. If operated at 100 - 150°C the sensor showed a fast and reversible response to hydrogen gas in air or in argon as shown in fig. 1. The sensitivity was of the order of a few ppm hydrogen in air[2].

The phenomenon was explained according to fig. 2[1,2]. Molecular hydrogen is dissociated at the Pd outer surface due to the catalytic activity of a Pd surface. The atomic hydrogen so formed will then diffuse through the Pd film to the $Pd-SiO_2$ interface. At both interfaces we will get an adsorption of atomic hydrogen. The adsorbed hydrogen will form a dipole layer at the metal interface, giving rise to a change in the effective work function of the metal. If oxygen is present the hydrogen will leave the structure as water vapor, else as hydrogen gas. Note that the unique property of Pd to dissolve very large amounts of hydrogen is not used here. The described effect works as well with platinum[3].

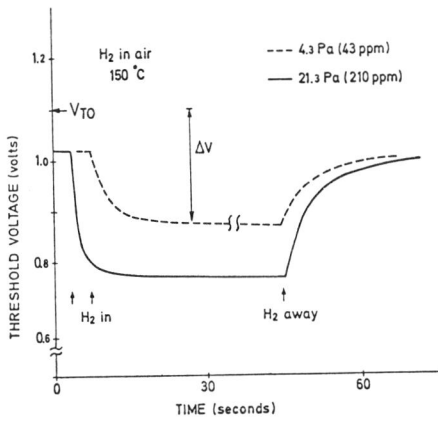
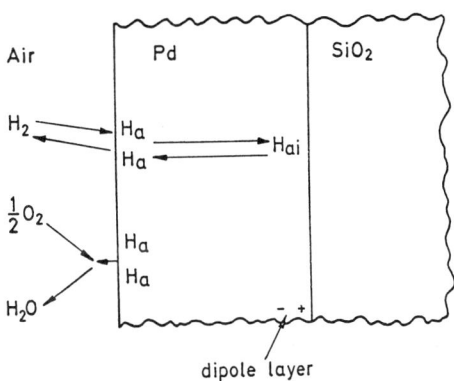

Fig. 1. Threshold voltage change of a Pd-gate MOS transistor in hydrogen containing air.

Fig. 2. Model of the hydrogen induced threshold voltage shift.

THE EFFECT OF SODIUM ON MERCURY WORK FUNCTION

A similar phenomenon as was described above have been observed in MOS capacitors with a mercury drop gate electrode[4]. By dissolving small amounts of sodium in the mercury a considerable shift in the MOS capacitor flat band voltage was observed as shown in fig. 3. Again the sensitivity is high. In fig. 4 we show the voltage shift versus sodium concentration. Note that as small sodium concentrations as 60 ppm gives half the maximum shift. This indicates that we again have to do with a surface effect, due to sodium adsorption in the mercury-silicon dioxide interface. This effect is then similar to the well known work function lowering of metal surfaces in vacuum when alkali metals are adsorbed on them. As in the vacuum case we observe quite large shifts, more than 2 volts.

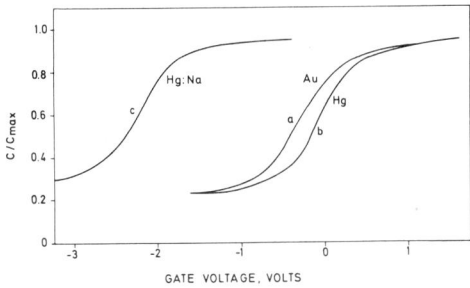

Fig. 3. CV-curves on MOS capacitors. The shift between b and c is 2.2 volts.

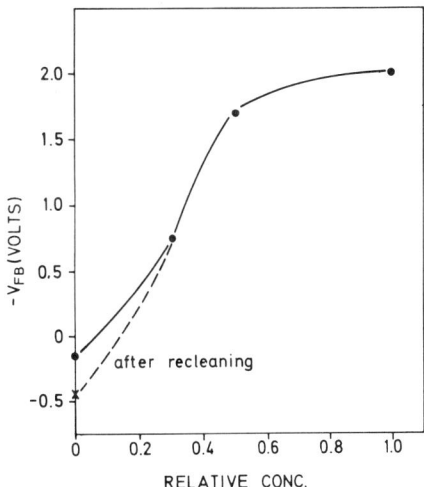

Fig. 4. Flat band voltage of Hg:Na gate MOS capacitors versus Na content. 1.0 correspond to about 0.002 Na atoms per Hg atom.

CHANGES IN ADHESION BETWEEN PALLADIUM AND SILICON DIOXIDE

Normally the adhesion between noble metals, like palladium or gold, and silicon dioxide is very poor. We found that it was possible to increase this adhesion by heat treatment to make photolitographic etching possible. It was, however, also very easy to lose the adhesion again[2]. Our observations are summarized in Table 1.

The observations were made on 100 nm thick evaporated Pd films on oxidized silicon. Adherence were judged according to the ability of the palladium film to withstand the photolitographic etching process.

TABLE 1   Effects of different treatments on Pd adhesion

TREATMENT	ADHESION	MODEL
Air $200^\circ C$ 10 min	increased	hydrogen extracted by oxygen
$N_2:H_2$ $500^\circ C$ 10 min	decreased	hydrogen introduction
Etching Aqua regia	decreased	hydrogen introduction through hydronium ions
Etching Aqua regia with excess HNO_3	constant	NO_3^- takes over oxidation role from H_3O^+

HYDROGEN INDUCED DRIFT IN PD GATE MOS DEVICES

Recent experiments indicates that we have an additional hydrogen induced voltage shift over what was discussed above in Pd-gate MOS devices. In fig. 5 we show the flat band voltage of a Pd-gate MOS capacitor versus time, when the gas composition around it is changed from oxygen with 3 ppm hydrogen to oxygen with 25 ppm hydrogen. We can see a fast voltage shift, corresponding to the effect demonstrated in

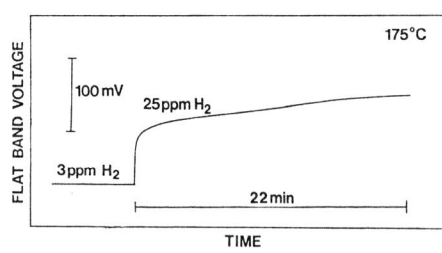

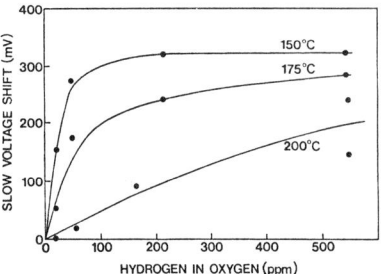

Fig. 5. Flatband voltage change when ambient is changed from 3 ppm to 25 ppm hydrogen in oxygen. The temperature is $175^\circ C$.

Fig. 6. Slow voltage shift versus hydrogen pressure at different temperatures.

fig. 2, followed by a very slow shift, similar in amplitude. If the gas composition is changing faster (in minutes) the latter effect will show up as an hysteresis in the voltage versus pressure curve. The phenomenon therefore can cause a lot of problems in gas sensing devices based on the principle described in fig. 1.

If only the slow shift is extracted from the measurements, we may study its pressure and temperature dependence, see fig. 6. The effect appears to saturate as a normal adsorption effect. For low pressures the voltage shift appears to be proportional to the hydrogen pressure, indicating H_2 adsorption, which is strange. In the case of the fast shift it is proportional to the squareroot of the hydrogen pressure indicating H adsorption, which is more reasonable. We cannot offer an explanation to the slow hydrogen induced voltage shift, unless it is coupled to the effect discussed below.

The sample used in the above measurement was made on n-type silicon, with 15 nm wet oxide and 100 nm thermally evaporated Pd.

THE EFFECT OF HYDROGEN ON ION DRIFT IN PD GATE MOS CAPACITORS

When investigating the above effect we discovered that hydrogen affects the ion drift in our MOS capacitors. We observed two effects, demonstrated in figs. 7 and 8.

Using a thin oxide device (the device used above, with 15 nm oxide) we observe an increase in the concentration of drifting ions (probably sodium ions) as a result of hydrogen adsorption. In fig. 7 we show a quasistatic measurement of the ion drift in the sample at 175°C. The full line is the quasistatic curve taken in oxygen by measuring the current through the structure during one cycle of a low frequency triangle voltage (0.0015 Hz). The first peak (a) represents a current of ions from the silicon to the metal. This peak occurs very early in voltage, indicating that the ions is loosely bound to the silicon surface. The valleys (c) represent the normal CV-curve of the MOS capacitor. The second peak (b) represents an ion current from the metal to the semiconductor. It occurs late in voltage and is not complete within the applied voltage, indicating strongly bound ions. The general behavior of this ion current is similar to what is usually observed for sodium ion drift in MOS structures. The dashed line shows a result of the same measurement done in 750 ppm hydrogen in oxygen. The curve as well as the sweep voltage has been shifted 0.6 volts, to compensate for the hydrogen induced voltage shift. We can clearly see that the number of moving ions has increased in the presence of hydrogen. The amount of drifting ions have increased from 7.5×10^{11} cm^{-2} to 10.3×10^{11} cm^{-2}. If we now return to pure oxygen again the amount of moving ions will slowly return to the original value (30 min.). The same phenomenon has also been observed in Pt gate samples[3].

The experiment described indicates that hydrogen in combination with a catalytic metal may release moving ions in the structure. One possible source of these ions could be inactive sodium, known to exist in silicon dioxide on silicon.

Using a thick oxide device (as above except that the oxide thickness is 100 nm) the measurement will be more sensitive to the actual voltages used. A similar measurement sequence as above on the thick oxide sample is shown in fig. 8. Here, the drifting ion density appears to decrease as a result of hydrogen adsorption. A closer look at the figure indicates that the second peak (b) has moved towards higher voltages in the presence of hydrogen, making the fraction of the peak swept less than before. The experiment indicates that adsorbed hydrogen increase the bonding between the moving ions and the metal.

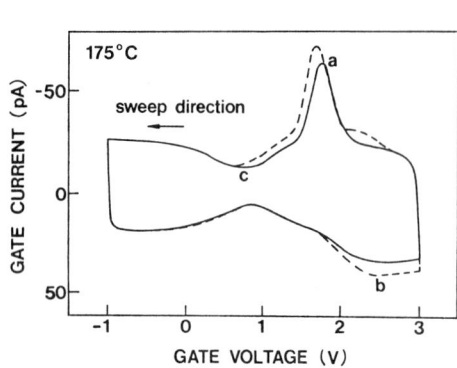 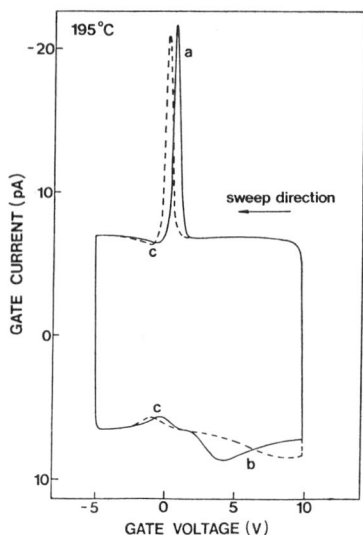

Fig. 7. Quasistatic current recording on thin oxide Pd gate MOS device showing ion drift, with hydrogen (---) and without hydrogen (——).

Fig. 8. Quasistatic current recording on thick oxide Pd gate MOS structure, with (---) and without hydrogen (——).

CONCLUSION

We have described a number of experiments, demonstrating that the metal-silicon dioxide interface may be very unstable in certain conditions. The observed phenomena can be used in the development of new devices, for example gas sensitive MOS transistors. Normally, however, most of these phenomena are unwanted and need to be investigated further to be understood and controlled. Such an effort will be important for the continuous search for more stable and reliable devices.

REFERENCES

1. I. Lundström, M.S. Shivaraman, C. Svensson and L. Lundkvist, Appl. Phys. Lett. 26, 55 (1975).
2. K.I. Lundström, M.S. Shivaraman and C.M. Svensson, J. Appl. Phys. 46, 3876 (1975).
3. I. Lundström and T. DiStefano, Sol. State Comm. 19, 871 (1976).
4. G.A. Corker and C.M. Svensson, J. Electrochem. Soc. 125, 1881 (1978).
5. M.S. Shivaraman and C.M. Svensson, J. Electrochem. Soc. 123, 1258 (1976).

FIELD EFFECT SPECTROSCOPY OF SEMICONDUCTOR-INSULATOR
INTERFACE STATES USING THIN FILM TRANSISTOR STRUCTURES

L.J. Brillson, F. Luo, and J. Wysocki
Xerox Webster Research Center, Webster, New York 14580

ABSTRACT

We present a new technique termed Field Effect Spectroscopy (FES) to probe semiconductor-insulator interface states optically. It consists of photoconductivity measurements of the semiconductor channel between source and drain of a gated thin film transistor. By varying gate voltage, one varies the relative contributions of bulk and interface states to overall device resistivity. We report results for various Al_2O_3:CdSe and SiO_2:CdSe interface structures.

INTRODUCTION

We have used the gate bias dependence of a semiconductor's resistivity in a thin film transistor (TFT) structure to probe semiconductor-insulator interface states. By varying the gate voltage and thereby the charge depletion within the thin (100-500Å) semiconductor channel, one can sweep the junction Fermi level through the interface state distribution and thereby alter the internal photoemission between insulator and semiconductor. These altered transitions can produce substantial changes in photoconductivity spectra of the semiconductor channel.

EXPERIMENTAL METHOD

Figure 1 illustrates schematically the bridge circuit used to measure surface photoconductivity of the TFT structure. Monochromatic light of energy $h\nu$ is focussed on the semiconductor channel between source and drain electrodes. Small changes in resistance $R_T(h\nu, V_G)$ at gate bias V_G are measured using a bridge configuration for maximum sensitivity. Bridge resistor $R_o \approx R_T$ to maximize dV/dR_T. Source-drain resistances for the various TFT's investigated ranged from 10^4 to $10^{10} \Omega$ and decreased by less than a factor of two under intense band gap illumination. Photoconductivity $\Delta\sigma$ spectra reflect ΔR_T as a function of $h\nu$ for a given V_G.

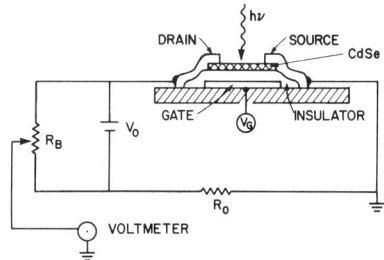

Fig. 1. Thin Film Transistor geometry in photoconductivity bridge circuit

RESULTS

Figure 2 illustrates FES spectra for a TFT consisting of 100Å CdSe channel, Au-In source and drain, and 2500Å Al_2O_3 insulator. An additional 2500Å Al_2O_3 overcoating isolated the CdSe from room ambient. The main spectral feature of each surface photoconductivity $\Delta\sigma$ vs. $h\nu$ curve is the absorption edge at 1.8eV, consistent with Eg = 1.74eV at room temperature. Several additional features appear at lower energy. The most pronounced is a photoconductivity increase at ~ 1.1eV and a quenching at 1.55eV. This structure bears a strong resemblance to surface photovoltage spectroscopy (SPS) features at 1.05eV and 1.60eV produced by Ar^+ bombardment of single crystal CdSe($10\bar{1}0$) surfaces in ultrahigh vacuum (UHV) which correspond to levels 1.05eV below the conduction band minimum (CBM) and 1.60eV above the valence band maximum (VBM) respectively.[1] Such defect structure is expected since the CdSe channel consists of polycrystalline grains. Additional structure appears at 0.7-0.8eV. All sub-band gap features are enhanced relative to the absorption edge under high depletion conditions (e.g., negative gate bias).

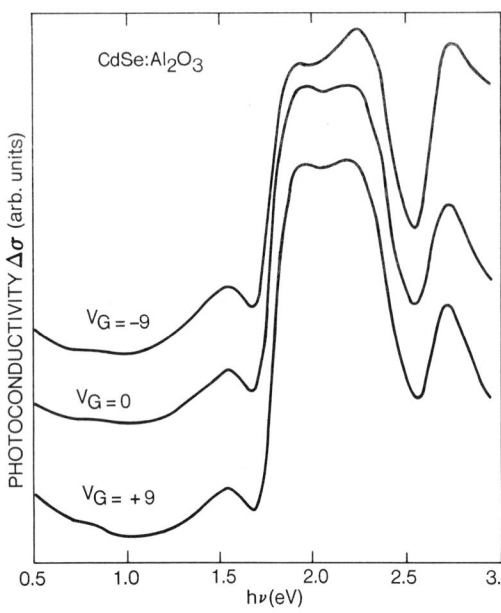

Fig. 2. Photoconductivity spectra of CdSe:Al_2O_3 TFT vs. gate voltage

Above Eg Fig. 2 shows additional features at 2.25, 2.55, and 2.70eV which do not correspond to any bulk or surface transitions involving CdSe alone. Using the same monochromator and detection scheme to measure surface and bulk photoconductivity spectra of cleaved single crystals of CdSe, we found only a monotonically decreasing response of $\Delta\sigma$ for E > Eg.[1] Gate voltage strongly affects the spectral intensities of these features but not their energies, suggesting a sharp density of states at the Al_2O_3:CdSe interface.

The interfacial nature of these states is confirmed by FES spectra of SiO_2:CdSe TFT's in Fig. 3, which exhibit different $\Delta\sigma$ structure for E > Eg. Here a 500Å CdSe channel, Au-In source-drain electrodes, and SiO_2 insulator comprise the TFT. In contrast to Fig. 2, these features above Eg exhibit only weak gate voltage dependence. Instead, the scale

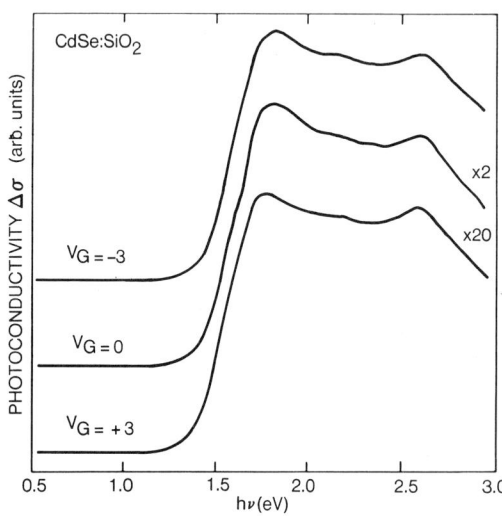

Fig. 3. Photoconductivity spectra of CdSe:SiO_2 TFT vs. gate voltage

changes in Fig. 3 show that gate bias produces a much stronger modulation of charge depletion within the semiconductor channel for the SiO_2 vs. Al_2O_3 TFT structure. This is indicative of a lower density of interface states and larger E_F movement within the CdSe band gap. Only weak structure appears for E < Eg. Studies using the same TFT structure but different source and drain electrodes exhibit first order changes for E < Eg only - due to diffusion of metal into the CdSe.

The difference in band structure as well as interface state density between Al_2O_3 and SiO_2 accounts for their contrast in FES response. Figures 4a and 4b illustrate schematically the interfaces between Al_2O_3 and CdSe and between SiO_2 and CdSe respectively. Band gap and electron affinity values for Al_2O_3[2,3] and SiO_2[4,6] are taken from optical absorption and photoemission measurements. The alignment of CdSe and insulator energy scales are accurate only to within a few tenths of eV due to possible dipole formation at the interfaces. Furthermore, the electron affinity of Al_2O_3 depends on preparation technique and may range up to 2.4eV for electron beam evaporation.[7] More recent analysis of the SiO_2 band gap yields values between 9 and 10eV.[8,9] For this vapor-deposited CdSe, the Fermi level E_F is several tenths of eV below the CBM. In Fig. 4a, the CdSe E_F is ~ 2.0-2.4eV above the VBM of Al_2O_3 - within the energy range of gate voltage-dependent features in Fig. 2. In contrast, the CdSe E_F is 3.7-4.2eV above the VBM of SiO_2. The above corrections to $\chi(Al_2O_3)$ and $Eg(SiO_2)$ only serve to refine this contrast. Thus, any optical transitions from insulator VBM to interface states near E_F for CdSe:SiO_2 interfaces fall outside the spectral range of these measurements.

Optical transitions from the Al_2O_3 VBM to the CdSe CBM can also contribute to an increase in photoconductivity. We attribute the $\Delta\sigma$ increase at $h\nu$ = 2.55-2.75eV to this process. As expected, the $\Delta\sigma$ increases under depletion conditions. However, the sharp $\Delta\sigma$ decrease at 2.25eV can only arise from Al_2O_3 VBM to interface states below the CdSe CBM, which quench the photoconductivity. Because of the sharpness of these features, the spectral energies do not shift with gate voltage. Assuming that the 2.55-2.75eV feature corresponds to a band-to-band transition, the 2.25 structure must be due to transitions from the Al_2O_3 VBM to interface states 0.3-0.5eV below the CdSe CBM (i.e., near the pinned Fermi level position). The $\Delta\sigma$ onset at 2.0eV may correspond similarly to transitions from interface states

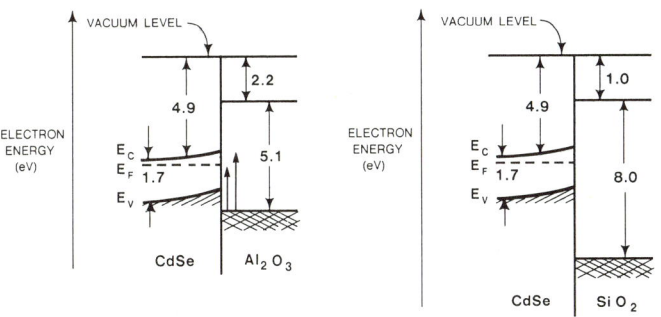

Fig. 4. Schematic energy band diagrams of a) CdSe:Al_2O_3 and b) CdSe:SiO_2 interfaces

below the CdSe VBM to the CdSe CBM or from interface states within the CdSe band gap levels above the CdSe CBM. FES spectra reveal changes in these features for Al_2O_3-CdSe TFT's exposed to atmosphere at various stages of fabrication, further confirming the interfacial nature of these transitions.

These interfacial features are altered when metal impurity atoms are intentionally diffused into the CdSe channel from the source and drain. Figure 5 illustrates the new $\Delta\sigma$ vs. $h\nu$ spectra for the CdSe:Al_2O_3 interface with In diffused throughout the CdSe channel. Again, structure appears at 0.65-0.8eV and at 1.1eV as in Fig. 2. However, the absorption edge clearly shifts 0.2eV to lower energy due to impurity level banding. In is known to produce electron donor levels at substitutional Cd Sites in CdSe.[10,11] The presence of In shifts the voltage-dependent peak structure 0.2eV to lower energy as well, thereby confirming that the feature involves a transition to a final state in the CdSe CBM. Figure 5 shows corresponding changes in the energy ranges 1.4-1.6eV and 2.0-2.2eV, both of which involve transitions to states several tenths of an eV below the CdSe CBM. In diffusion also increases the influence of gate voltage on charge depletion dramatically, suggesting a reduction in the density of interface states.

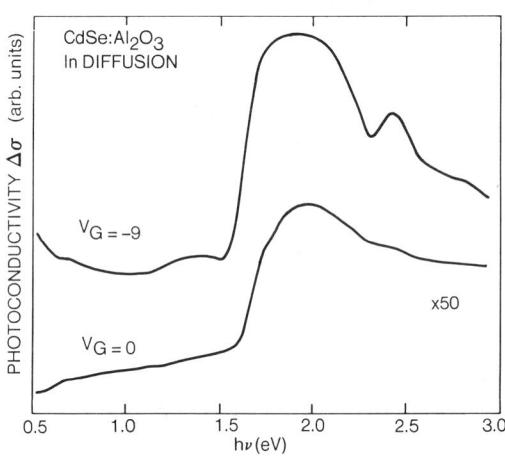

Fig. 5. Photoconductivity spectra of CdSe:Al_2O_3 TFT with In diffusion vs. gate voltage

These results demonstrate that FES technique is a useful technique for the optical measurement of states at semiconductor-insulator interfaces. FES enhances sensitivity to impurity and defect states at energies within the semiconductor band gap as well. Furthermore, the FES technique is uniquely suited to probing the electronic properties of gated semiconductor-insulator structures. Variations of this technique include monitoring photoconductivity as a function of gate voltage at constant $h\nu$ and modulating either $h\nu$ or V_G for added sensitivity.

REFERENCES

1. L.J. Brillson, Surf. Sci. <u>69</u>, 62 (1977).

2. A.M. Goodman, J. Appl. Phys. <u>41</u>, 2176 (1970).

3. T. Miyazaki, T. Edahiro, and J. Nakai, Oyo Butsuri <u>36</u>, 797 (1967).

4. W. Groth and H.Z. v.Weyssenhof, Naturforsch. <u>11A</u>, 165 (1956).

5. A.M. Goodman and J.J. O'Neill, Jr., J. Appl. Phys. <u>37</u>, 3580 (1966).

6. A.M. Goodman, <u>Proceedings of the Third International Conference on Photoconductivity</u>, ed. E.M. Pell (Pergamon Press, New York, 1969), p. 69, and references therein.

7. F.L. Schuermeyer, C.R. Young, and J.R. Blasingame, J. Appl. Phys. <u>39</u>, 1791 (1968).

8. T.H. DiStefano and D.E. Eastman, Solid State Commun. <u>99</u>, 2259 (1971).

9. S.T. Pantelides, in <u>Proceedings of the Conference on the Physics of SiO_2 and its Interfaces</u>, ed. S.T. Pantelides (Pergamon Press, New York, 1978) p. 80 and references therein.

10. A.V. Martinaitis, A.P. Sakalas, Z.V. Januskevicius, and J.K. Viscakas, Phys. Stat. Sol. <u>47</u>, 187 (1978) and references therein.

11. F.C. Luo, J. Vac. Sci. Technol. <u>16</u>, 1045 (1979).

Note added in proof:

The well-known optical response of Al_2O_3 at 5.1eV[2,3,6] reported for optical absorption and electron and hole internal photoemission experiments may be due to a high density of impurity or defect states located 5.1eV below the Al_2O_3 conduction band edge. Such electronic effects are a strong function of insulator preparation.[7] This designation of the initial state involved in the optical transitions depicted in Fig. 4a does not affect the interpretation of the spectral features above the absorption edge in Figs. 2 and 5.

THE PROPERTIES AND APPLICATIONS OF GaAs AND InP MIS STRUCTURES

D. L. Lile and H. H. Wieder
Naval Ocean Systems Center, Electronic Material Sciences Division,
San Diego, California 92152

ABSTRACT

MIS structures on GaAs and InP are of technological importance for high-speed IC and optoelectronic applications. This paper will attempt to relate the presently available understanding that exists on "free" surfaces of these materials to experimental measurements performed on MIS structures and insulated gate FETs and endeavor to project the consequences of these results for their device and circuit applications.

INTRODUCTION

The III-V compounds as a class are of considerable technological significance for device applications both because of their attractive transport characteristics as well as because of their favorable optical properties. Much of the potential device development and application of these materials has been hampered, however, by a lack of understanding, with a consequent inability to control their surface properties. This deficiency has had an extremely detrimental impact on the use of these semiconductors affecting not only the development of good ohmic and Schottky barrier contacts but also preventing the preparation of compatible dielectric/semiconductor structures for MIS digital and analog circuit applications. In contrast to Si, where a very good interface, low in localized energy states, may be prepared by oxidation, oxidation of a compound may well result in non-stoichiometry and the generation of defects due to the unequal rates of oxidation of the anion and cation species.

SURFACES OF GaAs AND InP

The starting point for a discussion of the properties of the surfaces of the III-V compounds is the observation that the clean unperturbed surfaces of these materials appear to be free of localized states in the bandgap; states at the surface being confined to energies beyond the band extrema due to surface reconstruction (1). This conclusion is based both on theoretical calculations as well as experimental studies of XPS, UPS, and photoluminescence on vacuum cleaved (110) surfaces of GaAs and InP (2). Real surfaces however will be at least partially oxidized and the effect of this on the expected properties is illustrated in Fig. 1 which shows the kind of behavior reported by Spicer, *et al*, on GaAs and InP during the initial stages of oxide formation (1). This should be contrasted with what is typically observed on Si where the addition of O_2 appears to remove the effect of surface trapping. The effect of oxygen is to shift the Fermi level position at the surface presumably as a result of the formation of extrinsic states at the interface. At large exposures, approaching what typically exists in a practical insulator-semiconductor system, the Fermi level on GaAs is seen to reside near mid-gap whereas on InP, for both n- and p-type material, the Fermi level is located near the CB minimum at the surface. It should be appreciated that the similar surface conditions on both n- and p-material are not surprising when it is considered that the differences in type at a doping level of 10^{17} cm^{-3} correspond to a change in just

1 atom in $\sim 10^5$ whereas a surface density of states of 10^{12} cm^{-2} correspond to approximately 1 state/10^2 surface atoms.

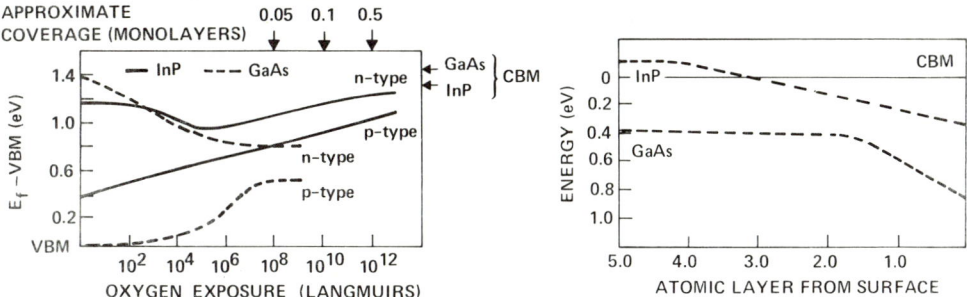

Fig. 1. Position of the Fermi level at the surface of GaAs and InP during exposure of a vacuum cleaved surface to a controlled amount of oxidation (After W. E. Spicer, et al (1)).

Fig. 2. Ideal anion vacancy energy level in InP and GaAs calculated as a function of its atomic layer distance from the surface. Only the highest (partially) filled level is shown (After Daw and Smith (4)).

During oxidation (or any process which releases sufficient energy to permit re-structuring of the surface) three effects are likely to occur. Firstly, surface relaxation will result in a reconstruction of the semiconductor surface; secondly, bonding between the O_2 and the anion and cation surface atoms will occur; and thirdly, a number of point defects in the form of vacancies, anti-site defects and vacancy complexes are likely to be created. Lucovsky and Bauer (3) have proposed that whereas in P bearing compounds, such as InP, following reconstruction the O_2 atoms can strongly chemisorb with the formation of a very stable double bond to the P atom in the case of those compounds containing As, the bonding is incomplete. A conclusion from such a model would be that the controlling species in a binary compound is the anion and that materials such as GaP and InP would be expected to have a higher density of saturated bonds and thus a lower surface trap density than, for example, GaAs and InAs. An alternative and perhaps supplemental model for the formation of the interface states on these materials is that during oxidation vacancies form due perhaps to selective depletion from the surface of both the anion and the cation. Daw and Smith (4) have performed a tight binding calculation for the bound state energy levels of ideal vacancies as a function of the distance of the defect from the surface and find that beyond the second atomic layer the results are much as for bulk vacancies with, for (110) GaAs, the As vacancy forming bound states ~ 1.1 eV and 0 eV above the VB edge and the Ga vacancy forming a bound state 0.4 eV above the VB maximum. As the vacancies are moved nearer the surface they shift in energy as schematically illustrated in Fig. 2 for the presumably dominant anion.

Daw and Smith report that increasing ionicity causes the bulk anion vacancy to move toward and eventually into the CB whereas the cation vacancy moves into the VB. Based on the results of Ref. (4) we have reproduced their data to obtain the approximate situation for InP shown in Fig. 2. The results of these calculations thus suggest that more covalent compounds such as GaAs may be expected to have deep levels due to the vacancies near the surface whereas in more ionic semiconductors the defects may be shallower or perhaps even completely within the bands.

Such surface defect levels, if present in sufficient density, will tend to control the position of the Fermi level at the surface E_F^*. In particular for GaAs where the above model predicts a distribution of deep states one would expect E_F^* to reside near mid-gap. This has in fact been observed to be the case by Spicer, *et al* (5) by means of photoemission measurements on the (110) surface of GaAs. They find that for vacuum-cleaved surfaces exposed to a wide variety of metals or oxygen the Fermi level is located $\sim$ 0.75 eV above the VB on n-type. These results are entirely consistent with what is typically observed in Schottky barrier studies on GaAs where barrier heights of $\sim$ 0.8 eV and 0.5 eV on n- and p-type material, respectively, are quite typical and suggest that the behavior of such metal semiconductor structures is dominated by the presence of surface states. In contrast to GaAs the results on InP indicate a Fermi level location near the CB edge. Particularly in the case of oxygen coverage Spicer, *et al* (5) observed a value of E_F^* within 0.3 eV of the CB on both p- and n-type material which is supported by Schottky barrier studies which show barrier heights on n-type material typically < 0.5 eV. The quantitative agreement between Schottky studies and barrier heights measured on purposely contaminated clean surfaces are not exact, however a strong qualitative correlation exists and does provide support for the idea of a common origin for both effects.

MIS STRUCTURES ON GaAs AND InP

Considerable technological significance is attached to the development of a compatible dielectric for use with the III-V compounds. Most effort in this area to date has been devoted to GaAs primarily because of the availability of this semiconductor in a relatively pure form. Despite wide variations in the methods of depositing the dielectric all data reported to date appear to be similar with variation occurring only in the resulting interpretation. Most of the data seems to indicate that in the absence of any applied bias the Fermi level at the interface of the GaAs and dielectric is located near mid-gap in close agreement with what is observed on the Schottky and free surface studies. This is, of course, completely consistent with the idea proposed in the preceding section where the defects at the interface with the oxides are no different to first order than those existing in the metal-semiconductor structure. Further consistency is seen in the results reported on InP MIS devices where once again apparently independent of the particular insulator or crystallographic orientation of the sample E_F^* resides close to the CB edge on both p- and n-material. The relative independence of these results on the specifics of the dielectric used would suggest that the interface properties are entirely determined prior to the deposition of the insulator. In any practical system the surface of the semiconductor will be covered with probably > 10 Å of thermal native oxide which, in light of the results of Spicer, *et al* (5), would be sufficient to completely determine E_F^*.

The quiescent surface potential of the semiconductor has at least a threefold significance as far as device applications are concerned. Firstly, it appears to determine to a large extent the quality of the Schottky barriers which may be formed. Secondly, it affects the ease of making low resistance ohmic contacts to the material. Finally, the position of the surface Fermi level determines, in part, the ease with which surface inversion may be achieved.

If the density of surface levels is sufficiently low (< $10^{11} cm^{-2} eV^{-1}$) over the energy region of interest then the quiescent surface potential becomes less important as all regions of the surface can be accessed by means of externally applied signals. In the case of GaAs most data available, obtained primarily by means of capacitance-voltage and conductance-voltage measurements on MIS diodes, gated van der Pauw galvanomagnetic measurements and three terminal transport measurements on MISFET structures indicate that the surface trap density is large; values well in excess of $10^{12} cm^{-2} eV^{-1}$ being typical over most of the gap (6). Such large surface state densities restrict the region of the gap accessible with applied surface fields to $\sim$ 0.5 eV thus precluding both the attainment of accumulation and inversion on this semiconductor. In contrast with GaAs it is possible on InP to move

the Fermi level at the surface across essentially the entire gap into the CB (7).
This means, for example, that on p-type InP it is possible to create a strong surface inversion of electrons as well as to strongly accumulate the surfaces of both
n-type and semi-insulating (SI) samples. An immediate consequence of this and one
with profound implications for the device and circuit applications of the III-V
compounds is that an enhancement-mode FET may be made on InP by inducing by means
of a voltage applied to an MIS gate a surface layer of electrons on what otherwise
would be either a p-type or SI substrate (8). Such devices have been made and demonstrated with field-effect mobilities well in excess of what can be achieved on Si.

A further parameter of importance as far as both device performance and an understanding of the physics of the interface are concerned relates to the time constant
for charging of the interfacial traps on these surfaces. The results of Meiners (9)
have shown that on GaAs MIS surfaces driven toward accumulation, the surface traps
appear relatively fast being able to respond to changes in induced carrier density
to frequencies in excess of 10^5 Hz. In depletion however, the states appear much
slower as is evidenced in the FET results reported by Lile (10) and Mimura (11).
The difference in speed of response between those states accessed for positive gate
bias and negative gate bias seems more than can be reasonably accounted for by
their location in energy in the gap although it is true that states deeper in
energy would be expected to respond more slowly. More likely is the explanation
often advanced in the early years of Si MOS development that slow trapping rates
($\tau \gtrsim 10$ msec) result from the traps being located in the dielectric rather than at
the semiconductor interface. If this is in fact the case then it would seem likely
that these slow traps could be located in the native oxide adjacent to the interface
rather than in the subsequently deposited dielectric. It is well known that in
general native oxides purposely grown on the III-Vs by either dry thermal or wet
chemical anodization are porous and often highly sensitive to water vapor and chemical etchants. Moreover thermal oxides are without exception extremely conductive.
A 10→20 Å intermediate layer of such an oxide, presumably highly non-stoichiometric
as a result of the preferential removal of one or other of the constituents from
the semiconductor surface would appear to be a likely candidate for the slow
trapping of carriers.

TECHNOLOGICAL IMPLICATIONS AND PROSPECTS

One of the main advantages of a III-V compound MIS technology is its potential for
making enhancement-mode transistors of wide dynamic range for direct coupled logic
at multigigabit data rates. To date the evidence suggests that because of a high
density of interfacial traps GaAs does not appear attractive for such an application. In contrast inversion and accumulation-mode enhancement MISFETs have been
demonstrated on InP and all available trap density data as well as the observation
of much lower recombination velocities on InP compared to GaAs (12) suggests that
InP may hold much promise for high-speed digital and analog ICs. The question still
remains unanswered however as to the preferable method of preparing the
dielectric (13).

One technique which seems to be getting fairly wide acceptance is pyrolysis or
chemical vapor deposition (CVD). This technique, analogous to vapor phase techniques used in epitaxial crystal growth results in layers with very good physical
and electrical bulk properties. SiO_2, Si_3N_4, and Al_2O_3 can all be grown by this
technique with a hard glass-like consistency resulting in layers compatible with
standard photolithographic device and IC processing. The disadvantage is that CVD
in its standard form requires a temperature somewhat in excess of 300°C for the
growth of SiO_2 and 600°C for Si_3N_4. Temperatures higher than 350°C are generally
believed to be unsuitable for InP and GaAs because of the potential problem of
their dissociation with the loss from the surface of the more volatile anion. To
overcome this problem various approaches to enhance the chemical reactivity of the

gases while at the same time retaining a low temperature have been tried including the use of a plasma discharge and photo-ionization by means of UV light. Once again these benefits are gained at the price of possible sample damage by the high energy ionized species but to date results using these techniques are encouraging. Fritzsche (14) using thermal CVD and Meiners (15) using plasma-enhanced CVD have both reported improved interface behavior using HCℓ in the reaction chamber during the initial stages of oxide growth on InP. HCℓ presumably acts as a reducing agent for the native surface oxide and thus although at this stage these observations and their interpretation must be considered tentative, they are at least consistent with our earlier statement concerning the possible role of the naturally present surface oxide in slow trapping on these compounds. Despite these uncertainties the fact remains that at least on InP relatively high gain microwave enhancement-mode MISFETs have been made which exhibit dynamic ranges far in excess of what is required for IC applications (< 5V). The inescapable conclusion is that in contrast with GaAs, InP has both a favorable quiescent surface potential and a sufficiently low density of fast states ($\tau \lesssim 10$ msec) over the surface potential range of interest to permit the fabrication of high performance FETs. Less encouraging is the irreproducibility that has been reported for these structures and the long term drift which may be associated with dielectric rather than interfacial defects. These are however exactly the same problems which plagued the early Si MOS technology and which were resolved only after a sufficient period of materials and device fabrication development.

REFERENCES

(1) W. E. Spicer, P. W. Chye, P. R. Skeath, C. Y. SU and I. Lindau, J. Vac. Sci. Technol., 16, 1422 (1979).
(2) R. A. Street, R. H. Williams and R. S. Bauer, J. Vac. Sci. Technol., 17, (1980).
(3) G. Lucovsky and R. S. Bauer, to be published in J. Vac. Sci. Technol. (1980).
(4) M. S. Daw and D. L. Smith, Phys. Rev., B20, 5150 (1979); M. S. Daw and D. L. Smith, Appl. Phys. Lett., 36, 690 (1980).
(5) W. E. Spicer, I. Lindau, P. Skeath, C. Y. SU and P. Chye, Phys. Rev. Lett., 44, 420 (1980).
(6) D. L. Lile and D. A. Collins, Thin Solid Films, 56, 225 (1979).
(7) L. G. Meiners, D. L. Lile and D. A. Collins, J. Vac. Sci. Technol., 16, 1458 (1979).
(8) L. G. Meiners, D. L. Lile and D. A. Collins, Electron. Lett., 15, 578 (1979).
(9) L. G. Meiners, J. Vac. Sci. Technol., 15, 1402 (1978).
(10) D. L. Lile, Solid-State Electron., 21, 1199 (1978).
(11) T. Mimura, K. Odani, N. Yakoyama, Y. Nakayama and M. Fukuta, IEEE Trans. on Electron Dev., ED-25, 573 (1978).
(12) H. C. Casey, Jr. and E. Buehler, Appl. Phys. Lett., 30, 247 (1977).
(13) H. H. Wieder, J. Vac. Sci. Technol., 15, 1498 (1978).
(14) D. Fritzsche, Electron. Lett., 14, 51 (1978).
(15) L. G. Meiners, to be published.

ACKNOWLEDGMENTS

The authors would like to acknowledge the assistance provided by D. Smith, W. E. Spicer, and L. G. Meiners for much helpful and stimulating discussion concerning the subject of this paper. Professor D. Smith is also to be thanked for making available to us preprints of his work prior to publication.

A STUDY OF THE ELECTRONIC STRUCTURE OF THE GaAs/NATURAL OXIDE INTERFACE

E. W. Kreutz and P. Schroll
Institut für Angewandte Physik, Technische Hochschule
Darmstadt, Schloßgartenstr. 7, D 6100 Darmstadt,
Fed. Rep. Germany

ABSTRACT

The electronic structure of the n (100) GaAs/natural oxide interface has been investigated by combined measurements of field effect, surface photoconductivity and thermally stimulated currents. Within the GaAs energy gap the density of interface states per energy interval consists of a continuous distribution with a minimum in the upper half of the energy gap separating a band of empty and occupied interface states. The separation of bulk and surface effects by differential methods allows to identify interface states energetically located within the bulk energy bands. The results are interpreted by a combination of different excitation mechanisms at the interface involving interface states, bulk states as well as oxide states. The response to physical and chemical influence discriminates between intrinsic and extrinsic states yielding a model of interface states describing satisfactorily the results.

INTRODUCTION

In relation to the group IV semiconductors the semiconducting III-V compounds have attracted increased attention as materials for high frequency low noise applications and for optoelectronic devices because of their favourable transport properties as high mobility, low diffusivity and large high field velocity. The preparation of good dielectric layers and high quality interfaces is an imperative technique for the development of a MOS or MIS technology for the semiconducting III-V compounds.

The present paper is dealing with combined measurements of field effect, surface photoconductivity, and thermally stimulated currents to derive the basic features of the natural oxide/GaAs interface and to enhance from our investigations considerably the understanding of the oxide/GaAs interface.

EXPERIMENTAL

The samples with a (100) plane as large surface were prepared from both epitaxially grown and bulk single-crystal GaAs. The bulk properties, the dimensions, and the mounting of the specimens including the cryostat with cooling and heating facilities have been described elsewhere.[1,2]

The remove residual damage the samples are chemically polished by etching either in HCl for the epitaxial wafers or in a mixture of HNO_3, HF, and distilled water (3:1:2) for the bulk material. The etching was stopped by flooding with distilled water, which was removed from the samples with filter paper. The oxidation state involved in the dissolution of the GaAs and the subsequent storage in laboratory air or in vacuum resulted in oxidized surfaces covered by a stable natural oxide layer[3-6] in the 20- to 30 Å thickness region with constant properties as a function of time. Possibly the chemical composition of the natural oxide grown during chemical etching and storage in laboratory air may be different as discussed in more detail by other authors.[6]

Pulsed field effect measurements with negative or positive step voltages were performed with a conventional circuit[2,7] minimizing the effect of the displacement current.[8] The arrangement and fabrication of current contacts, potential probes, and field plate have been reported.[1,2] The same sample geometry and bridge circuit have been used for the measurements of thermally stimulated currents and surface photoconductivity. Further details especially the optical system with the correction for the incident photon flux and the assumptions for the analysis of the data have been described elsewhere.[1,9,10]

RESULTS

The external electric fields pulsed with low frequencies cause positive or negative changes $\Delta\sigma_s$'s of surface conductivity σ_s (Fig. 1) with positive and negative voltages applied to the field plate. The surface condition and the induced charge Q_s govern the sign of the $\Delta\sigma_s$'s.[2] The bulk carrier concentration, the sample thickness, the intensity, and the duration of the electric field applied determine the magnitude of the $\Delta\sigma_s$'s.[1,2] The $\Delta\sigma_s$'s decay partially during and after applying the external voltages. The kinetics of the relaxation in $\Delta\sigma_s$ differ for the application and the removal of the electric field.[2] The magnitude of the relaxation in $\Delta\sigma_s$ depends on the sign, intensity, and duration of the electric field.[2]

Qualitatively similar $\Delta\sigma_s$'s are observed[11] during illumination with visible and UV photons (Fig. 1). After the illumination is stopped, the photon-induced $\Delta\sigma_s$'s partially decay leaving quasistable $\Delta\sigma_s$ at the illumination temperature. The

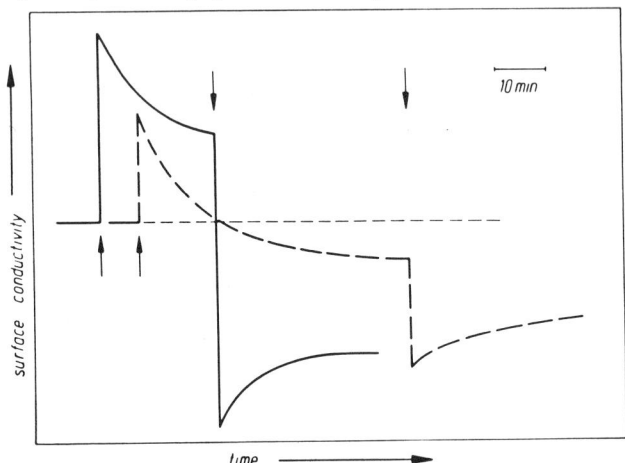

Fig. 1. Schematic time dependences of electrical (—— E = +1.15x10⁶ V/cm, T = 76.8 K) and optical (-- hν = 4.5 eV, T = 320 K) induced changes of surface conductivity. The marks indicate the beginning or the interruption of the external influence.

spectral response of the surface photoconductivity is plotted in Fig. 2 showing various structures in the energy range investigated. The sign and the magnitude

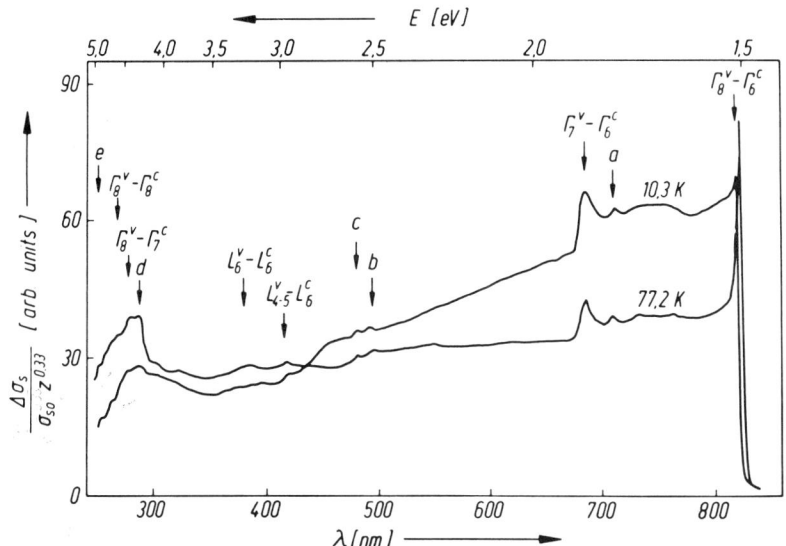

Fig. 2. Spectral response of relative surface photoconductivity at different temperatures. σ_{so} denotes the surface conductivity before illumination. $Z^{0.33}$ is the correcting factor[9] for the incidenting photon flux. The different peaks originate from bulk (labeled by band-to-band transitions) and surface (labeled by small letters) effects.

of the photon induced $\Delta\sigma_s$ depend on the photon flux, on the illumination temperature, on the photon energy, and on the surface condition.[9,11]

Starting with positive or negative $\Delta\sigma_s$'s the activation energies of the annealing processes have been evaluated from thermally stimulated current measurements which have been used to gauge the field effect measurements. Following Brown's procedure[12] the charge Q_{ss} captured by the interface states is determined in the energy gap. Q_{ss} is positive near the conduction band minimum and negative about mid-gap.[1]

Fig. 3 shows the density of semiconductor-oxide interface states proposed for the GaAs surfaces covered with natural oxide. The field effect measurements[1] yield a U-shaped distribution of interface states with a minimum in the lower half of the gap. The thermally stimulated currents in combination with the kinetics[2] of the relaxation of electrical and optical induced $\Delta\sigma_s$'s follow a broad peaking distribution of interface states superposed about midgap.[13] The contribution of bulk and surface (oxide [9]) effects to the surface photoconductivity via different excitation mechanisms as a function of temperature and photon energy (Fig. 2) claims a broad peaking distribution below the bulk valence band maximum (VBM).

DISCUSSION

Recently[14] we have reported on the chemical, structural, and electronic properties of InSb (110) surfaces. Because of the striking similarities in the features in the proposed density of interface states for GaAs surfaces (Fig. 3) and the corresponding results for InSb surfaces the density of interface states is discussed on the base of intrinsic and extrinsic states neglecting true oxide states.[9]

The analyses of LEED intensities have revealed that the clean GaAs (110) surface exhibits a structural rearrangement. Experimental data[14] and computational results[15] strongly suggest that the atomic displacements at the GaAs (110) surface penetrate at least three layers into the bulk. The rearrangement of surface atoms follows electronic rearrangements leading to the generally accepted model[16] of surface state bands: empty Ga derived surface states in the conduction band region and filled As derived surface states in the valence band region with the two kinds of surface state bands separated in energy by about the gap.

Changes of boundary conditions[17] cause a scattering of the Ga and As derived states from the bulk band region into the gap resulting altogether in a tailing distribution of intrinsic interface states (Fig. 3).[13] The foregoing considerations show that the spread in boundary conditions accounts for the $N_{ss}(E)$ spectrum. The larger the spread of the boundary conditions[17] the more the interface state is located in the middle of the gap, i.e. these are the states with the largest disturbance of their surrounding. Small fluctuations result in states near the band edges. Large fluctuations, however, confine the originating states into the middle of the gap. The nature of the source responsible of the gap state, i.e. the distortion of the chemical environment introduced by the source, determines the energy of the states. The concentration of the sources governs their density.

Surface defects (vacancies, anti-site defects, clusters etc.) also influence the atomic rearrangements and consequently the surface electronic structure as evidenced for cleavage defects on clean GaAs (110) surfaces.[18] The defects produce large strains within the surface. The resulting distortions in the arrangement of surface atoms probably introduce large changes of the boundary conditions and cause extrinsic surface states. For different sources of different concentration the density and the energy of the states are different resulting in a non-uniform density of states. The high density of extrinsic surface states originating from

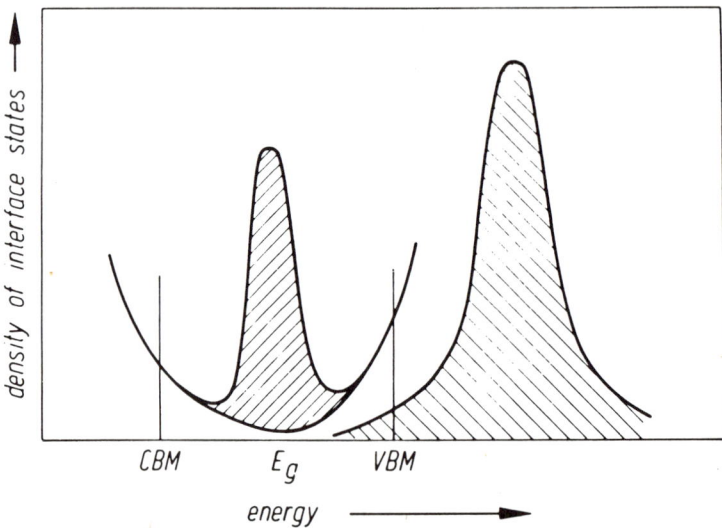

Fig. 3. Density of states for the GaAs/natural oxide interface.

the high concentration of defect introduced distortions degenerate by potential fluctuations or Coulomb interaction into peaking distributions in the gap superposed to the monotonic distribution. The experimental results show that the broad peak about 0.7 eV below the bulk conduction band minimum (CBM) is dominant for the GaAs surface covered with natural oxide (Fig. 3). These considerations are supported by a model proposed by other authors[19] for the formation of III-V semiconducting compound insulator interface states attributing the 0.7 eV state to a deficit of As which is initiated by oxygen uptake via the heat of condensation during the oxidation process. The missing anion gives rise to an acceptor level which is confirmed by the kinetics[2] of the relaxation in electrical and optical induced $\Delta\sigma_s$'s.

The peaking distribution (Fig. 3) degenerating with the bulk bands below VBM may be caused by filled As derived or defect introduced interface states. In the former case the oxidation favourably proceeds without breaking bonds. The oxide is connected to the GaAs surface by bonding through the excess electrons at the As atom removing no states from the band region into the gap. The peaking distribution at most shows a chemical shift by the bond distortions. In the latter case the oxidation favourably proceeds by defect generation.[19] The oxide is connected to the GaAs surface by a highly distorted interface layer assumingly introducing interface states within the valence band region. Ellipsometry,[20] RHEED,[21] backscattering measurements,[22] and AES-SIMS depth profiling[23] have shown the existence of such an interface layer which differs in chemical composition from the bulk and from the oxide overlayer. Calculations[24] for surface defects associated with one and two dangling bonds on either metallic and non-metallic atoms yield two clearly distinct groups of states where the upper one corresponds to centers with one dangling bond on the metallic atom and the lower one contains the levels of all the other possibilities.

[1] E. W. Kreutz and P. Schroll, Phys. Status Solidi (a) 53, 499 (1979).
[2] E. W. Kreutz and P. Schroll, Surf. Technology in press.
[3] A. J. Rosenberg, J. Phys. Chem. Solids 14, 175 (1960).
[4] F. Lukes, Surf. Sci. 30, 91 (1972).
[5] S. P. Murarka, Appl. Phys. Lett. 26, 180 (1975).
[6] B. Schwartz, CRC Crit. Rev. Solid State Sci. 6, 609 (1975).
[7] E. W. Kreutz and P. Schroll, Surf. Sci. 37, 410 (1973).
[8] G. G. E. Low, Proc. Phys. Soc. B 68, 1154 (1955).
[9] E. W. Kreutz and P. Schroll, to be published.
[10] E. W. Kreutz, Phys. Status Solidi (a) 13, 557 (1972).
[11] E. W. Kreutz and P. Schroll, Phys. Lett. 65 A, 65 (1978).
[12] W. L. Brown, Phys. Rev. 100, 590 (1955).
[13] E. W. Kreutz, Phys. Status Solidi (a) 56, 687 (1979).
[14] E. W. Kreutz, E. Rickus, P. Schroll, and N. Sotnik, Surf. Sci. 86, 794 (1979).
[15] e.g. Proc. VIth Conf. on the Physics of Compound Semiconductors and Interfaces, J. Vac. Sci. Technol. 16, No. 5 (1979).
[16] P. E. Gregory, W. E. Spicer, S. Ciraci, and W. A. Harrison, Appl. Phys. Lett. 25, 511 (1974).
[17] H. Flietner, Surf. Sci. 46, 251 (1974).
[18] W. E. Spicer, I. Lindau, P. E. Gregory, C. M. Garner, P. Pianetta, and P. W. Chye, J. Vac. Sci. Technol. 13, 780 (1976).
[19] W. E. Spicer, P. W. Chye, P. R. Skeath, C. Y. Su, and I. Lindau, J. Vac. Sci. Technol. 16, 1422 (1979).
[20] T. Ito and Y. Sakai, Trans. Inst. Electr. Engng. Japan 93 A, 11 (1973).
[21] E. W. Kreutz, Proc. VIth Intern. Vacuum Congr. and IInd Internat. Conf. Solid Surfaces Part 2, 445 (1974).
[22] J. P. Stromsheim, T. Olsen, and D. J. Ruzicka, Phys. Status Solidi (a) 39, 167 (1977).
[23] K. Watanabe, M. Hashiba, Y. Hirohata, M. Nishino, and T. Yamashina, Thin Solid Films 56, 63 (1979).
[24] D. Lohez, P. Masri, M. Lannoo, L. Soonckindt, L. Lassabatere, Proc. Conf. Phys. Semiconductor Surfaces and Interfaces, in press.

INTERFACE STATES IN GaAs/LaF$_3$ CONFIGURATIONS

A. Sher and Y.H. Tsuo
SRI International, Menlo Park, California 94025

J.E. Chern
College of William and Mary, Williamsburg, Virginia 23185

W.E. Miller
NASA Langley Research Center, Hampton, Virginia 23665

ABSTRACT

We have used a refined version of the well-established quasistatic and conductance methods to examine undoped GaAs samples with composite insulators. The insulator consists of thin native oxides $\approx$ 20 Å thick, followed by 500 Å of LaF$_3$. The resulting insulator capacitance is 500 nF/cm^2. With this large insulator capacitance, it is possible to cause the Fermi energy to scan over the principal interface state density peak that occurs at $\approx$ 0.95 eV above the valence band edge. The peak corresponds to an interface state density of 5×10^{12} eV^{-1}–cm^{-2}. Several subsidiary features are also observed. Part of the refinement involves the use of low-intensity light at photon energies above the band gap. The principal effect of the light is to shorten the response time of some interface states. This helps to separate the measured interface state profile into contributions from various origins. The shape of the observed hysteresis in C-V curves, with and without light, can be understood from the behavior of the interface states.

We have recently reported on a new class of insulators, LaF$_3$, for MIS structures,[1] a refined interface state measurement method,[2] and the application of this method to a study of the Si–SiO$_2$ interface.[2] Here the method is applied to the GaAs–LaF$_3$ interface and is extended so it can be used even if there is a small leakage current. The application of the new methods to GaAs is not complete, but tentative results can be understood by examining Figs. 1 and 2 along with Figs. 2 and 3 in Ref. 1. We shall first introduce each of these curves and then interpret some aspects of them.

In Fig. 1, a version of Fig. 2 of Ref. 1, in which the negative voltage variation of the capacitance of a sample with 500 Å of LaF$_3$ (GaAs-8) is followed to lower voltages, the capacitance scale is increased by a factor of ten, and the experiment is repeated for three light intensities shone through the front transparent contact. The dashed curves were taken with a negative ramp and the solid curves were taken with a positive ramp. The frequency for all curves was 1 kHz, and the magnitude of the ramp rate was 50 mV/s. The light was at wavelength 0.82 μm, and the maximum intensity, in the no-filter case, was 2 μW/cm^2. The labels 10^{-1} and 10^{-2} are the attenuation factors of the filters used to lower the light intensity used to take that particular data. The major effect of the light used comes from modifications of the response times of some of the interface states caused by the light-generated electron-hole pairs. This can be seen in Fig. 3 of Ref. 1, where the frequency dependence of the series capacitance C_s and the dissipation factor for zero gate bias are shown for these same light intensities. Notice that the peak in the dissipation factor (it follows the peak in G_p/ω for the interface states) and the corresponding change in the series capacitance shift to higher frequency as the light intensity is increased. Most importantly, the light brings the peak into a frequency range where its peak height can be measured. The position of the peak in G_p/ω identifies a particular interface state feature, and its height determines the interface state density associated with that feature.[3]

A quasistatic capacitance (QSC) measurement[3] determines the interface state density at a given bias voltage and the variation of the surface Fermi energy with applied voltage. Since the insulators used in the preliminary study are imperfect and leak, the data reduction is somewhat more involved than the usual Berglung method.[3,4] In the usual case (where there is no net current), the relation between the device capacitance C(V) as a function of applied voltage V and the surface potential ψ_s

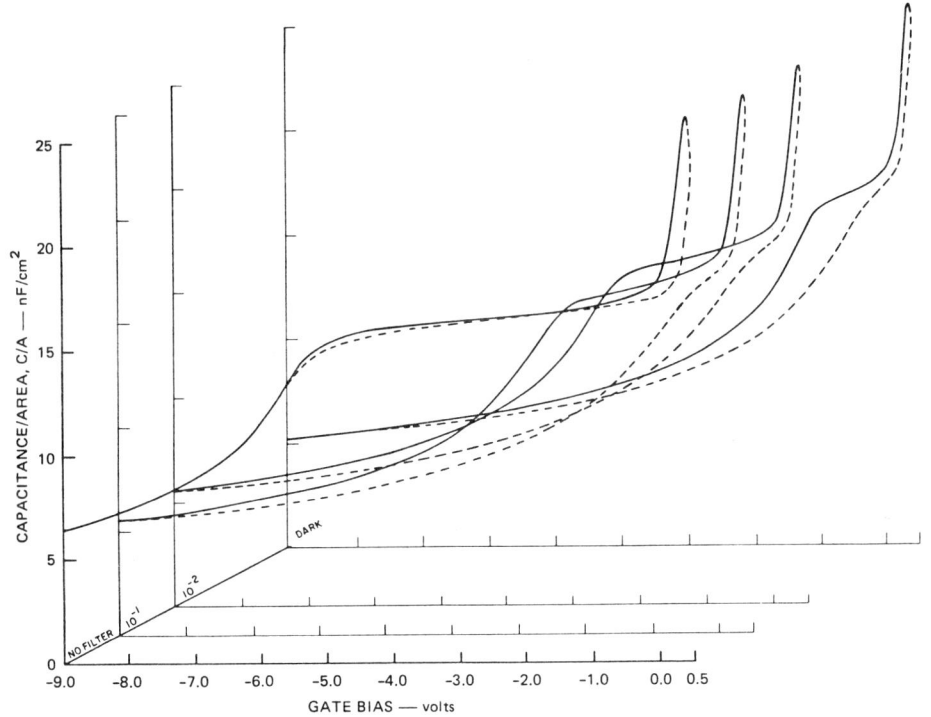

Figure 1 C-V characteristic for different light intensities of a GaAs sample with a 500 Å-thick LaF$_3$ insulator. The light intensity in the no-filter case was 2 μW/cm^2 at a wavelength of 0.82 μm. The labels 10^{-1} and 10^{-2} are the attenuation factors of filters used to reduce the light intensity.

$$\frac{d\psi_s}{dV} = (1 - C(V)/C_0) , \qquad (1)$$

where C_0 is the insulator capacitance and V_0 is the initial ramp voltage. If, in addition, a net current density, j, is present, then the change of the Fermi energy $E_F(x)$ with distance x away from the interface is given by the relation

$$\frac{dE_F(x)}{dx} = \frac{1}{\mu n(x)} j , \qquad (2)$$

where μ is the electron mobility, and $n(x)$ is the electron concentration at x. Equation (2) is correct if essentially all the current is carried by the majority carriers, in this case the electrons. The total series capacitance of the device $C(V)$ can be decomposed into contributions from the depletion/accumulation layer $C_{d/a}(V)$ the interface states $C_{ss}(V)$, and any inversion layer $C_i(V)$.

$$C^{-1}(V) = C_0^{-1} + (C_{d/a}(V) + C_{ss}(V) + C_i(V))^{-1} . \qquad (3)$$

If $C_{d/a}$, C_{ss}, or C_i is large compared to C_0, then it is evident [from Eqs. (1) and (3)] that $d\psi_s/dV$ becomes very small. In such cases, the surface potential is pinned as the gate bias is varied and does not change. When this pinning is caused by the inversion layer, it leads to the familiar limitation of the depletion layer thickness that causes the high-frequency capacitance to reach a minimum value in inversion. A slightly less familiar case results from the pinning of ψ_s by a large interface state density. The phenomena are, however, identical. We shall argue that this is precisely what happens in GaAs for the commonly used kinds of insulators. Notice that this pinning occurs more readily if C_0 is small.

When the surface potential is pinned and the magnitude of the gate voltage is increased further, all the additional voltage must drop (in the absence of leakage) across the insulator. If the insulator leaks, then [according to Eq. (2)] the Fermi level bends, predominately in the depletion region where n(x) is small. In this case, the Fermi energy is pinned on the interface states, but the surface potential is no longer pinned. If C_d is much smaller than C_0, then ψ_s approximates V and C_d is found experimentally to be proportional to $\psi_s^{-1/2}$ or $V^{-1/2}$ for low-current densities, just as one would expect for a thick depletion layer if no current were flowing.

A complete theory of the relation between the gate voltage and the surface potential in the presence of a leakage current is quite involved.[5] However, if the experimental result $C_d \sim \psi_s^{-1/2}$ is used as a guide to an approximation, one may assume that the functional form of the potential $\psi(x)$ through the space charge region is independent of the current. (The boundary conditions change with current, so the space charge is affected by a current.) Then, for $\psi_s < 0$, one has:

$$\frac{d\psi(x)}{dx} = -\frac{2}{\beta L_D} F(\beta\psi) , \quad ((4))$$

where L_D is the Debye length, $\beta = e/kT$, and

$$F(\beta\psi) \equiv (e^{\beta\psi} - \beta\psi - 1)^{1/2} . \quad ((5))$$

Then, using the relation

$$n(x) = n_i e^{[(E_F(x) - E_{Fo}) - (E_i(x) - E_{io})]/kT} , \quad ((6))$$

where E_i is the intrinsic Fermi level $[E_i(x) - E_{io} = -e\psi(x)]$ along with Eqs. (2), (3), and (4), yields

$$e^{(E_{FS} - E_{Fo})/kT} = 1 - \frac{j}{2(eDn_i/L_D)} \int_0^{(-\beta\psi_s)} \frac{e^y}{F(y)} dy , \quad ((7))$$

for $\psi_s < 0$, where D is the diffusion coefficient. Equation (7) provides a relation between the Fermi energy at the surface E_{Fs} and the measured variables ψ_s and j.

Thus, to determine the energy variation of the interface state density shown in Fig. 3, one first measures the dc current voltage characteristic to find j(V). Then, from a quasistatic capacitance measurement, $\psi_s(V)$ is determined by integrating Eq. (1). The integrating constant $\psi_s(V_0)$ is found using the well-established method of fitting the high-frequency capacitance variation with gate voltage to its theoretical curve.[3] The fact that the high-frequency capacitance fits the theoretical curve is a check on the validity of this procedure. Then, from $\psi_s(V)$ and j(V), one finds $E_{Fs}(V)$ from Eq. (7). Thus, the variation of the quasistatic capacitance with voltage can be translated into the functional dependence of the total interface state density with energy in the band gap. This whole procedure must be repeated for each light intensity.

Since several peaks in G_p/ω are observed for each gate voltage, the net interface state density determined by the quasistatic capacitance can be decomposed into component parts. We have observed at least three different time constants for some gate voltages. Only one of them corresponds to a large interface state density; the others are smaller than the principal features by factors of 10 to 100. A plot of the total interface state density, the main light-sensitive feature, and the difference between these two curves is plotted in Fig. 2. Until these measurements can be repeated on better samples, they should be used only as an indication of trends. They are offered here to indicate the kind of information that can be found from this method.

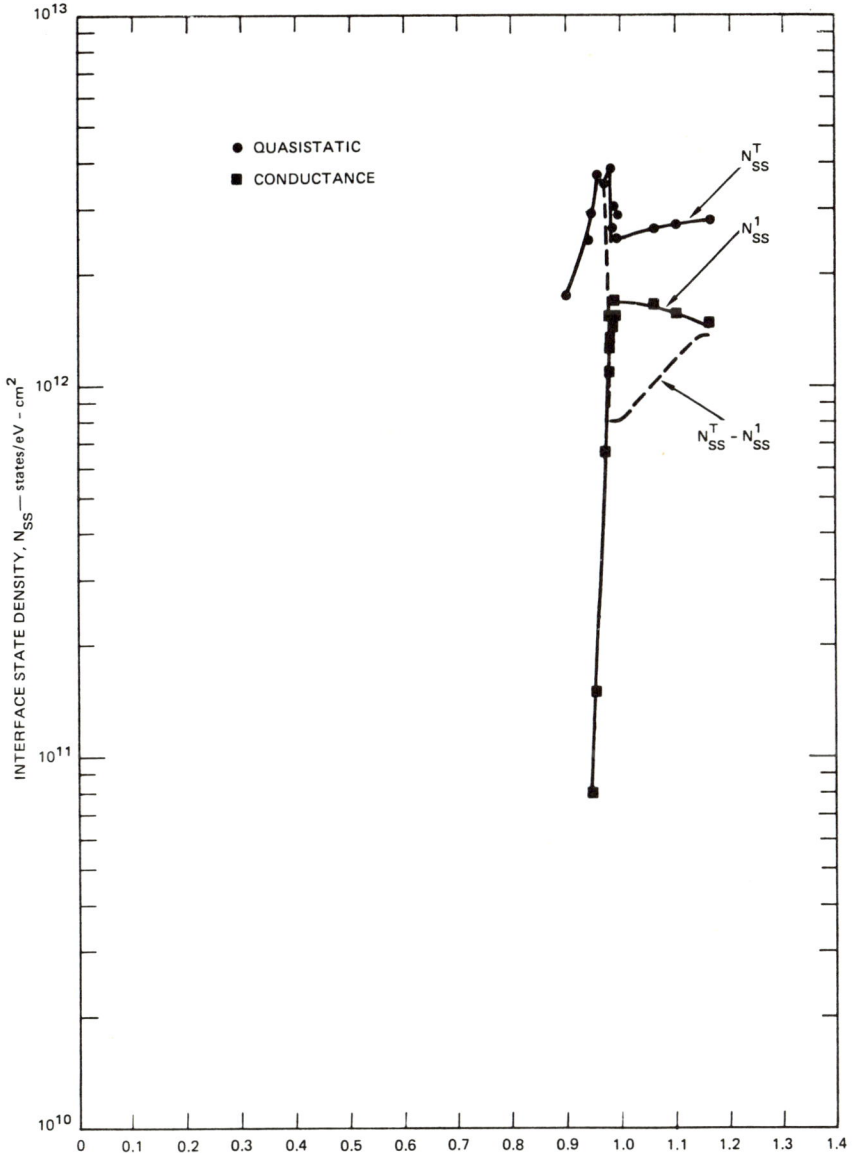

Figure 2 Interface state density of a GaAs sample with a LaF$_3$ insulator as a function of energy relative to the valence band edge. The dots are data taken in the light with no filter in a quasistatic measurement, the squares are data taken in a conductance-versus-frequency measurement, and the dashed curve is the difference between the others.

There are several noteworthy features in Fig. 3:

- The peak interface state density of approximately 4×10^{12} states/eV-cm^2 is quite large, corresponding to a capacitance C_{ss} of about 600 nF/cm^2. However large this number seems, it is small compared to those reported for other insulators on GaAs.[6,7] Thus, unless the insulator capacitance is very large by usual standards, the surface potential will be pinned.

- Only a small portion of the interface state density has been probed (from $\simeq 0.9$ to 1.16 ev, although the gate voltage was varied from 0.2 to -6 V. This occurs because the insulator leaks and the Fermi energy is essentially pinned, as discussed previously. However, if the effective insulator capacitance were not as large as those used here ($C_0 \simeq 500$ nF/cm^2), the Fermi energy would have moved even less. We should emphasize that finding a high-frequency capacitance which remains constant with gate voltage in depletion is insufficient evidence to deduce that a sample is inverted. We believe that several groups who have used thick oxides to prevent leakage have fallen into the trap of concluding their samples had inverted when they saw a flat voltage variation of the high-frequency capacitance.

- The general shape of the interface state density agrees with behavior reported by Spicer et al.[9] on several semiconductors using entirely different measurement techniques on quite different samples. The main peak of our curve GaAs falls at 0.96 eV, while theirs is placed at 0.75 eV. Until we examine additional samples, this difference should not be viewed as significant.

This brief survey is intended to introduce the measurement methods and some preliminary conclusions. It has undoubtedly raised more questions than it has answered, among them: the difference between the states that do and do not respond to light, what the interface state density does outside the energy range we have studied, and the origin of these states. We hope to answer these questions through continued studies of carefully prepared samples. The most important result is that the measured GaAs–LaF$_3$ interface state density peak is at least an order of magnitude lower than those reported for other insulators.

This work was supported in part by NASA Grant NSG-1385.

REFERENCES

1. A. Sher, W.E. Miller, Y.H. Tsuo, J.A. Moriarty, R.K. Crouch, and B.A. Sieber, Appl. Phys. Lett *34*, 799 (1979).

2. P. Su, A. Sher, Y.H. Tsuo, J.A. Moriarty, and W.E. Miller, Appl. Phys. Lett. (15 June 1980).

3. A. Gotzberger, E. Klausman, and M.J. Schultz, CRC Crit. Revs. in Solid-State Sci. *6*, 1 (1976).

4. C.N. Berglund, IEEE Trans. Elect. Dev. *ED-13*, 701 (1966).

5. J.R. MacDonald, Sol. State Electron. *5*, 11 (1962).

6. H.H. Wieder, Inst. Phys. Conf. Ser. *50*, 234 (1980).
   H.H. Wieder, J. Vac. Sci. Technol. *15*, 1498 (1978).

7. K. Yamasaki and T. Sugano, Appl. Phys. Lett. *35*, 932 (1979).

8. W.E. Spicer, P.W. Chye, P.R. Skeath, C.Y. Su, and I. Lindall, Inst. Phys. Conf. Ser. *50*, 216 (1980).

CHAPTER VII
DEFECTS AT INTERFACES

GENERATION OF INTERFACE STATES IN THE Si-SiO$_2$ SYSTEM BY PHOTOINJECTION OF ELECTRONS

Stella Pang, S.A. Lyon and Walter C. Johnson
Department of Electrical Engineering & Computer Science
Princeton University, Princeton, NJ 08544

ABSTRACT

The generation of interface states is observed in the Si-SiO$_2$ system when a photo-injected electron current is passed through the MOS structure. The increase in the density of interface states as a function of total charge passed can be fitted by assuming that there are about 10^{12} sites/cm^2 that can interact with the electron current, with a cross section of 10^{-18}-10^{-19} cm^2, and become interface states. Within the sample-to-sample variations, generation is independent of direction of current and of temperature (97°K-0°C). More than 50% of the states are stable for 6 months at room temperature but they anneal rapidly above 200°C leaving a residue of electrons trapped near the Si-SiO$_2$ interface.

INTRODUCTION

A number of investigators have measured electron trapping in the oxide layers of Metal-SiO$_2$-Si structures after large numbers of electrons have been passed through the SiO$_2$.[1-9] We are reporting the first quantitative study of the interface states generated by electron currents in MOS devices. Using UV light, electrons have been photoinjected from both the Si substrate and the Al field plate to produce an electron current through the oxide at moderate fields. Large numbers of interface states are generated at both 90°K and 0°C. We report here on the observed rate of generation of the states, the effect of bias polarity and temperature during injection, and the annealing properties of the states. Our data indicate that there are preexisting centers with a density of about 10^{12} cm^{-2} which are activated by the passage of electrons. The cross section for this activation process is measured to be 10^{-18}-10^{-19} cm^2. In addition, we note that the interface states produced by electron currents appear to be different from those generated by ionizing radiation. We observe that the former generate immediately at 90°K, while the latter do not appear until the sample is warmed if the irradiation is carried out at liquid-nitrogen temperature.[10-13]

SAMPLE PREPARATION AND EXPERIMENTAL TECHNIQUES

The samples used in our experiments were MOS capacitors with n-type (100) Si substrates of 5-10 ohm-cm resistivity. The oxide films were grown in dry O$_2$ with 3% HCl at 1000°C for 30 min. Semitransparent ($\sim$120Å) Al field plates were used to allow UV light transmission. Aluminum back contacts were used, and the devices were sintered in H$_2$ at 450°C for 30 min after Al evaporation. Bias-temperature-stress measurements[14] showed that sodium contamination was negligible.

The interface-state density was measured by a technique proposed by Jenq.[11] Two curves are taken at liquid-nitrogen temperature, one with electrons frozen into the interface states and the other with holes frozen into the states. The two

C-V curves have parallel portions that differ by a translation along the voltage axis owing to the different amounts of charge at the interface. The method includes all interface states except those so close to a band edge that they can emit their charge carriers during the short time required to record the C-V curve. Measurements requiring a few seconds allow states within ∼0.2 eV of a band edge to emit their carriers[15] at 90°K; thus, at this temperature, the Jenq technique includes those interface states lying within approximately the central 0.7-eV portion of the silicon bandgap. The number of interface states per unit area is given by $N_{ss} = C_{ox}\Delta V/e$, where C_{ox} is the oxide capacitance per unit area, ΔV is the translation between the two curves, and e is the magnitude of the electronic charge. The application of the method is illustrated by the low-temperature (97°K) C-V curves of Fig. 1. Each curve was taken at a ramp speed of 1.0 V/sec. On the downsweep, electrons are frozen into the interface states in the central region of the Si bandgap. At the leftmost portion of the sweep the sample is illuminated temporarily to supply holes to the interface, after which the upsweep is taken. The horizontal translation between the parallel portions of the curves occurring near $C/C_{ox} = 0.35$ provides the measure of the density of interface states. The shallow ramp of Curve 3 shows a lateral nonuniformity.

Ultraviolet light from a 1 kW xenon arc lamp, together with a field-plate bias providing an average oxide field of 1 MV/cm, produced the electron current in the oxide. A water filter and a Corning 7-54 glass filter were used to cut off the IR and visible light, allowing photons with energies in the range 3.2 - 4.7 eV to pass. The light was focused by an S1-UV quartz lens onto the semitransparent Al electrode. The photoinjection experiments were done at both liquid-nitrogen temperature and at 0°C.

GENERATION OF INTERFACE STATES

Figure 1 shows typical low-temperature (97°K) C-V curves obtained in the photo-injection experiments. Curve 1 shows the initial condition of a fresh sample. The number of interface states indicated by the Jenq technique is 2×10^{10} cm^{-2}. The sample was then biased, gate positive, to an average field of 1 MV/cm, and was exposed to the UV light while being held at 97°K. The resulting current density was about 1.3 μA/cm^2. Curve 2 shows the low-temperature C-V curves obtained after photoinjecting 5.4×10^{-2} C/cm^2 through the oxide. The interface density has increased to 2.8×10^{11} cm^{-2}. The photoinjection was continued for a total of 58 hrs, during which time a total charge of 0.24 C/cm^2 passed through the oxide. Curve 3 is the resulting low-temperature characteristic. The interface-state density has increased to 7.9×10^{11} cm^{-2}.

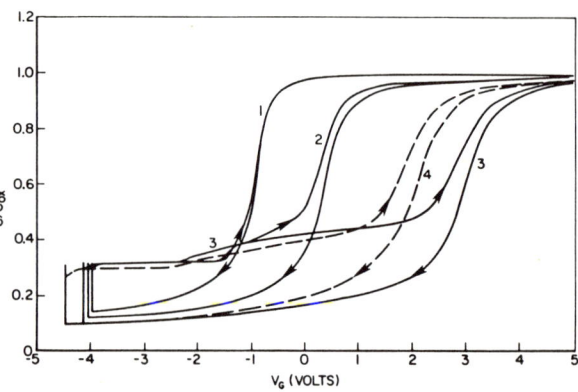

Fig. 1. C-V curves taken at 97°K, showing the effects of electron photoinjection from the substrate. Average field in the oxide was 1 MV/cm. The ramp rate was 1 V/sec. Curve 1: Fresh sample. Curve 2: After photo-injecting 0.054 C/cm^2. Curve 3: After photo-injecting 0.24 C/cm^2. Curve 4: After 4 days at room temperature.

Fig. 2. The number of interface states generated as a function of the charge passed through the oxide at 97°K. Injection was from the substrate. Average field in the oxide was 1 MV/cm. The solid curve is an exponential (see text).

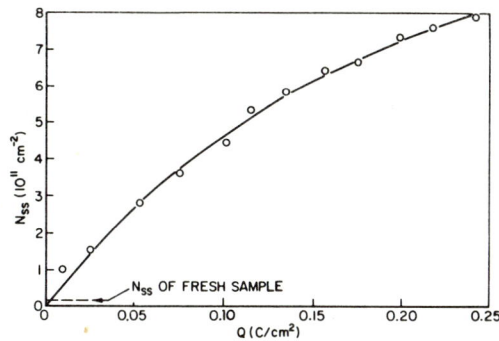

Figure 2 is a plot of the number of interface states as a function of charge passed through the SiO_2. The data indicate that initially the number of interface states generated is directly proportional to the total number of electrons passed through the oxide, but that there is a trend toward eventual saturation. The data can be fitted with an equation of the form:

$$N_{ss} = N_{sat}(1 - e^{-\sigma N_{inj}}) , \qquad (1)$$

where N_{ss} is the areal density of generated interface states, N_{sat} is the saturated density of interface states, σ is the cross section for the creation of an interface state by an electron, and N_{inj} is the number of injected electrons per unit area. Using $N_{sat} = 1.1 \times 10^{12}$ cm^{-2} and $\sigma = 8.7 \times 10^{-19}$ cm^2 we obtain the solid curve in Fig. 2, which shows excellent agreement with the data. The initial slope yields the product $\sigma N_{sat} = 9.6 \times 10^{-7}$. Different samples give slightly different values for σN_{sat}. Among the samples we have studied, the product σN_{sat} has not varied by more than a factor of two.

We also performed photoinjection experiments at 0°C. The initial interface-state generation was proportional to the amount of charge passed through the oxide. With a current density of 2.6 µA/cm^2 maintained for 5 hrs, the charge passed was 4.7×10^{-2} C/cm^2, and the interface-state density increased from 1.8×10^{10} cm^{-2} to 6.3×10^{11} cm^{-2}. The initial generation rate yields the product $\sigma N_{sat} = 2.1 \times 10^{-6}$.

With negative bias applied to the Al field plate, electrons were photoinjected from the gate into the insulator. With the smaller photoinjection barrier height, the electron current was larger than with positive field-plate bias. At 97°K, with a current of 4.4 µA/cm^2 maintained for 2 hrs, the charge passed was 3×10^{-2} C/cm^2, and the interface-state density increased from 2×10^{10} cm^{-2} to 3×10^{11} cm^{-2}. The initial generation rate yields the product $\sigma N_{sat} = 1.5 \times 10^{-6}$.

ANNEALING PROPERTIES OF THE INTERFACE STATES

The interface states generated at either liquid-nitrogen temperature or at 0°C with either polarity of bias showed partial annealing at room temperature over long periods of time. Curve 4 of Fig. 1 was taken at 97°K after the sample had been kept at room temperature for 4 days. The C-V curves have shifted to the left and the interface-state density has decreased to 6.2×10^{11} cm^{-2}. Typically, the interface state density decreased about 20% in a few days and about 40% in 5 months.

Fig. 3. Reduction of interface states by a set of isochronal anneals of 15 min duration each. A charge of 0.25 C/cm^2 had been photoinjected through the oxide at liquid-nitrogen temperature, and the sample had been kept at room temperature for 5 months prior to the annealing treatment.

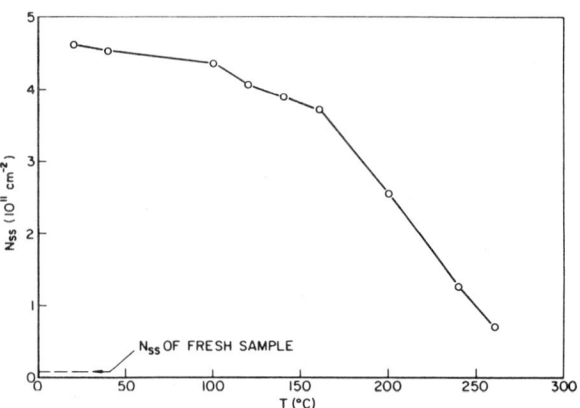

At higher temperatures, the generated states can be almost entirely annealed out. In preparation for an annealing experiment, a sample was photoinjected with 0.25 C/cm^2 with the field plate positive at liquid-nitrogen temperature, and showed an interface-state density of 7.9×10^{11} immediately after the photoinjection. With the sample open circuited and kept in the dark at room temperature for 5 months, the interface-state density was reduced to 4.7×10^{11} cm^{-2}. A set of isochronal anneals, of 15 min duration each, were then performed, with the results shown in Fig. 3. The interface-state densities plotted in that figure were, for the sake of consistency, obtained by cooling to liquid-nitrogen temperature and using the Jenq technique. The results of these measurements were confirmed by use of high-frequency and quasi-static C-V curves taken at room temperature. The results of Fig. 3 show that annealing of the generated interface states is rapid at temperatures above 200°C.

We find that a residue of negative charge ($\sim 10^{11}$ electrons/cm^2) remains trapped in the SiO$_2$ after annealing. Preliminary photocurrent-voltage[16,17] experiments indicate that the negative charge resides very close to the Si-SiO$_2$ interface. This result is similar to that obtained by Solomon[9] in Fowler-Nordheim tunneling experiments. The residual negative charge in our samples was stable in vacuum at 400°C.

CONCLUSIONS

We have made a quantitative study of the interface states generated by a photo-injected electron current in MOS devices. The production of interface states depends upon the total charge passed through the SiO$_2$ in a manner which suggests the interpretation that there are sites in the SiO$_2$ at or near the Si interface that can interact with (possibly capture) an electron to produce an interface state. We deduce a site density of about 10^{12} cm^{-2} and an interaction cross section of 10^{-18}–10^{-19} cm^2 at 90°K. The similarity between the density and cross section we measure and those observed for electron traps in SiO$_2$[7,8,9] indicate that they may be related and that these interface states may have contributed to the flat-band and threshold shifts observed by other authors.

The product of site density and cross section does not show a strong dependence on the temperature at which the photoinjection is conducted or on the direction of electron flow. The variation of σN_{sat} among samples photoinjected at one temperature is as large as the difference in the generation rates between 97°K and 0°C. Similarly, the values for σN_{sat} when electrons are photoinjected from the Al field plate (negative bias) fall within the sample-to-sample variations observed for positive bias.

We find that the electron-current-induced states anneal slightly at room temperature. After 6 months with the sample open circuited, however, more than 50% of the original interface states remain. At a temperature of 200°C the interface states anneal rapidly and they can be almost completely eliminated at 300°C. After annealing, a residue of negative charge remains trapped near the $Si-SiO_2$ interface similar to that observed in Fowler-Nordheim tunneling experiments.[9]

ACKNOWLEDGEMENT

This work was supported by the Defense Advanced Research Projects Agency through the Army Night Vision and Electro-Optics Laboratory.

REFERENCES

1. E.H. Nicollian, A. Goetzberger, and C.N. Berglund, Appl. Phys. Lett. 15, 174(1969).
2. E.H. Nicollian, C.B. Berglund, P.F. Schmidt, and J.M. Andrews, J. Appl. Phys. 42, 5654 (1971).
3. A. Ushirokawa, E. Suzuki, and M. Warashima, Jpn. J. Appl. Phys. 12, 398 (1973).
4. R.A. Gdula, J. Electrochem. Soc. 123, 42 (1976).
5. D.J. DiMaria, J.M. Aitken, and D.R. Young, J. Appl. Phys. 47, 2740 (1976).
6. T.H. Ning, J. Appl. Phys. 49, 5997 (1978).
7. T.H. Ning and H.N. Yu, J. Appl. Phys. 45, 5373 (1974).
8. D.R. Young, E.A. Irene, D.J. DiMaria, R.F. DeKeersmaecker, and H.Z. Massoud, J. Appl. Phys. 50, 6366 (1979).
9. P. Solomon, J. Appl. Phys. 48, 3843 (1977).
10. C.C. Chang, Ph.D. dissertation, Princeton University (unpublished - available from University Microfilms International, P.O. Box 1764, Ann Arbor, MI, 48106).
11. C.-S. Jenq, Ph.D. dissertation, Princeton University (unpublished - available from University Microfilms International, P.O. Box 1764, Ann Arbor, MI, 48106).
12. J.J. Clement, Ph.D. dissertation, Princeton University (unpublished - available from University Microfilms International, P.O. Box 1764, Ann Arbor, MI, 48106).
13. Genda Hu, Ph.D. dissertation, Princeton University (unpublished - available from University Microfilms International, P.O. Box 1764, Ann Arbor, MI, 48106).
14. E.H. Snow, A.S. Grove, B.E. Deal, and C.T. Sah, J. Appl. Phys. 36, 1664 (1965).
15. C.C. Chang and W.C. Johnson, IEEE Trans. Electron. Devices ED-24, 1249 (1977).
16. R.J. Powell and C.N. Berglund, J. Appl. Phys. 42, 4390 (1971).
17. D.J. DiMaria, J. Appl. Phys. 47, 4073 (1976).

REDUCED OXIDATION STATES AND RADIATION-INDUCED TRAP
GENERATION AT Si/SiO$_2$ INTERFACE*

F. J. Grunthaner, B. F. Lewis, R. P. Vasquez, and J. Maserjian
Jet Propulsion Laboratory, California Institute of Technology
Pasadena, California 91103, and
A. Madhukar
University of Southern California, Los Angeles, California 90007

ABSTRACT

Thin thermal SiO$_2$ films (< 80 Å) grown on Si (100) substrates are irradiated with electrons from 0 to 20 eV during *in situ* XPS measurements. Both oxide/vacuum surface states and Si(+3) species at the Si/SiO$_2$ interface are generated and allowed to relax during the course of the measurements. The results are correlated with the presence of a strained layer of SiO$_2$ (~ 20 Å) at the interface that we had previously reported.

INTRODUCTION

Considerable effort has been devoted to the characterization of the Si/SiO$_2$ interface using surface-sensitive electron-emission spectroscopies. These investigations have attempted to probe the detailed chemical structure of this interface and an overview of this work can be found in the recent reviews by Helms[1] and Blanc.[2] X-ray photoemission techniques (XPS)[3-5] have been particularly well suited for deducing local atomic environments. In our previous work,[3] we utilized high-resolution XPS and chemical-depth-profiling methods to determine the oxide structure as a function of distance from the Si/SiO$_2$ interface of thermally grown oxides. We concluded that the oxide was stoichiometric SiO$_2$ to within one monolayer of the single-crystal substrate. An abrupt chemical interface (~ 1 monolayer thick) between oxide and substrate was found with a mixed composition comprising Si$_2$O$_3$, SiO and Si$_2$O. We also observed a variety of SiO$_2$ signals throughout the oxide, and a significant change in their distribution upon approaching the interface. This was interpreted using a structure-induced charge-transfer model (SICT)[3,6] which suggests that the actual charge transfer between oxygen and silicon in an Si-O-Si bridging bond is determined by the bond angle. The bond-angle distribution is determined by the ring structure of the oxide network. The bulk amorphous oxide is primarily comprised of 144° bond angles similar to α-quartz, which indicates 6-member rings. In addition,

---

*This paper presents one phase of research performed at the Jet Propulsion Laboratory, California Institute of Technology, sponsored by the National Aeronautics and Space Administration, and, in part, by the Defense Advanced Research Projects Agency through the National Bureau of Standards, and the Defense Nuclear Agency through Harry Diamond Laboratory.

a significantly smaller number of bond angles at 120° (4-member rings) and 160° or greater (7-, 8-member rings) are also found. In the immediate region of the interface, however, we observed a strained layer induced by the lattice mismatch, which is largely composed of 4-member (120°) rings. This strain layer is between 15 and 30 Å in width and is largely controlled by details of the processing chemistry.

In this paper, we extend our previous work by examining the chemical changes introduced into the oxide structure by *in situ* electron irradiation experiments. We first demonstrate that electron- and x-ray-induced charging effects are quite controllable and well understood. We find that with electron energies greater than 7 eV, holes are generated by impact ionization. These holes can be trapped near the interface and lead to the creation of Si(+3) oxidation states through cleavage of strained Si-O-Si bonds. In the course of these electron-irradiation experiments, we also observe the creation of electron traps on the oxide/vacuum surface. We find that the generation and relaxation of both these surface traps and the Si(+3) species are strongly correlated with the proximity of the oxide surface to the strained oxide layer. Finally, we suggest a structural model to interpret the results and our previous observations. This model is viewed in light of previous defect models for silica and α-quartz.[7-10]

EXPERIMENTAL TECHNIQUES

The XPS spectrometer used in this work is a modified Hewlett-Packard 5950A spectrometer connected to an environmental chamber used for sample preparation.[3] For these experiments, the spectrometer was operated at a high resolution (better than 0.35 eV) and at a pressure less than 5×10^{-10} Torr. The spectrometer is equipped with a low-energy electron gun (flood gun) which can irradiate the samples with electrons of kinetic energy between 0 and 10 eV. The sample probe is electrically isolated to enable probe current measurement and independent bias of the sample. Maximum electron-current density to the sample in these experiments was about 10^{-5} amps/cm^2. A number of SiO$_2$ films of various thicknesses ranging from 12 to 86 Å were grown at 850°C on (100) silicon surfaces cleaned by a standardized procedure. Oxide thicknesses greater than 50 Å were measured ellipsometrically. Thickness below this value was generally determined by observed oxide/element ratios obtained from the XPS spectra.

CHARGING EFFECTS IN XPS

The presence of significant positive or negative charge on the oxide will change the oxide surface potential and thus cause a shift in the position of the oxide line relative to the silicon substrate line. The silicon substrate line reflects the silicon surface potential and therefore can shift at most by one eV relative to the Fermi level as dictated by the well-known field effect. Such shifts give rise to a potential gradient through the oxide and, consequently, also broaden the oxide spectral envelope. This behavior is illustrated in Fig. 1 for a 41 Å SiO$_2$ film on elemental silicon under various levels of electron irradiation. Because of the extended escape depth in SiO$_2$, the silicon substrate peak (at the right) can be clearly observed in the Si2p spectra. The apparent chemical shift of the oxide peak (at the left) relative to the substrate decreases with increasing negative bias, and the linewidth of the oxide increases proportionately. The silicon substrate line shifts by only a few tenths of an eV until it is pinned in strong inversion. This field-effect behavior is summarized in Fig. 2 for 49, 23, and 6 Å films of SiO$_2$ on Si. The condition wherein the charge on the oxide surface is essentially zero corresponds to a flatband condition with no potential gradient throughout the oxide films. This case results in the narrowest oxide signal. Consequently, by monitoring the width of the oxide line as a function of

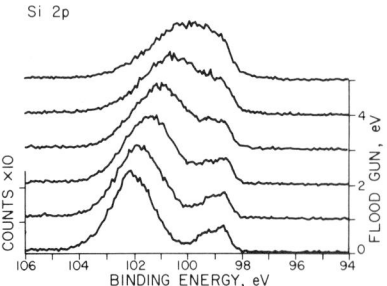

Fig. 1. XPS spectra of Si2p region of 41 Å oxide sample for various flood-gun voltages.

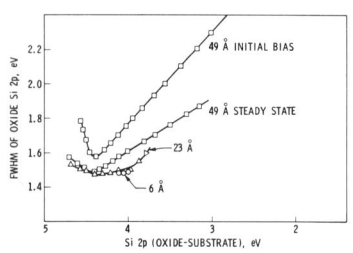

Fig. 2. Plot of the full width at half maximum (FWHM) of oxide Si2p peak vs apparent chemical shift of oxide line referenced to silicon substrate.

the electron-induced bias to the sample, the true chemical shift of the oxide relative to the substrate silicon can be obtained. For the thicker oxides, this linewidth minima corresponds to an oxide-to-substrate energy difference of 4.4 eV.

A number of more extensive charging experiments have been performed that cannot be reported here. These results have been analyzed in terms of the simple field effect as well as with the additional contribution of random-charge fluctuations. The behavior has been quite consistent to first order with the simple field-effect analysis. Only minor modifications are introduced by the random-charge fluctuation treatment.

ELECTRON-GENERATED TRAPS AT THE OXIDE/VACUUM SURFACE

The spectra of Fig. 1 are rather complex under strong electron-induced negative bias because of the interference of the substrate silicon line. Modification of the oxide surface potential causes an identical translation and broadening of the O1s photoelectron line. The effects of electron-induced bias experiments can therefore be readily monitored by following the position of the O1s line. In Fig. 3, the O1s centroid binding energy is plotted for various electron irradiation conditions. The initial start position for this 41 Å oxide is 533.2 eV with no flood-gun-generated current. Using the flood gun at successively increasing potentials, one observes at first a monotonic decrease in apparent binding energy, and at potentials above 7 eV, a reversal. Above 7 eV, the flood-gun current to the probe tends to saturate, but hole-generation by impact ionization can lead to a reduction of net oxide charge. In this experiment, the sample is irradiated with 10 eV electrons for several hours after which time the flood-gun source is switched off and the decay process is monitored. The initial negative charge decay process is quite rapid leaving a residual positive charge. During the first hours, positive charge is annihilated with a decay period of approximately two hours, after which a new rest binding energy is observed. This shift in binding energy of nearly 1 eV relative to the start condition indicates that electrons are trapped in oxide surface states which are in equilibrium with the substrate Fermi level. This experiment is summarized in the energy diagram of Fig. 4. In the flood-gun-on case, both shallow and deep states can be occupied. In the relaxed case, the shallow states quickly depopulate, but the deep states (estimated at 5.2 ± 0.5 eV below the oxide conduction band) are still occupied because of tunneling equilibration with the electrons in the substrate. This surface-state occupation leads to an observed oxide-to-substrate chemical shift of 3.7 eV as compared to the flatband case of 4.4 eV. Consideration of the oxide thickness and the appropriate tunneling

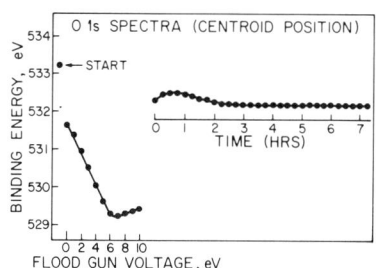

Fig. 3. Binding energy of centroid of oxygen 1s line for 41 Å oxide for various flood-gun-on conditions and during relaxation.

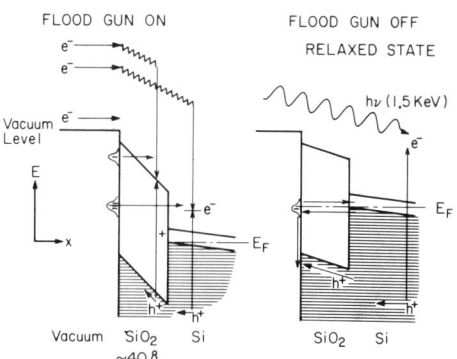

Fig. 4. Energy diagram for electron irradiation experiment.

rate leads to an interesting observation. If the 5.2 eV surface states had been present prior to electron irradiation, they would have been occupied and, consequently, the initial apparent chemical shift would have been 3.7 eV. Since the surface potential of the oxide is not pinned initially, we conclude that these states are induced by the electron flux to the oxide surface. To test this hypothesis, electron irradiations were performed at different biases with the oxide-to-substrate shift measured before and after each irradiation. If the shift is reversible, then no new states have been created by the electron irradiation. The data for a 41 Å oxide sample indicate that these experiments are reversible for low electron fluxes at flood-gun potentials below 0.5 eV. For potentials between 0.5 and 2.0 eV, this deep surface state is created. No further change occurs between 2 and 8 eV until hole generation begins in the oxide. For oxide samples thicker than 50 Å where tunneling communication with the substrate is negligible, the bias experiment is reversible for all potentials with a time constant of approximately one hour. This observation is consistent with dosimetry experiments which indicate that the x-ray-generated hole flux from the oxide and silicon is sufficient to annihilate the electron charge in the surface states within the time scale observed. For thinner oxides in the range of 20 - 30 Å, this state is populated but then decays on a time scale of several hours. Since tunneling would require a time scale of 10^{-3} seconds or less, we conclude that, for sufficiently thin oxides, the generated surface state is chemically annihilated. For very thin samples (< 20 Å), either the generation is low, or the chemical relaxation rate is so fast that no significant surface-state population can be observed.

We note that this same deep surface trap (5.3 eV) has been previously observed in electron irradiation of the vacuum-oxide surface.[11] Further, other investigators[5] of Si/SiO$_2$ interfaces have reported a chemical shift for the Si2p line of the oxide relative to the substrate of 3.8 eV. We attribute this observation to electron-generated oxide surface-state formation due to sufficiently large secondary electron fluxes in their experiment.

INTERFACE STATE GENERATION

Referring again to Fig. 4, we note that energetic electrons can create holes in the silicon substrate and oxide by impact ionization. With 10 eV electrons and the strong negative surface potential on the oxide, energetic holes from the substrate can be injected into the interfacial region of the SiO$_2$. In the extended 10 eV electron irradiation experiment discussed above, we can examine the XPS spectra to determine whether any chemical changes are being induced in the oxide interfacial region. During this irradiation, the Si2p spectra are essentially

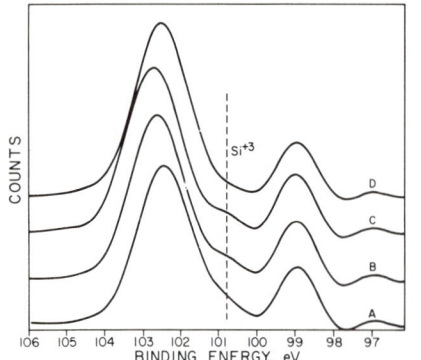
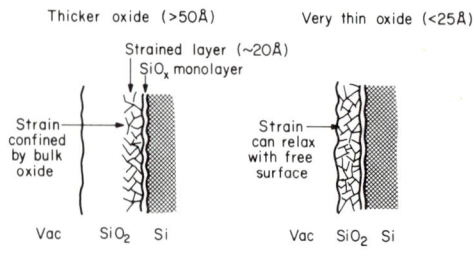

Fig. 5. Sequential decay of Si2p spectra (A to D) after 10 eV electron irradiation.

Fig. 6. Interaction of surface with strained SiO_2 transitional region.

obscured by charge broadening, but during the decay the oxide peak has recovered significantly. In Fig. 5, we plot sequential Si2p spectra for the initial period of the decay. The lower spectrum corresponds to the fourth spectrum taken after removal of the bias, while the uppermost curve corresponds to the 35th spectrum. Note the decay of the feature 1.9 ± 0.2 eV to the right of the oxide peak manifold. The final spectra are nearly identical to the oxide spectra prior to electron irradiation. This feature or state is not observed after irradiation with electrons with bias less than 7 eV. Exposure of a 41 Å sample to 20 eV electrons leads to nearly twice the state density obtained with 10 eV electrons for the same total flux. Irradiation of samples less than 30 Å in thickness does not give rise to such an observable feature. Irradiation of 50 Å samples gives a larger density of the state generation and a slower rate of decay than that obtained for 40 Å samples. The effect is further enhanced for 86 Å samples as compared to 50 Å thicknesses.

Detailed analyses of spectral assignments are beyond the scope of this paper. However, we note that further work indicates the interfacial state assigned to Si(+3) arises from the cleavage of 120° strained Si-O-Si bonds leading to a non-bridging oxygen and a singly occupied dangling orbital on a silicon atom which is bonded to three bridging oxygens. The observed binding energy suggests that the dangling Si orbital has been decorated by a labile hydrogen atom.

CONCLUSIONS

The thickness dependence found for both interface and oxide-surface-state generation and decay is consistent with a strain-related mechanism, as suggested by Fig. 6. As noted in the introduction, the presence of this strained interfacial region has been demonstrated in our previous work.[3] With a thicker oxide, a maximum strain exists in the interfacial region which is confined by the bulk oxide, and the release of this strain energy can assist in the generation of the Si(+3) state by a trapped hole. With very thin oxides, some of this strain energy is removed by the free surface, making generation of Si(+3) states less favorable and allowing both oxide surface and interface states that are generated to be annihilated more easily by structural relaxation. This interpretation further suggests that for oxides of device thickness, a substantial Si(+3) state density can be generated and be stable over extended times. This predicted relation to the strained interfacial layer may offer a significant insight into radiation-hardening processes, especially since we have already observed major differences in the amount of strain with process changes. Some preliminary results of current investigations have already borne out that such a relation exists—that is, radiation-hardening has been seen to improve with reduced amounts of 120° bond-angle strain.

Finally, we have demonstrated that electron-induced bias experiments offer new information in the XPS study of the Si/SiO$_2$ interface. It offers a more comprehensive understanding of trapping and charging effects, which otherwise could introduce artifactual results, and allows *in situ* manipulation of the interface under study.

ACKNOWLEDGMENTS

The authors wish to thank C. Akers, who performed the data reduction, C. Butler, for technical support, D. Lawson, who fabricated the samples, M. Brandenberg for preparation of the manuscript, and P. J. Grunthaner and J. A. Wurzbach for stimulating discussions.

REFERENCES

1. C. R. Helms, J. Vac. Sci. Technol. 16, 608 (1979).

2. J. Blanc, Proceedings of the Topical Conference on Characterization Techniques of Semiconductor Materials and Devices, Electrochem. Soc. Proc. 78-3, 100 (1978).

3. F. J. Grunthaner, P. J. Grunthaner, R. P. Vasquez, B. F. Lewis, J. Maserjian, and A. Madhukar, J. Vac. Sci. Technol. 16, 1443 (1979); Phys. Rev. Lett. 43, 1683 (1979).

4. R. S. Bauer, J. C. McMenamin, R. Z. Bachrach, A. Bianconi, L. Johansson, and H. Petersen. Inst. Phys. Conf. Ser. 43, 797 (1979).

5. A. Ishizaka, S. Iwata, and Y. Kamigaki, Surf. Sci. 84, 355 (1979).

6. R. N. Nucho and A. Madhukar, Phys. Rev. B 21, 1576 (1980).

7. D. L. Griscom, in *The Physics of SiO$_2$ and Its Interfaces*, S. Pantelides, Editor, Pergamon Press, New York, 1980, p. 232.

8. N. F. Mott, Adv. Phys. 26, 363 (1977).

9. G. N. Greaves, Phil. Mag. B37, 447 (1978).

10. G. Lucovsky, Phil. Mag. B39, 513 (1979).

11. J. M. Fanet and R. Piorier, Appl. Phys. Lett. 25, 183 (1974).

STUDIES OF ELECTRON-BEAM RADIATION AND HYDROGENATION
EFFECTS ON Si-SiO$_2$ INTERFACE AND SiO$_2$ BY XPS

Takeo Hattori, Takashi Totsuka and Toshihisa Suzuki
Department of Electrical Engineering
Musashi Institute of Technology
1-28-1 Tamazutsumi, Setagaya-ku, Tokyo 158, Japan

ABSTRACT

Si-SiO$_2$ system were studied by XPS without depth profiling. Possible defect distribution are deduced from the analysis of the experimental data. The effects of electron-beam irradiation and annealing on the Si 2p photoelectron spectra were not observed in most cases. Si 2p photoelectron spectra from Si-SiO$_2$ interface was not identified in the present study.

INTRODUCTION

Electron-beam lithography has been used to fabricate micron and submicron devices[1]. Energetic electrons used to achieve good image resolution create damages on the surface of SiO$_2$, in the bulk of SiO$_2$ and at the Si-SiO$_2$ interface. These damages have been shown to contribute to the increase in positive and neutral trap densities in the bulk of SiO$_2$ and to the increase in the interface state densities[2,3]. The relation of interface states with chemical structures of Si-SiO$_2$ system have been studied by various surface sensitive techniques[4]. However, the relation of these traps with chemical structures remain unanswered. In this paper we report the result of our study using XPS to discuss distribution of defects in the bulk of SiO$_2$. In order to detect the small change in photoelectron spectra produced by electron-beam irradiation, depth profiling techniques using argon ion bombardment can not be used in the present case because of the appreciable change produced by ion knock-on effects[5].

DEFECT DISTRIBUTION IN Si-SiO$_2$ SYSTEM

Defects in Si-SiO$_2$ system and Si atoms in the nonstoichiometric transition region at Si-SiO$_2$ interface are considered to contribute to Si 2p photoelectron spectra whose binding energy is between those of Si and SiO$_2$. One of the possible defect distribution in SiO$_2$ are schematically shown in region II of Fig. 1. n_1 and n_s are volume density of Si atom in the transition region I and Si substrate. n_4 is surface density of Si atom contributing to NI. Here, NI expresses numbers of photoelectrons arising from three regions in Fig. 1. In order to avoid the change in signal intensity due to the change in the sensitivity of the detecting system and that due to contamination, intensity ratio (NI/NS) is used for the quantitative analysis. Here, NS expresses numbers of photoelectrons arising from Si substrate. The expression of the intensity ratio (NI/NS) for three regions are

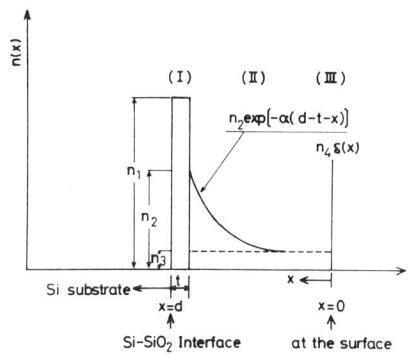

Fig. 1

REGION	n(x) & (NI/NS)
(I) Si-SiO$_2$ interface	$n(x) = n_1$, $(NI/NS) = (n_1 \Lambda_t / n_s \Lambda_s)[\exp(t/\Lambda_t) - 1]$
(II) Defects in SiO$_2$	$n(x) = n_2 \exp[\alpha(x - d + t)] + n_3$ $(NI/NS) = (n_2 \Lambda_e / n_s \Lambda_s) \exp(t/\Lambda') [\exp((d-t)/\Lambda_e) - 1]$ $+ (n_3 \Lambda_o / n_s \Lambda_s) \exp(t/\Lambda_t) \exp(d/\Lambda_o)$ $\Lambda_e = \dfrac{\Lambda_o}{1 - \alpha \Lambda_o} \qquad \Lambda' = \dfrac{\Lambda_o \Lambda_t}{\Lambda_o + \Lambda_t}$
(III) Defects at the surface of SiO$_2$	$n(x) = n_4 \delta(x)$ $(NI/NS) = (n_4 / n_s \Lambda_s) \exp(t/\Lambda_t) \exp(d/\Lambda_o)$

Table 1

Fig. 2 $\Lambda_o = 25$ A and $\Lambda_s = 23$ A are determined by Flitsch et al.[9]

Dopant	Orientation	Resistivity (Ωcm)
p-type	(100)	7 – 10
	(111)	10 – 20
n-type	(100)	0.009 – 0.018
		2 – 4
	(111)	1.0 – 1.7

Table 2

Oxidation Condition	Annealing	Number in Fig. 4
800°C Dry O$_2$	No annealing	1
	30min. annealing in N$_2$ at 800°C	2
	30min. annealing in N$_2$ at 1000°C	3
1000°C Dry O$_2$	No annealing	4
	30min. annealing in N$_2$ at 1000°C	5
800°C Wet O$_2$	No annealing	6
	30min. annealing in N$_2$ at 800°C	7
	30min. annealing in N$_2$ at 1000°C	8
1000°C Wet O$_2$	No annealing	9
	30min. annealing in N$_2$ at 1000°C	10

Table 3

listed in Table 1. Here, it is assumed that thickness of the transition layer and SiO_2 are both uniform. Λ_s, Λ_t, Λ_o in Table 1 are escape depth in Si substrate that in transition layer and that in SiO_2, respectively. Here, it is assumed that values of escape depth, Λ_t and Λ_o are not affected by the existence of defects discussed above because of the small amounts of the defects.

EXPERIMENTAL DETAILS

Dopants and orientation of Si substrates used for the thermal oxidation are listed in Table 2. Oxidation and annealing conditions used are listed in Table 3. Oxidized wafer thus obtained were cut in square shape with area of 16 mm^2 for the study by XPS. Direct and indirect irradiation were done by using scanning electron microscope JSM-V3 in the dose range from 10^{-6} to 2×10^{-4} C/cm^2. Indirect irradiation was done through 5000 Å thick aluminum film supported by Ni mesh. The transmittance of Ni mesh is 78.8%. Electron dose is calibrated with Faraday cup. Dose in the case of indirect irradiation is made equal to the numbers of electron incident on the aluminum film. The accelerating voltage unless specified is 20 kV. The radius of electron-beam is 0.7 µm. The electron-beam is scanned with spacing of 8 µm. After electron-beam irradiation, annealing was done in N_2 at 440 °C for 30 minutes or in the mixture of 90% N_2 and 10% H_2 for 30 minutes. X-ray excited photoelectron spectra were measured by using Dupont ESCA 650B spectrometer by using $K\alpha_{1,2}$ radiation of magnesium. The details of the XPS spectrometer are described elsewhere[6]. In order to analyze small change, the photoelectron spectra are smoothed by using Fourier transform technique[7].

EXPERIMENTAL RESULTS AND DISCUSSIONS

Dependence of intensity ratio (NI/NS) on the oxide film thickness are shown in Fig. 2, where NI expresses photoelectrons arising from SiO_2. From this figure, electron escape depth in Si and SiO_2 are determined such as to explain experimental data. Namely, by using the value of 2.255 g/cm^3 for the dendity of SiO_2 made by dry oxidation[8], the values of 32 Å and 25 Å are obtained for Λ_s and Λ_o, respectively. A curve calculated by using the values of 23 Å and 25 Å for Λ_s and Λ_o[9] is also shown in this figure. In the following, quantitative discussions are made by using Λ_s = 32 Å and Λ_o = 25 Å.

A typical example of indirect electron-beam irradiation effects with total dose of 10^{-5} C/cm^2 on the Si 2p photoelectron spectra and the annealing effects in N_2 at 440 °C for 30 minutes after irradiation are shown in Fig. 3. As can be seen in Fig. 3, increase in NI by irradiation and a decrease in NI by annealing were observed. These spectra can be separated into three lines. Higher binding energy line located at around 103.6 eV is associated with SiO_2 and its line shape is determined by refering to the line shape for 400 Å thick SiO_2 film. Lower binding energy line having intensity of NS located at 99.3 eV is associated with silicon substrate and its line shape is determined by refering to the line shape for chemically etched Si surface. Intermediate binding energy line having intensity of NI is thus determined. Intensity ratio (NI/NS) exhibit thickness dependence as shown in Fig. 4. Specimens used for the present experiment were prepared at various conditions listed in Table 3. Intensity ratio (NI/NS) for specimens prepared at 1000°C and those annealed at 1000°C fit to a straight line quite well irrespective of the oxidation condition. This straight line can be expressed by $\exp(d/\Lambda_e)$ with $1/\Lambda_e$ = 0.048 $\simeq$ $1/\Lambda_o$, which leads to two posssible explanation. One is related with uniform distribution of defects in SiO_2. Another is related with defects on the surface of SiO_2. If we assume that the main contribution to this dependence is uniform distribution of defects in SiO_2, the percentage of Si atoms contributing to this spectra having intermediate binding energy is 2.7. The specimens prepared at 800 °C and those annealed at 800 °C

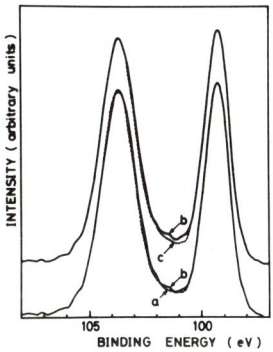

Fig. 3

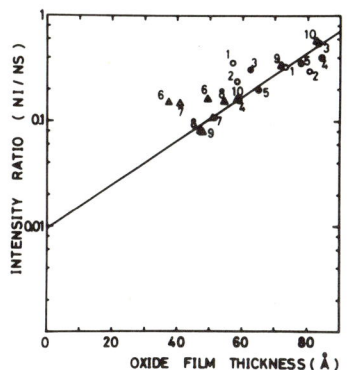

Fig. 4 numbers in the figure are listed in Table 3.

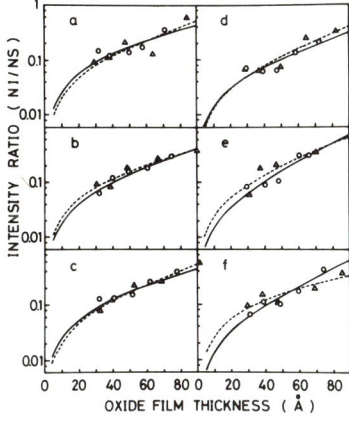

Fig. 5 oxide films by dry oxidation (o), oxide films by wet oxidation (△)

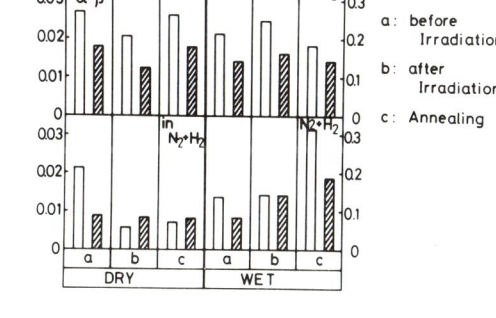

Fig. 6

deviate from this straight line suggesting the existence of interface dependent defects illustrated in Fig. 1 in addition to the uniformly distributed defects in SiO_2 In the following, detailed discussion on the oxide films grown by dry and wet oxidation at 800 °C and annealed in N_2 or in the mixture of 90% N_2 and 10% H_2 at 800 °C for 30 minutes are given. The dependence of intensity ratio (NI/NS) on oxide film thickness are shown for two cases in Fig. 5. One case is intended to see the effects of electron-beam irradiation   (see figure"b" in Fig. 5) and the effects of annealing of defects in N_2 at 440 °C for 30 minutes after irradiation(see figure"c"). Another case is intended to see the effects of electron-beam irradiation on the defects(see figure"e") and the effects of annealing of defects in the mixture of 90% N_2 and 10% H_2 at 440 °C for 30 minutes after irradiation(see figure"f"). In these two cases, the experimental data does not follow $\exp(d/\Lambda_e)$ dependence, but follow $B[\exp(d/\Lambda_e) - 1]$ dependence described in Table 1, where $B = (n_2\Lambda_s/n_s\Lambda_s)$. This means that in the case of straight line in Fig. 4 uniformly distributed defects should be considered, while in the case of Fig. 5 interface dependent defects should be considered. The solid and dotted curves in Fig. 5 are obtained by the least squares fit of the expression $B[\exp(d/\Lambda_e) - 1]$ to experimental data. These solid and dotted curves correspond to dry and wet oxidation, respectively. From the values of Λ and B thus determined, α and β are defined by the following relations; $\alpha = (1/\Lambda) - (1/\Lambda_o)$, $\beta = B(\Lambda_s/\Lambda_e)(n_{so}/n_{so})$, where n_{so} expresses the density of Si atoms in SiO_2. The values of α and β thus determined are shown in Fig. 6. It is quite difficult to find out the effects of electron-beam irradiation and annealing on the values of α and β except one figure"f" in Fig. 5. Namely, the oxide film made by wet oxidation are irradiated with electron dose of 10^{-5} C/cm^2 and then annealed in the mixture of N_2 and H_2 to result in the appreciable increase in the value of α. This suggests that dangling bonds are produced in SiO_2 by electron-beam irradiation and resultant dangling bonds are satisfied by hydrogenation. However, the value of β is not affected also in this figure "f".

CONCLUSION

Si-SiO$_2$ system were studied by XPS without depth profiling. Possible defect distribution in SiO$_2$ are deduced from the analysis of the experimental data. The effects of electron-beam irradiation and annealing were not observed in most cases. Si 2p photoelectron spectra from Si-SiO$_2$ interface was not identified in the present study.

ACKNOWLEDGEMENTS

The authors wish to express their gratitude to Mr. Hideaki Noguchi for his valuable discussions, taking some of the photoelectron spectra, and numerical calculations, to Mr. Takashi Enomoto, Mr. Hiroyuki Saito and Mr. Yukimoto Hisajima for their help in numeical calculations.

REFERENCES

1. H. N. Yu, A. Reisman, C. M. Osburn and D. L. Critchlow, IEEE J. Solid-State Circuits, SC-14, 240 (1979).
2. J. M. Aitken, ibid, p. 294 (1979).
3. G. A. Scoggan and T. P. Ma, J. Appl. Phys., 48, 294 (1977).
4. S. Pantelides ed., Proc. Intl. Topical Conf. SiO$_2$ and Its Interfaces, (Yorktown Heights, 1978) Pergamon Press.
5. T. Hattori, Surf. Sci. 86, 555 (1979).
6. T. Hattori, Thin Solid Films, 46, 47 (1977).
7. G. K. Wertheim, J. Electron Spectrosc. Rel. Phenom. 6, 239 (1975).
8. R. P. Donovan, Fundamentals of Silicon Integrated Device Technoloty, Prentice-Hall, ed. R. M. Burger and R. P. Donovan.
9. R. Flitsch and S. I. Raider, J. Vac. Sci. Technol., 12, 305 (1975).

A MICROSCOPIC MODEL FOR THE Q_{ss} DEFECT
AT THE Si/SiO$_2$ INTERFACE

Gerald Lucovsky
North Carolina State University, Raleigh, NC 27650

and

D. J. Chadi
Xerox Palo Alto Research Center, Palo Alto, CA 94304

ABSTRACT

This paper presents a model for the local atomic structure at the positively charged Q_{ss} defect at the Si/SiO$_2$ interface. The defect center is shown to consist of a threefold-coordinated oxygen atom, bonded to three neighboring silicon atoms in a pyramidal configuration. The net positive charge of this center is distributed over this entire group of atoms.

INTRODUCTION

In spite of extensive progress in device fabrication technology, residual defect densities of the order of 10^{11} cm^{-2} can be present in the immediate vicinity of a Si/SiO$_2$ interface. These defects, which play an important role in determining the electronic properties of the interface, have generally been classified phenomenologically through their electronic properties. Nishi[1] reported three strong esr signals at the Si/SiO$_2$ interface and designated them as P$_a$, P$_b$ and P$_c$. Poindexter, Caplan and their co-workers[2,3] have shown that only one of these three, the P$_b$-center is an intrinsic bonding defect, a threefold-coordinated Si-atom, Si^{III}, with three Si-neighbors. The unpaired electron responsible for the esr response, occupies a p-rich orbital that is normal to the interface for an oxidized [111] Si-face[2,3], and is canted, as expected on the basis of this assignment, for a [100] face[4,5]. Poindexter and co-workers further demonstrated: (i) that the densities of the P$_b$-centers and interface trapping states, N_{st}, behaved similarly with respect to various annealing and oxidation cycles, (ii) that the density of centers responsible for a fixed positive space charge within the oxide layers and designated as Q_{ss}[6] did not correlate with the systematic and coupled changes in P$_b$ and N_{st}, and finally, (iii) that E'-centers, consisting of two neighboring Si^{III}, each with three oxygen neighbors and "sharing" a single and unpaired electron[7], were not observable; hence, their density was less than 10^{10} cm^{-2}.

On the basis of (iii), they concluded that the E'-centers, which carry a net positive charge, could not possibly be the main source of Q_{ss} which is generally in excess of 10^{11} cm^{-2}. This is not surprising since E'-centers are only observed in bulk a-SiO$_2$ following intense irradiation[7]. The correlated behavior of P$_b$ and N_{st} further suggests that these two centers may be associated with the same Si^{III} centers[2-5]. Finally, the lack of any correlation between Q_{ss} and P$_b$ (or equivalently N_{st}) demonstrates that Q_{ss} is not related in any simple way to the P$_b$ environment. This paper proposes a new atomic model for the Q_{ss} center, based on a threefold-coordinated and positively charged oxygen center, O^{III}, the so-called oxonium ion[8]. This proposal for Q_{ss} derives from a defect assignment for bulk a-SiO$_2$ in which the 604 cm^{-1} Raman line is assigned to the local cluster containing the O^{III} atom[9].

BONDING DEFECTS IN a-SiO$_2$

Recently there has been considerable interest in the nature of intrinsic bonding defects in both chalcogenide and oxide glasses. Well-annealed bulk chalcogenide glasses display no esr signals; however, an esr signal may be produced by optically pumping a sample held at low temperature[10]. The build-up of this esr signal is accompanied by a decrease in the photoluminescence efficiency, and an increase in the near infrared absorption just below the optical gap. Street and Mott[11], Mott, Davis and Street[12] and Kastner, Adler and Fritzshe[13] have shown that this behavior is explained by a model in which the intrinsic defects in these glasses are oppositely charged pairs of over and under-coordinated chalcogen atoms, so-called valence alternation pairs or VAP's. Each member of the pair has a spin-paired electron configuration, hence no dark esr. However, trapping of an optically-generated charged carrier at either member of the pair leads to a neutral center with an unpaired spin and hence the photo-induced esr described above. One of us (GL) has shown how this defect model can be extended to a-SiO$_2$[9] wherein the corresponding bonding defects are onefold and threefold-coordinated oxygen atoms, C_1^- and C_3^+ in the notation developed by Kastner et al[13]. One electron energy diagrams for these centers have been published elsewhere[9,14] and will be discussed below. These have been constructed using the approach of ref. 13. Referring now to the shorthand notation, the letter signifies the atom species, C for chalcogen including oxygen (we shall later use T for a tetrahedrally-coordinated atom, Si); the subscript gives the coordination and the superscript the charge state relative to the normal bonding environment which is taken to be neutral. This charge state index is therefore a charge associated with the number of valence electrons in the particular configuration minus the charge associated with the valence electrons of the neutral atom. It is not the effective ionic charge of any particular atom, but is the integrated net charge in the immediate vicinity of a particular atom site. It is convenient to expand the notation and include in parenthesis the nature and number of the nearest neighbors. Hence, in a-SiO$_2$, where all bonds are assumed to be heteropolar in character, the normally-bonded oxygen is C_2°(2T) and the defects identified above are C_1^-(1T) and C_3^+(3T). In the C_2°(2T) configuration, four of the six valence electrons of the neutral oxygen atom are in non-bonding orbitals, and two electrons participate in covalent σ-bonding with the two silicon neighbors. Later on we shall discuss an alternative bonding picture that includes dative π-bonding using p_π orbitals of the oxygen (the donor in the dative bond) and d_π orbitals of the silicon (the acceptor). In the σ-bonding configuration the bond charge is displaced toward the oxygen atom giving the Si-O bond a partial ionic character; however, as discussed above, and in the spirit of the Mott-Street[11] and Kastner-Adler-Fritzshe[13] models, the charge superscript is taken to be zero. In the C_1^-(1T) configuration the oxygen atom starts out with seven valence electrons ($2s^2 2p^5$), rather than six, so that the excess charge relative to the normal valence state is -1. One of the seven valence electrons is utilized in a covalent σ-bond with its neighboring silicon. The oxygen orbital in this bond is nominally pure p. The remaining six electrons go into non-bonding orbitals, two into a low-lying doubly occupied 2s-like state, and the remaining four into two higher lying and degenerate p-states with π-symmetry[14]. These $2p_\pi$ electrons enter into dative $d_\pi p_\pi$ bonds donating electron density to otherwise empty d-orbitals of the silicon neighbor[14]. Finally, the C_3^+(3T) center starts out with five electrons; hence the charge of +1. Two of these five electrons are non-bonding, whereas the remaining three enter into σ-bonds with the three silicon neighbors. The charge in these σ-bonds is displaced toward the oxygen atom, so that the net charge of +1 is distributed over the entire defect cluster consisting of the threefold-coordinated oxygen atom and its three silicon neighbors. The bonding geometry at this center is pyramidal with the oxygen atom at the apex and the three silicon atoms at the base[9,14].

Kastner et al[13] pointed out that the formation of a charged-defect pair does not decrease the total number of covalent bonds, so that the energy for the creation of a defect pair is low (∼1-2eV) and consequently, the number of defect-pairs in a well-annealed glass could be high. For example, the number of defects, as estimated from an interpretation of weak features in the vibrational spectra of a-SiO$_2$ is ∼10^{19} cm^{-3} [9]. Transport studies, in particular the scattering mechanism for the electron mobility, also suggests a similarly high density of neutral defect centers[15].

Greaves[16] has proposed an alternative model in which the defect pairs in a-SiO$_2$ are C_1^-(1T) and T_3^+(3C). Lucovsky[9] has argued against this model on two counts: (i) it requires the breaking of a heteropolar bond which has considerable covalent character and its replacement by a weaker ionic bond, and (ii) the Raman signature of a T_3^+(3C) center would consist of two-polarized defect modes, not the single mode which is observed. Before turning to the question of defects near the Si/SiO$_2$ interface, we point out that Gee and Kastner[17] have found intrinsic defect luminescence in a-SiO$_2$. A comparison of the features in both the excitation and luminescence spectra with the corresponding features in a-As$_2$S$_3$ has been taken by them as a demonstration that the intrinsic bonding defects in a-SiO$_2$ are similar in character to the valence-alternation-pair defects of a-As$_2$S$_3$[13].

THE LOCAL ATOMIC STRUCTURE OF Q_{ss}

The density of Q_{ss} defects in the vicinity of the Si/SiO$_2$ interface is typically of the order of 10^{11} cm^{-2}. Since these defects are distributed over a distance of approximately 100Å, the volume density is of the order of 10^{17}/cm^3. We assume that the defect configuration is associated with an atomic configuration in which a single silicon or oxygen atom has a valence other than that required for the normal bonding in a-SiO$_2$. This gives two candidate defect configurations, T_3^+(3C) and C_3^+(3T). We have argued that T_3^+(3C) is not a characteristic bonding defect in a-SiO$_2$; however, experiments have shown that T_3^+(3C) centers can be formed following irradiation and that this center is a constituent of the E'-complex[7]. In contrast, C_3^+(3T) centers are present in pristine a-SiO$_2$ and their number increases upon irradiation as evidenced by the behavior of the 604 cm^{-1} Raman mode[9]. We have argued above that C_3^+(3T) centers occur in pairs with C_1^-(1T) so that their positive charge is compensated by their negatively charged partner. Hence, in order for the C_3^+(3T) center of a VAP to contribute to Q_{ss} the charge on the companion C_1^-(1T) must be injected in the crystalline Si substrate on which the oxide is grown, or into the metal layer of an MOS structure. This is not likely since the companion center, a $C_1°$(1T) has an unpaired spin [13] and would therefore contribute to an esr signal which is not observed. Alternatively, pair formation may not necessarily occur near the interface due to deviations from stoichiometry and C_3^+(3T) centers may be produced without their conjugate charged partner.

ELECTRONIC STRUCTURE OF C_3^+(3T)

The pyramidal geometry of C_3^+(3T) as deduced from the Raman activity, suggests that the bonding at this center is predominantly through the oxygen p-orbitals. This local environment is to be contrasted with that of the isoelectronic neutral nitrogen center in Si$_3$N$_4$ [18], in which the nitrogen atom, $P_3°$(3T) in the notation used above (P ≡ pnictide atom), is at the center of an equilateral triangle defined by its three Si-neighbors. The atomic states of nitrogen in this planar bonding configuration are separated into two groups; (i) three sp^2 hybrids that form σ-bonds with the neighboring silicons, and (ii) one p-orbital that is perpendicular to the σ-bonding plane. This p-orbital is involved in a π-bonding configuration in which the nitrogen atom is the donor, and three neighboring silicon atoms are acceptors. The orbitals on the silicon atom used in this bond must have d-symmetry in order to give a positive overlap with the nitrogen p-orbital, e.g., d_{xy}, d_{xz} or d_{yz} depending upon the choice of axes. The existence of the planar geometry for the $P_3°$

site in Si_3N_4 implies that the additional bond strength due to the use of sp^2 orbitals for the σ-bonds, and the availability of the third p-orbital for a dative π-bond, is more than enough to compensate for the promotional energy necessary to form the hybrids. The 2s-2p separation in oxygen is much larger than in nitrogen, ~16.7 eV in oxygen as compared to 10.2 eV in nitrogen, and it has been suggested[14] that this is the reason that threefold-coordinated and positively charged oxygen sites have a pyramidal rather than a planar geometry, i.e., that the promotion energy can not be recovered through the increased bond strength of the sp^2 hybrids or the availability of dative π-bonds. π-bonding in the pyramidal geometry would require the use of the 3d-states of oxygen. These are too high in energy and hence are not available, so that the bonding at the C_3^+(3T) center is by σ-bonds.

There are a number of arguments put forth that the local atomic structure in a-SiO_2, in particular, the relatively large bond angle at the oxygen center, $C_2°$(2T), is indicative of substantial π-bond character[18]. This bonding is dative in character, involving the doubly occupied oxygen p-states, those which have an otherwise nonbonding π-symmetry for a 180° bond angle, and unoccupied 3d-orbitals of the silicon atom. The 3d-states that have a symmetry appropriate for dative bonding are d_{3z^2-1} and $d_{x^2-y^2}$ [18,19]. In spite of these associations between local atomic structure and the inferred importance of the silicon 3d-states, most calculations of the energy bands of α- and β-quartz, β-cristobalite and a-SiO_2 [20-22] have used a basis that is restricted to only s and p-states of both the silicon and oxygen atoms. The quality of these band structure calculations is determined by comparisons with various optical spectra which are not very sensitive to a small 3d-admixture. Revesz[23] has pointed out the importance of the 3d-states in the SiO_2 structure and recent calculations by Fowler and his co-workers[24-26] have used a mixed basis which includes participation of the silicon 3d-states. Silicon 3d-character is found by them in the uppermost valence bands, which in the calculations using only s and p-basis functions are simply derived from oxygen $2p_\pi$ states. The mixing reported in refs. 24-26 is here taken to be an indication of the $d_\pi p_\pi$ bonding discussed above.

The energy of the C_3^+(3T) center relative to the top of the valence band will depend on the basis set used to calculate the electronic spectrum of the host network. For example, if the basis set includes only silicon 3s and 3p states, then it is likely that the calculated energy of the C_3^+(3T) center will place it below the top of the valence band. Alternatively, if the basis set includes silicon 3d states, then the energy of the highest lying valence band will be lowered, with the other valence bands also downshifted. Experience with mixed basis sets has also shown that the optical gap will not be changed appreciably. However, the energy of the C_3^+(3T) defect which involves a silicon 3s and 3p states will not be changed and may then lie above the top of the valence band; i.e., within the forbidden energy gap of SiO_2. We suggest the following calculations: (i) a band structure calculation for the β-cristobalite structure using a basis set including the silicon 3d-states, essentially what has already been done by Fowler and his co-workers[24-26], and (ii) a calculation for a C_3^+(3T) defect incorporated into that structure.

SUMMARY

Application of the valence-alternation-pair model to a-SiO_2 predicts that the dominant positively charged defect center is C_3^+(3T). In this paper we associate this center with Q_{ss}. A calculation of the energy of this defect center should include 3d-states for the calculation of the valence band states of the SiO_2 host material.

ACKNOWLEDGEMENT

One of the authors (GL) is pleased to acknowledge partial support for this research under ONR Contract No. 0014-79-C-0133.

REFERENCES

1. Y. Nishi, Japan J. Appl. Phys. $\underline{10}$, 51 (1971).
2. E. H. Poindexter, E. R. Ahlstrom and P. J. Caplan, in The Physics of SiO_2 and Its Interfaces ed. by S. T. Pantelides (Pergamon, New York, 1978), p.227.
3. P. J. Caplan, E. H. Poindexter, B. E. Deal and R. R. Razouk, J. Appl. Phys. $\underline{50}$, 5847 (1979).
4. P. J. Caplan and E. H. Poindexter, this conference.
5. E. H. Poindexter, P. J. Caplan and J. J. Finnegan, this conference.
6. B. E. Deal, J. Electrochem. Soc. $\underline{121}$, 1980 (1974).
7. D. L. Griscom, in Ref. 2, p. 260.
8. F. A. Cotton and G. Wilkinson, Advanced Inorganic Chemistry, (Wiley, Interscience, New York, 1972), p. 407.
9. G. Lucovsky, Phil. Mag. $\underline{B39}$, 518 (1979).
10. S. G. Bishop, U. Strom and P. C. Taylor, Phys. Rev. Lett. $\underline{34}$, 1346 (1975); $\underline{36}$, 543 (1976).
11. R. A. Street and N. F. Mott, Phys. Rev. Lett. $\underline{35}$, 1293 (1975).
12. N. F. Mott, E. A. Davis and R. A. Street, Phil. Mag. $\underline{32}$, 961 (1975).
13. M. Kastner, D. Adler and H. Fritzshe, Phys. Rev. Lett. $\underline{37}$, 1504 (1976).
14. G. Lucovsky, J. Non Cryst. Solids $\underline{35}$ & $\underline{36}$, 825 (1980).
15. G. Lucovsky, Phil. Mag. $\underline{B39}$, 531 (1979).
16. G. N. Greaves, Phil. Mag. $\underline{B37}$, 447 (1978).
17. C. M. Gee and M. Kastner, Phys. Rev. Lett. $\underline{42}$, 1765 (1979).
18. J. E. Huheey, Inorganic Chemistry, 2nd Edition, (Harper and Rowe, New York, 1978), p. 713.
19. C. J. Ballhausen and H. B. Gray, Molecular Orbital Theory, (W. A. Benjamin, New York, 1964), pp. 107-128.
20. S. T. Pantelides and W. A. Harrison, Phys. Rev. $\underline{B13}$, 2667 (1976).
21. J. R. Chelikowsky and M. Schluter, Phys. Rev. $\underline{B15}$, 4020 (1977).
22. D. J. Chadi, R. B. Laughlin and J. D. Joannopoulos, Ref. 2, p. 55.
23. A. G. Revesz, Phys. Rev. Lett. $\underline{27}$, 1578 (1971).
24. P. B. Schneider and W. Beall Fowler, Phys. Rev. $\underline{B17}$, 7122 (1978).
25. E. Calabrese and W. Beall Fowler, Phys. Rev. $\underline{B17}$, 2888 (1978).
26. W. Beall Fowler, P. M. Schneider and E. Calabrese, Ref. 2, p. 70.

EPR DEFECTS AND INTERFACE STATES ON OXIDIZED (111) AND (100) SILICON

P. J. Caplan and E. H. Poindexter
US Army Electronics Technology and Devices Laboratory (ERADCOM)
Fort Monmouth, New Jersey 07703

B. E. Deal and R. R. Razouk
Research and Development Laboratory, Fairchild Camera and Instrument Corp.
Palo Alto, California 94304

ABSTRACT

The P_b center, previously assigned to oriented $\cdot Si \equiv Si_3$ on (111) wafers, has a complex spectrum on (100). The different anisotropy of the (100) signal clearly indicates that P_b reflects crystal structure. It can be resolved into two centers. One has near-axial symmetry with $g_1 = g_{//}$ 60° off the (100) axis toward (110). The g-values are like P_b (111), and suggest $\cdot Si \equiv Si_3$ oriented to fit (100) face structure. The second (100) center has a triaxial g-ellipsoid with g_1 30° off the (100) axis toward (110). Unassignable as any known defect in Si or SiO_2, the second center is consistent with the partially oxidized defect $\cdot Si \equiv Si_2O$, plausible on (100), but structurally awkward and not observed on (111). The $\cdot Si \equiv Si_3$ center on (100) may be selectively reduced with respect to the other center by suitable annealing. For rapidly cooled oxidized (100) wafers, the midgap interface state density D_{it} is about equal to P_b spin concentration, and both quantities are about 1/3 the values on (111). This is evidence of common origins of interface state defects and EPR centers. Several other EPR centers are observed with augmented n-doping or implantation of iron.

INTRODUCTION

An improved structural model of the Si/SiO_2 interface in oxidized silicon wafers has become more important as technology moves toward submicrometer device elements.[1] Electron paramagnetic resonance (EPR) has been shown to have promise in this area.[2] In this paper, our earlier study of (111) wafers is extended to include (100) wafers, doping variations, and ion implantation effects. Both electrical and EPR analyses have been compared. This present work concentrates on the midgap interface state density D_{it} and the trivalent silicon EPR signal P_b. Other EPR signals, from P_a, P_c, D, and E' centers,[3-5] were observed with n-doping or iron implantation.

EXPERIMENTAL DETAILS

A discussion of wafer sample preparation and analytical procedures has been given previously.[2] The only important additions were the use of ion implantation for interface doping with Fe, and especially careful selection and alignment of (100) wafers for good EPR signal-to-noise ratio with these weak signals. Unless otherwise noted, samples were n- and p-type, 4-6Ω-cm wafers for C-V measurements, and 4x20 mm wafer slices, 100Ω-cm for EPR. Most EPR runs were at 295°K.

RESULTS AND DISCUSSION

Nature of EPR P_b Centers on (100) Wafers

The P_b center has a complex nature and seems to have a connection with the inherent defect structure of the Si/SiO_2 interface. In our previous study,[2] the simple anisotropic P_b signal from (111) wafers was tentatively assigned to trivalent silicon broken bonds in the $\cdot Si \equiv Si_3$ configuration, with the odd orbital perpendicular to the interface. On (100), however, the P_b signal is quite different. A selection of the observed anisotropic (100) signals is shown in Fig. 1. The signal is visibly resolved into 2, 3, or 4 components. This signal anisotropy has one very important meaning even without further analysis: it indicates that the P_b signal is responding to interface crystallography.

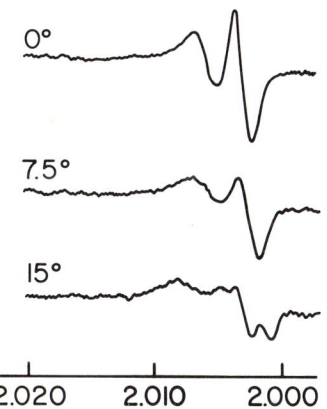

Fig. 1. EPR signals from P_b centers in oxidized (100) silicon wafers.

The anisotropy map for P_b (100) is shown in Fig. 2. The simplest resolution of the pattern requires two EPR centers, noted by solid and dashed lines. These two centers will be called P_{b0} and P_{b1}; their deduced g-value ellipsoid parameters, and values for P_b from (111) wafers, are summarized in Table I. The near-identity of principal g-values for P_{b0} (100) and P_b (111) strongly suggests their equivalence, with the P_{b0} (100) ellipsoid oriented to fit Si orbital arrangement on (100), rather than perpendicular, as on (111). Thus P_{b0} (100) is assigned to $\cdot Si \equiv Si_3$. The second component, P_{b1} (100), has a triaxial ellipsoid whose g-values do not resemble any of the well-known centers found in either Si or SiO_2. By crystallographic symmetry, P_{b1} is consistent with $\cdot Si \equiv Si_2 O$. However, no assignment will be made on the basis of the limited evidence presently available.

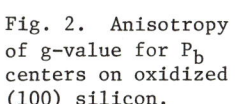

Fig. 2. Anisotropy of g-value for P_b centers on oxidized (100) silicon.

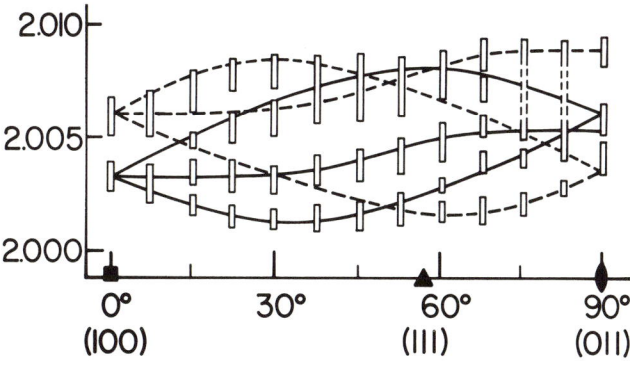

The dual P_b (100) center was examined in several sets of sample wafers, to seek possible distinction of P_{b0} and P_{b1} under different chemical procedures. Some wafers showed a much reduced P_{b0} component. This aided a more confident resolution of the g-anisotropy map, and reinforced the separate identity of P_{b0} and P_{b1}.

Table I. EPR g-values for P_b centers in oxidized (111) and (100) silicon wafers. The value with H_o normal to the wafer face is denoted g_0.

	(100) ·Si≡Si$_3$	(100) ·Si≡Si$_2$O(?)	(111) ·Si≡Si$_3^2$
g_0	2.0060	2.0032	2.0013
g_1	2.0015, 60° off (100)	2.0012, 30° off (100)	2.0012 ⊥ (111)
g_2	2.0080	2.0076	2.0086
g_3	2.0087	2.0052	2.0086

A model of the Si/SiO$_2$ interface on (100) silicon derived from these EPR results is shown in Fig. 3. The more well-established ·Si≡Si$_3$ center is shown with two different orientations. The speculative ·Si≡Si$_2$O center is included. It is apparent from the figure that it is fairly natural to derive the latter center as part of the advancing oxide front. In contrast, an analogous partially oxidized structure is much more difficult to envision on (111), since awkward oxygen penetration beneath the first silicon line is required.

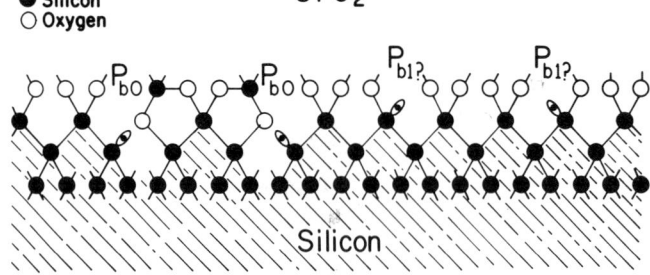

Fig. 3. Suggested model of Si-SiO$_2$ interface on (100) silicon.

Comparison of P_b and D_{it} in Oxidized Wafers cooled in Dry O_2

Variations in annealing and cooling are known to play a substantial role in oxide charges.[6] These variations were observed to have strong effects on EPR signals as well. Fig. 4 shows the relation between P_b and midgap D_{it} for samples oxidized in dry O_2 and cooled rapidly in O_2. Both (100) and (111) samples, n- and p-type, are shown; and although scatter is evident in the data, a clear quantitative correlation seems to exist between P_b and D_{it}. This correlation persists through the range of oxidation temperatures, 800-1200°C. The concentration of P_b centers is about equal to the midgap D_{it} for both (100) and (111) wafers. A ratio of about 1:3 between (100) and (111) holds for D_{it} and P_b, for both n- and p-type samples, although both quantities are lower in p-type. (As will be seen later, this was apparently due to processing anomalies, rather than the dopant.)

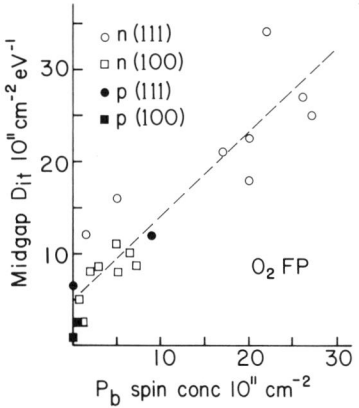

Fig. 4. Midgap D_{it} vs. P_b spin density on (100) wafers oxidized in dry O_2 with fast pull in O_2.

Inert Gas Anneals

The annealing of thermally grown silicon dioxide films in inert ambients, particularly nitrogen, can effect a reduction in oxide charge densities. Differences have been reported between annealing in nitrogen and in argon.[1] For our study, test wafers were oxidized at 1000 or 1200°C for 360 or 60 min respectively, and then annealed in situ in nitrogen or argon for 10 or 60 min at the oxidizing temperature. For N_2 anneals, the correlations between D_{it} and P_b were found to have different slopes for n- or p-type silicon. The cause of this difference is not clear, although it may reflect a more complex interface chemistry in N_2-annealed wafers. For argon-annealed wafers, a higher level of interface states was observed with short anneal time (10 min) and a lower level for longer anneal time (60 min) as compared to nitrogen annealed samples for 10 and 60 min, respectively. There was apparently no difference between n-type and p-type wafers for argon anneals.

Dopant Variation

Variations in dopant type and concentration were studied to discern any effects on P_b or D_{it}, and to fix their apparent correlation more accurately. Previous data have usually been taken on silicon samples with different resistivities, to operate in the region most suitable for EPR or C-V techniques. For this study, n- and p-type (111) wafers, 25, 50, and 100Ω-cm, were oxidized at 1000°C for 360 minutes in dry O_2 and cooled in dry O_2 or N_2 for both EPR and C-V tests.

No trends were noted for either P_b or D_{it}. Before hydrogen anneal D_{it} was 33 ± 4 (10^{11}cm^{-2}eV^{-1}) for oxygen pulls, and 21 ± 3(10^{11}cm^{-2}eV^{-1}) for nitrogen. P_b signals likewise showed no dopant concentration effects. For oxygen pulls, P_b concentration was 14 ± 1(10^{11}cm^{-2}), and for nitrogen, 10 ± 1(10^{11}cm^{-2}). The average ratio of midgap D_{it} to P_b center concentration is a little more than 2:1. This is probably more nearly correct than the approximate equality noted in other parts of this investigation. Densities of N_{it} and P_b should not necessarily be exactly equal, even if P_b centers were associated with all interface states on a 1:1 basis. The observed P_b signal draws only from levels occupied by a single electron at zero applied bias, and may be affected by band-bending at the interface. The midgap D_{it}, however, is deconvolved from the voltage-swept C-V and reflects a different surface potential.

Besides the invariant P_b, a different EPR signal emerged in some samples. Upon rerun at 77°K, the P_a center at g=1.999 appeared in the n-doped wafers, with largest signal in 25Ωcm. Its presence only in n-doped samples suggests donor and/or conduction electrons. A brief HF-HNO$_3$ etch is required to eliminate P_a. which fixes its location in the first few micrometers of silicon below the interface. Large phosphorus pile-up near the interface has been detected by Auger spectroscopy.[7]

Incorporation of Metallic Impurities

Metallic impurities with bandgap energy levels are a potential direct source of interface states.[8] Moreover, their presence may disturb the lattice and create or modify other defect centers. To examine such effects, iron was implanted either in bare (111), n-type wafers to a depth of 450 Å, or in oxidized wafers (1000 Å SiO$_2$) at the interface, in order to produce similar interface iron concentrations in the two cases. The dose was $10^{13} \pm 10\%$ (cm^{-2}). Following implantation, the bare silicon wafers were oxidized to a final SiO$_2$ thickness of 1000 Å. All oxidations were carried out in a dry O_2 ambient at 1000°C for 120 min, followed by a 10 min anneal and a slow pull (2 min) in nitrogen. Selected samples were further annealed for 60 min at 1000°C in a 10% hydrogen-in-nitrogen ambient.

Four EPR species were observed at 77°K (Fig. 5): P_b from interface ·Si≡Si$_3$; P_c

from neutral Fe; a damage signal D at g=2.0055 from broken-bond clusters within the silicon bulk; and E' from $\cdot Si\equiv O_3$ in the damaged SiO_2. Both damage signals and P_b vanish after H_2 annealing, though P_c increases (presumably due to reduction to the visible neutral state Fe^o).[9] These samples generally had a very high D_{it} ($>50\times10^{11} cm$-eV^{-1}) before H_2 anneal, and the usual correlation between D_{it} and P_b was maintained throughout. The implanted iron did not have a clear effect on D_{it}, P_b, or P_c. The latter result is surprising in view of the large interface implant concentration. Since both implanted and inherent bulk Fe must be reduced to Fe^o for EPR visibility, different annealing efficiencies in the two regions may be the explanation. Furthermore, despite the abundance of implanted iron, it is no more than comparable to the overall bulk contamination in typical device-grade silicon.

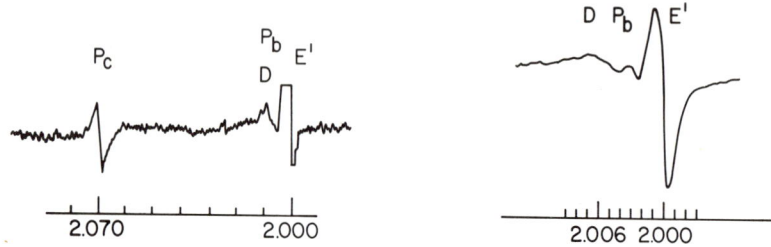

Fig. 5. EPR centers in (111) iron-implanted silicon wafers.

CONCLUDING REMARKS

The consistent behavior of the EPR P_b signal on both (111) and (100) silicon reinforces its assignment to $\cdot Si\equiv Si_3$. Additional correlations of P_b and D_{it} were observed in this study, which strengthen the idea of their possible common origins. The P_b-D_{it} correlation was found to be independent of doping type or level, but showed a possibly significant difference in the case of nitrogen annealing/cooling as compared to argon or oxygen. Finally, EPR was shown to serve as a convenient analytical detector for several other device-pertinent defects in silicon wafers.

REFERENCES

1. B. E. Deal in Semiconductor Silicon 1977, H. R. Huff and E. Sirtl, Ed. (Electrochemical Society, Princeton, NJ, 1977) 276.
2. P. J. Caplan, E. H. Poindexter, B. E. Deal, and R. R. Razouk, J. Appl. Phys. 50, 5847 (1979).
3. Y. Nishi, Jpn. J. Appl. Phys. 10, 52 (1971).
4. B. P. Lemke and D. Haneman, Phys. Rev. Lett. 35, 1379 (1975).
5. D. L. Griscom, E. J. Freibele, and G. H. Sigel, Jr., Solid State Commun. 15, 479 (1974).
6. R. R. Razouk and B. E. Deal, J. Electrochem. Soc. 126, 1574 (1979).
7. S. A. Schwarz, C. R. Helms, W. E. Spicer, and N. J. Taylor, J. Vac. Sci, Technol. 15, 227 (1978).
8. A. Goetzberger, E. Klausmann, and M. J. Schulz, CRC Crit. Rev. Solid State Sci. 6, 1 (1976).
9. P. J. Caplan, J. N. Helbert, B. E. Wagner, and E. H. Poindexter, Surface Sci. 54, 33 (1976).

CHARACTERISTIC DEFECTS AT THE Si-SiO$_2$ INTERFACE

N. M. Johnson, D. K. Biegelsen, and M. D. Moyer
Xerox Palo Alto Research Center, Palo Alto, California 94304

ABSTRACT

The Si-SiO$_2$ interface possesses defects which may be considered characteristic of the thermal oxidation process. Electronic defects introduce a broad peak in the interface-state distribution which is centered ~0.3 eV above the silicon valence-band maximum. Furnace anneals remove these levels yielding the generally observed U-shaped distribution. The ESR interface defect was evaluated over a range of annealing temperatures ($\leq$600 C). The spin signal rapidly decays in MOS structures annealed above 250 C. An anneal in atomic deuterium at 230 C completely annihilated the spin center, which could not be recovered with vacuum anneals up to 600 C.

INTRODUCTION

It has been established that the Si-SiO$_2$ interface possesses electronic defects which may be considered characteristic of the thermal oxidation process. Electrical measurements on MOS devices reveal a continuous distribution of interface states that extends throughout the silicon forbidden energy band and a random spatial distribution of fixed positive charge situated in the oxide near the interface. For interfaces obtained by thermal oxidation without a subsequent anneal, the interface-state distribution is dominated by a broad peak centered approximately 0.3 eV above the silicon valence-band maximum[1,2] This peak may be considered the signature of an inherent interface defect. Theoretical studies have shown that such a peak can arise from silicon dangling bonds at the interface.[3] Electron spin resonance (ESR) reveals the presence of a paramagnetic center, designated as the P_b center, at the Si-SiO$_2$ interface which has been identified as a trivalent silicon defect bonded to three silicon atoms, with the dangling bond oriented normal to the interface on (111) silicon.[4] The spin density correlates with the density of (midgap) interface states but not with the annealing behavior of the fixed positive space charge. In this paper the nature of this correlation is examined more closely with specific attention given to the annealing behavior of the characteristic interface-state distribution and low-temperature ($\leq$600 C) annealing of the spin center.

SAMPLE PREPARATION AND MEASUREMENT TECHNIQUES

Specimens for both electrical and ESR measurements were prepared from single-crystal silicon wafers. The wafers were oxidized in dry O$_2$ at 1000 C. Specimens which received no further thermal processing are designated "as oxidized." Other specimens received a 1000-C anneal in either argon or nitrogen in order to evaluate the effect of high-temperature anneals. All specimens were rapidly cooled to room temperature in the furnace ambient. MOS structures were formed by vacuum evaporating aluminum films with an RF-induction heated source; this avoided the generation of radiation damage in the SiO$_2$ layer.

Electrical measurements were performed on MOS capacitors fabricated on (100)-oriented p-type silicon. Deep-level transient spectroscopy (DLTS) and the quasistatic capacitance-voltage (QSCV) technique were used to measure electronic defect levels at the Si-SiO$_2$ interface. The DLTS measurement was performed in the constant-capacitance mode, which has distinct advantages for data analysis.[1] The experimental and analytical techniques for obtaining the interface-state distribution from CC-DLTS measurements are discussed elsewhere.[1,5] The QSCV technique provides the interface-state distribution over the central portion of the semiconductor bandgap.[6]

ESR measurements were performed on (111)-oriented p-type silicon wafers, with both surfaces polished. Specimens for ESR did not receive a high-temperature anneal. Room temperature ESR absorption spectra (X band) were measured at a nonsaturating power level and digitally recorded. Spin densities and g values were calculated by comparison with a weak pitch standard and DPPH.

EXPERIMENTAL RESULTS

In this section, results are presented from electrical and ESR measurements of characteristic defects at the thermally oxidized silicon surface. The interface-state distribution in as-oxidized and annealed MOS capacitors was measured by DLTS and QSCV techniques. ESR was used to investigate low-temperature (≤ 600 C) annealing of the interface spin center.

Interface-state distributions are shown in Fig. 1 for as-oxidized and annealed MOS capacitors. The distributions were obtained from the QSCV technique. For the as-oxidized sample the interface-state distribution displays a prominent peak located approximately 0.3 eV above the silicon valence-band maximum. On (100)-oriented silicon, the peak density is $\sim 1 \times 10^{12}$ eV^{-1} cm^{-2}. A furnace anneal in argon at the temperature of oxidation results in an overall reduction of the interface-state density across the silicon bandgap, with the characteristic peak remaining the dominant feature. The addition of a standard forming-gas (15% H_2, 85% N_2) anneal, performed after aluminum deposition, results in a substantial reduction of the interface-state density and essentially complete removal of the characteristic peak. Remaining is the U-shaped interface-state distribution which is generally observed in fully processed MOS test devices.

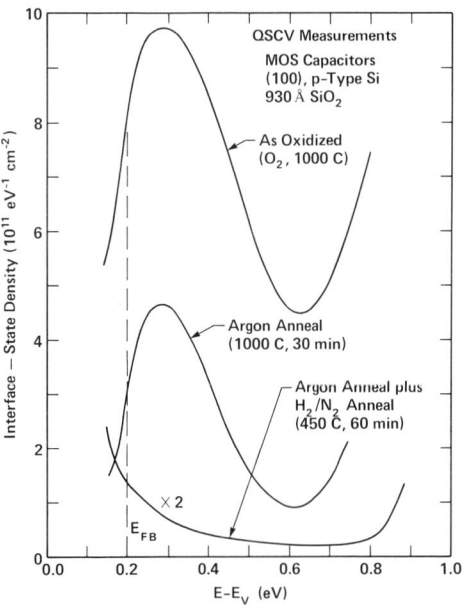

Fig. 1. Interface-state distributions in as-oxidized and annealed MOS capacitors.

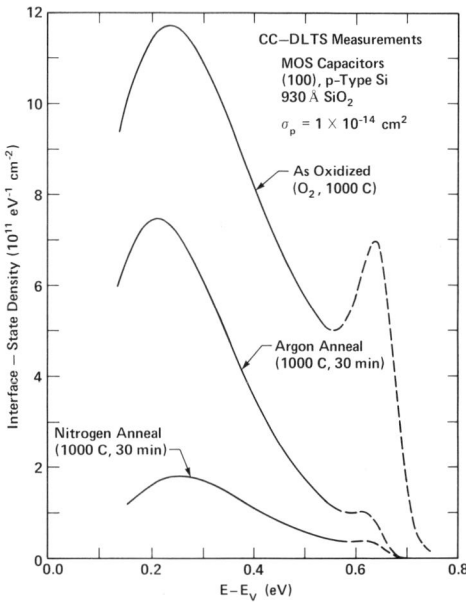

Fig. 2. Effect of high-temperature anneals on the characteristic interface-state distribution.

The effect of the annealing ambient on the interface-state distribution is shown in Fig. 2. The distributions were measured by CC-DLTS. In a previous study it was shown that the cross section for hole capture, σ_p, is essentially constant and equal to 1×10^{-14} cm^2 for the characteristic interface-state distribution on oxidized (100)-oriented silicon.[2] The results in Fig. 2 show that a high-temperature post-oxidation anneal in either argon or nitrogen reduces the interface-state density over the measured energy interval, with nitrogen more effective as an annealing ambient. The characteristic peak clearly dominates the interface-state distribution in the CC-DLTS measurement, with the peak located approximately 0.25 eV above the silicon valence-band maximum. Near the silicon midgap the DLTS measurement is influenced by minority-carrier surface generation and bias-dependent trap occupation statistics and therefore does not yield the interface-state distribution; this is signified by the dashed-line segments in the distributions.

The ESR absorption spectrum for the interface defect center is shown in Fig. 3. The specimen consisted of (111)-oriented silicon, with both surfaces polished, which was oxidized in dry O_2 at 1000 C to yield an SiO_2 thickness of 2000 Å. The spectrum was obtained with the magnetic field H_o normal to the Si-SiO$_2$ interface and was averaged over five scans to improve the signal-to-noise ratio. The dominant spin signal has a g-value of 2.0015 ± 0.0001 (for $H_o \perp$ interface). This is in agreement with results from previous studies of the Si-SiO$_2$ interface.[4] The interface ESR signal in Fig. 3 corresponds to a spin density of $6 (\pm 1) \times 10^{11}$ spins/cm^2. Also identifiable in the absorption spectrum is a weak signal with an isotropic g-value of 2.0057. By systematically varying the ratio of edge-to-surface area, it was demonstrated that this signal originates from the edges of the specimen and can be ascribed to dangling bonds on the rough-cut and exposed edges of the silicon wafer.

The effect of isochronal anneals on the interface spin density is shown in Fig. 4. For this study the thermally oxidized (111)-oriented silicon wafers were coated with 1000 Å of aluminum. The MOS specimens were annealed in vacuum ($\sim 10^{-6}$ Torr) for 15 min. at the specified temperatures. After annealing, the aluminum films were chemically etched for ESR measurements, in order to prevent loading of the microwave cavity. In specimens annealed with aluminum, the spin density decreases rapidly to zero for anneal temperatures above 250 C. As a means of assessing the role of the aluminum film, a single specimen which had been annealed with aluminum at 250 C was further annealed without aluminum. As shown in Fig. 4, temperatures above 400 C were required to significantly affect the spin density, and temperatures above 500 C were required to completely annihilate the spin signal. In a third experiment a specimen, with the aluminum removed, was annealed in atomic deuterium at 230 C for a time sufficient to completely remove the spin signal. The specimen was then vacuum annealed at successively higher temperatures from 230 C to 600 C. As shown in Fig. 4, the P_b spin signal was not regained for anneal temperatures up to 600 C.

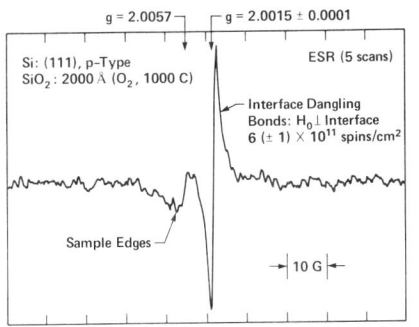

Fig. 3. ESR absorption spectrum for the P_b defect center at the Si-SiO$_2$ interface.

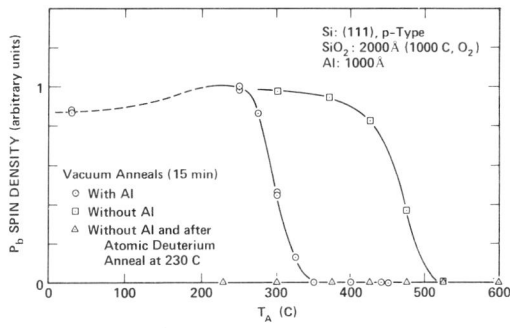

Fig. 4. Isochronal annealing of the P_b spin center in as-oxidized and deuterium-annealed specimens.

The effect of low-temperature isochronal anneals on the interface-state distribution is shown in Fig. 5. The distributions were measured by the QSCV technique on (100)-oriented silicon specimens. On MOS capacitors the vacuum anneals reduce the interface-state density nearly uniformly over the silicon bandgap. After a 400-C anneal, the density of the characteristic peak has decreased by approximately 60%.

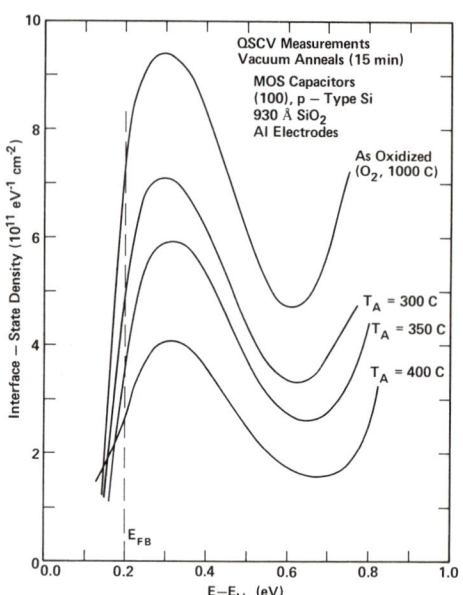

Fig. 5. Isochronal annealing of the characteristic interface-state distribution in as-oxidized MOS capacitors. The QSCV results are unreliable for energies that are nearer to the valence bandedge than the bulk Fermi energy E_{FB}.

DISCUSSION AND CONCLUSIONS

With integrated-circuit quality MOS devices, it is generally found that the Si-SiO$_2$ interface possesses a U-shaped continuum of interface states as illustrated by the bottom curve in Fig. 1. Such distributions are obtained by the combination of a high-temperature post-oxidation anneal and a low-temperature post-metallization anneal. As shown in Fig. 1, the as-oxidized interface does not possess a U-shaped distribution. Rather, the defect-level continuum is dominated by a broad peak centered approximately 0.3 eV above the silicon valence-band maximum. A high-temperature anneal results in a reduction of the interface-state density over the entire measured energy interval (see Figs. 1 and 2). However, the low-temperature post-metallization anneal completely removes the characteristic peak, yielding the U-shaped distribution in Fig. 1. For aluminum films on SiO$_2$ it has been proposed that the low-temperature anneal releases residual hydrogen at the metal-oxide interface which subsequently diffuses to the Si-SiO$_2$ interface where it reacts with silicon dangling bonds.[7,8] However, this mechanism has been considered speculative because the exact chemical nature of the interface defect has not been identified.

The results presented in Fig. 2 were obtained by CC-DLTS and provide an independent absolute measurement of the interface-state distribution for comparison with the results from the QSCV technique shown in Fig. 1. Both methods clearly establish the existence of the characteristic peak in the interface-state distribution. The CC-DLTS measurement further provides the cross section for majority carrier capture at the interface defects.[1,5]

The P_b interface spin center has been identified with trivalent silicon, and the spin density varies with silicon crystal orientation and oxidation/annealing conditions in a manner similar to interface states.[4] In this study, low-temperature isochronal anneals were used to obtain detailed information on the spin center for comparison with electrical measurements. The ESR measurements were conducted on (111)- rather than (100)-oriented silicon because of the higher spin density.[4] Figure 4 presents the first detailed information on the low-temperature annealing behavior of the interface spin center. For aluminum coated oxides, the P_b spin density decreases rapidly for anneal temperatures above 250 C. This behavior is attributable to the composite MOS structure, as evidenced by the significantly different response for samples annealed without aluminum. With the identity of the interface defect, the above low-temperature annealing mechanism may be applied as follows: spin-center annihilation during anneals with aluminum is caused by the release of atomic hydrogen at the aluminum-SiO_2 interface which subsequently diffuses through the oxide layer and reacts with trivalent silicon defects to form stable non-paramagnetic silicon-hydrogen bonds. The role of atomic hydrogen in removing the P_b spin signal is further established with the results from the atomic deuterium anneal. Exposure of the oxidized silicon to atomic deuterium at 230 C completely annihilated the P_b center, and the signal was not recovered with subsequent isochronal anneals up to 600 C. This is the most direct evidence to date that subjecting MOS devices to a low-temperature anneal results in the hydrogenation of silicon dangling bonds at the Si-SiO_2 interface.

The results in Fig. 5 provide information on low-temperature annealing of interface states for comparison with the ESR results. Specifically, isochronal vacuum anneals of MOS capacitors reveal that even after a 400-C anneal (for 15 min.) a significant interface-state density remains, with the characteristic peak reduced in density by only ~60%. Figure 1 illustrates that a conventional 450-C furnace anneal (for 60 min.) produces the U-shaped distribution. However, the electrical measurements were performed on (100)-oriented silicon and therefore do not permit a direct comparison with the ESR data; the requisite electrical measurements on (111)-oriented silicon are in progress. The above results suggest that, while both the electronic interface defect and the P_b spin center are removed at low temperatures, the annealing processes are not identical. Further study is required to establish the nature of the correlation between these characteristic interface defects.

ACKNOWLEDGEMENT

The authors express their appreciation to E. H. Poindexter and P. J. Caplan for helpful discussions. They also thank H. Parker, R. Lujan, and N. Latta for assistance with sample preparation. The work was supported by the U.S. Army Research Office.

REFERENCES

1. N. M. Johnson, D. J. Bartelink, and M. Schulz, The Physics of SiO_2 and its Interfaces, ed. S. T. Pantelides (Pergamon, New York, 1978), pp. 421-427.

2. N. M. Johnson, D. J. Bartelink, and J. P. McVittie, J. Vac. Sci. & Technol. 16, 1407 (1979).

3. R. B. Laughlin, J. D. Joannopoulos, and D. J. Chadi, Phys. Rev. B (to be published).

4. P. J. Caplan, E. H. Poindexter, B. E. Deal, and R. R. Razouk, J. Appl. Phys. 50, 5847 (1979).

5. N. M. Johnson, Appl. Phys. Lett. 34, 802 (1979).

6. A. Goetzberger, E. Klausmann, and M. Schulz, CRC Crit. Rev. 6, 1 (1976).

7. P. Balk, Ext. Abstr., Electronics Div., Electrochem. Soc. 14, 237 (1965).

8. E. Kooi, Philips Res. Repts. 20, 578 (1965).

IMPURITY SEGREGATION AT THE Si/SiO$_2$ INTERFACE*

R. W. Barton, J. Rouse, S. A. Schwarz, and C. R. Helms
Stanford Electronics Laboratories, Stanford University,
Stanford, California 94305 USA

ABSTRACT

We have used Auger sputter profiling to examine the distribution of impurities at the Si/SiO$_2$ interface. Impurities studied include phosphorus, a high concentration dopant in silicon, and chlorine, commonly added as HCl to the oxidation ambient. Both impurities are found to segregate to the Si/SiO$_2$ interface, a phenomenon only recently accounted for in models of integrated circuit processing. After describing the use of the Auger technique in the quantitative determination of impurity profiles, we will demonstrate how simple segregation models can be used to predict the dependence of these profiles on parameters such as oxidation time, temperature, and doping level.

INTRODUCTION

Interface segregation, an important phenomenon in fields such as metallurgy and heterogeneous catalysis, is becoming increasingly important in the study of thin films. In its more familiar context, segregation has been used to explain temper embrittlement in metal alloys as well as the activity of heterogeneous alloy catalysts (1). Segregation is in fact expected when a multicomponent system reaches equilibrium at the interface between two bulk phases (2). Since one component will generally have a lower free energy in the interface region, that component will preferentially segregate to that region. In semiconductors, however, where the distribution of dopants and other impurities is an important determinant of device behavior, segregation effects have only recently been considered (3,4). These effects are expected to become increasingly important as device dimension decrease.

We have used Auger sputter profiling to measure the distribution of phosphorus and chlorine at the Si/SiO$_2$ interface. (3) We find that both impurities tend to segregate to the interface. The significance of this observation for the case of phosphorus, an n type dopant in silicon, is illustrated in Fig. 1. The distribution has been expected to be influenced by oxidation: phosphorus is piled-up in front of the moving interface because it is less soluble in SiO$_2$ than it is in silicon. However, because phosphorus is a relatively fast diffuser, this pile-up is expected to extend for at least a micron into the silicon, and to be no more than 1.1 times more concentrated at the interface than in the bulk (5). This contrasts with our Auger sputter profiles, which reveal large concentrations (up to ten times the bulk level) occurring within a 50Å region at the Si/SiO$_2$ interface.

Further studies have revealed that this pile-up near the interface is an interface segregation effect which occurs in addition to the "normal redistribution"

---

*This work is supported by the Defense Advanced Research Projects Agency, Contract No. DAAB07-77-6-2684

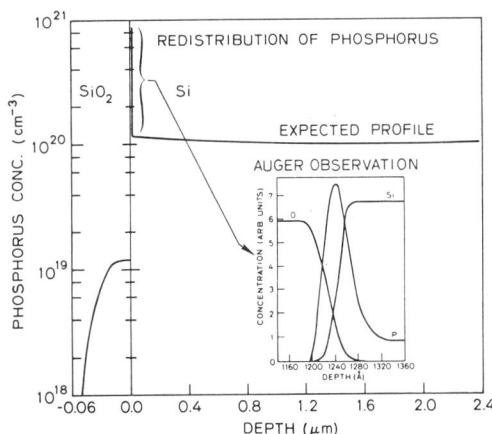

Fig. 1. A comparison of phosphorus distributions near a moving Si/SiO$_2$ interface. The "expected" profile results from bulk diffusion calculations, the narrow peak at the interface represents the results of Auger sputter profiling.

of phosphorus, and is not associated with the actual movement of the Si/SiO$_2$ interface. For instance, the concentration of phosphorus increases during early oxidation times - at a rate that cannot be accounted for by mere redistribution from the SiO$_2$. Furthermore, if a thin (<100 Å) oxide is annealed, the interface concentration will increase further until it reaches a maximum or "saturated" level, characteristic of measurements on thicker oxides. Longer oxidations or post-oxidations anneals do not alter the saturated distribution. The phosphorus is therefore thought to be segregated, or attracted, to the interface by a lower chemical potential in that region.

Large amounts of phosphorus can exist in the saturated pile-up, up to 7×10^{14} atoms per square centimeter. Future models of dopant distribution should account for this segregation, and include its dependence on processing parameters such as oxidation time, temperature, and bulk doping level.

Similarly large amounts of chlorine, an additive in the oxidation ambient, are found at the Si-SiO$_2$ interface (see Fig. 2). An understanding of the reasons for these segregations will enhance our understanding of the structure of the interface itself, and may explain the dependence of other physical properties, such as oxidation rate, on the presence of large amounts of phosphorus and chlorine.

EXPERIMENTAL

Our chemical concentration profiles are obtained by Auger sputter profiling with a Varian 2730 Auger Spectrometer. Profiles are obtained by monitoring the heights of characteristic Auger spectra while simultaneously sputtering with a neon ion beam. Several experimental factors must be considered before these profiles are quantitatively interpreted in terms of chemical concentration and interface width.

Concentrations are generally determined by comparing the peak to peak heights of the Auger derivative spectra. The phosphorus LVV spectrum, for instance, is compared in height to the LVV spectrum of the silicon substrate. These ratios must be corrected by a spectral sensitivity factor, determined from calibrated standards. We use Si or SiO$_2$ samples ion implanted with a known dose of impurity as our standard. Care is taken to insure that the incident electron current is low enough to prevent effects such as desorption, decomposition, and diffusion in the sample. Typically we use 4.5 kev electrons at currents of 10 microamps, rastered over a 600 micron square area. We find our concentration measurements to be reproducible to +/- 10%.

Our depth scale is determined by the sputter rate, which is measured by the time taken in sputtering through an SiO$_2$ film of known thickness. The sputter rates in Si and SiO$_2$ are not expected to be different by more than 5 or 10 percent. The depth resolution is determined to a certain extent by the electron escape depth, but, primarily, by the mixing effect of the sputter ion beam. In previous work, we have measured how profiles are broadened by sputter ions of different mass, energy, and incidence angle (6); neon, for instance, at a sputtering energy of 1 kev, will broaden the profile as much as 30 Å.

Several chemical concentration profiles, such as shown in Fig. 2, were obtained by the above methods. Phosphorus concentrations as high as 1.2×10^{21} cm^{-3} have been measured in the interface of samples doped in the bulk at concentrations of 1.2×10^{20} cm^{-3}. Chlorine peak concentrations as high as 3×10^{20} cm^{-3} have also been measured. Both pile-ups are typically 40 Å in width, with a relative displacement that associates the phosphorus primarily with the silicon side of the interface and chlorine with the SiO$_2$ side.

Since our profiles are significantly broadened by the mixing of the ion beam, the actual profile may actually be quite a bit narrower and proportionately more concentrated than shown. The integrated concentration expected to be an accurate indication of the amount of impurity at the interface. Typical values for phosphorus are 7.0×10^{14} cm^{-3}, measured after an 800° oxidation, which, if concentrated in one atomic layer at the interface, would comprise 50% of that layer.

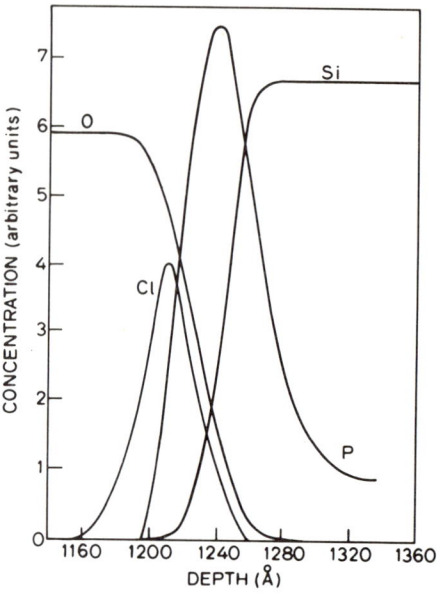

Fig. 2. Auger sputter profiles of oxygen, "free" silicon, phosphorus, and chlorine in the region of the region of the Si/SiO$_2$ interface. Si/SiO$_2$ interface. SiO$_2$ is on the left, Si on the right. The depth scale originates on the surface of SiO$_2$.

SEGREGATION MODELING

Impurities will segregate to an interface when they have a lower chemical potential in that region. Several physical mechanisms might be used to account for this effect: charged impurities will be attracted by interface electric fields, impurities of odd sizes will be attracted by strain fields, and impurities with different bonding geometries will be attracted if they can reduce their electronic energy. All of these cases, however, can be treated thermodynamically in a simple way by assuming that the impurity is held at the interface in a separate chemical phase with a lower standard state free energy. In the limit of low concentration, its profile will be determined by a Boltzmann relation such as:

$$C_I = C_B \exp(-\Delta\mu°/kT) \tag{1}$$

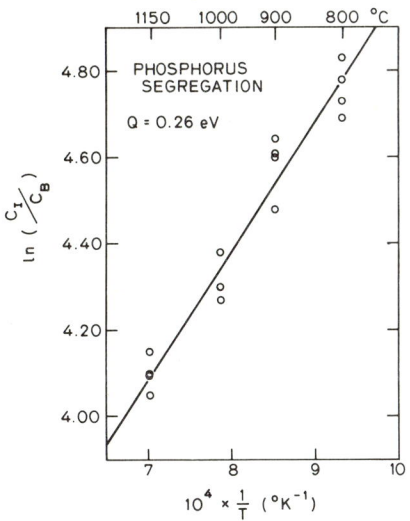

Fig. 3. The temperature dependence of phosphorus interface segregation. The interface concentrations are determined by integrating the Auger sputter profile and assuming that the excess phosphorus concentration resides in a single atomic layer at the interface.

where $\Delta\mu^\circ$ represents the chemical potential difference between bulk and interface and includes both the heat and entropy of segregation.

This relationship has in fact been verified in our Auger sputter profiles of samples that had been post-oxidation annealed at a series of different temperatures. We find a reversible dependence of phosphorus interface concentration on temperature: The lower temperatures being associated with higher concentrations at the interface. The width of the profiles is not significantly affected by temperature. Fig. 3 shows the expected temperature dependence as an Arhenius plot, verifying again that the phosphorus distributes itself in equilibrium with the interface. The slope of the graph indicates that the heat of phosphorus segregation, Q, is 0.26 electron volts.

As indicated previously, in early oxidation times the segregated layer is observed to increase in concentration at a rate faster than that expected by redistribution from the SiO_2. Before the interface reaches its saturated or equilibrium concentration it presumably fills with phosphorus by diffusion out of the bulk of the silicon as well as the SiO_2. We have modeled this diffusion process by assuming that the interface represents initially a square well in the chemical potential of phosphorus. The model then uses common values of phosphorus difffusion coefficient and Si/SiO_2 interface velocity to predict the amount of time necessary to fill this well to its saturated value. Our diffusion model accurately predicts the relationship between total interface concentration and oxidation time. It also determines phosphorus concentrations in the silicon near the interface, which are depleted while diffusion proceeds towards the interface. Fig. 4 compares the prediction of the diffusion model to an actual Auger sputter profile.

Because our Auger sputter profiles are broadened by the mixing effect of the ion beam, our knowledge of the relative depth and width of the potential well remains incomplete. We find that adjustment of these parameters in our diffusion model has no significant effect on its results, their being dependent only on the total volume of the potential well. The Arhenius plot of Fig. 3 is useful because it defines a heat of segregation, but the depth of the chemical potential well also depends on an as yet undetermined entropy of segregation. Nevertheless, we have shown that knowledge of the amount of phosphorus in a segregated layer saturated at one oxidation temperature is sufficient to predict the amount that will appear there after other oxidation times and temperatures.

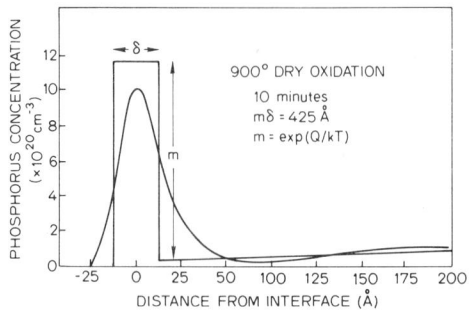

Fig. 4. A comparison between diffusion theory and experiment for an as yet "unsaturated" phosphorus profile that had been oxidized at 900°C for ten minutes. The theory predicts the area under the profile as well as the depletion in the silicon that accompanies diffusion from the bulk.

SUMMARY

Phosphorus and chlorine are found to segregate to the Si/SiO_2 interface in equilibrium with the structural properties of the Si/SiO_2 interface, and in a manner that depends predictably upon temperature and processing time. The segregated concentration is large enough to merit serious consideration in future studies of thin films, and may even yield valuable information about the structural properties of the Si/SiO_2 interface itself.

REFERENCES

(1) See for example: Interface Segregation, W. C. Johnson and J. M. Blakely, Eds., ASM, Metals Park, Ohio, 1979.

(2) See for example: P. Wynblatt and R. C. Ku, Surf. Sci. 65, 511 (1977).

(3) R. W. Barton, S. A. Schwarz, W. A. Tiller, and C. R. Helms in Thin Film Interfaces and Interactions, J. E. E. Baglin and J. M. Poate, eds., Electrochemical Society, Princeton, N. J., 1980; S. A. Schwarz, R. W. Barton, C. P. Ho, and C. R. Helms, to be published in J. Electrochem. Soc. (1980).

(4) M. M. Mandurah, K. C. Saraswat, C. R. Helms, and T. I. Kamins, to be published in J. Appl. Phys.

(5) B. E. Deal, A. S. Grove, E. H. Snow, and C. T. Sah, J. Electrochem Soc. 117, 308 (1965).

(6) S. A. Schwarz and C. R. Helms, J. Vac. Sci. Technol. 16, 781 (1979).

INVESTIGATION OF HYDROGEN AND CHLORINE AT THE SiO_2/Si INTERFACE

I. S. T. Tsong, M. D. Monkowski and J. R. Monkowski
The Pennsylvania State University, University Park, PA 16802

P. D. Miller, C. D. Moak, B. R. Appleton and A. L. Wintenberg
Oak Ridge National Laboratory, Oak Ridge, TN 37830

ABSTRACT

Silicon oxides thermally grown in H_2O, O_2, HCl/O_2 and Cl_2/O_2 ambients were analyzed, via $^1H(^{19}F,\alpha\gamma)^{16}O$ nuclear reaction and SIMS, for the presence of hydrogen. In addition, those oxides grown in HCl/O_2 and Cl_2/O_2 ambients were analyzed with SIMS for the presence of chlorine. The SIMS data show that the hydrogen levels in these oxides were below the limit of detection for nuclear reaction experiments. The $^{35}Cl^+$ depth-profiles show that chlorine is enriched at the SiO_2 interface for the HCl/O_2 grown oxides while it is more evenly distributed in oxide bulk in the Cl_2/O_2 grown samples.

INTRODUCTION

It has long been speculated that hydrogen (or water) is incorporated into thermally grown SiO_2 films on silicon during the oxidation process and this hydrogen contaminant plays an important role in the interface properties [1-3]. However, apart from the measurements by Beckmann and Harrick [4] using infrared internal reflection spectroscopy, direct evidence of the existence of hydrogen has been lacking. Recent development of nuclear reaction techniques such as $^1H(^{19}F,\alpha\gamma)^{16}O$ and $^1H(^{15}N,\alpha\gamma)^{12}C$ [5,6] show great promise for absolute determination of hydrogen in solids, but search for hydrogen at the SiO_2/Si interface has produced negative results [7]. One problem associated with nuclear reaction experiments is beam-induced hydrogen mobility [7,8,9] which could conceivably drive the hydrogen away from the interface. Since this beam-induced migration is thought to be a thermal effect, we have, in the present work, undertaken to perform ^{19}F nuclear reaction experiments on samples cooled to $-40°C$ to determine the hydrogen concentration and distribution in thermally grown SiO_2. In addition, we have also used the secondary ion mass spectrometry (SIMS) technique to obtain hydrogen depth profiles on the same samples.

It is well known that the presence of chlorine during the oxidation of silicon produces several beneficial effects in the electrical characteristics of MOS devices. As a result, numerous studies have been carried out on the role of chlorine incorporation in SiO_2 films and the results have been extensively reviewed recently by Monkowski [10]. In the present study, we have determined the depth-profiles of chlorine using SIMS in several SiO_2 films grown in HCl/O_2 and Cl_2/O_2 ambients under a variety of conditions. These results are compared with data from previous work and their significance are discussed.

EXPERIMENTAL

The $^1H(^{19}F,\alpha\gamma)^{16}O$ resonant nuclear reaction experiments were carried out at the Oak Ridge National Laboratory Tandem Van der Graaff accelerator. The experimental set-up was similar to that described previously [9,11]. The sample holder was in contact with a liquid nitrogen reservoir via a copper shroud such that the samples could be cooled by conduction. After several hours of cooling, a final temperature of -40°C was reached. Depth profiling was accomplished by raising the beam energy (in the lab frame) from 16.4 MeV to 17.4 MeV, equivalent to a depth of about 0.4 μm in SiO_2. A second strong resonance at 17.56 MeV lab energy limits the useful range to this particular depth.

The SIMS depth profiles were measured using a newly developed SIPS-SIMS scanning ion probe. The sputtered-induced photon spectrometry (SIPS) part of the apparatus was not used in this series of experiments. To perform hydrogen analysis using SIMS, special care was made to keep the hydrogen partial pressure low during analysis and the Ar^+ ion beam was mass analysed so that no proton component was allowed to reach the target. Typical pressure in the target chamber during analysis was 5×10^{-9} torr. The 7 keV Ar^+ beam was focussed to 100 μm in diameter with a current density of 6 mA cm^{-2}. Raster-gating technique [12] was used to achieve maximum depth resolution. To overcome charging of the SiO_2 surface, a 2 keV electron beam focussed to a 3 mm diameter spot was directed on to the target surface. Neutralization was complete with an electron beam current of about 40 μA. These neutralization conditions were very similar to those described by Magee and Harrington [13].

RESULTS

^{19}F Nuclear Reaction

Four SiO_2 films grown in (a) H_2O at 900°C for 40 mins., (b) dry O_2 at 1100°C for 50 mins., (c) $HCl(6\%)/O_2$ at 1150°C for 20 mins., and (d) $Cl_2(0.7\%)/O_2$ at 1150°C for 20 mins. were depth-profiled. Throughout the 0.4 μm depth being probed, the γ-ray counts never rose above background which was determined with the beam on and no target. Our calculation indicated that the level of hydrogen throughout the bulk must be less than $\sim 10^{20}$ atoms cm^{-3} and if the hydrogen was concentrated at the interface, then its density must be below 2×10^{14} atoms cm^{-2} since the depth resolution of the ^{19}F beam was ~ 200 Å.

SIMS

In addition to the above four SiO_2 films, four other films were examined by SIMS. Samples (e), (f) and (g) were grown in $HCl(6\%)/O_2$ ambient at 1100°C for 10 min., 35 min. and 55 min. respectively, and sample (h) in $Cl_2(5\%)/O_2$ at 1100°C for 10 min.

Fig. 1 shows the $^{16}O^+$ and $^{30}Si^+$ depth profiles of the $HCl(6\%)/O_2$ (1150°C) oxide, i.e., sample c, under conditions of complete and incomplete neutralization. It is clear that if the surface is not completely neutralized by the electron beam, the $^{16}O^+$ and $^{30}Si^+$ profiles become distorted. Before profiling $^1H^+$ or $^{35}Cl^+$ for each sample, an $^{16}O^+$ profile was always taken to see if complete neutralization was in effect.

The $^1H^+$ depth profiles for the eight samples are shown in Figs. 2(a) and 2(b). The hydrogen concentration was calibrated by a single-crystalline Si wafer implanted with a known dose of protons. This would serve as an approximate

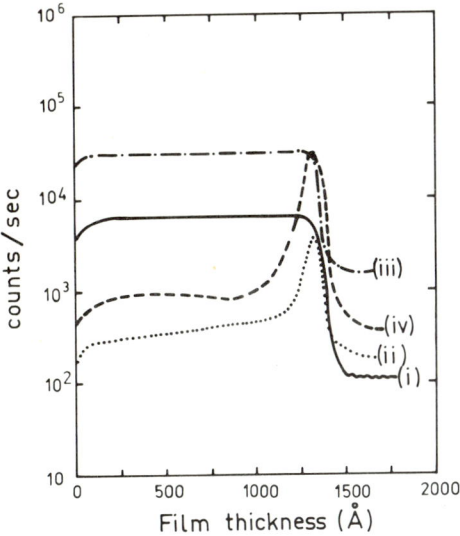

Fig. 1 SIMS depth profiles of $^{16}O^+$ in SiO_2 under (i) complete charge neutralization and (ii) incomplete charge neutralization. Depth profiles of $^{30}Si^+$ in SiO_2 under (iii) complete and (iv) incomplete charge neutralization.

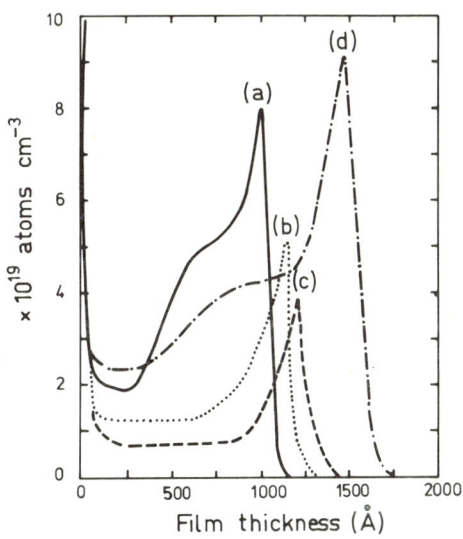

Fig. 2(a) SIMS $^1H^+$ depth profiles of SiO_2 films grown in (a) H_2O at 900°C, (b) dry O_2 at 1100°C, (c) HCl(6%)/O_2 at 1150°C and (d) Cl_2 (0.7%)/O_2 at 1150°C.

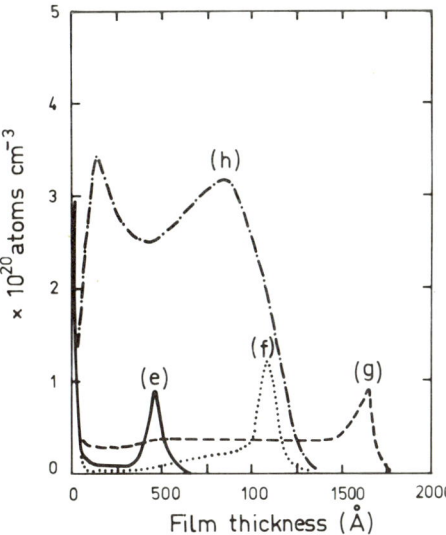

Fig. 2(b) SIMS $^1H^+$ depth profiles of SiO_2 films grown in HCl(6%)/O_2 at 1100°C for (e) 10 mins., (f) 35 mins., and (g) 55 mins., and film (h) grown in Cl_2(5%)/O_2 at 1100°C.

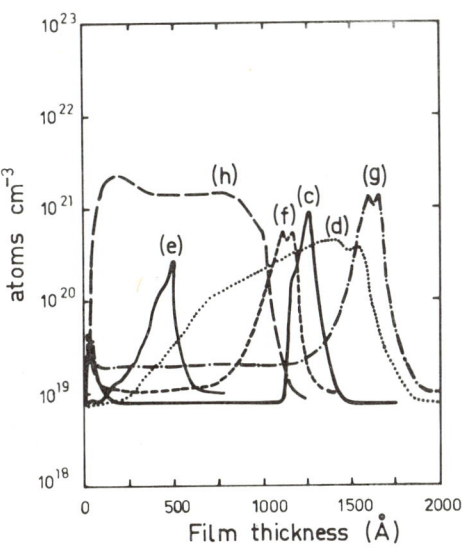

Fig. 3 SIMS $^{35}Cl^+$ depth profiles of SiO_2 films grown in HCl(6%)/O_2 (c) 1150°C, 20 mins., (e) 1100°C, 10 mins., (f) 1100°C, 35 mins., (g) 1100°C, 55 mins., and films grown in (d) Cl_2 (0.7%)/O_2, 1150°C, 20 mins., and (h) Cl_2(5%)/O_2, 1100°C, 10 mins.

calibration since a proton implanted SiO_2 standard was not available. Allowing for oxygen enhancement effect in the SiO_2 in which the $^{30}Si^+$ signal is 20 times higher than in Si, we obtain a calibration factor of 1×10^4 counts s^{-1} for 1×10^{21} H atoms cm^{-3} in SiO_2. The neutralizing action of the electron beam also caused electron-stimulated desorbed (esd) $^1H^+$ to form a background $^1H^+$ signal. The $^1H^+$ profiles shown in Fig. 2 all have this esd background subtracted. The esd background was measured periodically during the depth-profiling process by simply blocking off the ion beam momentarily.

The $^{35}Cl^+$ profiles are shown in Fig. 3. The chlorine concentration was calibrated by Rutherford backscattering performed on samples e, f and g, at Brookhaven National Laboratory. The consistency between the RBS and the SIMS data is extremely good, with all three samples yielding a calibration factor of $(8.4 \pm 0.1) \times 10^{17}$ atoms cm^{-3} per c/s at a sputtering rate of 3.1 ± 0.1 Å s^{-1}.

DISCUSSION

Hydrogen

The fact that hydrogen was not detected in the ^{19}F nuclear reaction experiments is consistent with previous observations by Benenson et al. [7] using the ^{15}N reaction. However, from the SIMS data, it is clear that the reason for the lack of hydrogen is due to the low levels of hydrogen concentration in samples a, b, c and d, all $\sim 10^{19}$ atoms cm^{-3}, thus falling below the limit of detection for nuclear reaction.

SIMS measurements show an enrichment of hydrogen at the SiO_2/Si interface in all the samples. The highest level of hydrogen concentration occurs in sample h, thermally grown in 5% Cl_2/O_2 ambient. The hydrogen is distributed throughout the bulk in this sample, with an average concentration of $\sim 3 \times 10^{20}$ atoms cm^{-3}.

Chlorine

The SIMS $^{35}Cl^+$ depth profiles of the oxides grown in HCl/O_2 ambients (samples c, e, f and g) agree very well with previous SIMS data obtained by Deal et al. [14]. The chlorine shows enrichment at the SiO_2/Si interface as well as on the immediate surface of SiO_2.

The $^{35}Cl^+$ depth profiles in the Cl_2/O_2 oxides (samples d and h) show that the chlorine is more evenly distributed in the oxide rather than simply piling up at the interface. This finding closely resembles that of van der Meulen et al. [15] who used Rutherford backscattering to show that under similar growth conditions the additive species Cl_2 results in higher, more evenly distributed chlorine levels in the oxide as opposed to the occurrence of highest chlorine concentration at the SiO_2/Si interface for the HCl oxides.

A feature quite apparent in the $^{35}Cl^+$ profiles is the presence of a sharp peak at the SiO_2/Si interface as well as a diffuse buildup within the bulk oxide extending from a maximum near the interface. The development with increasing oxidation time of these two aspects of the profile can be followed in the series (e), (f), (g). At 10 min., the diffuse buildup is present only as a slight shoulder, while at 35 and 55 min. it appears as a prominent peak. The effect of increased chlorine partial pressure can be seen in (d) and (h) as the chlorine extends further away from the interface into the oxide, while the interface peak becomes only a shoulder in the profile of the $Cl_2(5\%)/O_2$ film.

These developments can be explained as due to the incorporation of chlorine into the SiO_2 network within the bulk, and into a separate phase at the SiO_2/Si interface. Observations of such an interfacial phase have been described by Monkowski et al. [16]. Support for this hypothesis is found in the $^1H^+$ profiles (Figs. 2a and b) which appear to coincide more with the network chlorine than with the interfacial phase.

The total amount of chlorine in the film (determined by integrating the area under the $^{35}Cl^+$ profiles) in the HCl(6%)/O_2 series is quite linear with oxidation time. This trend, despite the existence of chlorine in two disparate phases, indicates that chlorine incorporation is limited by a reaction with silicon at the interface as suggested by earlier work [17]. This implies that the shape of the network chlorine profile is likely due to the relative increase in Cl_2 partial pressure at the growth interface as the transport of O_2 becomes diffusion limited.

ACKNOWLEDGEMENT

This work was supported in part by the National Science Foundation under grant DMR-7809767 and by the U.S. Department of Energy, Division of Basic Energy Sciences under Contract No. W-7405-eng-26 with Union Carbide Corporation. The authors thank Dr. H. W. Kraner for performing the RBS measurements.

REFERENCES

[1] A. G. Revesz, The Physics of SiO_2 and Its Interfaces, Ed. S. T. Pantelides, Pergamon Press (1978) p. 222.
[2] C. M. Svensson, The Physics of SiO_2 and Its Interfaces, Ed. S. T. Pantelides, Pergamon Press (1978) p. 328.
[3] A. G. Revesz, J. Electrochem. Soc. 126, 122 (1979).
[4] K. H. Beckmann and N. J. Harrick, J. Electrochem. Soc. 118, 614 (1971).
[5] D. A. Leich and T. A. Tombrello, Nucl. Instrum. Meth. 108, 67 (1973).
[6] W. A. Lanford, H. P. Trautvetter, J. F. Ziegler and J. Keller, Appl. Phys. Lett. 28, 566 (1976).
[7] R. E. Benenson, L. C. Feldman and B. G. Bagley, Nucl. Instrum. Meth. 168, 547 (1980).
[8] D. A. Leich, T. A. Tombrello and D. S. Burnett, Earth Planet. Sci. Lett. 19, 305 (1973).
[9] G. J. Clark, C. W. White, D. D. Allred, B. R. Appleton and I. S. T. Tsong, Phys. Chem. Minerals 3, 199 (1978).
[10] J. Monkowski, Solid State Tech. 22, 58 (1979) and Solid State Tech. 22, 113 (1979).
[11] G. J. Clark, C. W. White, D. D. Allred, B. R. Appleton, C. W. Magee and D. E. Carlson, Appl. Phys. Lett. 31, 582 (1977).
[12] C. W. Magee, W. L. Harrington and R. E. Honig, Rev. Sci. Instrum. 49, 477 (1978).
[13] C. W. Magee and W. L. Harrington, Appl. Phys. Lett. 33, 193 (1978).
[14] B. E. Deal, A. Hurrle and M. J. Schulz, J. Electrochem. Soc. 125, 2024 (1978).
[15] Y. J. van der Meulen, C. M. Osburn and J. F. Ziegler, J. Electrochem. Soc. 122, 284 (1975).
[16] J. Monkowski, J. Stach and R. E. Tressler, J. Electrochem. Soc. 126, 1129 (1979).
[17] J. R. Monkowski, J. Stach and R. E. Tressler, Proc. 30th IEEE Electronic Components Conf., San Francisco (April 1980), p. 61.

SURFACE-POTENTIAL DEPENDENCE OF EPR CENTERS AT THE Si/SiO$_2$ INTERFACE

E. H. Poindexter, P. J. Caplan, and J. J. Finnegan
US Army Electronics Technology and Devices Laboratory (ERADCOM)
Fort Monmouth, New Jersey 07703

N. M. Johnson, D. K. Biegelsen, and M. D. Moyer
Xerox Palo Alto Research Center
Palo Alto, California 94304

ABSTRACT

Quantitative correlation of midgap D_{it} with the EPR P_b center ·Si≡Si$_3$ on oxidized (111) silicon suggests a possible contribution to D_{it} by this Si^{III} defect. An external voltage might thus affect the population of the spin center. We have now observed a variation in EPR amplitude of this center with an electric field normal to the interface during the measurement. A bias variable between +15 and -10 volts was applied to a 1000 A, 1 cm^2 aluminum electrode deposited on the oxide of a (111) silicon wafer. The EPR amplitude declined reversibly about 25% between 0 and -4 volts. In situ C-V measurements of surface potential indicate that the responsive component of the P_b center is just below the Fermi level in 50Ωcm p-type silicon. Details of further tests and suggested structural and electrical models will be discussed.

INTRODUCTION

The paramagnetic P_b defect center at the Si/SiO$_2$ interface of oxidized silicon wafers has been observed to be correlated with interface trap density D_{it} over a variety of material and processing variations.[1-3] The P_b spin concentration is quantitatively about equal to midgap D_{it}. The P_b center itself has been tentatively identified as a trivalent silicon defect, mainly ·Si≡Si$_3$, oriented to fit the respective crystallographic directions of non-bonded Si orbitals at the Si/SiO$_2$ interface of (111) or (100) wafers.[2] Because of the time-honored supposition that Si^{III} is a major source of interface states, it is of immediate interest to determine the exact nature of the connection (if any) between P_b and D_{it}.

Several possible situations can be imagined: (a) Si^{III} itself is a chargeable electrical trap, and a direct source of D_{it}; (b) Si^{III} is an intrinsic feature of the P_b center, but D_{it} arises from ancillary charges in the center; or (c) the P_b-D_{it} relation is indirect, or even coincidental. Important evidence on this question should be obtained from the behavior of the EPR signal with a DC potential applied normal to the Si/SiO$_2$ interface. In case (a) above, it might be expected that the shift of the P_b level(s) with respect to the Fermi level at the interface would cause a change in the occupancy of the center, varying between 0, 1, or 2 electrons as the interface goes from accumulation to depletion under different surface potentials. Thus a decisive effect on P_b amplitude would strongly support case (a). A weak effect would support (b), and no effect, (c).

EXPERIMENTAL DETAILS

Sample preparation for voltage-controlled EPR is a delicate compromise between serious difficulties. Leakage-prone large capacitors (1 cm^2) were required for useful EPR signal-to-noise ratio; these were fabricated on single-crystal (111) silicon wafers, p-type (boron doped), 20-50Ωcm. For electrical contact to the silicon substrate with minimal cavity loading, one edge was ion implanted with a high dose of boron prior to oxidation. Wafers were oxidized in dry O_2 at 1000 C for oxide thickness of 1980 Å, then rapidly cooled in the O_2 ambient. This procedure maximizes P_b. Silicon dioxide was selectively removed from the wafer back and from the boron-implanted substrate-contact region prior to vacuum deposition of an aluminum thin film. The aluminum thickness was kept to less than 1000 Å, to minimize cavity loading. Evaporation was performed with an rf induction-heated source to avoid radiation damage in the SiO_2. The Al layer was selectively etched to define the gate electrode and substrate contact. These oxidation and processing procedures have been shown to produce and retain electronic defects which may be considered characteristic of the thermally oxidized silicon surfaces.[3] The sample structure is shown in Fig. 1.

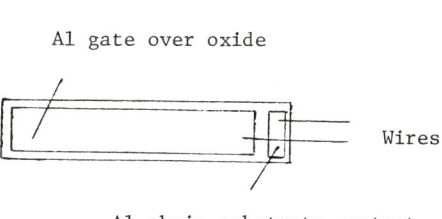

Fig. 1. Wafer sample for study of EPR P_b centers with applied DC bias.

The capacitors were electrically evaluated with conventional current and C-V techniques. Current measurements identified large-area low-leakage devices suitable for EPR. The high-frequency C-V curve is shown in Fig. 2. The C-V results were used to compute the silicon surface potential and intersection of the Fermi energy with the Si/SiO_2 interface.[4]

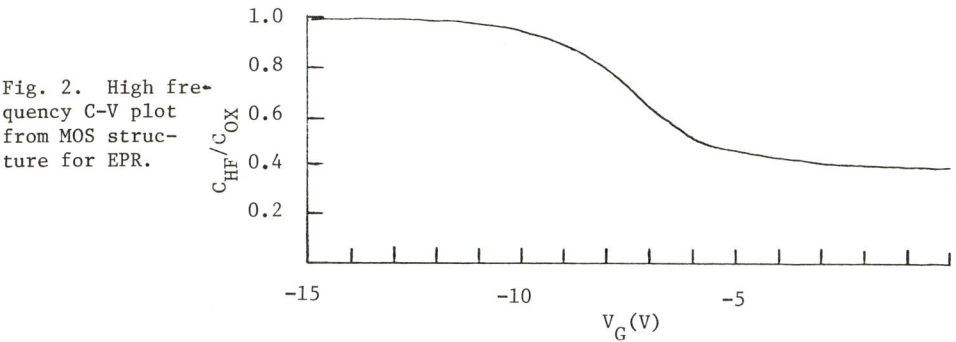

Fig. 2. High frequency C-V plot from MOS structure for EPR.

EPR at 295°K was observed on a Varian x-band spectrometer. The EPR signal was much weaker than our previous studies, which were typically made with a group of 5 samples oxidized on both sides. So it was necessary to use signal averaging, and to minimize spurious resonances. A freshly etched (resonance-free) silicon strip served as a neutral support for the test sample, and as a mount for the biasing wires. Pressure contacts were formed between the wire ends and the aluminum oxide plate and ohmic-contact plate, respectively. The fine wires did not load the cavity.

The sample assembly was suspended in the EPR cavity from the bottom of a quartz tube without an enclosing container. Applied DC potential was varied between -10V and +15 volts on the oxide plate. The leakage current with -10V was typically 1.5 ma, and the voltages indicated were not exceeded to avoid breakdown. The wafer was oriented with H_0 perpendicular to the crystal face, but the anisotropy of the P_b signal on a (111) face was confirmed.

RESULTS AND DISCUSSION

EPR signal amplitudes were measured at three microwave power levels: 2, 10, and 30 mw. The observed amplitude (in arbitrary units) for 10 mw power is plotted in Fig. 3. A large, reproducible change was observed between 0 and -4 volts. This suggests a broad spin level just beneath the zero-bias Fermi level in our 50Ωcm p-silicon.

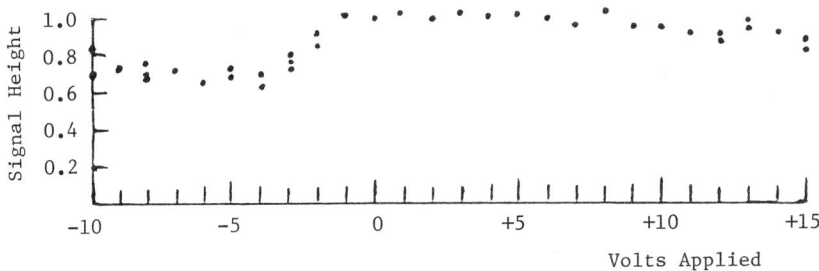

Fig. 3. EPR signal amplitude of P_b centers in MOS structure under applied DC bias.

This result, however, is misleading. When the microwave power was varied, the behavior of P_b was quite different. This indicated EPR saturation. A simple relation for signal height as a function of power P is $S=AP^{1/2}(1+KP)^{-1}$. The saturation parameter is K, and is proportional to the EPR relaxation time T_1. The spin concentration is A. The common procedure for quantitative comparison of spin concentrations is operation at low powers, $KP\ll 1$, so $S\sim P^{1/2}$. In our case, however, in the region of useful signal-to-noise, this is not true; amplitude data must be corrected for saturation to yield the true spin concentration. The saturation parameter K as a function of applied voltage was derived from the data at three power levels, and is shown in Fig. 4. The relaxation parameter K varies by a factor of 3 between 0 and -4 volts bias.

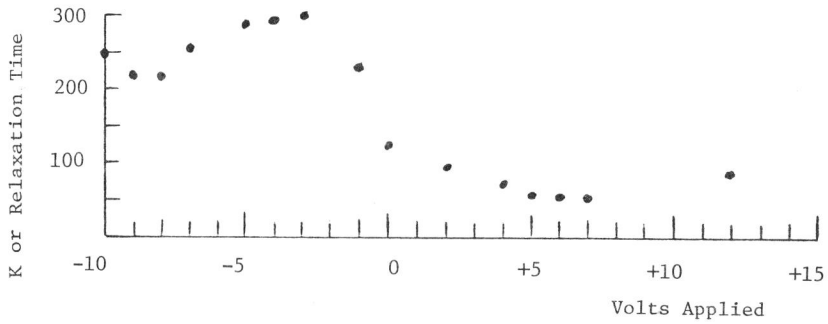

Fig. 4. Relative EPR relaxation time of P_b centers under applied DC bias.

The variation of K with applied voltage is a very significant finding. A readily reversible effect, stronger with positive bias, suggests coupling of P_b spins with mobile electrons attracted to the interface in the inversion regime. Simple spin-spin relaxation requires interacting spins to be within ~10 Å of each other, i.e., much closer than the tunneling limit. The P_b electrons are thus in close communication with the silicon. It cannot be stated whether the relaxation effect arises from mobile conduction electrons, or from electrons trapped near the $\cdot Si \equiv Si_3$ defect. The latter situation is case (b), defined in the introduction: Si^{III} as an intrinsic component of a host site for band-gap charge traps. The apparent close coupling of P_b with silicon electrons has another important meaning. It reinforces other lines of reasoning which place P_b at or very near the interface (i.e., etching experiments; the orientation dependence of P_b, uniquely acceptable at the interface, but very implausible in either the bulk Si or SiO_2).

The occupancy and energy level structure of P_b spin centers are of great interest. Saturation-corrected spin concentration is related to surface potential (E_F-E_V) in Fig. 5. It is emphasized that this is an uncertain result and is presented mainly to illustrate the theory of case (a), above. If Fig. 5 were indeed correct, it can be interpreted with two species of spin centers: (1) a donor level D just above the valence band, normally doubly occupied at zero bias: (2) an amphoteric center with donor level A_1 a little below midgap, and conjugate acceptor level A_2 near the conduction band. These levels are sketched in Fig. 6. At high negative bias A_1 and A_2 are empty; but a tail of D is above the Fermi level and thus singly occupied, yielding an EPR signal. With electron accretion at -4v bias, A_1 crosses the Fermi level, becomes singly occupied, and has an EPR signal. Finally, at high positive bias, A_2 gains an electron, which pairs with the A_1 electron, reducing the EPR signal. It is noteworthy that a similar amphoteric model has been invoked for optically-stimulated EPR in hydrogenated amorphous Si.[5]

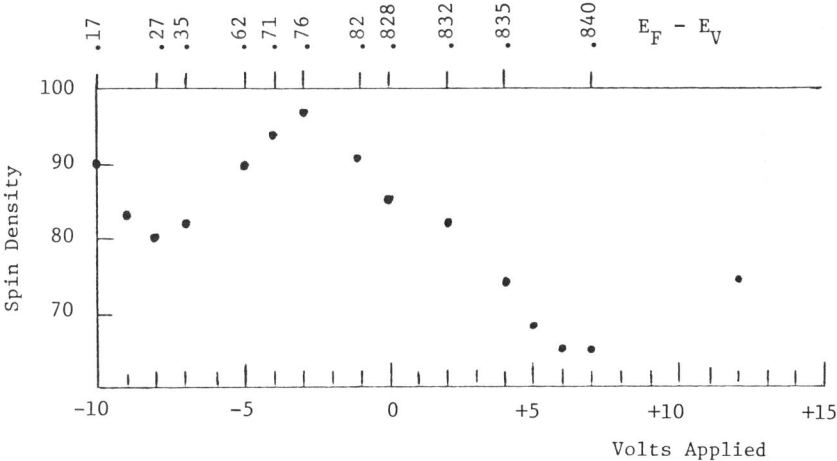

Fig. 5. Apparent P_b spin concentration under applied DC bias.

CONCLUDING REMARKS

Electrical control of a spectroscopic signal from the Si/SiO_2 interface is of considerable interest and reduces the likelihood of case (c), mentioned earlier, for the role of Si^{III} at the interface. The amplitude of the P_b ($\cdot Si \equiv Si_3$) EPR signal in p-doped silicon wafers has been observed to vary as a function of applied voltage.

Most or all of the effect is due to a change in electron relaxation time. The shorter time occurs with positive applied bias, and may reflect interaction of P_b electrons with interface electrons in the inversion regime of p-type silicon. The relaxation effect shows that P_b centers are in close proximity (10 Å) to interface electrons. Not only does this substantiate the interface location of P_b, but it also suggests that Si^{III} may be part of a host site for interface charge traps (case (b), above), even if it does not itself act as a trap.

A number of experimental problems prevent confident deduction of net voltage-dependent spin concentration at this time. Sample burn-out precluded application of high potentials in the interesting regions near band-gap edges. The large relaxation effect over-shadows any spin population change. Nonetheless, this sharp change in electron T_1 indicates that the interface is usably uniform despite leakage currents. Improved technique may allow better measurement of spin concentration and possible discrimination between cases (a) and (b) for the role of $\cdot Si \equiv Si_3$ in regard to interface states.

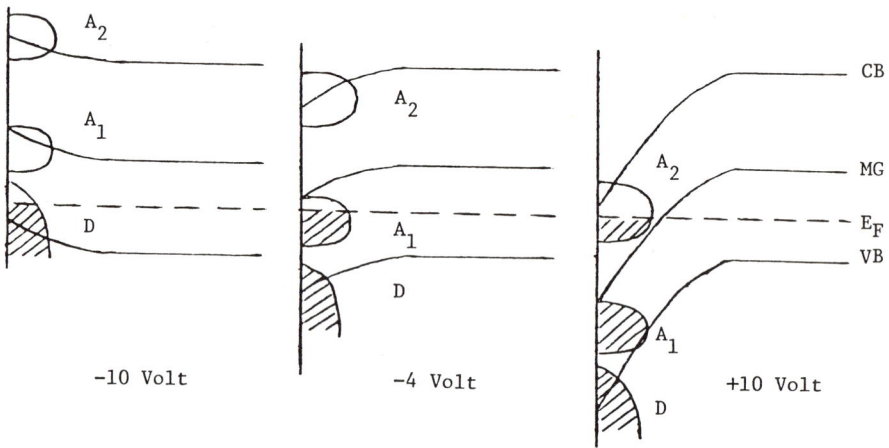

Fig. 6. Hypothetical spin-level structure and occupancy of P_b centers.

ACKNOWLEDGMENT

A portion of this research was supported at Xerox by the US Army Research Office.

REFERENCES

1. P. J. Caplan, E. H. Poindexter, B. E. Deal, and R. R. Razouk, J. Appl. Phys 50, 5847 (1979).
2. P. J. Caplan, E. H. Poindexter, B. E. Deal, and R. R. Razouk (this conference).
3. N. M. Johnson, D. K. Biegelsen, and M. D. Moyer (this conference).
4. A. Goetzberger, E. Klausmann, and M. Schulz, CRC Crit. Rev. 1, 1 (1976).
5. R. A. Street and D. K. Biegelsen, Solid State Commun. 33, 1159 (1980).

ELECTRON BEAM INDUCED DEFECTS AT Si-SiO$_2$ INTERFACE

E. Rosencher, A. Chantre and D. Bois
Centre National d'Etudes des Télécommunications
Centre de Microélectronique de Grenoble
B.P. 42   38420 Meylan, France

ABSTRACT

Electron beam induced defects at Si-SiO$_2$ interface are studied using high frequency C(V), conductance and DLTS measurements. A set of levels near E_c-80 meV with a very low electron capture cross section (10^{-19} cm^2) is found to emerge from a low density continuum. Discrepancies with other works are discussed in terms of surface potential fluctuations.

INTRODUCTION

The investigation of electron beam effects on MOS structures is of physical interest, since such radiations are likely to cause bonds at the Si-SiO$_2$ interface to break. It is also of technological interest due to the growing importance of energetic environment in the integrated circuit technology (microlithography, plasma deposition, metallisation,...). A lot of studies have been devoted to such effects in the literature (1-3), but using unreliable capacitance measurements ; moreover, only high irradiation doses were used, probably due to the lack of sensitivity of these measurements. We report here an experimental investigation of electron beam induced defects at the Si-SiO$_2$ interface using Deep Level Transient Spectroscopy (DLTS) and conductance measurements ; low doses were used to avoid complicated phenomena (defect interactions, aggregations,...), and allow first stage introduction kinetics to be analyzed.

EXPERIMENTAL DETAILS

State-of-the-art n-type MOS capacitors (N_D = 1.5x10^{15} cm^{-3}, (100) surface orientation) have been used in this work. 1500 Å thermal oxides were grown in dry O$_2$ at 1000 °C for 90 min with 2 % HCl, followed by a 30 min *in situ* anneal. 0.1 mm^2 Al dots were then evaporated, and the structures annealed at 400 °C in N$_2$. The interface state density Nss was controlled to be in the 10^9 eV^{-1} cm^{-2}.

25 keV electron beam irradiations were performed in a Scanning Electron Microscope, with both MOS electrodes grounded. Each sample received increasing doses, from 10^{-8} C cm^{-2} up to 10^{-5} C cm^{-2}. The beam was kept unfocused during the exposures, and scanned troughout the structure, in order to achieve homogeneous irradiations. 30 min annealing at 50 °C was performed to stabilize the Qss charge before interface state characterization. This first stage annealing was controlled by *in situ* C(V) measurements.

After each exposure, the interface state density was analyzed using three methods : high frequency C(V) measurements (Terman), conductance vs voltage G(V) and frequency G(ω)/ω (4), and transient capacitance DLTS (5,6). These experiments were performed on a fully computer-controlled setup, using a lock in system where the phase is monitored by on line computer ; this set up allows 10^9 and 10^{10} states eV^{-1} cm^{-2} to be detected through G(V) and DLTS measurements respectively. DLTS spectra were recorded at a constant equilibrium capacitance, i.e. the bias voltage was adjusted in order to keep the steady-state capacitance constant throughout the temperature scan. However, the release of trapped electrons was actually measured through capacitance transients (maximum amplitude less than 5 % of the total capacitance). The Fermi level was thus kept over $E_c-0.4$ eV at the $Si-SiO_2$ interface, i.e. in the depletion regime.

RESULTS

Figure 1 shows the $N_{ss}(E)$ spectra at increasing electron doses, as deduced from Terman method, i.e. comparison between experimental and theoretical high frequency C(V) curves (4). The effect of the irradiation on the interface state density is not clear on this figure, which even shows a decrease of N_{ss} at low doses. For the highest investigated doses, these data are in rough agreement with previously published works (2). However, such C(V) measurements lead to erroneous results, because of the surface potential (ψ_s) fluctuations (7). Moreover, these spectra were found to be highly frequency dependent, which is known to introduce virtual features in high frequency measurements. We have used conductance measurements to demonstrate these points (8).

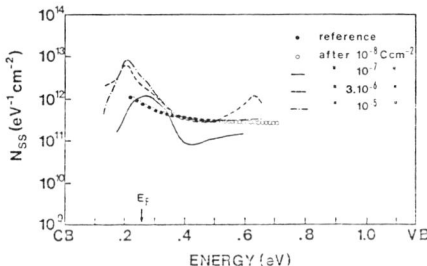
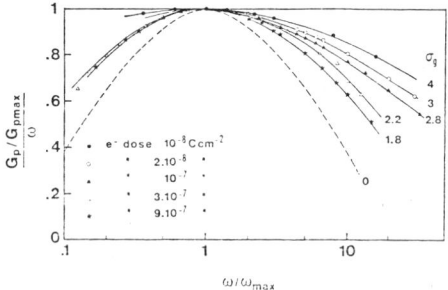

Fig. 1. Interface state distribution obtained by high frequency C(V) technique at increasing electron doses

Fig. 2. Normalized $G(\omega)/\omega$ curves at different irradiation doses for the same sample (T=300 °K, f=.1-200 kHz)

G(V) curves recorded at room temperature and 10 kHz indicate N_{ss} values of 3×10^9 eV^{-1} cm^{-2} near midgap for the non-irradiated samples. This result is in agreement with DLTS measurents, which demonstrate that the interface state density before irradiation lies below 10^{10} eV^{-1} cm^{-2}. The large discrepancy between these low values and the 5×10^{11} eV^{-1} cm^{-2} given by C(V) measurements can be easily understood if the presence of large surface potential fluctuations is taken into account. Figure 2 shows the $G(\omega)/\omega$ curves at different irradiation doses for the same sample. They are compared to the theoretical curves calculated for a continuous density-of-states with various surface potential fluctuations (4). It can be seen that large fluctuations are present before irradiation : the experimental curve can be fitted assuming a variance σ_g = 4 kT (T=300 K). Such fluctuations are known to affect C(V) curves in a similar manner as interface states, leading to apparent N_{ss} values much larger than the real ones (7,9).

As the electron dose increases, the ψ_s fluctuations decrease significantly (a factor of two), as demonstrated in Fig. 2. This is surprising since fixed charges homogeneously deposited by the beam should not lead to a decrease in the absolute value of the ψ_s fluctuations. Moreover, as the density of interface states probed by the conductance technique (within 100 meV above midgap) is found to be in the few 10^{10} eV^{-1} cm^{-2} range, the corresponding increase of interface state capacitance C_{ss}

cannot be responsible for a noticeable screening effect in the potential fluctuations through the relation (4)

$$\sigma_g = q\, kT\, \sigma_q/(C_{ox} + C_{si} + C_{ss}) \qquad (1)$$

where σ_q is the variance of the Gaussian oxyde charge distribution, C_{ox} and C_{si} the oxyde and silicon capacitance respectively. One must then assume that such fixed charges are more likely to be created in the regions where less fixed charges preexisted in the material. This can only be understood if the defects responsible for those fixed charges interact in such a way that each introduces some kind of forbidden region around it. Whatever the real physical mechanism leading to the decrease in the ψ_s fluctuations, this decrease accounts for the initial reduction of the Nss values observed in Fig. 1.

The evolution of the DLTS spectrum as a function of irradiation dose is shown in Fig. 3. As mentioned previously, no noticeable signal is observed on the non irradiated sample. The effects of the electron beam appear as two main features on the DLTS curves : a low temperature peak A, and a broad response B extending up to room temperature. A large temperature shift vs delay times is found for peak A, as shown in Fig. 4. The behaviour of B is different : while no change is observed up to 280 K, the room temperature peak decreases at increasing delay times, with almost no shift in its position. The large temperature shift of peak A can be understood assuming a small electron capture cross section σ_n for the corresponding levels. Indeed, σ_n can be determined by measuring this T shift for various DLTS signal values (10,11). σ_n appears to remain constant in the 10^{-19} cm^2 range for the levels giving a DLTS response between 100 K and 150 K. This low value was confirmed by a direct determination based on the analysis of the change in DLTS peak height as a function of injecting pulse width. Further experiments are under progress, which are intended to clarify the energy and temperature dependence of this capture cross section. It is then possible to deduce an energy range for peak A : one finds a set of levels located around E_c-80 meV. The enlargement of the DLTS peak on the high temperature side implies an extention of this shallow density of states over about 50 meV on the high energy side. Figure 3 shows that the maximum of peak A shifts towards lower temperature at increasing electron doses, i.e. the levels tend to move into the conduction band. Such behaviour is expected for a set of interacting defects.

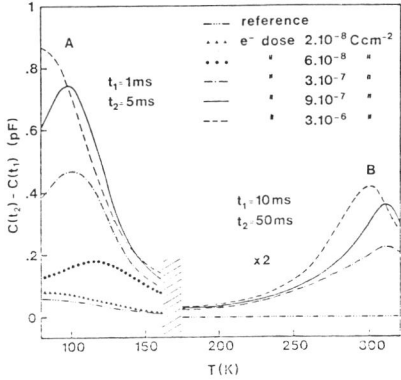

Fig. 3. DLTS spectra obtained at a constant equilibrium capacitance for increasing electron doses using two sets of delay times (t_1, t_2).

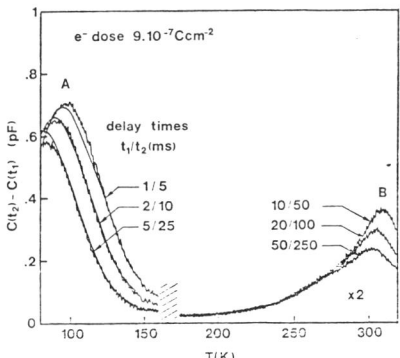

Fig. 4. DLTS spectra at a fixed irradiation dose (9x10^{-7} C cm^{-2}) for different delay times.

The density of states corresponding to the broad response B can be easily determined by simply dividing the DLTS signal by kT and converting the temperature scale into an energy scale (6), which is valid only for a slowly varying density of states distribution. Such a treatment leads to a low (10^{10} eV^{-1} cm^{-2}) almost constant density down to Ec-300 meV, slightly increasing towards midgap. The decrease of the DLTS signal at high temperature is likely to be due to the cut off associated with the window rate crossing down the Fermi level at the interface. Figure 5 shows the introduction rate of the set of shallow levels (DLTS peak height vs dose). It turns out that the kinetics is well accounted for by a logarithmic function. The kinetics of introduction of B was found to roughly follow a (dose)$^{0.7}$ law.

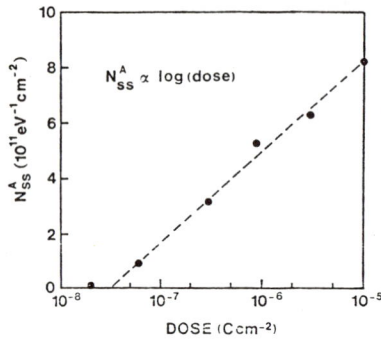

Fig. 5. Introduction rate of electron beam induced shallow levels A. The interface state density at the maximum of the DLTS peak is plotted vs dose.

DISCUSSION - CONCLUSION

DLTS measurements show that low energy electron irradiations create two kinds of defects at the Si-SiO$_2$ interface (in the upper half gap). The first which are greatly dominant, introduce a rather sharp set of levels below the conduction band, with a very low electron capture cross section. This low value can be interpreted assuming either a repulsive centre, or a centre with a large lattice relaxation, i.e. with an elastic barrier against electron capture. Further measurements have to be carried out before this point can be clarified. The other type of defects introduce a very broad continuum extending over the whole upper half gap, with some increase towards midgap.

These results are very different from those previously published in the literature : a broad band was generally found near E_c - 300 meV (2). This disagreement can be partly attributed to the presence of surface potential fluctuations which have been analyzed in this work, but were generally forgotten in others. These ψ_s fluctuations lead to erroneous interface state densities as obtained through either high frequency or quasi-static (9) C(V) measurements. In addition, due to the low sensitivity of these measurements, high irradiation doses were commonly used in the past (above 10^{-5} C cm^{-2}). Such doses lead to interface state densities large enough to produce secondary reactions between the defects. Indeed, even for the low doses investigated in the present work (from 10^{-8} to 10^{-5} cm^{-2}), the DLTS peak was found to shift significantly with the dose.

The introduction rate of the shallow levels has been found to be logarithmic. Such kinetics are generally characteristic of non mobile defects interacting with each other : one must assume that the defects are non mobile after their formation, and that each one is surrounded by some "forbidden" volume where no other defect can be created (14). This interpretation is in fair agreement with our experimental observation of the surface potential fluctuations decreasing at increasing irradiation doses, as described in the text. These shallow levels are likely to be associated with simple primary defects introduced by the electron beam : one possible candidate would be a broken bond at the Si-SiO$_2$ interface (12). Finally, as the samples were zero biased during the irradiations, it is likely that only excitonic transport in SiO_2 followed by a dissociation at the $Si-SiO_2$ interface could be responsible for the creation of interface states (3,13). Further experiments are in progress to investigate the effects of polarisation during irradiation on the nature and creation rate of these defects.

ACKNOWLEDGEMENT

We are indebted to Prf. J.C. Pfister and Dr A. Nouailhat for fruitful discussions.

REFERENCES

1. R. Hezel, Solid St. Electron. 22, 735 (1979)
2. G. A. Scoggan and T. P. Ma, J. A. P. 48, 294 (1977)
3. P. S. Winokur and M. M. Sokoloski, A. P. L. 28, 627 (1976)
4. A. Goetzberger, E. Klausmann and M. Schulz, CRC Critical Review, 1 (Jan. 1976), and references listed therein
5. D. V. Lang, J. A. P. 45, 3023 (1974)
6. N. M. Johnson, D. J. Bartelink and M. Schulz, Proc. of Int. Conf. of the Physics of SiO_2 and its Interfaces, Yorktown Heights (1978)
7. R. Castagne et A. Vapaille, C. R. Acad. Sc. Paris 270, série B, p. 1347, 1970
8. M. R. Boudry, A. P. L. 22, 530 (1973)
9. V. A. Gergel, Sov. Phys. Semicond. 13, 385 (1979)
10. M. Schulz and N. M. Johnson, Solid State Comm. 25, 481 (1978)
11. E. Klausmann, Int. Phys. Conf. Ser. n° 50, p. 97 (1979)
12. R. B. Laughlin, J. D. Joannopoulos and D. J. Chadi, Proc. of Int. Conf. of the Physics of SiO_2 and its Interfaces, Yorktown Heights (1978)
13. Z. A. Weinberg and G. W. Rubloff, RADC Interim Report - TR - 78 - 115, May 1978
14. E. Mercier, G. Guillot and A. Nouailhat, Phys. Rev. B 20, 1678 (1979)

CHAPTER VIII
DEVICE PHYSICS

Anomalous gate current on avalanche hot electron injection
in MOS structures

Kikuo Yamabe* and Yoshio Miura
Cooperative Laboratory VLSI Technology Research Association,
Takatsuku, Kawasaki, 213, Japan

ABSTRACT

Hot electron injection current in MOS structures is studied by avalanche technique. The gate current markedly increases with injected electron density below the avalanche threshold gate voltage, in spite of an increase in electron trapping in SiO_2. This anomalous gate current can be explained by assuming that hot electrons are able to tunnel into the SiO_2 conduction band through defect centers generated by hot electron injection.

INTRODUCTION

Hot electron injection into thermal grown gate oxide has been investigated in recent years because of its important effects on instabilities of MOS devices, such as flatband voltage shift and transconductance degradation [1]. In order to study the instabilities due to injected electrons, the avalanche injection method is widely used because of the simplicity in the MOS capacitor structures required. Some experimental results observed by this technique are electron trapping in gate oxide and interface state generation due to hot electron collision with Si-SiO_2 interface [2]. Nicollian et al. [2] reported that flatband voltage shift due to electron trapping equals DC bias applied by feedback circuit to MOS capacitor to keep injection current constant. These results suggest that the injection current monotonically decreases as electron trapping in the oxide progresses. Results obtained by the present authors do not support the model proposed by Nicollian et al.: the gate current increased with injected electron density, N_{inj}, especially in the near avalanche region, which will be defined in a later section. This anomalous gate current can be explained by assuming that hot electrons can tunnel into the SiO_2 through intermediate centers of trivalent silicon in the oxide damaged by avalanche injection.

SAMPLE PREPARATION

Sample were prepared by oxidizing 0.3 Ωcm (100) oriented P-type silicon wafers in dry oxygen at 1000°C to form SiO_2 layers with a thickness of about 790 Å. The oxidation was followed by an *in situ* annealing in nitrogen at 1000°C for 15 minutes.

---
* Present address: Integrated Circuit Laboratory, Toshiba Research and Development center; 72, Horikawacho, Saiwai-ku, Kawasaki-city, Kanagawa, 210, Japan

Al was evaporated onto the SiO$_2$ layer from an electron gun source. After Al deposition, the wafers were annealed in nitrogen at 450°C for 20 minutes to eliminate the radiation damage introduced during electron gun deposition of Al. The devices were attached to TO-5 headers.

EXPERIMENTAL TECHNIQUE

The complete capacitance - voltage (C-V) characteristics of the virgin MOS capacitor are measured at high (1 MHz) and low (17 Hz) frequencies. Hot electrons are injected from P-type silicon substrate into SiO$_2$ layer by the avalanche injection technique using a constant voltage pulse. Pulse frequency, f, and width, T_w, are 200 kHz and 2 μsec, respectively. During hot electron injection, injection gate current, I_g, and the number of injected electrons, Ne, that is, $Ne = \frac{1}{q}\int_0^t I_g \, dt$, are measured. The pulse gate voltage is periodically interrupted to automatically measure the flatband voltage shift, ΔV_{FB0}, with high frequency in order to monitor the trapped electrons built up in the SiO$_2$ and the generated interface states. After a given number of hot electrons are injected, the capacitor is reconnected to the C-V measuring

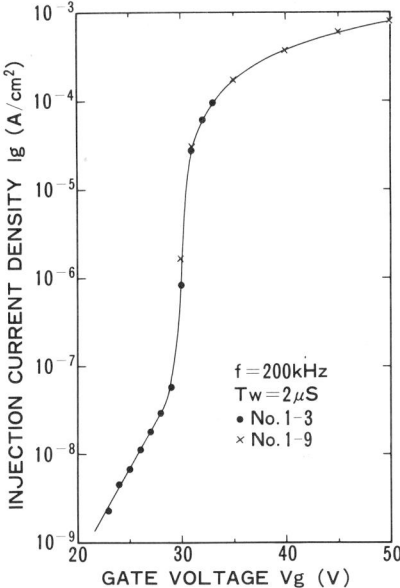

Fig. 1. I_g-V_g characteristics upon hot electron injection in the virgin MOS capacitors

apparatus and complete C-V curves at high and low frequencies are determined. From these C-V curves, the generated interface state density, ΔN_{SS}, is calculated. The flatband voltage shift due to the generated interface stage, ΔV, is given by, $\Delta V = q \Delta N_{SS}/C_{ox}$, where C_{ox} is the oxide layer capacitance per unit area and q is the electron charge.

EXPERIMENTAL RESULTS

It is well known that avalanche multiplication can be induced by applying a voltage pulse that tends to invert the Si surface in an MOS capacitor. The electrons thus generated in the depletion region gain energy from the field as they drift towards the interface. Most of these electrons contribute to forming the surface inversion layer, while a small fraction with sufficient energy to surmount the interface barrier can be injected into the SiO$_2$ layer. Most of the injected electrons contribute to the gate current, and only a small portion are trapped causing a shift in the flatband voltage of MOS capacitors.

Figure 1 shows a typical plot of averaged injection gate current, I_g, as a function of the gate voltage in virgin MOS capacitor. Note that there is a rapid enhancement of gate current around V_g = 30 V. At V_g = 30 V, the Si surface field is about 0.58 MV/cm. This value is consistent with that of the avalanche breakdown threshold field in silicon. Therefore, this gate current enhancement is due to the rapid multiplication of the total number of electrons arriving at the interface. There is also a current tail extending to the gate voltage well below 30 V. This tail can be explained by the fact that the avalanche multiplication of electrons in Si is not abrupt. This tail is called "near avalanche region" in the following. For an injection gate voltage above 30 V, the gate current saturates because most of the

electrons generated by avalanche multiplication form the surface inversion layer, which weakens the electric field in the depletion region. The avalanche injection gate current is greatly affected by the flatband voltage shift due to electron trapping in the oxide as described by Nicollian et al. (2).
After the injected hot electrons have generated interface states, the flatband voltage shift due to the trapped electrons in the gate oxide, ΔV_{FB1}, is given by

$$\Delta V_{FB1} = \Delta V_{FB0} + |\Delta V|, \quad (1)$$

where ΔV_{FB0} is the net flatband voltage shift measured with high frequency and ΔV is the flatband voltage shift due to the generated interface states.

Figure 2 shows ΔV-N_{inj} characteristics with the injection pulse condition as a parameter. This figure shows that the generated interface state density increases not only with N_{inj} but also with V_g.

Figure 3 is obtained by combining the ΔV-N_{inj} characteristics shown in Fig. 2 with the experimentally obtained ΔV_{FB0}-N_{inj} characteristics, based on Eq. (1). This figure shows that injected electron trapping efficiency increases with pulse height. Such a tendency cannot be explained by an oxide field dependence for a capture cross section in an electron trapping center in the oxide (3). It is assumed that the hot electron injection generates new electron trapping centers in the oxide as a result of damage induced by hot electron injection. The model proposed by Nicollian et al. (2) gives an indication that I_g -

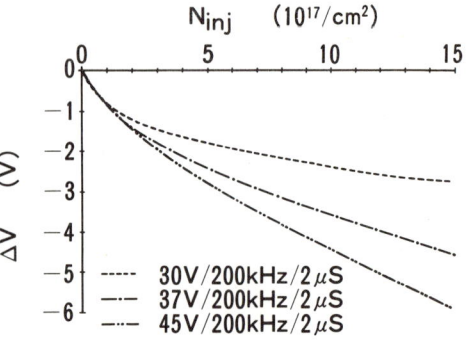

Fig. 2. ΔV-N_{inj} characteristics with V_g as parameter

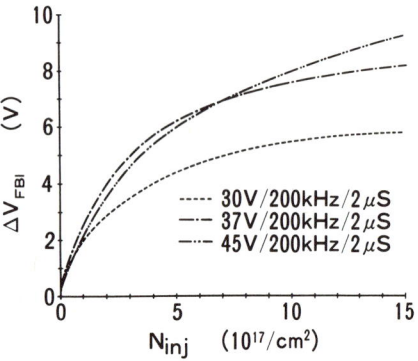

Fig. 3. ΔV_{FB1}-N_{inj} characteristics with V_g as parameter

(V_g - ΔV_{FB1}) characteristics become independent of the cumulative injected electron density. However, the gate current increases with N_{inj}, especially in the near avalanche region as shown in Fig. 4. This increasing current during electron injection is referred to as anomalous gate current.

DISCUSSION

There are two possible reasons which can explain the anomalous gate current in the near avalanche region.

First is energy barrier lowering at the interface. Even if V_g is modified by ΔV_{FB0} instead of ΔV_{FB1}, anomalous gate current shown in Fig. 4 cannot be explained; the modification includes both the electron trapping effect and the interface state generation effect from the injected MOS capacitors, so that the potential distribution at the Si surface of the injected MOS capacitor is equivalent to that in the virgin MOS capacitor. Then, positive charge at the interface or in the oxide near the interface weakens the oxide field in the neighborhood of the interface rather than strengthens it, and cannot lower the interface barrier.

Second is an increase in electrons arriving at the interface. Considering the enhancement of the etching rate of an Si surface layer after avalanche hot electron injection (4), it is possible to consider that electron multiplication factor, M_n, increases in the surface depletion region. The following experiment was performed in order to estimate this increase using an n^+-p junction. The stress bias with 200 kHz pulse frequency, 2 μsec pulse width and a reverse voltage which is slightly higher than the PN junction breakdown voltage, is applied to the junction. The average reverse current caused by the application of stress bias increases at most six times as much as that in virgin PN junction. This increase is not enough to explain the anomalous gate current shown in Fig. 4.

Now, a new mechanism is proposed to explain the anomalous gate current as a consequence of negation of the above two possibilities.

The extent of hot electron injection damages is shown by ΔV. Figure 2 shows that ΔV not only increases with N_{inj} but also depends on V_g. These results are interpretted as follows.

SiH bond at the interface is broken by hot electron injection, resulting in Si˙ + H. The trivalent Si atom at the Si surface, Si˙, bonded to three neighbor Si atoms, is generally considered to form interface state. By analogy with the interface state generation mechanism by hot electron injection, it is probable to consider that the SiH bond in the oxide near the interface is broken by hot electron injection resulting in Si˙. Laughlin et al. (5) theoretically found that the trivalent Si in the oxide bonded to three oxygens gave rise to a state between the Si and SiO_2 conduction bands minimums. In addition, SiH concentration in the oxide increases sharply towards the interface (6). This would help confirm the concept that, in the oxide, a large number of Si˙ are generated in the vicinity of the

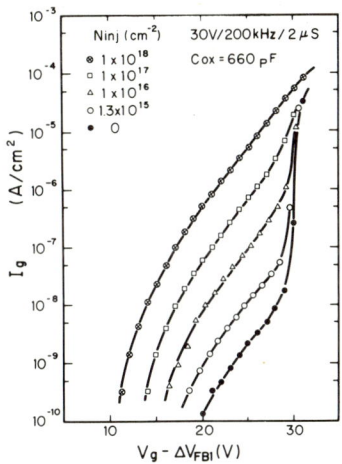

Fig. 4. Enhancements of gate current in the near avalanche region during avalanche hot electron injection

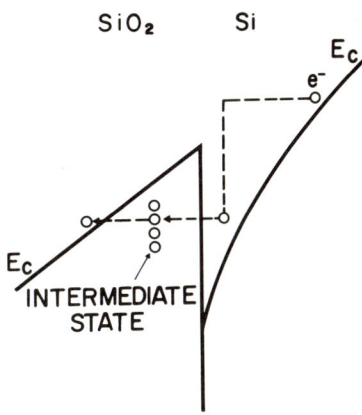

Fig. 5. Anomalous hot electron injection model in the injected MOS capacitor by tunnel process through the intermediate centers

interface by avalanche injection, and hot electrons can tunnel into the conduction band of SiO_2 through these intermediate Si˙ levels as shown in Fig. 5, which explain the anomalous gate current on hot electron injection.

CONCLUSION

Hot electron injection current was studied by avalanche technique in MOS structures. The gate current markedly increase in the near avalanche region with cumulative injected electron density, in spite of an increase of trapped electron density in the oxide. By analogy with the interface state generation mechanism that injected hot electron generates interface state due to the formation of trivalent silicon Si˙ at the interface, it is considered that trivalent Si may also be generated in the oxide by breaking the SiH bond. Therefore, anomalous gate current is explained by assuming that hot electrons can tunnel into the SiO_2 conduction band through intermediate centers of Si˙ in SiO_2 damaged by avalanche injection.

ACKNOWLEDGEMENT

The authors wish to thank Dr. Y. Tarui for his helpful suggestions and K. Taniguchi for his critical reading of this manscript.

REFERENCES

(1) T. H. Ning, Hot electron emission from silicon into silicon dioxide, Solid-St. Electron. 21, 273 (1978)
(2) D. R. Young, J. Appl. Phys., October 1979.
   E. H. Nicollian, C. N. Berglund, P. F. Schmidt, and J. M. Andrews, Electrochemical charging of thermal SiO_2 films by injected electron currents, J. Appl. Phys. 42, 5654 (1971)
(3) K. Yamabe and Y. Miura, submitted for publication in J. Appl. Phys.
(4) Y. Miura and K. Yamabe, submitted for publication in Electron. Lett.
(5) R. B. Laughlin, J. Joannopoulos, and D. J. Chadi, The Physics of SiO_2 and its Interfaces, edited by S. T. Pantelides, Pergamon Press, New York, Chap. VI, pp321-327 (1978)
(6) A. G. Revesz, The role of hydrogen in SiO_2 films on silicon, J. Electrochem. Soc. 126. 122 (1979)

NOISE FROM MOS TRANSISTORS AT WEAK AND UNIFORM INVERSION

A.A. WALMA

Centre d'Etudes d'Electronique des Solides, associé au C.N.R.S.
Université des Sciences et Techniques du Languedoc
34060 MONTPELLIER CEDEX, FRANCE

It has been suggested that negative U-centres exist near the Si-SiO$_2$ interface and that they provide the slow states from Si-Technology[1]. In equilibrium the bonds will form and break with a spectrum of relaxation times. This led to the idea of noise measurements for a spectroscopy of these bond-states[2]. Surface related methods (C-V plots, conductance, low temperature and low frequency techniques) are not able to detect paired states. Although a wealth of data exists for 1/f noise in MOS transistors, they usually have been carried out at strong inversion or above pinch-off. This implies non-linearity, hot electrons and non-ohmic behaviour. The proposed theories cannot explain coherently the noise as a function of gate--voltage[3]. Discrepancies arise with p and n layers, particularly with respect to temperature. In order to reconcile conflicting experimental data, a transition region with a strong electron-phonon coupling appears to be an interesting alternative[3]. In this paper the unwanted effects from a (badly understood) drain region and drain bias are avoided.

An on-chip double transistor configuration has been used with a high source resistance. For resistances of the order of 10^{10} Ω, uniform band-bending can be obtained up to canal resitances in the mΩ range. The uniformity is important for the final results. Experimentally, the high resistances pose a problem. Not only because of the amplifiers but also with respect to a steady signal. The measurements were carried out by conventional means without using a FFT.

A typical noise result at a fixed frequency of 80 Hz is given in Fig. 1 for a source resistance of 10^7 Ω. The sharp increase at about 8 V drain-bias is caused by T_1 going into saturation. This is confirmed by the I-V characteristic. For voltages smaller than 8 V the output senses T_2 (T_1 is now strongly inverted). This gives the well known pattern for a MOS in pinch-off.

The spectra near this sharp increase are plotted in Fig. 2. A real uniform bending is expected up to about 100 kΩ for the channel resistance (Details will be presented elsewhere). These results are a little unexpected, see discussion. Their magnitude follows the canal resistance. This is confirmed by Fig. 1. At for instance 7.95 V the noise exceeds the plateau (at 80 Hz) by an order of magnitude. The roll-off is caused by a system capacity.

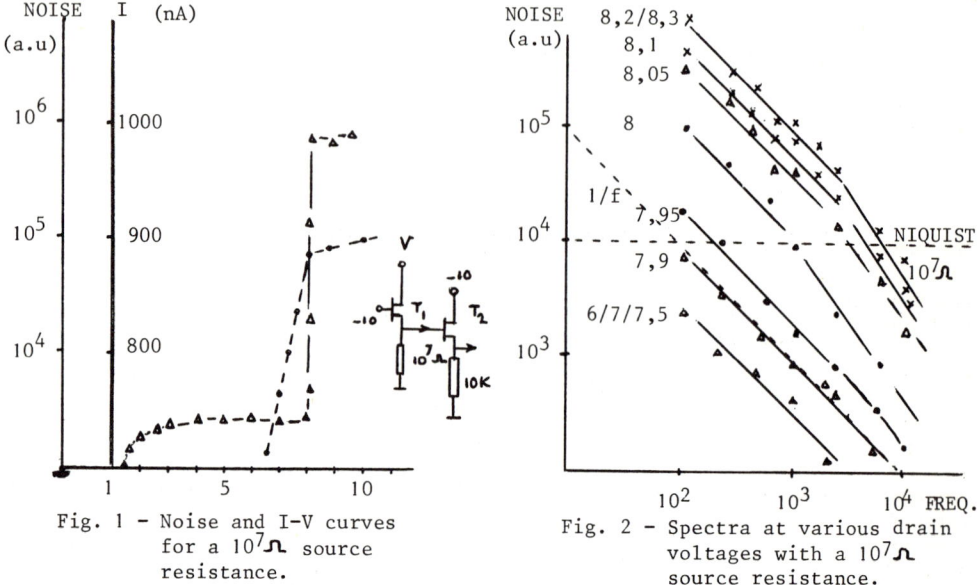

Fig. 1 - Noise and I-V curves for a $10^7 \Omega$ source resistance.

Fig. 2 - Spectra at various drain voltages with a $10^7 \Omega$ source resistance.

For increasing source resistances, a similar behaviour is observed up to about $10^8 \Omega$. After that a drastic change occurs both in the noise versus drain voltage and the spectrum. A typical result for a reverse biased diode in the source ($\sim 10^{10} \Omega$) is shown in Fig. 3. This pattern has been observed in various devices. The peak occurs shortly before pinch-off. It will be shown elsewhere that this distance increases for higher frequences.

Some typical results for the spectra near the peak are given in Fig. 4. The origins of the "G-R bump" in the tail of the spectra is not clear for the moment.

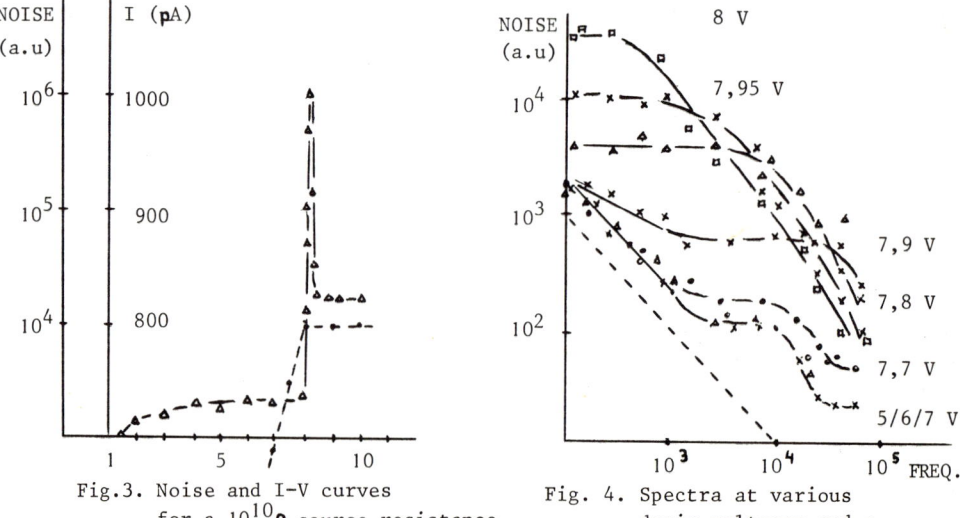

Fig. 3. Noise and I-V curves for a $10^{10} \Omega$ source resistance.

Fig. 4. Spectra at various drain voltages and a source resistance of $10^{10} \Omega$.

It is of minor interest for our discussion however. At low voltages, T_1 is strongly inverted and the spectra are determined by T_2. This is why they saturate at about 7 V (Plateau in Fig. 3). The abrupt change into a G-R type of spectrum is thought to be related to the occurrence of the peak, see discussion. The white levels at the higher frequencies provide us with no additional information since they deal with T_2 and the output impedance.

DISCUSSION

The noise from a uniformly inverted channel is unexpectedly high. It can be shown that with no current passing the noise is white and violates in certain devices the Fluctuation Dissipation theorem. This is therefore remarkable since it implies that the Einstein relationship does not hold for a system relatively near equilibrium[4]. These are strong concepts in physics, which suggests that it is dealt with a rather special system. The non-Gaussian and non-Markovian behaviour of an electron in a random potential then becomes an attractive alternative. This is not unreasonable since it has been shown[5] by means of drifting Na^+ ions to the interface that the 1/f noise increases drastically for a hopping process.

The noise magnitude from Fig. 2 deserves attention. At 7.9 V the channel is still relatively strong inverted. At 100 Hz the noise level has a current spectral density of $1.6 \cdot 10^{-27}$ A^2 Hz^{-1}. Comparing this with Hooge's formula yields for $\alpha/N = 1.6 \cdot 10^{-13}$. For moderately weak inverted channels N is of the order of 10^{15}. This yields an unusual value for α (Bulk Si yields $\alpha \sim 10^{-3}$).

If one accepts the usual trapping explanations for 1/f noise, than the increase in the density of states deeper in the gap (as is suggested by the spectra of Fig. 2) is not in accord with the results from surface related methods (C-V, conductance etc). It is consistent however with the pairing idea (deep donor states).

For very weak (uniform) inversion, the peak poses a problem in Fig. 4. Initially[6] it was attributed to a localized bandtail and a mobility shoulder. Numerous spectra suggested a more straight forward explanation. It will be shown elsewhere that a parasitic capacity could give rise to a peaking behaviour. There are some problems with this kind of explanation however. (i) The turn-over frequency for the 7.95 V spectrum in Fig. 4 for example, yields a capacity of about 10 pf, which is rather high (ii) The plateau's tend to disagree with first oder measurements, thus violating the F-D theorem (iii) The peak tends to dissappear with substrate bias (iv) There still remains the question how 1/f noise can become white noise (this should have a physical reason). In other words, even if one accepts the parasitic capacity idea, the amorphous interpretation still remains an attractive one.

If however the spectra reflect a physical mechanism, then Fig. 4 suggests that each carrier density has its own relaxation time.

It can be concluded that in both cases the statistical and physical aspects of a bond breaking idea are promising. It complicates a theoretical noise analysis considerably however[3].

REFERENCES

1. N.F. Mott, Adv. Phys., 26, 363, 1977
2. P.W. Anderson, Journal de Physique, 37, 339, 1976
3. A.A. Walma, 1/f Symposium, Orlando, USA, 1980
4. A.A. Walma, Sem. Dyn. Theory Phys. Syst., Udine, Italy, 1979
5. R.F. Voss, 1/f Symposium, Tokyo, Japan, 1977
6. A.A. Walma, A. Touboul, ESSDERC'78, Montpellier, France, 1978

POLYMERIZED LANGMUIR FILM MIS STRUCTURES

K. K. Kan, M. C. Petty and G. G. Roberts
Department of Applied Physics and Electronics, University
of Durham, England

ABSTRACT

MIS structures have been prepared using Langmuir films of suitably substituted diacetylene monomer. These monolayers are then polymerized when they are on the semiconductor substrates. The resulting films are of high quality, show little dispersion and have high breakdown strengths. Admittance data are described for MIS diodes based on p-type InP and n-type GaP.

INTRODUCTION

Biophysicists and biochemists have long been aware of the usefulness of the Langmuir trough technique to form molecular species into accurately defined super molecular structures. However, it is only recently that the real device potential for multilayer films based on ordered Langmuir monomolecular layers has been convincingly demonstrated. For example, the thinness and perfection of films of conventional Langmuir-Blodgett materials coupled with their excellent insulating qualities have resulted in interesting MIS transistor (1) and photovoltaic structures (2). Other applications have been reported in the areas of integrated optics and electron beam lithography. The scope of the technique has also been considerably widened by the demonstration (3) that aromatic molecules with their extended π-electron orbitals can be handled on the Langmuir trough.

The particular dielectrics used in previous investigations have quite low melting points and therefore are unlikely to form the basis of practical devices. We have therefore turned our attention to more stable materials and have, in particular, studied suitably substituted diacetylenes. Enkelmann et al (4) have discussed the polymerization of such molecules using u.v. light when chains containing conjugated double and triple bonds are formed. The inset to Fig. 1 illustrates both the monomeric and the polymeric forms of the material (abbreviated 12-8 diacetylene) whose properties are described in this paper. It will be seen that the end groups of the monomer become the side groups in the polymer and therefore play a vital role in packing the monomeric units into a reactive configuration. We report for the first time the a.c. and d.c. properties of these polymerized diacetylene multilayers and discuss their performance as passivating layers in MIS structures based on InP and GaP.

PREPARATION OF DIACETYLENE LANGMUIR FILM STRUCTURES

The techniques required to obtain high quality multilayers have been described else where (1) and therefore only a few of the essential details are mentioned here: The substituted diacetylene *monomer* (synthesized by Dr. D. Bloor, Q.M.C., London) was dissolved in chloroform and spread onto the surface of water obtained from a Millipored purification system. This aqueous subphase contained 2.5×10^{-4} M $CdCl_2$ and its pH at $17°C$ was adjusted to be in the range 6.2 to 6.4 (rather higher than for conventional fatty acid films). The surface area of the monolayer was varied by using a motor driven glass fibre barrier of constant perimeter; optimum

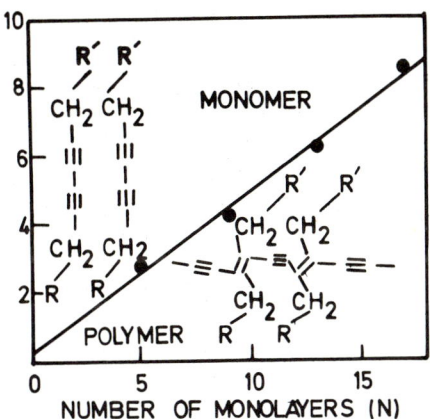

FIG.1. INVERSE CAPACITANCE V.S NUMBER OF MONOLAYERS FOR LANGMUIR FILM OF SUBSTITUTED DIACETYLENE WHOSE MONOMERIC AND POLYMERIC FORMS ARE SHOWN IN THE INSETS.

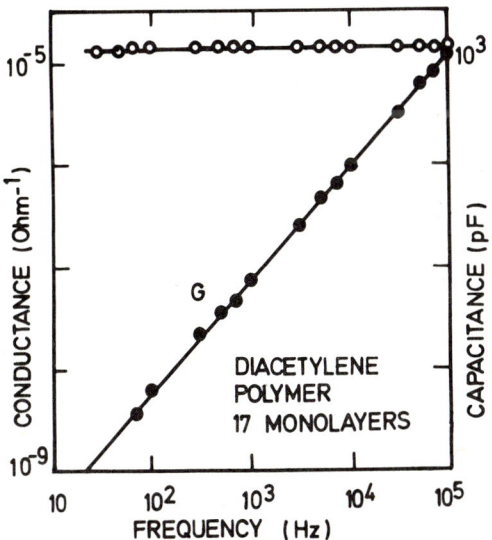

FIG. 2. CONDUCTANCE AND CAPACITANCE CURVES FOR MIM STRUCTURE INCORPORATING 17 MONOLAYERS OF DIACETYLENE POLYMER.

conditions for building up monolayers were established by studying the pressure-area isotherms. An electronic feedback system was used to maintain the surface pressure at 1.5×10^{-2} Nm^{-1} during the deposition process. After the required number of monomer monolayers had been deposited the film was immediately polymerized by exposure to an intense 300nm ultra-violet source. Multilayers deposited onto aluminised glass slides showed a distinct colour change to red during this process. The polymerized films were stored for two days in a desiccator under a low pressure of nitrogen before gold top electrodes were deposited by thermal evaporation. To minimise possible damage to the film the samples were cooled to about -100°C during the evaporation procedure. The carefully selected glass slide substrates were ultrasonically cleaned and vacuum coated with aluminium before Langmuir film deposition; the bulk p-type InP substrate preparation prior to coating involved etching in a fresh solution of 1% bromine in methanol. However, good adhesion to our n-type epitaxial GaP surfaces was obtained after simply refluxing the semiconductor for one hour in isopropylalcohol vapour.

DIACETYLENE POLYMER FILM CHARACTERISTICS

Figure 1 shows the reciprocal capacitance of an Au-diacetylene polymer - Al structure as a function of the number of deposited layers N. The slope of the straight line gives the dielectric thickness (d_M/ε_M) of each monolayer while the intercept yields similar information about the interfacial layer. We thus calculate the following: $(d_M/\varepsilon_M) = 1.33$nm; $d_{ox}/\varepsilon_{ox} = 0.45$nm. It is reasonable to assume that the substituted 12-8 diacetylene film has a similar dielectric constant to that of arachidic acid. Taking $\varepsilon_M = 2.5$ then gives $d_M = 3.3$nm, a value in good agreement with a simple calculation of the molecule chain length based on the configuration shown in Fig. 1.

Conductance and capacitance measurements were made using phase-sensitive techniques (Ortholoc 9502 Bridge). Typical data are shown in Fig. 2 for an MIM sandwich structure incorporating 17 monolayers of diacetylene polymer. The capacitance displays virtually no dispersion throughout the frequency range 70Hz to 100kHz. The conductance varies according to the equation $G \propto \omega^n$ with n of the order of unity, signifying good insulator properties. D.C. measurements (not shown) confirmed that the polymerized Langmuir films were very resistive ($\rho \sim 4.0 \times 10^{11}$ Ω cm at a field strength of 5×10^5 V cm^{-1}); linear plots of log current versus square root of voltage were independent of the polarity of the electrode thus implying a barrier limited conduction process (Poole-Frenkel effect) through the bulk of the film.

MIS CHARACTERISTICS

p-type InP

Current theories and experimental evidence (5) suggest that Fermi level pinning occurs at the surfaces of III-V semiconductor compounds; this arises due to extrinsic surface states localized at surface imperfections and stoichiometric defects; their energy levels are thought to be independent of the characteristics of the adatoms so that, if an energetic process is used to deposit the insulator on the semiconductor, the natural oxide might be expected to play only a minor role. It is instructive therefore to compare admittance data for MIS structures prepared using relatively high temperature deposition techniques with those for structures fabricated using the low temperature Langmuir-Blodgett method. Such an investigation has been carried out by Sykes et al (6) using p-type InP and conventional fatty acid salt Langmuir films. Their C/V data show both accumulation and depletion regions and contrast markedly with earlier reports describing an inverted semiconductor even at zero gate bias; their G/V data are also distinctive in that they exhibit structure not previously seen with p-type material.

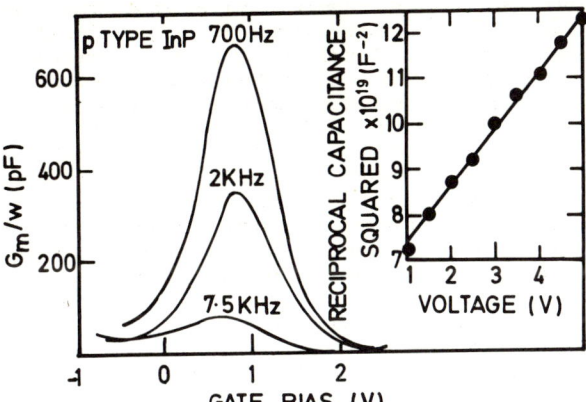

FIG.3. MEASURED CONDUCTANCE VERSUS VOLTAGE FOR p–TYPE InP COATED WITH 25 MONOLAYERS OF DIACETYLENE POLYMER; THE INSET SHOWS THE MEASURED CAPACITANCE AT 70 KHz.

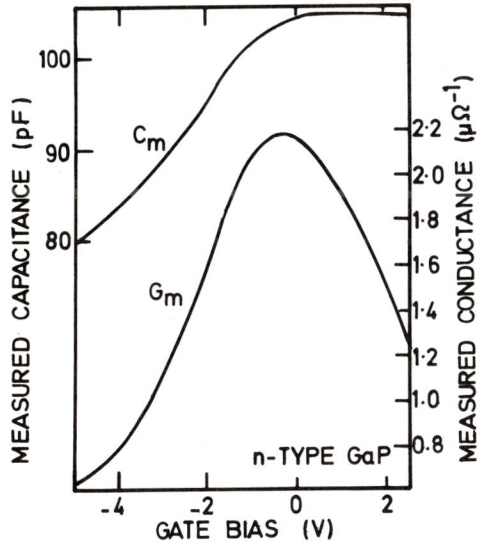

FIG.4. MEASURED ADMITTANCE CHARACTERISTICS AT 100 KHz FOR n–TYPE GaP COATED WITH 25 MONOLAYERS OF DIACETYLENE POLYMER.

Figure 3 shows the almost identical data we have obtained for p-type InP coated with 25 monolayers of 12-8 diacetylene polymer. The inset to this diagram confirms the linear relationship between C^{-2} and V expected in a depletion situation. A notable feature which is not shown in the figure is the small degree of hysteresis for normal ramp speeds. Previous efforts to fabricate MIS structures using polymers deposited in conventional ways have always been plagued with large hysteresis effects.

n-type GaP

Few attempts have been made to characterize the interface properties of GaP and compatible insulating layers; such information is of interest, however, for electroluminescent devices might be constructed using GaP MIS structures to which an alternating voltage is applied (7). Our preliminary results for an n-type epitaxial GaP film (on an n^+ substrate) coated with 25 monolayers of 12-8 diacetylene polymer are shown in Fig. 4. The capacitance in forward bias falls a factor of three short of the insulator value indicating that the surface does not accumulate with gate voltages less than breakdown. In other respects the C-V and G-V curves (both obtained in the dark) are fairly conventional and show little hysteresis.

CONCLUSION

We have demonstrated for the first time that it is possible to construct useful MIS structures by first depositing monomeric Langmuir films and then polymerizing the monolayers when they are on the semiconductor substrate. The resulting films are of high quality, show little or no dispersion and have high breakdown strengths and may therefore find application in practical devices.

ACKNOWLEDGEMENTS

We should like to thank T. Fok, J. P. Lloyd and R. W. Sykes for useful discussions, and D. Bloor for supplying the diacetylene monomer material.

REFERENCES

(1) G. G. Roberts, K. P. Pande and W. A. Barlow, InP/Langmuir Film MISFET, Solid St. Elect. Devices 2, 1969 (1978).

(2) I. M. Dharmadasa, G. G. Roberts and M. C. Petty, CdTe/Langmuir Film Photovoltaic Structures, Electronics Letts. 16, 201 (1980).

(3) G. G. Roberts, T. M. McGinnity, W. A. Barlow and P. S. Vincett, A.C. and D.C. Conduction in Lightly Substituted Anthracene Langmuir Films, Thin Solid Films 68, 223 (1980).

(4) V. Enkelmann, R. J. Leyner and G. Wegner, Investigation of Topochemical Solid State Polymerization of a Diacetylene by X-ray Methods and Brillouin-Spectroscopy, Makromol Chem, 7, 180 (1979).

(5) W. E. Spicer, Nature of Interface States at III-V Insulating Interfaces, in Insulating Films on Semiconductors, Inst. Physics Conf. Series 50, 216 (1980).

(6) R. W. Sykes, G. G. Roberts, T. Fok and D. T. Clark, p-type InP/Langmuir Film MIS Diodes, Solid St. and Elect. Devices Vol. 127, Issue 3, 137-139 (1980).

(7) C. N. Berglund, Electroluminescence using GaAs MIS Structure, Appl. Phys. Lett., 9, 441 (1966).

MOS WEAROUT AND BREAKDOWN STATISTICS

D. Wolters, T. Hoogestyn and H. Kraaij
Philips Research Laboratories, 5600 MD Eindhoven, The Netherlands

ABSTRACT

The statistical behaviour of MOS failures due to electric breakdown of the insulator has been studied. We measured the cumulative fraction of failures P(t) at a constant field depending on stress time (t) and the cumulative fraction of failures P(E) while ramping the field E.
We found that

$$\ln\ln(1-P(t))^{-1} \propto \ln(t)$$

and

$$\ln\ln(1-P(E))^{-1} \propto E - E_{max}$$

Where E_{max} is an experimentally found upper limit of the field E above which hardly any capacitor survives. The expressions found can be qualitatively explained assuming electron injection and subsequent charging of the dielectric.

INTRODUCTION

It was suggested by Kristiansen[10] that Weibull distribution functions might be used for the description of cumulative failure characteristics. These were found applicable to our measurements provided that the appropriate variables were chosen.

EXPERIMENTAL

Measurements of time dependent and field dependent breakdown were performed on thin oxide layers (370 – 700 Å). Circular windows of area 0.2 mm^2 were etched out of thick oxide (5000 Å). Then the thin oxide was grown in a double-walled silica tube, resistance-heated at 950°C in an ambient of dry oxygen, in some cases mixed with HCl. The thick oxides were grown at 1100°C under the same conditions. The outer chamber of the tube was flushed continuously with HCl-containing oxygen to avoid contamination. TSIC and TVS checks[1] on MOS capacitor samples indicated mobile ion concentrations of about $2 \times 10^{10}/cm^2$. Electrodes were made by electron gun evaporation of high-purity Al or depositon of poly-Si by LPCVD-techniques. The electrodes always overlapped the windows completely. Post metallisation annealing for 30 min at 450°C was carried out in wet N_2 for Al and in forming gas for poly electrodes. Contacting was performed by pressing 96 spring-loaded pins on the contacts at some distance from the windows in order to avoid damage of the thin oxide layer. Standard measurements were performed at 300°C under N_2 flush. A constant voltage for wearout or a ramping voltage for breakdown (ramprate 0.25 V/sec) was applied to the contacts by 96 current detectors changing their logical state when the current exceeded 5 µA. The state of the detectors was scanned every 3 ms and the information was fed into a computer. In addition large-sample (400) breakdown measurements were performed on an automatic stepper and tester (Fairchild tester 500 c) with a ramp rate of 20 V/s at room temperature in open air.

RESULTS

Wearout

Data for wearout (defined as the failure at a constant field strength as a function of time) are plotted in fig. 1. The scales are chosen so as to approximate Weibull probability paper. In fact $\ln \ln (1 - P(t))^{-1}$ is plotted against $\ln(t)$. $P(t)$ is the fraction of failures at time (t). The use of probability paper and the methods of extreme value statistics are given by Gumbel[2]. The fact that the lines are straight suggests the relationship

$$\ln\ln (1 - P(t))^{-1} = n \ln \left(\frac{t}{\bar{t}}\right) \qquad (1)$$

or

$$P(t) = 1 - \exp\left(-\left(\frac{t}{\bar{t}}\right)^n\right) \qquad (2)$$

Where $\bar{t}$ is the value at $\ln\ln (1 - P)^{-1} = 0$ i.e. $P = 1 - 1/e = 0.63$ and n corresponds to the slope of the curves. The varied parameters are given in the caption to the figures. The large difference (on the logarithmic scale) is striking and in agreement with the trend of the corresponding curves published in literature (see the review article of Osburn[3]). The reliability is improved by using polygates and HCl containing oxygen. Important for the results is the linear relationship between $\ln\ln (1 - P(t))^{-1}$ and $\ln(t)$. This relationship has been found by other authors also[4,5]. For comparison data from Osburn[6] have been replotted in fig. 2 (upper curve).

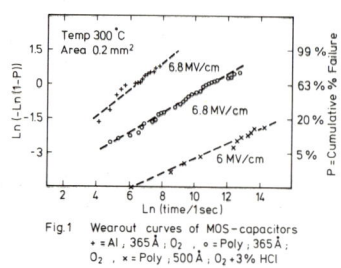

Fig.1 Wearout curves of MOS-capacitors
+ = Al, 365 Å ; O_2 , ○ = Poly, 365 Å ;
O_2 , × = Poly, 500 Å ; O_2 + 3% HCl

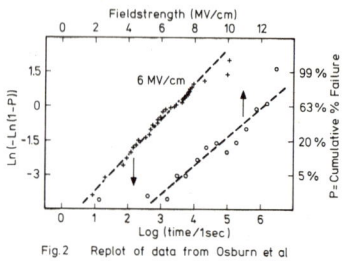

Fig.2 Replot of data from Osburn et al

Breakdown

Plotting the function $\ln\ln (1 - P(E))^{-1}$ against the applied field strength E shows that a substantial fraction of the total population can be fitted to a straight line. This is shown in fig. 2 (lower-curve) for a replot of data from Osburn[7] and in fig. 3 for samples from the same runs as used in fig. 1.
It can be seen in fig. 3 that at least half of each population can be fitted to a straight line. At a high field strength there is a deviation for the polygated sample which is due to a rather small fraction of capacitors. The behaviour was also found with some other runs but it was not typical. This deviation at a high field strength may be due to poor contacting and was not further investigated.
A striking feature of the curves in fig. 3 is that the straight lines can be prolonged to intersect at a high field strength E_{max} for $P(E) = 99.9\%$. This means that the chance of surviving E_{max} is very small (<0.1%) irrespective of the distribution of the particular population. It was found that the values of E_{max} may scatter over a somewhat wider range than suggested by the results in fig. 3. E_{max} is further dependent on measuring temperature, substrate doping, capacitor area and contaminations of the oxide.

The deviation from the straight lines in fig. 3 at a low field strength was found to be typical for a large number of runs. The explanation for this deviation could be a breakdown mechanism differing from the main mechanism or a subpopulation with different properties. To investigate this, a much larger sample of capacitors was measured. Test capacitors of area 0.09 mm² were made in a standard MOS process with HCl containing oxygen and poly-gates. Several runs were tested and regularly two or three straight lines showed up in the breakdown curve. An example is given in fig. 4a where three subpopulations can be distinguished (in most cases only two can be distinguished clearly). These three subpopulations are re-plotted and normalised to the sample size of the subpopulation in fig. 4b. It may be seen that here again the subpopulation curves can be prolonged to a common intersection point E_{max} at $P(E) = 99\%$. The occurrence of E_{max} was found for a large number of runs. The value of E_{max} varied between 9 and 10 MV/cm for over 50 runs with a sample size of 400 capacitors. It is assumed that a certain value of E_{max} is a typical property for all breakdown curves. The general expression for $P(E)$ can be given in terms of this E_{max} and the 63% value $\overline{E}$.

$$P(E) = 1 - \exp(-\exp m(\frac{E - \overline{E}}{E_{max} - \overline{E}})) \tag{3}$$

where $m/(E_{max} - \overline{E})$ is the slope of the curve and m is the value of $\ln\ln (1 - P(E))^{-1}$ where the various curves intersect. $\overline{E}$ is the field strength at which $P(E) = 1 - 1/e = 0.63$ which can be approximately taken as the mean field strength when the sample size is very large[2]. It can be objected that E_{max} is dependent on the capacitor area (A) so that a physical interpretation of the value is not possible at all. It must be seen as a practical value which can be determined with a much higher precision than the usual values determined from a histogram for the breakdown frequency against field strength. It was found that $E_{max} \propto -\ln(A)$, hence, for very small areas, E_{max} tends to increase theoretically to very large values. For large values of A, E_{max} will become very small.

The significance of E_{max} is, that for the very-low-field defects, the sample seems to "know" that the 99% point is at E_{max}. Probably there is a field strength E_{max} for which the probablity of survival is very low regardless of whether the modal capacitor breaks down at a high or a low field strength.

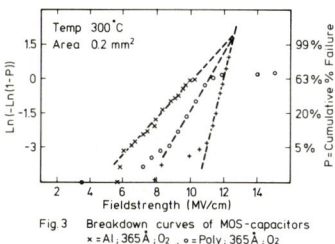

Fig. 3 Breakdown curves of MOS-capacitors
x = Al ; 365 Å ; O₂ , o = Poly ; 365 Å ; O₂
+ = Poly ; 500 Å ; O₂ + 3% HCl

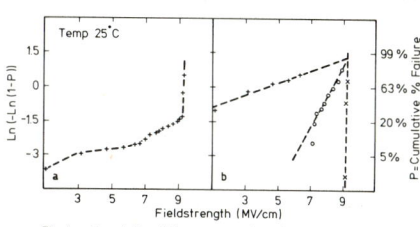

Fig. 4 Cumulative failure curves for MOS-capacitors
d_{ox} = 700 Å ; Poly gate ; a) Total population
b) 3 subpopulations

DISCUSSION

The distribution functions found belong to the field of extreme value statistics (Gumbel[2]) and it is beyond the scope of this paper to derive the exact statistical parameters. The occurrence of several populations which converge to a same value E_{max} in the case of breakdown and the presence of parallel curves in the case of wearout give an indication of the basic mechanism responsible for the breakdown. We shall, here, give only the qualitative arguments for this mechanism based on the observation that insulators stressed under high fields show charging effects. The quantitative description will be given elsewhere.

Charging of dielectrics by injected electronic charges has been found by numerous authors (Walden[8], Young et al.[9]). The charging effect is measured by monitoring the flat band voltage during the injection period. The magnitude of the flat band voltage shift can be of the order of volts. Rosenberg et al.[5] have shown that an increase in the applied field strength of this magnitude (MV/cm) can shorten the life, defined at a certain failure rate, by 7 orders of magnitude. We want to emphasize that the charging process, if present, should always have a severe effect on the life of MOS capacitors. It is also observed that charging and wearout have the same logarithmic time law (see table I). Adopting the charging model for the field dependent breakdown it may also be seen that the field dependence of both processes is the same.

TABLE I: Comparison of field and time dependence of charging and failure statistics.

Charging	$Q(t,E) \propto E$	$Q(t,E) \propto \ln(t)$
Failure statistics	$\ln\ln(1-P)^{-1} \propto E$	$\ln\ln(1-P)^{-1} \propto \ln(t)$

Summarizing, charging increases the local fields above a critical breakdown value. The maximum breakdown fields can be seen as theoretical points where the applied fields are equal to the critical ones. The slopes in the wearout and breakdown curves can be explained by means of this model. The quantitative details will be given elsewhere.

CONCLUSIONS

Breakdown and wearout curves of MOS capacitors are presented. The use of exponential distribution functions has been demonstrated. The physical interpretation of the found curves will be given in detail elsewhere.

ACKNOWLEDGMENTS

The authors are indebted to W. Voncken for the permission of presenting the data of fig. 4. Discussion and critical revising of the manuscript by J. Verwey are gratefully acknowledged.

REFERENCES

1. J.F. Verwey, Inst. Phys. Conf., Ser. No. **50**, 62 (1979).
2. E.J. Gumbel (1957), Statistics of Extremes, Columbia University Press, New York.
3. C.M. Osburn, Journ. of Solid State Chem., **12**, 232 (1975).
4. D.L. Crook, Technical Digest, IEEE – IEDM, Washington, D.C., 444 (1978).
5. S.J. Rosenberg, D.L. Crook and B.L. Euzent, IEEE Trans. on Electron Devices, ED-26, No.1 48 (1979).
6. C.M. Osburn and E. Bassous, J. Electrochem. Soc., **122**, 89 (1975).
7. C.M. Osburn and D.W. Ormond, J. Electrochem. Soc., **119**, 591 (1972).
   C.M. Osburn and D.W. Ormond, J. Electrochem. Soc., **119**, 597 (1972).
8. R.H. Walden, J. Appl. Phys., **43**, 1178 (1972).
9. D.R. Young, Inst. Phys. Conf., Ser. No. **50**, 28 (1979).
10. K. Kristiansen, Vacuum, **27**, 227 (1977).

EFFECT OF PREPARATION METHODS ON PERFORMANCE OF
MOS PHOTOVOLTAIC SOLAR CELL

Fouad Abou-Elfotouh and Mohammad Al-Mass'ari
Physics Department, University of Riyadh
Riyadh, Saudi Arabia

ABSTRACT

A study of the structure variation and performance of p-type Si MOS solar cell, due to various oxidation and annealing conditions was carried out. From SIMS, XPS and C-V measurements, an improvement in the structure order, and thining of the transition layer width were found in devices prepared by growing oxide films in oxygen at 10^{-6} Torr followed by annealing in hydrogen or nitrogen at 450°C. Under these conditions an increase of values of open circuit valtage Voc (0.65 V) and fill factor Ff(0.72) were obtained. In devices prepared by oxidation in air or boiling nitric acid, Voc improves after high temperature ($\geqslant$600 C) annealing only, but supression of current was observed.

INTRODUCTION

Evaluation of the behaviour and properties of the surface and interface states of nSi-SiOx system were the subject of an extensive studies using a wide range of MOS devices. However a complete control of the properties of this interface is still lacking. It is known (ref.1.3) that the photovoltaic yield of MOS solar cell is strongly dependent on the action of the interface states. In this paper we present a study of the composition and structure variations of the interface and the performance of p-type silicon MOS solar cell due to various oxidation and post-oxidation heat treatments involved in the cell fabrication.

EXPERIMENTAL DETAILS

Sample Preparation

The (111) surface of p-type silicon wafers (0.35 mm thick, and 0.4 ohm cm resistivity) were mechanically polished to optical finish on one side (cutting marks were removed from the other side), chemically cleaned, and etched for 10 - 20 seconds in H NO_3/HF mixture 3 / 1 by volume. Oxide films of thickness between 20A° and 150A° were immediately grown by dipping into boiling nitric acid, heating in air at temperatures between 600°C and 900°C or by heating in oxygen at pressure 10E-6 Torr. at 600°C. Post oxidation annealing was carried out in the temperature range 200 to 850°C in air, oxygen at pressure 10^{-5} Torr, dry hydrogen or dry nitrogen.

Device Fabrication

In a completed MOS solar cell, the back ohmic contact was prepared by vacuum evaporation of Au at a vacuum of 10^{-5} Torr on the unpolished surface (after removing the oxide layer). The front contact was made in the form of fingers covering about 10-15 precent of the surface.

Structure and Performance Determination

The two surface analysis techniques SIMS and XPS were employed to investigate the chemical composition of the resulting interface during the growth of the oxide film, after its formation, and after annealing under various conditions. Depth profiles were obtained to characterize the transition layers. The fixed oxide charges Qox were estimated from the flat band voltage of the high frequency capacitance-voltage measurements. Cell performance was determined from measurements of open circuit voltage Voc and fill factor Ff.

RESULTS AND DISCUSSIONS

The chemical species and oxidation states detected by SIMS analysis of the oxide layer grown in oxygen at pressure of 10E-6 Torr are Si^{+++}, Si^{++}, Si^{+}, SiO, Si_2, SiO_2 Si_2O, Si_3, Si_2O_5, SiO_4, Si_3O, Si_2O_3, Si_4, Si_3O_2, Si_2O_4, Si_4O, Si_3O_3. and Si_5. Oxide films prepared by different methods showed similar spectra in addition to O, O_2, SiH, and $SiOH$. The mass peaks of O and O_2 in samples oxidized in air were 3 to 4 times higher than those obtained in samples oxidized in boiling nitric acid. Depth profiles of O_2, SiO_2, SiO_4, Si_2O_3, and Si_2O_5 before and after annealing in hydrogen are shown in Fig.1.

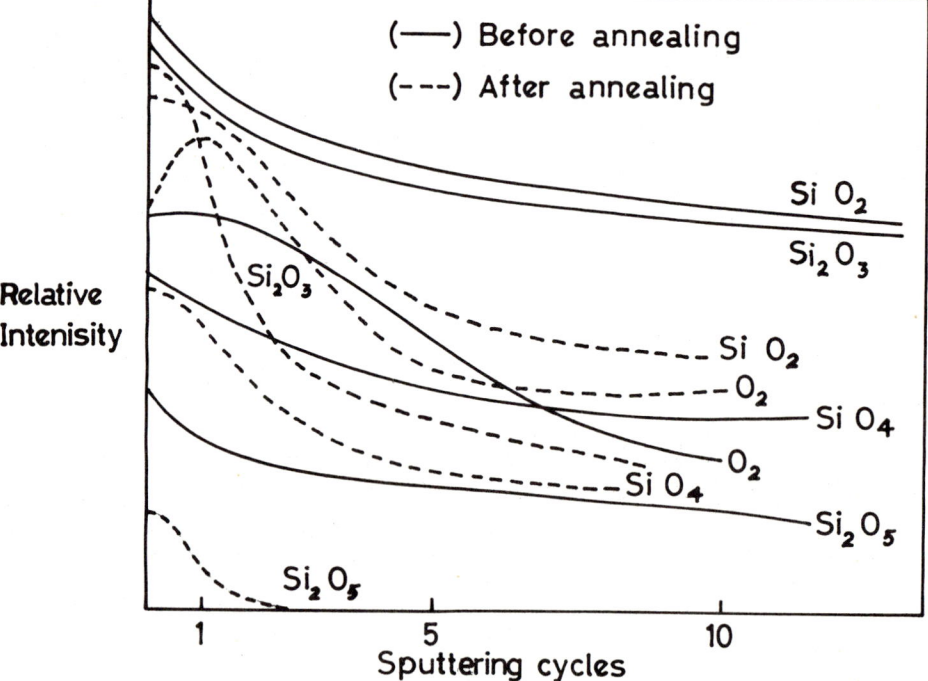

Fig.1 Depth profiles by SIMS using 3Kev ion beam and 45 uA/cm^2 current density, for oxide grown in oxygen and annealed in hydrogen (t_{ox} = 100 A).

The profiles before annealing showed a slow decreasing of all peaks, and a strong oxide interface due to annealing. Most of the SixOy oxides ($x \geqslant 1$) were almost disappeared from all samples after high temperature ($\geqslant 800°C$) annealing for periods longer than 30 minutes. Oxygen accomulation was observed at the oxide-silicon interface particularly in samples oxidized in air or oxygen. This accomulation was increased by annealing in hydrogen or nitrogen, while vacuum annealing did not show appreciable effect. Figure 2 shows the total number of fixed oxide charges as a function of oxidation and annealing conditions.

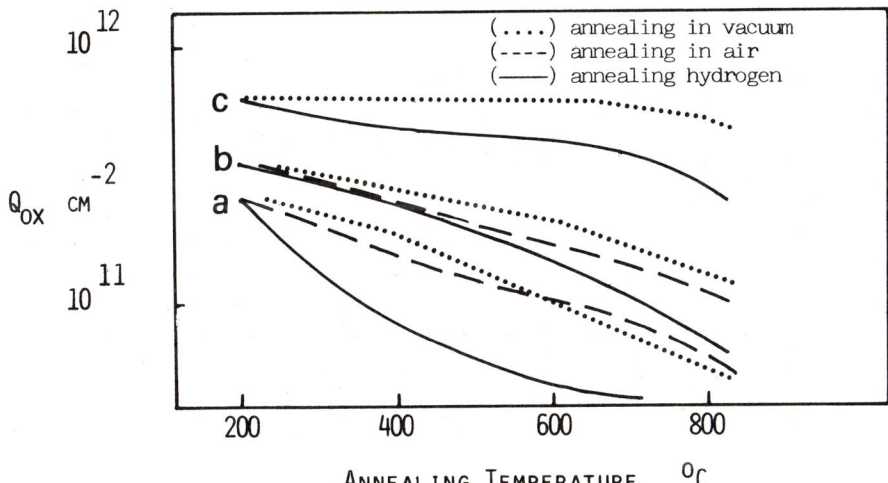

FIG.2 Fixed oxide charges versus annealing temperature for devices oxidized in HNO_3 (c), air (b), oxygen (a).

The highest values of Qox was observed in samples oxidized in boiling nitric acid while the minimum was found in samples oxidized in oxygen. All annealing treatments were able to reduce Qox in samples oxidized in air or oxygen. On the other hand annealing in vacuum did not affect Qox which was reduced only at temperatures higher than 600°C in samples oxidized in boiling nitric acid. Figure 3 shows the changes in open circuit voltage Voc with oxidation condition and annealing temperatures for samples annealed in hydrogen.

The highest value of Voc and Ff was obtained for samples oxidized in oxygen after annealing in hydrogen or nitrogen at 450°C (at this condition the transition layer width was less than 8A°). In samples oxidized in air and boiling nitric acid a current supression started after annealing at temperatures higher than 400°C. This is responsible for the reduction in the Ff value although Voc was increasing. After annealing in hydrogen, the maximum values of the conversion efficiency η and Ff was maintained upto values of the oxide thickness t_{oxm} as high as 55 A° in devices oxidized in oxygen at low pressure. In devices oxidized in air t_{oxm} was limited to 40 A° only while t_{oxm} was only 25A° for samples oxidized in boiling nitric acid. It was shown (ref.4) that t_{oxm} could be increased to more than 70 A°. The results also demonstrated that the thickness of the transition layer as defined in (ref.5) was always decreasing at the same condition for increased Voc. At these conditions an improved structure order in the oxide layer was observed, and molecular oxygen was diffusing to the Si matrix from the interface.

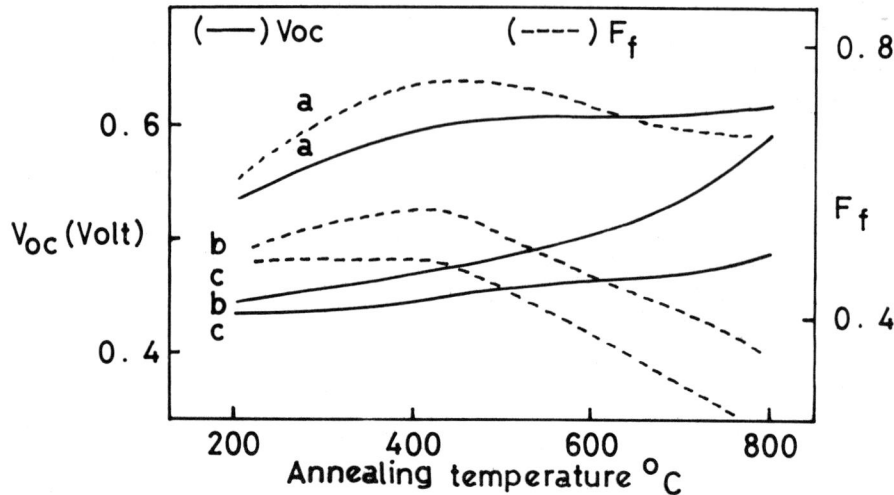

Fig.3 Open circuit voltage Voc and fill factor F_f versus temperature of annealing in dry hydrogen for samples oxidized in: oxygen(a), air(b) HNO_3(c).

CONCLUSION

The characteristic performance of MOS solar cell could be controlled by following preparation procedure and heat treatment that can minimize the density of interface states and transition layer thickness, and improve the structure order at the interface. Appropriate action of the interface states should be also achieved (ref.4). Therefore a reliable model for silicon-silicon oxide interface that can explain the interface states properties in terms of the structure variations is needed. A combination between the oxygen model for interface (ref.6) and Deal's model will be successful to explain most of our results as well as those of others. The details of such model will be discussed in another publication (7).

REFERENCES

(1)   P.Viktorovitch, G.Kamarinos, and P.Even, Effect of insulating film and interface states upon photovoltaic efficiency of an MIS structure, C.R. Acad. Sci. (France) 283 B,119 (1976).

(2)   J.Chewchun, R. Singh, and M.A. Green, Theory of metal - insulator - semiconductor solar cells, J.Appl. Phys. 48, 765 (1977).

(3)	P.Viktorovitch and G.Kamarinos, Improvement of the Photovoltaic efficiency of a metal - insulator-semiconductor structure, Influence of interface states, J.Appl.Phys. 48, 3060 (1977).

(4)	F.A.Abou-Elfotouh and M.A.Al-Mass'ari, Operational Characteristics and structure of the $Si-SiO_2$ interface of MOS solar cell, Inst. Phys. Conf. Ser. No. 50, 174 (1980).

(5)	C.R. Helms, W.E.Spicer, and N.M.Johnson, New Studies of the $Si-SiO_2$ interface using Auger sputter profiling, Solid state commun. 25, 673 (1978).

(6)	Y.C.Cheng, Electronic states at the silicon-silicon dioxide interface, Progress in Surface Sci, 8,181 (1977).

(7)	F.A.Abou-Elfotouh, studies on MOS silicon schottky barriers, submitted to the 27th national symposium of the American Vacuum Soc., 14 - 17 Oct. 1980.

LIST OF PARTICIPANTS

AHMED KHAIRY ABOULSEOUD
UNIVERSITY OF ALEXANDRIA
DEPARTMENT OF ENGINEERING
4 RASMY STREET, GLIM.
ALEXANDRIA, EGYPT

MUSTAFA ABUSAID
NORTH CAROLINA STATE UNIVERSITY
RALEIGH, NC 27650

M. A. AL-MASS'ARI
UNIVERSITY OF RIYADH
PHYSICS DEPARTMENT
PO BOX 2455
RIYADH, SAUDI ARABIA

JUDY ANDREWS
UNC-CHAPEL HILL
CHAPEL HILL, NC 27514

GEORGE W. ARNOLD
SANDIA NATIONAL LABORATORIES
P.O. BOX 5800, DIVISION 5112
ALBUQUERQUE, NM 87185

D. E. ASPNES
BELL LABS
600 MOUNTAIN AVENUE
MURRAY HILL, NJ 07974

BILL ATKINS
ELECTRICAL ENGINEERING DEPARTMENT
NORTH CAROLINA STATE UNIVERSITY
RALEIGH, NC 27650

CHESTER L. BALESTRA
TEXAS INSTRUMENTS, INC.
P. O. BOX 225936, MS 118
DALLAS, TX 75265

GERARD BARBOTTIN
IBM FRANCE
224, BLD., J. KENNEDY - BP #58
91102 CORBEIL-ESSONNES CEDEX
FRANCE

DIRK J. BARTELINK
XEROX CORPORATION
PALO ALTO, CA 94304

ROGER W. BARTON
STANFORD UNIVERSITY
STANFORD, CA 94305

ROBERT S. BAUER
XEROX PARC
3333 COYOTE HILL ROAD
PALO ALTO, CA 94025

JOE BAUGHEN
TELE-TYPE CORPORATION
555 TOUHY
SKOKIE, IL 60077

BURHAN BAYRAKTAROGLU
UNIVERSAL ENERGY SYSTEMS, INC.
39195 PLAINFIELD ROAD
DAYTON, OH 45432

DAVID L. BEAM
IBM
P. O. BOX 12195
RESEARCH TRIANGLE PARK, NC 27709

CHUCK BENNETT
PHYSICS DEPARTMENT
NORTH CAROLINA STATE UNIVERSITY
RALEIGH, NC 27650

STEVEN B. BIBYK
CASE WESTERN RESERVE UNIVERSITY
UNIVERSITY CIRCLE
CLEVELAND, OH 44106

NATHAN BLUZER
WESTINGHOUSE ATL - MS 3525
P. O. BOX 1521
BALTIMORE, MD 21203

HAROLD EDWIN BOESCH, JR.
HARRY DIAMOND LABS
2800 POWDER MILL ROAD
ADELPHI, MD 20783

DANIEL BOIS
CNET
4 CHEMIN DES PRE'S
MEYLAN, FRANCE

JOSE LUIS BOLDER
DEPARTMENT OF PHYSICS & ASTRONOMY
UNC-CHAPEL HILL
CHAPEL HILL, NC 27514

J. R. BREWS
BELL LABS
600 MOUNTAIN AVENUE
MURRAY HILL, NJ 07974

LEONARD J. BRILLSON
XEROX WEBSTER RESEARCH CENTER
800 PHILLIPS ROAD - W114
WEBSTER, NY 14580

M. H. BRODSKY
IBM - T. J. WATSON RESEARCH CENTER
P. O. BOX 218
YORKTOWN HEIGHTS, NY 10598

PHILIP J. CAPLAN
U. S. ARMY
FORT MONMOUTH, NJ 07703

JON CARNES
PHYSICS DEPARTMENT
NORTH CAROLINA STATE UNIVERSITY
RALEIGH, NC 27650

CHI CHANG
UNIVERSITY OF CALIFORNIA
ELECTRONICS RESEARCH LAB
BERKELEY, CA 94720

ALAIN CHANTRE
CNET
4 CHEMIN DES PRE'S
MEYLAN, FRANCE

JOSEPH CHIANG
ELECTRICAL ENGINEERING DEPARTMENT
NORTH CAROLINA STATE UNIVERSITY
RALEIGH, NC 27650

WAI-YIM CHING
UNIVERSITY OF MISSOURI
KANSAS CITY, MO 64114

MARK A. CHONKO
MOTOROLA, INC.
3501 ED BLUESTEIN BLVD.
AUSTIN, TX 78721

CHARLES CLAHASEY
MIT
DEPARTMENT OF ELECTRICAL ENGINEERING
CAMBRIDGE, MA 02139

LARRY COOPER
OFFICE OF NAVAL RESEARCH
CODE 427
EARLINGTON, VA 22217

JAMES H. CRAWFORD, JR.
PHYSICS & ASTRONOMY
UNC-CHAPEL HILL
CHAPEL HILL, NC 27514

JOHN L. CROWLEY
LOCKHEED PALO ALTO RESEARCH LAB
3251 HANOVER
PALO ALTO, CA 94304

WALTER E. DAHLKE
LEHIGH UNIVERSITY
SHERMAN FAIRCHILD LAB, BLDG. 161
BETHLEHEM, PA 18015

TIMIR DATTA
DEPARTMENT OF PHYSICS & ASTRONOMY
UNC - CHAPEL HILL
CHAPEL HILL, NC 27514

BRUCE E. DEAL
FAIRCHILD R & D
4001 MIRANDA AVENUE
PALO ALTO, CA 94304

R. F. DEKEERSMAECKER
K. UNIVERSITY LEUVEN
ESAT LABORATORY KARD.
HAVERLEE, BELGIUM

DONELLI J. DIMARIA
IBM CORPORATION
P. O. BOX 218
YORKTOWN HEIGHTS, NY 10598

CHARLES DOZIER
NAVAL RESEARCH LABS
WASHINGTON, DC 20375

ARTHUR H. EDWARDS
LEHIGH UNIVERSITY
BETHLEHEM, PA 18015

DANIEL L. ELLSWORTH
COLORADO STATE UNIVERSITY
DEPT. OF ELECTRICAL ENGINEERING
FT. COLLINS, CO 80524

DAVID EMIN
SANDIA NATIONAL LABS
P. O. BOX 5800, DIV. 5151
ALBUQUERQUE, NM 87185

L. FARAONE
LEHIGH UNIVERSITY
FAIRCHILD LAB - 161
BETHLEHEM, PA 18015

FRANK J. FEIGL
LEHIGH UNIVERSITY
FAIRCHILD LAB - 161
BETHLEHEM, PA 18015

DOMINQUEZ FERRARI
C.S.I.C.
INSTITUTO DE ELECTONICA DE
  COMMUNICACIONES
SERRANO, 144 MADRID, SPAIN

GILL FOUNTAIN
NORTH CAROLINA STATE UNIVERSITY
RALEIGH, NC 27650

W. BEALL FOWLER
LEHIGH UNIVERSITY
BETHLEHEM, PA 18015

ALEXANDRE FRANCOIS
CNET
BAGNEUX, FRANCE

H. FRENZEL
RWTH
D-5700 AACHEN
FEDERAL REPUBLIC OF GERMANY

FRANK L. GALEENER
XEROX PALO ALTO RESEARCH CENTER
3333 COYOTE HILL ROAD
PALO ALTO, CA 94304

CAROLINE M. GEE
MASSACHUSETTS INST. OF TECHNOLOGY
DEPT. OF PHYSICS, ROOM 13-2134
CAMBRIDGE, MA 02139

RANDY GLATTFELDER
GTE MICRO CIRCUITS DIVISION
3883 NORTH 28th AVENUE
PHOENIX, AZ 85017

J. M. GOLIO
NORTH CAROLINA STATE UNIVERSITY
RALEIGH, NC 27650

RONALD D. GRAFT
U. S. ARMY NIGHT VISION LAB
FT. BELVOIR, VA 22060

A. J. GRANT
MOD, ROYAL SIGNALS & RADAR
  ESTABLISHMENT
ST. ANDREWS ROAD
MALVERN, WORCS. WR 14 3PS

RONALD W. GRANT
ROCKWELL INTERNATIONAL
1085 CAMINO DOS RIOS
THOUSAND OAKS, CA 91360

DAVID L. GRISCOM
NAVAL RESEARCH LAB
CODE 6570
WASHINGTON, DC 20375

FRANK J. GRUNTHANER
JET PROPULSION LAB
4800 OAK GROVE DR. MAIL STOP 198-231
PASADENA, CA 91103

CIRO FALCONY-GUAJARDO
LEHIGH UNIVERSITY
BETHLEHEM, PA 18017

ROBERT B. HAMMOND
LOS ALAMOS SCIENTIFIC LAB
E-10, MS 430
LOS ALAMOS, NM 87545

FREDDIE L. HAMPTON
WESTINGHOUSE ELECTRIC
ADVANCED TECH. LABS, BOX 1521
BALTIMORE, MD 21203

TAKEO HATTORI
MUSASHI INSTITUTE OF TECHNOLOGY
1-28-1 TAMAZUTSUMI
SETAGAYA-KU, TOKYO, JAPAN

FRANK HERMAN
IBM RESEARCH LABORATORY - K33/281
5600 COTTLE ROAD
SAN JOSE, CA 94193

THOMAS W. HICKMOTT
IBM T. J. WATSON RESEARCH CENTER
P.O. BOX 218
YORKTOWN HEIGHTS, NY 10598

MASATAKA HIROSE
DEPARTMENT OF ELECTRICAL ENGINEERING
HIROSHIMA UNIVERSITY
HIROSHIMA 730

GUY HOLLINGER
INST. DE PHYSIQUE NECLEAIRE
43, BLVD DU 11 NOVEMBRE 1918
69621 VILLEURBANNE, FRANCE

CHUNG S. HSU
MOSTEK
1215 W. CROSBY ROAD
CARROLLTON, TX 75006

FU-LONG HSUEH
FAIRCHILD LAB - 161
LEHIGH UNIVERSITY
BETHLEHEM, PA 18015

XI-YI HUANG
UNIVERSITY OF HAWAII
DEPARTMENT OF ELECTRICAL ENGINEERING
HONOLULU, HAWAII 96822

K. HUEBNER
SEKTION PHYSIK DER WILHELM
PIECK UNIVERSITÄT
ROSTOCK, UNIVERSITATSPLATZ 3
GERMAN DEMOCRATIC REPUBLIC

HOWARD HUFF
SIGNETICS
811 E. ARGUES AVENUE
SUNNYVALE, CA 95052

H. L. HUGHES
U. S. NRL
CODE 6816
WASHINGTON, DC 20375

ROBERT C. HUGHES
SANDIA NATIONAL LABS, DIVISION 5152
P. O. BOX 5800
ALBUQUERQUE, NM 87185

GABRIEL IGWE
ELECTRICAL ENGINEERING
NORTH CAROLINA STATE UNIVERSITY
RALEIGH, NC 27650

EUGENE A. IRENE
IBM - T. J. WATSON RESEARCH CENTER
P. O. BOX 218
YORKTOWN HEIGHTS, NY 10598

FLOYD J. JAMES
UNC - CHAPEL HILL
CHAPEL HILL, NC 27514

BRUCE K. JANOUSEK
THE AEROSPACE CORPORATION
P.O. BOX 92957
LOS ANGELES, CA 90009

J. D. JOANNOPOULOS
M.I.T.
DEPARTMENT OF PHYSICS
CAMBRIDGE, MA 02139

NOBLE M. JOHNSON
XEROX CORPORATION
3333 COYOTE HILL ROAD
PALO ALTO, CA 94304

ROBERT L. JOHNSON
U. S. AIR FORCE
AFWAL/AADR
WRIGHT-PATTERSON AFB
DAYTON, OHIO 45433

RODNEY JONES
UNC - CHAPEL HILL
CHAPEL HILL, NC 27514

STEVEN JOST
PRINCETON UNIVERSITY/ITT EOPD
7635 PLANTATION ROAD
ROANOKE, VA 24019

VIKRAM J. KAPOOR
CASE WESTERN RESERVE UNIVERSITY
DEPARTMENT OF ELECTRICAL ENGINEERING
CLEVELAND, OH 44106

S. KAR
DEPARTMENT OF ELECTRICAL ENGINEERING
INDIAN INSTITUTE OF TECHNOLOGY
KANPUR-208016, INDIA

MARK KELLAM
UNC - CHAPEL HILL
CHAPEL HILL, NC 27514

M. K. KIM
ELECTRICAL ENGINEERING DEPARTMENT
NORTH CAROLINA STATE UNIVERSITY
RALEIGH, NC 27650

EVERETT E. KING
NAVAL RESEARCH LAB
CODE 6816
WASHINGTON, DC 20375

C. T. KIRK
MIT LINCOLN LAB
RM B-192, P. O. BOX 73
LEXINGTON, MA 02173

NICHOLAS KLEIN
DEPARTMENT OF ELECTRICAL ENGINEERING
TECHNION
HAIFA, ISRAEL

MARJORIE KLENIN
PHYSICS DEPARTMENT
NORTH CAROLINA STATE UNIVERSITY
RALEIGH, NC 27650

E. W. KREUTZ
INSTITUTE FUR ANGEWANDTE PHYSIK
FACHBEREICH 5, SCHLOBGARTENSTR 7
61 DARMSTADT

L. M. LAMBERT
PHYSICS DEPARTMENT
COOK BUILDING A-405
UNIVERSITY OF VERMONT
BURLINGTON, VT 05406

DIETRICH W. LANGER
AVIONICS LAB - AFWAL/AADR
WRIGHT-PATTERSON AFB
DAYTON, OH 45433

PETER LANYON
SERI
1617 COLE BLVD.
GOLDEN, CO 80401

ROBERT B. LAUGHLIN
BELL LABS
600 MOUNTAIN AVENUE
MURRAY HILL, NJ 07971

YOUNG WON LEE
NORTH CAROLINA STATE UNIVERSITY
RALEIGH, NC 27650

YEU-PYNG LIAW
ELECTRICAL ENGINEERING DEPARTMENT
NORTH CAROLINA STATE UNIVERSITY
RALEIGH, NC 27650

DEREK L. LILE
NAVAL OCEAN SYSTEMS CENTER
271 CATALINA BLVD.
SAN DIEGO, CA 92106

YUNG-TAO LIN
TEXAS INSTRUMENTS, INC.
P. O. BOX 225936
DALLAS, TX 74265

N. O. LIPARI
IBM-T.J. WATSON RESEARCH CENTER
P. O. BOX 218
YORKTOWN HEIGHTS, NY 10598

JIANN LIU
PHYSICS DEPARTMENT
NORTH CAROLINA STATE UNIVERSITY
RALEIGH, NC 27650

JUAN LLABRES
CSIC
INSTITUTE FISICA ESTADO SOLIDO
CANTOBLANCO, MADRID, SPAIN

GERALD LUCOVSKY
PHYSICS DEPARTMENT
NORTH CAROLINA STATE UNIVERSITY
RALEIGH, NC 27650

STEPHEN A. LYON
PRINCETON UNIVERSITY
B 428 ENGINEERING QUAD
PRINCETON, NJ 08544

ANUPAM MADHUKAR
UNIVERSITY OF SOUTHERN CALIFORNIA
EXPOSITIONS PARK
LOS ANGELES, CA 90007

S. K. MALIK
IBM CORPORATION
390 S. ROAD
POUGHKEEPSIE, NY 12602

AZZAM MANSOUR
PHYSICS DEPARTMENT
NORTH CAROLINA STATE UNIVERSITY
RALEIGH, NC 27650

J. L. MARTINEZ
INSTITUTO DE CIENCIAS, UAP
APDO. POSTAL J-48
PUEBLO, MEXICO

JOSEPH MASERJIAN
JET PROPULSION LAB
4800 OAK GROVE DR., MAIL STOP 198-231
PASADENA, CA 91103

BRUCE D. MCCOMBE
NAVAL RESEARCH LAB
CODE 6800
WASHINGTON, DC 20375

DANIEL MCGRATH
TEXAS INSTRUMENTS
BOX 225936, MAIL STATION 119
DALLAS, TX 75265

STEPHEN W. MCKNIGHT
NAVAL RESEARCH LAB
WASHINGTON, DC 20375

F. BARRY MCLEAN
HARRY DIAMOND LABS
2800 POWDER MILL ROAD
ADELPHI, MD 20783

WILLIAM G. MEYER
BELL LABS
2315 LASALLE DRIVE
WHITFIELD, PA 19609

DALE MILLER
LAWRENCE LIVERMORE LABORATORY
P. O. BOX 808, L-156
LIVERMORE, CA 94550

UMESH KUMAR MISHRA
LEHIGH UNIVERSITY
FAIRCHILD LAB #161
BETHLEHEM, PA 18015

JOE R. MONKOWSKI
PENN STATE UNIVERSITY
ELECTRICAL ENGINEERING DEPARTMENT
UNIVERSITY PARK, PA 16802

MICHAEL D. MONKOWSKI
PENN STATE UNIVERSITY
226 STEIDLE
UNIVERSITY PARK, PA 16802

CHARLES MORRISON
ELECTRICAL ENGINEERING DEPARTMENT
NORTH CAROLINA STATE UNIVERSITY
RALEIGH, NC 27650

TOMMY MYERS
PHYSICS DEPARTMENT
NORTH CAROLINA STATE UNIVERSITY
RALEIGH, NC 27650

DONALD K. NICHOLS
JET PROPULSION LAB
4800 OAK GROVE DRIVE
PASADENA, CA 91103

JEAN-JACQUES NIEZ
CENG - LETI/MEA
AVENUE DES MARTYRS - BP 85X
38041 GRENOBLE CEDEX, FRANCE

KAZUNORI OHNISHI
DEPT. OF ELECTRONICS ENGINEERING
NIHON UNIVERSITY, CHIYODA-KU
TOKYO, 101, JAPAN

KUNIICHI OHTA
NIPPON ELECTRIC CO., LTD.
1753 SHIMONUMABE, NAKAHARA-KU
KAWASAKI 211, JAPAN

SIEGFRIED OTHMER
NORTHROP RESEARCH & TECHNOLOGY CTR.
ONE RESEARCH PARK
PALOS VERDES PENINSULA, CA 90274

M. A. PAESLER
HARVARD UNIVERSITY
DIVISION OF APPLIED SCIENCES
9 OXFORD STREET
CAMBRIDGE, MA 02138

STELLA PANG
PRINCETON UNIVERSITY
DEPARTMENT OF ELECTRICAL ENGINEERING
PRINCETON, NJ 08544

S. T. PANTELIDES
IBM - T. J. WATSON RESEARCH CENTER
P. O. BOX 218
YORKTOWN HEIGHTS, NY 10598

D. D. PATEL
IBM CORPORATION
P. O. BOX 12195
TRIANGLE PARK, NC 27709

MARTIN C. PECKERAR
WESTINGHOUSE ELECTRIC CORP.
BALTIMORE, MD 21203

ROBERT PFEFFER
U. S. ARMY ELECTRONIC TECH. & DEVICES
   LABORATORY
8 STANDISH DRIVE
OCEAN, NJ 07712

EDWARD H. POINDEXTER
3 ROSEDALE TERRACE
HOLMDEL, NC 07733

WILLIAM B. POLLARD
DEPARTMENT OF PHYSICS
ATLANTA UNIVERSITY
ATLANTA, GA 30314

ROGER W. PRYOR
PITNEY BOWES
BOX 6050
NORWALK, CT 06852

DENNIS D. RATHMAN
M.I.T. LINCOLN LAB, M.S.E. 0124A
244 WOOD STREET
LEXINGTON, MA 02173

SHANG-YUAN REN
UNIVERSITY OF MISSOURI
DEPARTMENT OF PHYSICS
1110 East 48th STREET
KANSAS CITY, MO 64110

A. G. REVESZ
COSMAT LABS
BOX 115
CLARKSBURG, MD 20734

PETER D. RICHARD
CHEMISTRY DEPARTMENT
NORTH CAROLINA STATE UNIVERSITY
RALEIGH, NC 27650

ELIEZER DAVID RICHMOND
NAVAL RESEARCH LAB
WASHINGTON, DC

DANIEL RICHMOND
432 CORY HALL
UNIVERSITY OF CALIFORNIA
BERKELEY, CA 94720

SERGE RIGO
G.P.S. DE L'E.N.S. UNIVERSITE
PARIS 7, 2 PLACE JUSSIEU
TOUR 23, 24 3E STAGE
PARIS (5E) FRANCE

G. G. ROBERTS
DEPARTMENT OF APPLIED PHYSICS
SCIENCE LABS, SOUTH ROAD
DURHAM, DH1 3LE, UNITED KINGDOM

CHRIS RODDY
PHYSICS DEPARTMENT
NORTH CAROLINA STATE UNIVERSITY
RALEIGH, NC 27650

PETER ROITMAN
NATIONAL BUREAU OF STANDARDS
WASHINGTON, DC 20234

RON RUDDER
PHYSICS DEPARTMENT
NORTH CAROLINA STATE UNIVERSITY
RALEIGH, NC 27650

JERZY RUZYLLO
PENNSYLVANIA STATE UNIVERSITY
UNIVERSITY PARK, PA 16802

ALI SAFAVI
ELECTRICAL ENGINEERING DEPARTMENT
NORTH CAROLINA STATE UNIVERSITY
RALEIGH, NC 27650

C. T. SAH
UNIVERSITY OF ILLINOIS
403 POND RIDGE LANE
URBANA, IL 61801

DALE SAYERS
PHYSICS DEPARTMENT
NORTH CAROLINA STATE UNIVERSITY
RALEIGH, NC 27650

JAN SCHETZINA
PHYSICS DEPARTMENT
NORTH CAROLINA STATE UNIVERSITY
RALEIGH, NC 27650

O. F. SCHIRMER
INSTITUT FUR ANGEWANDTE
FESTKOERPER PHYSIK
D78 FREIBURG, WEST GERMANY

A. G. SCHREINER
CHEMISTRY DEPARTMENT
NORTH CAROLINA STATE UNIVERSITY
RALEIGH, NC 27650

BERTRAM SCHWARTZ
BELL LABS
600 MOUNTAIN AVENUE
MURRAY HILL, NJ 07090

GARY P. SCHWARTZ
BELL LABS
600 MOUNTAIN AVENUE
MURRAY HILL, NJ 07974

CHOI SEONG SAO
UNC-CHAPEL HILL
CHAPEL HILL, NC 27514

ARDEN SHER
SRI INTERNATIONAL
333 RAVENSWOOD AVENUE
MENLO PARK, CA 94025

GLENN SHIRLEY
MOTOROLA, INC.
5005 E. MCDOWELL ROAD
PHOENIX, AZ 85008

MARVIN SILVER
DEPARTMENT OF PHYSICS & ASTRONOMY
UNC-CHAPEL HILL
CHAPEL HILL, NC 27514

MAYRANT SIMONS
ROUTE L, BOX 12194
RESEARCH TRIANGLE PARK, NC 27709

W. LEE SMITH
LAWRENCE LIVERMORE LAB
P. O. BOX 808
LIVERMORE, CA 94550

RALPH J. SOKEL
SANDIA NATIONAL LABS
P. O. BOX 5800
ALBUQUERQUE, NM 87185

HERMAN J. STEIN
SANDIA NATIONAL LABS
DIVISION 5112
ALBUQUERQUE, NM 87185

G. STEPHENS
IBM CORPORATION
P. O. BOX 12195
RESEARCH TRIANGLE PARK, NC 27709

T. SUDA
INSTITUTE OF VOCATIONAL TRAINING
1960 AIHARA, DAGAMIHARA-SHI
KANAGAWA, JAPAN 229

TAKUO SUGANO
DEPARTMENT OF ELECTRONIC ENGINEERING
UNIVERSITY OF TOKYO
3-1 HONGO 7 CHOME, BUNKYO-KU
TOKYO, 113, JAPAN

YUAN-CHEN SUN
SOLID STATE ELECTRONICS LAB
155 ELECTRICAL ENGINEERING
UNIVERSITY OF ILLINOIS
URBANA, IL 61801

CHRISTEN M. SVENSSON
IBM, LINKOPING UNIVERSITY
S-581 83 LINKOPING
SWEDEN

MING L. TARNG
RCA CORPORATION
DAVID SARNOFF RESEARCH CENTER
PRINCETON, NJ 08540

YVES THENOZ
THOMSON CSF
38120 ST. EGREVE, FRANCE

STEPHEN TITCOMB
LEHIGH UNIVERSITY
SHERMAN FAIRCHILD LAB #161
BETHLEHEM, PA 18015

IGNATIUS S. T. TSONG
PENN STATE UNIVERSITY
MATERIALS RESEARCH LAB
UNIVERSITY PARK, PA 16802

DERWEI TU
ELECTRICAL ENGINEERING DEPARTMENT
NORTH CAROLINA STATE UNIVERSITY
RALEIGH, NC 27650

JOSEPH J. TZOU
SOLID STATE ELECTRONICS LAB
155 ELECTRICAL ENGINEERING
UNIVERSITY OF ILLINOIS
URBANA, IL 61801

RONALD UTTECHT
IBM
ESSEX JUNCTION, VERMOUNT

PENNING DE VRIES
TWENTE UNIVERSITY OF TECHNOLOGY
P. O. BOX 217
FINSCHEDE, THE NETHERLANDS

BENJAMIN H. VROMEN
IBM CORPORATION
P. O. BOX 390
POUGHKEEPSI, NY 12602

ALLEN A. WALMA
CEES USTL
PLACE E BATAILLON
MONTPELLIER-CEDEX, FRANCE

XUN WANG
DEPARTMENT OF PHYSICS
THE UNIVERSITY OF WISCONSIN
MILWAUKEE, WI 53201

M. WHITE
MOD, ROYAL SIGNALS & RADAR
    ESTABLISHMENT
ST. ANDREWS ROAD
MALVERN, WORCS., WR 14 3PS

HARRY H. WIEDER
NAVAL OCEAN SYSTEMS CENTER
271 CATALINA BLVD.
SAN DIEGO, CA 92152

C. KENNETH WILLIAMS
NORTH CAROLINA STATE UNIVERSITY
DANIELS HALL
RALEIGH, NC 27650

CARL W. WILMSEN
COLORADO STATE UNIVERSITY
ELECTRICAL ENGINEERING DEPARTMENT
FT. COLLINS, CO 80523

HORST R. WITTMAN
U. S. ARMY RESEARCH OFFICE
P. O. BOX 12211
RESEARCH TRIANGLE PARK, NC 27709

D. R. WOLTERS
PHILLIPS LABORATORIES
RESEARCH LAB WBT
EINDHOVEN

C. K. WONG
NORTH CAROLINA STATE UNIVERSITY
RALEIGH, NC 27650

G. B. WRIGHT
OFFICE OF NAVAL RESEARCH
ARLINGTON, VA 22217

JAMES A. WURZBACH
JET PROPULSION LAB
4800 OAK GROVE DR., MAIL STOP 189-1
PASADENA, CA 91103

KIKUO YAMABE
TOSHIBA RESEARCH & DEVELOPMENT CTR.
IC LAB 72 HORIKAWACHO
SAIWAI-KU 210
KAWASAKI-CITY, KANAGAWA, JAPAN

DONALD R. YOUNG
IBM CORPORATION
P. O. BOX 218
YORKTOWN HEIGHTS, NY 10598

A. R. M. ZAGHLOUL
CAIRO UNIVERSITY
ELECTRICAL ENGINEERING DEPARTMENT
CAIRO, EGYPT

AUTHOR INDEX

AboulSeoud, A. K., 177
Abou-Elfotouh, F., 353
Agius, B., 167
Ahrenkiel, R. K., 212
Al-Mass'ari, M., 353
Anderson, G., 212
Appleton, B. R., 321
Arnold, G. W., 112
Aspnes, D. E., 197
Bachrach, R. Z., 221
Baglee, D. A., 191
Balk, P., 246
Barton, R. W., 316
Bauer, R. S., 221
Bayraktaroglu, B., 207
Berberian, M., 24
Bibyk, S. B., 117
Biegelsen, D. K., 311, 326
Bois, D., 331
Brillson, L. J., 221, 265
Butler, S. R., 127, 142
Caplan, P. J., 306, 326
Chadi, D. J., 301
Chantre, A., 331
Chern, J. E., 280
Ching, W. Y., 63, 73
Dahlke, W. E., 232
Daugherty, M. J., 217
Deal, B. E., 306
DiMaria, D. J., 1
Dominguez, E., 250
Dunlavy, D., 212
Edwards, A. H., 59
Ellsworth, D., 152
Emin, D., 39
Falcony-Guajardo, C., 127
Feigl, F. J., 127, 142
Ferry, D. K., 191
Ferrer, F., 250
Finnegan, J. J., 326
Fowler, W. B., 59, 97, 142
Frenzel, H., 246
Galeener, F. L., 77
Gee, C. M., 132

Gibbs, G. V., 92
Ginley, D. S., 147
Gómez, E., 157
Grant, R. W., 202
Greve, D. W., 232
Griscom, D. L., 97
Grunthaner, F. J., 290
Gualtieri, G. J., 197
Hammond, R. B., 212
Hattori, T., 255, 296
Helms, C. R., 316
Henderson, D. J., 107
Herman, F., 107
Hickmott, T. W., 227
Hill, W. A., 202
Hirose, M., 255
Hollinger, G., 87
Hoogestyn, T., 349
Huang, X., 44
Hübner, K., 82
Hughes, R. C., 29
Janousek, B. K., 217
Ji L., 34
Johnson, N. M., 311, 326
Johnson, R. L., 207
Johnson, W. C., 285
Kan, K. K., 344
Kapoor, V. J., 117
Kashat, I., 19
Kasowski, R. V., 107
Kastner, M., 132
Klein, N., 19
Kowalczyk, S. P., 202
Kraaij, H., 349
Kreutz, E. W., 275
Langer, D. W., 207
Laughlin, D. H., 191
Laughlin, R. B., 68
Lehmann, A., 82
Lewis, B. F., 290
Lile, D. L., 270
Liu, F., 44
Llabrés, J., 250
Lora-Tamayo, A., 250

Lora-Tamayo, E., 250
Lucovsky, G., 301
Luo, F., 265
Lyon, S. A., 285
Maggiore, C., 212
Martinez, J. L., 157
Maserjian, J., 290
Masoud, S., 177
McKinley, G. T., 77
McKnight, S. W., 137
Mier, M. G., 207
Miller, P. D., 321
Miller, W. E., 236, 280
Miura, Y., 336
Moak, C. D., 321
Monkowski, J. R., 321
Monkowski, M. D., 321
Moyer, M. D., 311, 326
Ngai, K. L., 44
Ni, X., 34
Ohring, M., 162
Osaka, Y., 255
Othmer, S., 49
Pang, S., 285
Payo, A., 250
Peercy, P. S., 147
Petty, M. C., 344
Pfeffer, R., 162
Poindexter, E. H., 306, 326
Rathman, D. D., 142
Razouk, R. R., 306
Ren, S.-Y, 73
Revesz, A. G., 92
Rigo, S., 167
Roberts, G. G., 344
Rochet, F., 167
Rosencher, E., 331
Rouse, J., 316
Ruzyllo, J., 54

Sakano, S., 255
Sakurai, T., 241
Schirmer, O. F., 102
Schoolar, R. B., 217
Schroll, P., 275
Schumann, L., 82
Schwartz, B., 197
Schwartz, G. P., 181, 197
Schwarz, S. A., 316
Sher, A., 236, 280
Sokel, R. J., 29
Srour, J. R., 49
Stein, H. J., 147
Stotlar, S., 212
Straboni, A., 167
Studna, A. A., 197
Stutius, W., 77
Su, P., 236
Sugano, T., 241
Suzuki, T., 296
Svensson, C. M., 260
Totsuka, T., 296
Tsong, I. S. T., 321
Tsuo, Y. H., 236, 280
Vasquez, R. P., 290
Waldrop, J. R., 202
Walma, A. A., 341
Wang, Y., 34
Wieder, H. H., 270
Wilmsen, C. W., 152, 191
Wintenberg, A. L., 321
Wolkenberg, A., 122
Wolters, D., 349
Wurzbach, J. A., 172
Wysocki, F., 265
Yamabe, K., 336
Zhang, L., 34

RAYMOND H. FOGLER LIBRARY